SO-AAD-292

Weyerhaeuser Environmental Books

WILLIAM CRONON, EDITOR

*Weyerhaeuser Environmental Books explore human relationships
with natural environments in all their variety and complexity.
They seek to cast new light on the ways that natural systems affect
human communities, the ways that people affect the environments of
which they are a part, and the ways that different cultural conceptions
of nature profoundly shape our sense of the world around us.*

GEORGE PERKINS MARSH

Prophet of Conservation

DAVID LOWENTHAL

FOREWORD BY WILLIAM CRONON

University of Washington Press Seattle & London

George Perkins Marsh, Prophet of Conservation has been published with the assistance of a grant from the Weyerhaeuser Environmental Book Endowment, established by the Weyerhaeuser Company Foundation, members of the Weyerhaeuser family, and Janet and Jack Creighton

George Perkins Marsh, Prophet of Conservation is a new biography. It has its roots in but wholly supersedes the author's earlier biography, *George Perkins Marsh, Versatile Vermonter* (Columbia University Press, 1958).

Copyright © 2000 by the University of Washington Press
Printed in the United States of America

All rights reserved. No part of this publication may be reproduced or transmitted in any form or by any means, electronic or mechanical, including photocopy, recording, or any information storage or retrieval system, without permission in writing from the publisher.

Library of Congress Cataloging-in-Publication Data

Lowenthal, David.
George Perkins Marsh, prophet of conservation / David Lowenthal ; foreword by William Cronon.
p.   cm. — (Weyerhaeuser environmental books)
Rev. ed. of: George Perkins Marsh. 1958.
Includes bibliographical references (p.   ).
ISBN 0-295-97942-9 (alk. paper)
1. Marsh, George Perkins, 1801–1882.   2. Statesmen—United States—Biography.   3. Conservationists—United States—Biography.   4. United States—Environmental conditions.   5. United States—Foreign relations—1861–1865.   6. United States—Foreign relations—1865–1898.   I. Lowenthal, David. George Perkins Marsh.   II. Title.   III. Series.   IV. Weyerhaeuser environmental books.

E415.9.M185 L6 2000
973.5'092—dc21
[B]
99-055465

The paper used in this publication is acid-free and recycled from 10 percent post-consumer and at least 50 percent pre-consumer waste. It meets the minimum requirements of American National Standard for Information Sciences-Permanence of paper for Printed Library Materials, ANSI Z39.48–1984.

CONTENTS

ILLUSTRATIONS

MAPS

PLATES

following p. 102

ILLUSTRATIONS

Foreword: Look Back to Look Forward

WILLIAM CRONON

IF I WERE ASKED TO NAME THE THREE BOOKS BY
American authors that have had the greatest impact on environmental
politics and on the struggle to build more responsible human relations
with the natural world, I would have little hesitation about my top can-
didates. One, surely, would be Rachel Carson's *Silent Spring*, published in
1962, that powerful indictment of pesticides and environmental arrogance which arguably launched the modern environmental movement.
A second would be Aldo Leopold's *A Sand County Almanac*, published
without much fanfare in 1949 but gradually emerging as the single
most influential defense of wilderness and a new "land ethic" for Americans in the closing decades of the twentieth century. Both these books
are reasonably familiar to modern environmentalists. But there is one
other, published well over a century ago in the midst of the Civil War,
that has had at least as great an influence in shaping the course of
conservation history in the United States and elsewhere. Although
Thoreau's *Walden* and John Muir's writings on the High Sierra are now
far better known, in fact neither had anything like the political impact
of a book that is today little read even by those who still remember

it: George Perkins Marsh's remarkable *Man and Nature*, first published in 1864.

It takes a real act of historical imagination to understand just how profoundly *Man and Nature* reshaped American attitudes toward the environment in the decades after its publication. At a time when the United States was moving at breakneck speed to industrialize and develop the national economy by exploiting its wealth of natural resources to the fullest, Marsh's was a lonely voice cautioning against the risks of careless growth. He did so in a most unlikely way. Long concerned about the possible effects of deforestation on his native New England, he began to look to world history as a tool for trying to imagine how environmental degradation might affect the United States. Having spent years as an ambassador from the United States to Turkey and Italy, he knew the Mediterranean basin better than most Americans of his day; and as one of his nation's most gifted linguists, fluent in some twenty languages (most of them self-taught), he was able to wander the past landscapes of Greece, Rome, and other ancient civilizations by exploring their histories in their original tongues. Looking back across the many centuries of European history and asking how these remarkable nations had risen to such grandeur only to collapse into barbarism, he came up with an unexpected but potent answer: the great civilizations of antiquity had disappeared, he said, because they abused the natural environments on which their own prosperity depended. Combining the role of historian with that of prophet, Marsh concluded his survey of the ancient past by looking forward to what it might portend for the future of his own nation. His message was a chilling one: if Americans failed to learn from the frightening environmental examples of long-vanished civilizations, their young nation could all too easily suffer the same fate.

Just how had environmental degradation contributed to the decline and fall of Rome and other ancient societies? Marsh's answer to this question was subtle and complicated, but in his mind one explanation stood out above all others: deforestation. By cutting their woodlands too aggressively, the ancients had brought upon themselves a host of environmental misfortunes. Erosion accelerated to disastrous levels, eventually undermining the soil fertility on which agriculture and other human activities depended. Watersheds, freed from the entangling web of roots beneath the surface of the ground, lost their ability to hold the precipitation that fell on them from the heavens. As a result, the flow of water became more erratic, making both droughts and floods more common. Whereas formerly, streams had flowed year round, now they carried too much water in times of abundance and too little in times of scarcity. Springs began to dry up. Even patterns of climate might possibly have changed, rains becoming scarcer without the evaporative powers of forests to sustain them (though Marsh was less convinced of this than many of his contemporaries). The result was a landscape dryer, wetter, hotter, colder, with less soil and lower fertility, than had existed before human axes and human fires had done their dangerous work. Environmental disaster led willy nilly to economic and social disaster, and the mainstays of civilization were weakened as a result.

Although not all of Marsh's diagnoses have been fully confirmed by subsequent scientific research, in the main his arguments have held up remarkably well. And about their effects on nineteenth-century contemporaries who read his book, there can be no question whatsoever. *Man and Nature* encouraged the passage of the 1873 Timber Culture Act to encourage settlers on the Great Plains to plant trees as a way of increasing rainfall. Far more important to the history of conservation, though, was the role the book played in laying the groundwork for protecting American forests. Its arguments helped buttress efforts in New York State during the 1870s and 1880s to create a permanent "forest reserve" in the Adirondack Mountains as a way of protecting the watershed of the Erie Canal and the Hudson. The result was the creation in 1885 of Adirondack State Park, at over 700,000 acres the largest wildland park east of the Mississippi River. This in turn laid the groundwork for the passage by Congress in 1891 of the Forest Reserves Act, giving the President the power to set aside tracts of forest land at the headwaters of rivers supplying major American cities with water. The law's justification was straight out of the pages of George Perkins Marsh's book, and it created the national forests as we know them today: no small achievement for a book that too few modern environmentalists have ever bothered to read.

In 1958, David Lowenthal published what has remained the standard biography of Marsh, but it was in many ways a book before its

time. Written just before environmentalism began to emerge as one of
the most important political movements of the late twentieth century,
it arrived on the scene at a time when too few readers had even heard
of Marsh, let alone recognized his crucial significance to the history of
American conservation. The book's title didn't help much: *George Perkins
Marsh: Versatile Vermonter*. One can certainly understand Lowenthal's difficulty in trying to figure out how to describe a man who had served in
Congress, been one of his nation's chief diplomatic representatives in
Europe, made major contributions to the development of nineteenth-
century linguistics, led the movement to create more sustainable fisheries in New England . . . *and* just happened to write the founding text
of the modern conservation movement. But although it was widely
praised, won prizes, and led to a new edition, edited by Lowenthal, of
Marsh's *Man and Nature*, too few people ever read this important biography. Lowenthal went on to a distinguished career as one of the leading historical geographers and intellectual historians of his generation,
but his first book was never reprinted and has received less attention
than both the book and its subject deserve.

I was delighted, therefore, when David Lowenthal responded en-
thusiastically to my suggestion that we reprint his pioneering biography of George Perkins Marsh in the Weyerhaeuser Series. I was even
more delighted when Lowenthal said he would like to look over the old
book and see whether perhaps it should be updated. None of us knew
at the outset just what this might mean, but as time went it became clear
that Lowenthal wanted to revise his earlier book quite dramatically in
light of recent scholarship, and that he was willing to revisit archives
not just in the United States but in Europe as well to gather a wealth of
new documents that had not been readily accessible to him at the time
he first wrote the book. As he accumulated more and more documents,
both primary and secondary, it became increasingly clear that the book
he would be giving us would not be a reprint at all, but would go well
beyond what is ordinarily even called a "revision"; it would in fact be a
new biography.

What Lowenthal makes clear in this biography is just how far
ahead of his time George Perkins Marsh truly was. Marsh expressed the
values and concerns of his own generation, his remarkably supple and
wide-ranging mind cast backward and forward in time to rethink the

long sweep of human history even as he sought to reimagine what our environmental future might be. Lowenthal does a superb job of showing us how the many facets of this very complicated man all contributed to his special vision. Marsh's astonishing facility with languages opened the window on past civilizations, giving him his opportunity as both historian and prophet. His extensive involvement with domestic and international politics, whether as a member of the House of Representatives or as an ambassador to Italy and Turkey, meant that his scholarship was never merely academic but always pointed out toward the wider world of affairs. Devoted to his home state of Vermont, he was also one of the most widely traveled Americans of his generation, giving him a perspective that was local and global at the same time. His early work as a fish commissioner in Vermont gave him direct experience with what a later generation might call "ecological restoration." It also encouraged his belief that science and enlightened self-interest really could make a difference in reversing damage done by past environmental misuse. Marsh offered his contemporaries an almost apocalyptic vision of what might happen to America if it repeated the environmental errors of past civilizations, but he combined this dark prophecy with a deeply optimistic faith that disaster could be averted if only people responded in time.

Marsh's message remains no less valuable today than it was when he first offered it in 1864. Our hope is that the time has come when environmentalists and other readers of this biography will finally recognize just how important its subject truly is. Few Americans have contributed more dramatically than Marsh to the protection of the natural environment not just in the United States but the world as well. Although the dense, challenging prose of *Man and Nature* will never attract as many readers as Rachel Carson or Aldo Leopold, there can be no question that George Perkins Marsh himself remains a pioneer of conservation whose legacy continues to benefit us all. Those who read David Lowenthal's fine biography will finally be able to understand and appreciate what a remarkable man Marsh was, and what an extraordinary difference he made to the natural ecosystems that continue to sustain us in no small part because of what he wrote and did.

Preface

THE LIFE OF GEORGE PERKINS MARSH SPANNED MOST
of the nineteenth century. On few aspects of his era did he leave no
mark. Lawyer, farmer, manufacturer, congressman, diplomat par excellence, Marsh was the broadest scholar of his day. He was at home
in twenty languages, became America's prime master of Scandinavian
and English literature and linguistics, made signal advances in comparative philology, helped to found and foster the Smithsonian Institution, spearheaded corporate railroad curbs and irrigation control, was
a wonted arbiter of public taste in art and architecture, shone fresh light
on the history of everyday life. Above all, his ecological insights pioneered alertness to human impacts on the earth, inspiring conservation
zeal in his day and in ours. Next to Darwin's *On the Origin of Species*,
Marsh's *Man and Nature* of 1864 was the most influential text of its time
to link culture with nature, science with society, landscape with history.

Its influence endures. A centenary reprint in 1965 spurred the environmental crusade launched in 1970 by Earth Day devotees of Rachel
Carson and Aldo Leopold. *Man and Nature* persists in public demand,
successive crises ever rekindling its relevance. As Marsh has been held
"the last person to be individually omniscient in environmental mat-

ters," so his text remains, in Lewis Mumford's phrase, "the fountain-head of the conservation movement."[1]

In the range of his efforts Marsh exemplified the yeasty and pragmatic ferment of his time. His history and geography are most recalled—the ecological wisdom, the stunning survey of human impacts on nature, the agenda for resource husbandry set forth in *Man and Nature*. But the clue to Marsh's polymath scholarship lies in his variety of ventures, wealth of experience, and many-sided, passionate involvement in the life of his times.

In 1941 the geographer Carl Sauer issued a plea for a biography of Marsh, a venture seconded by the historian Merle Curti. I became a student of both. Marsh had sought above all to fathom natural history with a human focus and to plumb human history with a natural focus. It was to explore the striking convergence of these two goals that drew me, fifty years ago, to trace Marsh's career.[2]

Could a youngster know enough of life to chronicle a long by-gone career? Some of my 1950s' mentors felt immaturity would unduly impair my biographical dissertation. A half century later, I'm unsure that age fits me better for rewriting Marsh's life. But the years have lent me some new skills, time to reflect, and greater hindsight. Beyond being able to correct errors and repair omissions, I've relished the unique chance to become reimmersed in the manifold concerns of an eminent, ardent, quirky, larger-than-life figure, by turns aloof and engagé, droll and austere.

Marsh's place in the nineteenth-century pantheon is more exalted today than fifty years back. His feats as scholar and diplomat, his pithy views on famed men and events, and the wealth of data bearing on every aspect of his career make his life an invaluable gloss on much of American history, enhanced by European aperçus.

Fifty years have made Marsh's concerns yet more salient. His devotion to tangible signs of the past, to popular science, to women's rights, to the career and meaning of words, to archaeology and site protection, to public architecture and the arts—these resonate more than ever today. And the conservation cause he pioneered is now global. His environmental insight is less famed than that of Henry Thoreau, John Muir, or Frederick Law Olmsted, but Marsh's is at least as vital to present needs.

Lewis Mumford, who restored Marsh to public memory in the 1920s and 1930s, held my 1958 subtitle "Versatile Vermonter" belittling; "it reduces a man of world stature & world influence to a mere provincial worthy," he upbraided me. "Just a *Vermonter*—like Calvin Coolidge?" Well, there are worse epithets (and less laudable presidents). And Marsh's insights and career owed much to his being a Vermonter. That state's reputation for stubborn individualism, for local community, and for on-the-job longevity all echo Marsh—like him, as Tom Bassett noted of land-grant college founder Justin Morrill's forty-four-year congressional career, a lot of Vermonters never really retire.[3]

Marsh also fits Vermont's repute for frugal probity. Asked what his congressional elections had cost him, he replied, "Not so much as a glass of wine" or any favors to constituents: *"In my State such things are not necessary."* A like seemliness led Dorothy Canfield Fisher's emigrant grandfather to conceive Vermont as national memory: "Keep it just as it is, with Vermonters managing just as they do—so the rest of the country could come in to see how their grandparents lived."[4] His idea presaged the national park created in Woodstock in 1998 to honor Marsh himself. But Marsh's stature and import far transcend his Vermont links; the epithet "versatile" did scant justice to his polymath talents.

Every career, like all history, must be assessed afresh by each generation. New data surfaces; insights alter; readers expect other ways of chronicling a life. Five specific challenges crucially reshaped my work on Marsh, making this in most ways a new book.

New primary sources. Archival data by and bearing on Marsh, already copious fifty years ago, have been much augmented. The papers of Hiram Powers, the neoclassical sculptor famed for his *Greek Slave*, are of prime import. Marsh and Powers were lifelong friends; their letters, found in Florence in the 1970s and now at the Archives of American Art, offer new insights on art and public life in America and Italy. Other new troves include the Crane Family Papers (Marsh's wife's family) in the New York Public Library, additions to Marsh papers at the University of Vermont, at the Vermont Historical Society, at Columbia University, and archival sources in Italy for Marsh's scholarship and diplomatic work in the 1860s and 1870s, the crucial decades of Italian unification.

New historical insights. Fifty years' work has amplified, revised, and often overturned received views on every enterprise Marsh engaged in. New insights revolutionize studies of language, race, national identity, religion, and nature in America and Europe. On Vermont society and culture, the Smithsonian Institution, antebellum American politics, European and Middle Eastern affairs, language theory and practice, and social and natural history, research throws fresh light on Marsh's views and influence. In particular, a celebratory view of the Risorgimento as a popular national crusade, previously conventional wisdom among historians of Italy, has given way to awareness of Italian unification as a largely elite and deeply flawed venture.

New biographical expectations. What readers expect of life stories changes apace. Ancient and medieval chroniclers dwelt on eternal truths or stressed momentous personal conversions. Biographers of Boswell's time limned bizarre eccentricities. Nineteenth-century hagiographers aimed to canonize their subjects, Lytton Strachey and Leon Edel to expose their defects or self-delusions, post-Freudians to ferret out their sexual fixations.[5] "Oh!" lamented fellow envoy and historian George Bancroft to Marsh in 1869, "these children and biographers who never leave in the dark what belongs there."[6] Even if curious about what remains hidden in Victorian shades, a biographer need not pander to current taste for sordid trivia, let alone succumb to deconstructors who deny a life any coherent meaning.

Marsh himself shunned brooding and autobiography alike. Finding "continued retrospection hurtful," he seldom spoke of his early childhood. "The best ordered life is that which is least haunted by its own past," he declared. "We must cast no backward looks at ourselves," he warned a niece, "and must strive to lead an objective, not subjective, life."[7] Only in correspondence was Marsh's subjective life revealed. "How little, after all, we knew this man!" exclaimed a Vermont neighbor when a volume of Marsh's letters appeared six years after his death. "We met him often, on the streets, at church, in our houses, this portly, grave, courteous man, this most impressive and instructive talker, when on rare occasions he talked freely to any but his intimate friends. But we felt that none save those friends knew, if they knew, what lay behind those spectacles in that large brain."[8]

Yet Marsh the historian saw personal privacy at odds with the demands of posterity: "Precisely the disclosures we shrink most from making with respect to ourselves, and the outspoken expressions we are shyest in using, attract us most in the life of distant ages. Refined and sensitive persons destroy their family letters, and are reluctant to record their names in albums," or on monuments. Marsh realized that personal mementos, even vainglorious graffiti, revealed more than any other verbal traces.[9]

We are fortunate that Marsh and many others heeded that precept. On returning to America, a year after his death in 1882, his widow gathered letters from friends the world over for a projected life by Samuel Gilman Brown (son of Francis Brown, president of Dartmouth when Marsh studied there).[10] When Brown died in 1885, Caroline Marsh took over what he had barely begun, compiling a first volume of *Life and Letters* (up to their departure for Italy in 1861). She cut, sanitized, and but scantily annotated his letters and sought to conceal many identities. Yet her perseverance, insight, and vivid portrayal of Marsh's passions and prejudices are invaluable. Honoring his "abhorrence of exaggeration or even concealment," she strove with much success not "to represent him greater or better than he was."

A paralytic stroke in March 1888 aborted Caroline's second volume. Then living in Scarsdale, New York, with daughters of her nephew Alexander B. Crane who had been much with the Marshes in Italy, Caroline rallied only enough for reading and desultory writing; she died in October 1901.[11]

Except for Sister Mary Philip Trauth's perceptive 1957 survey of Italo-American diplomacy during Marsh's ambassadorial tenure in Italy, no substantial study of Marsh appeared before my 1958 book. But for summary sketches, little on Marsh has emerged since.[12]

The altered biographer. In 1994 Michael Holroyd began revising the book on Lytton Strachey he had published in the 1960s. As he read what he had written, he was dismayed by his egregious errors and abysmal prose. "What a naive, opinionated young fool I was," he told an audience. But his rewriting was stymied at every turn. "What a naive, opinionated old fool I must have now become!" he thought.[13] Holroyd's new edition needed to be finer, more mature, *different*; surely he could improve on the first. But in the end he chose to hew to the muscular essence of the original.

So too my initial plan for Marsh. I would keep the syntax and structure of my younger self; altering it would violate the flow as well as the content and framework of the narrative. I would confine textual changes to words and phrases that now seemed obscure or jarring or anachronistic. But as I went on, more and more of the original seemed not just outdated but deeply flawed. The lapse of time, far longer than Holroyd's, made much obsolete. It was not enough just to correct errors; I had to reconsider histories, reassess motives and outcomes, revise and reverse judgments. This forced me to unravel, even jettison, earlier work and add new matter in its stead.

Environmental understanding. Eulogies on Marsh's death judged *Man and Nature* of less moment than his language studies. Only since the 1930s have his insights and warnings on environmental degradation been accorded more consequence. Marsh's Woodstock birthplace is now a national historical park, designed to convey the story of American conservation. That designation attests his pioneering primacy.

Mounting public anxiety has broadened and deepened concerns over human damage to the earth. The perfervid polemics of today's environmental causes echo Marsh's apocalyptic tone. But they lack his basic optimism, grounded in Enlightenment faith in progress. Technological advance spurred him to advocate public stewardship of natural resources. Marsh was convinced nature must be controlled; once brought under human dominion, the environment required ceaseless care, lest nature and society alike decay. As he put it, once man ceased to be nature's master, he became nature's slave. Marsh understood the critical distinction, blurred by today's ecocentrists, between "first nature"—the pristine environment prior to human impress—and "second nature"—the world as irrevocably overlain by human agency. My final chapter suggests how Marsh's environmental insights may be of use to us, who inherit the earth whose devastations he so vividly portrayed and deplored.

It would be an error to enlist Marsh in support of any current environmental credo. He was a man of his time; his perceptions like his concerns may yet inspire us, but they are bound to be anachronistic.

Today's main environmental fears—global warming, ozone-layer depletion, radioactive emissions—were unheard of in his time or, for that matter, even in Rachel Carson's time a century later.

Yet Marsh's resonance remains potent; he faced human dilemmas that strike us as both familiar and novel. His was a life of rare felicity in its amalgam of chance and fulfillment, furthered by a family network even then unusual. His environmental insights emerged from his political, entrepreneurial, and diplomatic careers. These callings gave him the occasions to observe and reflect on an awesome range of physical and social milieus.

A century ago Carlyle's dictum that "History is . . . the biography of great men" was engraved on the west corridor of the great hall of the Library of Congress. It is now out of fashion. Marsh himself stressed that humble and unsung lives were as deserving of memory as those of the great, and collectively of far more consequence for both human and earth history. Yet for the insights he signally added to our world view Marsh's own life is unusually worth study, as I hope to show in what follows.

Money and the lack of it occupied much of Marsh's time and energy. To appreciate the sums that concerned him, the dollar figures in the text should be multiplied by twenty.[14]

Acknowledgments

Marsh's many-sidedness compelled my extensive reliance on countless others. Carl Sauer and John Leighly introduced me to Marsh's *Man and Nature*, Merle Curti to his historical insights, Lewis Mumford to his import as a social reformer. My gratitude goes first to them, next to Wisconsin mentors Fulmer Mood, Vernon Carstensen, and Richard Hartshorne, and to John Kirtland Wright. All were hugely generous with time and wisdom.

My Vermont debts have multiplied over time. In Woodstock, Rhoda and Frank Teagle long cheered me on, David Donath (Billings Farm Museum and Woodstock Foundation) and Janet Houghton (Billings Family Archives and National Park Service) were ever helpful, Laurance Rockefeller and Franklin and Polly Billings warmly hospitable. Rolf Diamant and Nora Mitchell at the nascent Marsh-Billings-

Rockefeller National Historical Park lent every support, as did NPS staff in Boston and Washington. I owe earlier Woodstock help to Elizabeth French Hitchcock, John H. McDill, Margaret Johnson, Loren R. Pierce, and E. E. Wilson.

In Burlington, special collections staff at the Bailey/Howe Library of the University of Vermont were peerless collaborators. Connell Gallagher smoothed my archival way; Elizabeth H. Dow and Angie Chapple-Sokol proved indefatigable sleuths on countless topics, persons, and places. Letters to and from Marsh's first wife's family arrived in the Marsh Papers at the eleventh hour through the kindness of John Hungerford Cox and Karen Perez, as did a Marsh manuscript on the history of engravings prepared for the painter G. P. A. Healy. For hospitality I am indebted to Daniel Gade, Kristin Peterson-Ishaq, and above all to Alice and Tom Bassett; Tom's kindly corrective judgment has guided my Marshiana searches over half a century. Lyman Allen, Leon W. Dean, Paul Evans, Gladys Flint, John Huden, Laura Loudon, Sidney Butler Smith, and Frank Spaulding previously gave valued help.

In Montpelier, State Archivist D. Gregory Sanford and Barney Bloom of the Vermont Historical Society found items not available when Harrison J. Conant, Elizabeth W. Niven, Arthur Wallace Peach earlier aided me.

Judy Throm, William Cox, and Eva Crider at the Archives of American Art, and Paul Theerman, Marc Rothenberg, and Pamela Henson at the Smithsonian Institution aided my review of old and alerted me to newly acquired materials; George Gurney helped with the history of Capitol sculpture. I am indebted to Antoinette J. Lee for assistance on architecture and other Washington matters beyond enumeration. Kathleen McDonough at the Library of Congress and John D. Stinson at the New York Public Library kindly located and copied Marsh correspondence, as did Tony Appel at Yale University Medical Library and Jane Lowenthal at Columbia University. Staff at the British Library, Senate House and University College London libraries, Turin's Archivio di Stato, Museo del Risorgimento, and Biblioteca della Provincia di Torino, and at scores of American and British libraries responded equably and fruitfully to my often rushed and baffling queries.

Research leads and hospitality in Alpine Europe came from Geneviève Heller-Racine and Carl Krummel in Switzerland, and above all from Anna Somers Cocks and Umberto Allemandi, whose home and connections in Turin were generously opened to me. Piedmontese prodigal with aid included Guido Gentile (Soprintendenza Archivistica per il Piemonte e Valle d'Aosta), Guido Sertorio (Arsenale di Artiglieria), Isa Massabò-Ricci (Archivio di Stato), and Mariella Tagliero (Fondazione Centro Culturale Valdese, Torre Pelice). Lavish with help and hospitality were Rinaldo Merlone in Piòbesi and Lodovico Sella and his staff at the Fondazione Sella in Biella. Roberta Ferrazza (Soprintendenza Beni Artistici di Firenze) unearthed the locale and later annals of Marsh's Florentine home, the Villa Forini. Luisa Quartermaine was a dedicated collaborator, supplier of Italian materials, and ever benign yet scrupulous critic. Richard Bosworth merits more than can be repaid for two detailed scrutinies, at short notice, of my Italian chapters.

Tamara Whited shared ideas and leads on French forestry, Marcus Hall on Italian forestry and conservation, Richard Judd on nineteenth-century New England environmental awareness, and Randolph Roth on Vermont society and religion, notably Masons and Anti-Masons. Philip Stott read my environmental chapters and gave invaluable leads on the history and rhetoric of ecology. I am grateful for Michael Williams's advice on forest history and persistent encouragement. Billie Lee Turner, Robert Kates, Mike Heffernan, Maurie Cohen, and Martin Rudwick let me explore Marsh's environmental insights with responsive audiences. Volker Welter helped on Patrick Geddes and Lewis Mumford's "discovery" of Marsh, Donald Dahmann on the reception of *Man and Nature* in America, Graeme Wynn on Marsh's influence in New Zealand, Richard Wunder on Marsh's relations with Hiram Powers, John Wolter on J. G. Kohl, Gillian Clarke and Wilhelm Hortmann on German arcana, Kenneth Olwig on Marsh's Scandinavian contacts, Joe Crystal on Florida's Naval Live Oaks Preserve, Eric Brown and Elihu Lauterpacht on boundary arbitrations, James E. Sherow on Marsh's repute among environmental historians, Jeanne Kasperson for unearthing the Russian translation of *Man and Nature*, Samantha Evans for searching the Charles Darwin Correspondence, and Suzanne Fleischman for assessing Marsh's flights of linguistic fancy. Dextrous patience by Catherine Pyke, Elanor McBay, and Nick Mann, of University College London's geography drawing office, transmuted my confused cartography into this book's maps.

Mary Alice Lamberty Lowenthal's sunny patience, unflappable expertise, and responsive enthusiasm assembled deluges of Marsh materials in coherent, legible, and virus-free form, produced innumerable almost timely drafts, and sustained unflagging devotion to the crotchety concerns of my cranky Yankee.

To Bill Cronon I am grateful for proposing this book, spurring it on, condoning delays in its completion, and inspiring me to revise linkages between Marsh's environmental and other careers. Leila Charbonneau has been an exemplary copy editor and Julidta Tarver a model and enthusiastic in-house editor, offering positive stimulus at every stage.

For advice and assistance I have also been indebted to Selig Adler, Wayne Andrews, Richard Beck, Saul Benison, Jan Broek, Francis Brown, Laura Burgess, John Christie, Andrew Clark, Francelia Clarke, Aurelia Crane, Caroline E. Crane, Hunter Dupree, Robert Durden, Ruth Miller Elson, Wilma Fairchild, Ralph Flanders, Frank Freidel, John C. Greene, Charles Griffin, Jack Harrison, Einar Haugen, Charles B. Hitchcock, Norman Hood, Sexson Humphreys, Donald Innis, Edward C. Kirkland, Mary Raymond Lambert, Carl Lokke, John Lowenthal, Max Lowenthal, Edmunds Lyman, James Malin, Howard Marraro, Bernard Noble, Marsh Pitzman, Robert Reynolds, Chalmers Roberts, Ishbel Ross, Margaret Sheppard, Elisabeth Shoemaker, Dorothy Swart, William L. Thomas, Jr., Glyndon Van Deusen, and J. Russell Whitaker.

Permission to quote manuscript materials was kindly given by the American Philosophical Society (Isaac Hayes Papers); Archives of American Art (Hiram Powers Papers, Miner K. Kellogg Papers); Billings Family Archives, Woodstock, Vt.; Boston Public Library (Charles Folsom MSS, C. C. Jewett Papers); Annmary Brown Memorial Library, Brown Univ. (Rush C. Hawkins Papers); Calais Free Library, Calais, Maine (J. S. Pike Papers); the Syndics of Cambridge University Library (Charles Darwin Papers); Columbia College Archives; Columbia Univ. Rare Books & MSS Library (Park Benjamin Coll., Charles Stewart Daveis Corr., Willard Fiske Papers); Dartmouth College Archives (College records, G. P. Marsh Papers, John K. Wright Papers); Harper & Bros. Archives, New York; Harvard College Archives (C. C. Felton

MSS); Harvard Univ. Herbarium (Asa Gray Coll.); Harvard Univ. Library (E. L. Godkin Papers, James Russell Lowell Papers, Charles Eliot Norton Papers, Charles Sumner Papers); Henry E. Huntington Library (Francis Lieber MSS); Library of Congress (Letter Book 1843–49, John M. Clayton Papers, C. S. Daveis Papers, Hamilton Fish Corr., Galloway-Maxcy-Markoe Papers, Francis Lieber Corr., S. F. B. Morse MSS, W. T. Sherman Papers, E. G. Squier Coll.); Massachusetts Historical Society (George Bancroft MSS, R. H. Dana, Jr., MSS, Edward Everett Papers, Norcross MSS, Robert C. Winthrop Papers); Minnesota Historical Society (C. C. Andrews Papers); New-York Historical Society (Albert Gallatin Papers, Hiram Powers MSS); New York Public Library (Crane Family Papers, Ford Coll.); New York State Library, Albany (Franklin B. Hough Papers); Univ. of Pennsylvania, Van-Pelt Dietrich Library (Henry C. Lea Papers); Ricasoli Archives, Brolio, Tuscany; Univ. of Rochester Library (W. H. Seward Papers, Lewis H. Morgan Papers); Fondazione Sella, Biella, Piedmont; Smithsonian Institution Archives (Corr. Rec'd, Regents' minutes, Spencer F. Baird Corr., Joseph Henry Papers); U.S. National Agricultural Library (Franklin B. Hough Papers); U.S. National Archives (Dept. of State appointment papers, Turkey and Italy legation corr., instructions, and despatches); Univ. of Vermont, Bailey/Howe Library (Jacob Collamer Papers, George P. Marsh Coll., James Marsh Coll., Karen Perez Coll.); Vermont Historical Society (Benjamin F. Bailey Papers, Marsh MSS); Yale Univ. Library (Knollenberg Coll.); Yale Medical School Library (Elisha Perkins letterbooks).

George Perkins Marsh
Prophet of Conservation

1. *Woodstock and the First Watershed*

We have a tradition in our family that sometime in the past century,
one of our ancestors crept out of the ground in Lebanon, Connecticut,
and as soon as he got money enough together to run away with,
came up to Queechy Vermont, where we have remained ever since.

Marsh, speech at Forefathers' Day, Middlebury, Vt., 1859[1]

GEORGE PERKINS MARSH WAS BORN IN WOODSTOCK, Vermont, "the true date of my happy nativity being the Ides of March 1801." So he told a friend who had wrongly antiquated him back into the eighteenth century.[2] It was a bitter winter, this first of the new century. Typhus raged in Woodstock village along the terraced flats of the Ottauquechee River, usually known as the Quechee. Across the river, the family of Charles Marsh, Esquire, Woodstock's leading lawyer and U.S. District Attorney for Vermont, escaped the lice-borne malady. Six years later the family left their riverside cottage for a brick mansion a hundred yards north at the foot of Mount Tom.

Here, overlooking the village opposite, Charles Marsh brought up a large family, entertained visiting dignitaries, and managed his farm, the choicest in the township. He owned much of Mount Tom, and his well-tilled fields and meadows covered the whole of the rich intervale in the great bend of the Quechee north of Woodstock. The new road from Royalton, a turnpike in which Marsh held a half interest, skirted his house and entered Woodstock on the Quechee bridge Marsh had built in 1797. Across this bridge to Elm Street, which he had likewise laid out, Charles Marsh would walk to his law office.

From the summit of Mount Tom, treeless since a devastating October fire in George's very early childhood, the boy could survey his entire cosmos. The windswept summit afforded an unobstructed view of the village five hundred feet below. To the west, the main range of the Green Mountains was dark with spruce, hemlock, and white pine. Seven miles to the east lay Hartford, the home of George's grandfather, Colonel Joseph Marsh. A few miles farther downstream, the Quechee entered the broad Connecticut River, spanned seven miles upstream by a bridge linking Norwich, Vermont, where young Marsh would briefly teach school, and Hanover, New Hampshire, the site of his alma mater, Dartmouth College.

This was no static panorama but one in ceaseless flux, rapidly being transformed by the forces George Marsh would so memorably limn in *Man and Nature*. Thirty years of clearing and planting had converted the wooded lower hills surrounding Woodstock into field and pasture. On higher, steeper slopes the forest was also receding, as demands for fuel and the effects of pioneer profligacy took their toll. The runoff of rain and snow on denuded hillsides sped erosion and depleted once abundant supplies of fish and game. Frequent floods washed out bridges and mill sites.

From his first years, Marsh was acutely aware of such metamorphoses. "Too sterile ungrateful and cold to furnish food and shelter even to the frugal and hardy Indian," the Green Mountains had remained for the most part "an untenanted and untenantable wilderness," he wrote half a century later.[3] "Born on the edge of an interminable forest," as he termed it, he himself saw much of it cut down for timber, fuel, and the making of fuller's soap from potash, then in huge demand for manufacturing woolen cloth, especially in England. In Woodstock "every family made its own soap, preparing also its own lye," Marsh recalled, from "the numerous potash factories supplied with material from the 'clearings' then going on."[4]

Such clearings had an impact on the fish in local streams. Sent to school in 1811 at Royalton, fifteen miles north, the ten-year-old fisherman found the White River "a very different ichthyological province from that of Woodstock." Lower-lying than the Quechee, the White River valley had been earlier and more thoroughly deforested. There Marsh caught eels and gathered freshwater clams "never found in forest streams but only [in] those cleared and cultivated."[5]

Birds were also affected. Commenting on their nocturnal migrations along natural channels now become thickly settled, Marsh in *Man and Nature* recalled a boyhood village

at the junction of two valleys, each drained by a mill stream, where the flocks of wild geese which formerly passed, every spring and autumn, were very frequently lost, as it was popularly phrased, and I have often heard their screams in the night as they flew wildly about in perplexity as to the proper course. Perhaps the village lights embarrassed them, or perhaps the constant changes in the face of the country, from the clearings then going on, introduced into the landscape features not according with the ideal map handed down in the anserine family, and thus deranged its traditional geography.[6]

Many such changes, Marsh later stressed, were at least for some time irreversible. The fierce fire that had denuded Mount Tom in his childhood consumed humus as well as trees: "the rains of the following autumn carried off much of the remaining soil, and the mountain-side was nearly bare of wood for two or three years afterward." Although Mount Tom was soon again thickly wooded, "the depth of mould and earth is too small to allow the trees to reach maturity." No trees over six inches in diameter survived, and seedlings would go on dying "until the decay of leaves and wood on the surface, and the decomposition" of rock beneath, had "formed, perhaps hundreds of years hence, a stratum of soil thick enough to support a full-grown forest."[7]

In fact, Mount Tom was replanted by Frederick Billings in the 1880s with exotics—Norwegian spruce, European larch, and white ash; thus only fifty years after Marsh's somber augury in *Man and Nature*, it was well clad in mature growth.[8] Yet Marsh's insight that the intensity

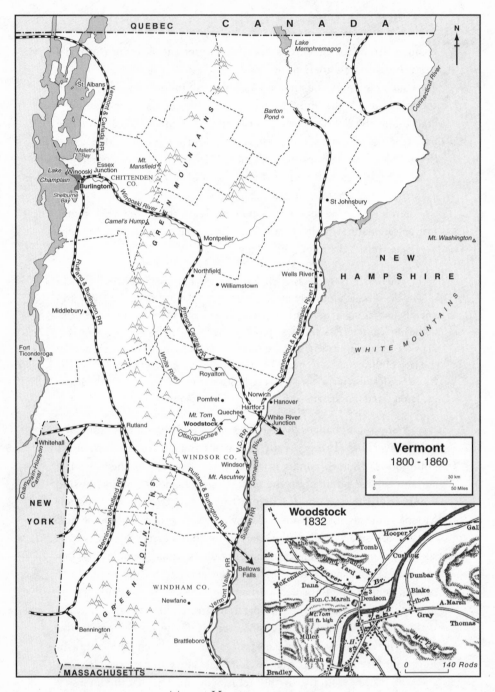

Map 1. Vermont, 1800–1860

of human-induced damage might long delay natural recuperation remains valid. And in this instance, as Marsh often came to stress, it was not nature alone but nature aided by human artifice that had restored Mount Tom.

*R*evolutionary Vermont usually calls to mind Ethan Allen and the Green Mountain Boys, brash, unschooled, ultrademocratic; but the Marshes were emphatically not Vermonters of this type. Indeed, for years they strove not to be Vermonters at all. They sought instead to set up their own Puritan commonwealth between the Green Mountains and the White Mountains, a pious, orderly, Dartmouth-focused New Connecticut. Their political arcadia unrealized, the gentry of eastern Vermont contented themselves with inculcating genteel virtues in their offspring. In frontier Vermont, George Marsh was brought up a sober Calvinist patrician.

For a long time the Green Mountains separated two hostile worlds. To the west, from Bennington and Rutland north through the Champlain Valley, came western Connecticut backwoodsmen. Dissenters from Congregational conformity, some were revivalists, others freethinkers. In orthodox eyes, western Vermont was notorious as the abode of atheists who "chuse to have no Sabbath—no ministers—no religion—no heaven—no hell—no morality."[9]

East of the mountains, settlers came from conservative central and coastal Connecticut. Orthodox in religion and Federalist in politics, many were educated and prosperous. The likes of the Marshes along the upper Connecticut and its tributaries were noted for "puritanical gravity, that shrewdness and Connecticut peddler's air, which enables them to drive a lucrative business in the humblest and most unpromising pursuits." But an 1840s visitor also found them honest and "punctual to a fault." In eastern Vermont roads were better than those to the west, fences in good repair, houses neat and weatherproof. Yale's president Timothy Dwight smugly judged, in 1803, that "steadiness of character, softness of manners, a disposition to read, respect for the laws and magistrates, are all extensively predominant in this region."[10]

Joseph Marsh, George's grandfather, exemplified such traits. With his mother, wife, nine children, and several brothers' and cousins' families, Marsh had moved north from Lebanon, Connecticut, in 1772. Hartford, their Vermont destination, was one of the Connecticut valley townships chartered by New Hampshire Governor Benning Wentworth at the close of the French and Indian Wars a decade before. In those wars Joseph Marsh had ably served. A man of substance, he was already a town proprietor before he came to Hartford, which throve on timber, potash, and wheat.

He rapidly gained further influence. Colonel of militia, convention delegate, Connecticut Valley spokesman, he was chosen lieutenant governor of the newly independent republic of Vermont in 1778. But misgivings about the radical Green Mountain Boys led his "Dartmouth College party," headed by Dartmouth president Eleazar Wheelock, to seek separate autonomy, or at least Vermont hegemony, for the upper Connecticut Valley. They failed. In 1784 the towns east of the river rejoined coastal New Hampshire, and the Marsh-Wheelock party capitulated to the Allens. Though they saw the Allens and their ilk as "friends of Hell," they were "prepared to make a pact with the devil himself" to protect their properties against New Hampshire and New York. The pact reflected little faith on either side, Ethan Allen terming the Marsh faction "a Petulent, Petefoging, Scribling sort of Gentry, that will keep any Government in hot water till they are Thoroughly brought under by the Exertions of authority." [11]

Joseph Marsh again became Vermont's lieutenant governor in 1787–89. Although radical Jeffersonians for the most part controlled the state (admitted to the Union in 1791), conservative Federalists long dominated its eastern counties. As chief justice of the Windsor County Court until 1795, Joseph Marsh was legendary for his capacious memory, logical acuity, and equable temper. Meticulous in dress and manner, he remained until his death in 1811 "a perfect Federalist gentleman . . . of the pure Washingtonian school," recalled a grandson, "and trained his children in it." [12] And trained his grandchildren in it too, recalled George Perkins Marsh near the end of his own life. [13]

Charles, eighth of Joseph Marsh's twelve children, and a graduate of Dartmouth and of Judge Tapping Reeve's famed law school in Litchfield, Connecticut, opened his Woodstock law practice in 1789—the first lawyer there, the third in Windsor County. Charles Marsh is said to have arrived in Woodstock without a shilling in his pocket beyond the price of his first fifty acres;[14] but with his undoubted talents and an influential father heading the county court, success and prestige came quickly.

In 1790, Marsh and his wife, Nancy Collins, moved from the Eagle Hotel into the frame cottage at the foot of Mount Tom, known for the next half century as "Mr. Marsh's hill."[15] Here Charles, Jr., was born in 1790, and Nancy Marsh died after giving birth to Ann Collins in 1793. Five years later Charles married Susan Perkins Arnold, also recently widowed, who brought with her a two-year-old daughter, also named Susan. Charles and Susan had five more children: Lyndon Arnold, 1799; George Perkins, 1801; Joseph, 1807; Sarah Burrill, 1809; and— the eldest son having died—another Charles, in 1821.

As a lettered man and trained advocate, Charles Marsh was at first an anomaly. Newly arrived from Connecticut, Charles's cousin Jeremiah Mason found Vermont's courts "badly organized and usually filled with incompetent men. Most of the members of the bar were poorly educated, and some of vulgar manners and indifferent morals." In Windham County's Newfane, just south of Woodstock, legal credentials scarcely mattered. "I certainly knew very little law," Mason recalled, "but that was the less necessary as my opponents knew not much more, and the judges I addressed none at all." Worse yet, "a large portion of the inhabitants were new settlers and poor, and of course not desirable clients." Mason departed to hang out his shingle in New Hampshire, where he hoped to find lawyers who were gentlemen, judges scholars, and clients well-heeled.[16]

Charles Marsh was made of sterner stuff—or perhaps he concurred with state supreme court judge Royall Tyler, who termed Vermont "a good place for lawyers" *because* "all the rogues and runaways congregated" there. Woodstock became the county seat in 1790, Marsh planning the new courthouse. For a decade he monopolized Woodstock's legal business, and for many years led the Windsor County bar. Severity and hot temper made him widely feared; for punching a legal opponent in court, Marsh was scathingly rebuked by his father, sitting as judge. He thereafter behaved with icy restraint, excoriating his foes in a low-pitched voice that compelled close attention. He browbeat ju-

ries, too. In one defense Marsh agreed with the prosecutor that his
clients were "poor and mean, wicked and criminal"; they ought to be
hanged. But, he warned, the jury must heed nothing but the evidence;
were they swayed in the slightest by his clients' reputation, "you will as
richly deserve the state prison as they deserve the gallows." Marsh won
the case.[17]

During the War of 1812, when antiwar Federalists briefly controlled Vermont, Charles Marsh was twice elected to public office. Each
time he courted public odium. In the Vermont Council of Censors,
which met every seven years to revise the state constitution, Marsh
urged a Senate elected by men of property to offset the "hasty, incon-
siderate, violent rabble" in the Assembly. In the U.S. Congress, Marsh
took a grim pride in unpopular stances. Backing a bill to raise congressional pay from six dollars a day—a bill disastrous to several politicians—he had the temerity to demand ten dollars! Like his son George
a generation later, Charles Marsh was pilloried for legislative arrogance.
Yet he remained staunchly republican; "how indignant my old Federal
father was," George Marsh later recalled, when England at the Congress
of Vienna in 1815 "abandoned the peoples of Europe to the tyranny of
priests & princes."[18]

Charles Marsh gained no public renown. Better schooled and more
brilliant than his popular father, he did not make the most of his talents.
Echoing Joseph's political and social views, Charles lacked the tact,
even the will, to make them palatable to others. He was never eager for
office; hauteur ensured he never long retained it.[19] Despite habitual
self-deprecation—inherited by and intensified in his son George—
Charles Marsh saw himself inferior only to God. To most mortals he
was clearly superior. Confident of never lying or acting unjustly, he
pitilessly pilloried anyone he thought did so. His portrait shows a man
one would not care to gainsay: a spare six-footer with a narrow, deeply
lined face and thin-lipped, derisive smile.

Reared in a stern religion, Charles Marsh remained a strict Calvinist. He helped found Woodstock's Congregational Society and gave the
land for its meeting house. A stalwart of the American Bible, Education,
and Colonization societies, Board of Foreign Missions, and Society for
the Promotion of Temperance, Marsh never deviated from their aims.

Yet he was besieged by doubt as to their efficacy and his own moral worth. It was like him to argue for banning liquor traffic by blaming himself for having "sold spirits to a person who under its influence had abused his family, and consequently been imprisoned." In grieving the untimely death of his eldest son, Marsh found himself "dumb before my Judge. . . . 'How shall I answer Him for one of ten thousand of my transgressions?'"[20] Rectitude was the mainstay of Marsh's piety; he taught his children to fear God not as a loving Father but as a righteous Judge.

Charles Marsh was gentler at home than in the courtroom, but a tyrant nonetheless. George's father was as quick to condemn stupidity as to denounce a lapse from probity; a young Marsh ought to know as much as possible about everything. Thoroughness was a lesson George learned early. Watching his father inspect the building of their new house, he "never forgot the stern rebukes administered to the laborers when their work was found to be ignorantly or carelessly done."[21] Slow and negligent learners took a back place in the family circle; George was impelled by more than mere curiosity to absorb the encyclopedia from the age of five. His family viewed him as a paragon because he yearned to know everything from mechanics to needlework; precocity aroused admiration, exemplary conduct exempted him from criticism. At an age when most children are learning the alphabet, George Marsh was well on the way to becoming a pedant.

His rigorous childhood training left lasting traces. Marsh ever reveled in encyclopedic knowledge; versatility was a passion for this polymath in myriad diverse realms. He hungered for data of all sorts. And he forever interjected recondite lore into letters and essays, sometimes to illustrate a point, often simply as self-indulgent fun. Marsh never ceased finding virtue in facts, or delighting in knowing so many of them. In the home of Charles Marsh this had paid well.

George's mother, unlike his father, was brought up in an ambience of quixotic fervor. Susan Perkins was born in Plainfield, Connecticut, in 1776, one of the nine children of Dr. Elisha ("Terrible Tractor") Perkins, an innovative and enterprising physician. To make ends meet for his large family, Dr. Perkins boarded local academy students and ran a lucrative business in breeding and trading mules. But his main hopes lay in improving on Franz Mesmer's by then discredited "animal magnetism" and on the "electrical" effects of metals that Luigi Galvani applied to nerves and muscles. In the early 1790s Perkins found rheumatic and other pains relieved and cures effected by stroking his patients with a polished knife blade or keeping a lacquered metal comb in their hair, "the most valuable discovery ever made," he assured his daughter and her first husband, for which an exclusive monopoly "would make me & mine as rich as we ought to wish to be." Since knives and combs yielded small profit, Perkins peddled pairs of three-inch gold- and silver-painted curative magnetic rods, pointed at one end and rounded at the other. In 1796 he secured a patent for his "tractors" from Congress in Philadelphia; George Washington among other leaders acquired a set. While Mesmer's method was said to succeed only with the "lower classes," Perkins's tractors were especially popular among "men of science and respectability."

Dr. Perkins was a entrepreneurial healer. A million and a half sets of his tractors, whose materials cost one shilling, were sold for five guineas a pair. Connecticut's Medical Society denounced Perkins's "bare-faced impositions" and expelled him as a user of nostrums; Perkins and his son Benjamin rebutted with testimonials from scores of statesmen and ministers and cited two thousand cures, notably of infants and horses.[22] In London, Benjamin set up a Perkinistic Institute, satirized by Vermont poet Thomas Green Fessenden:

> With powers of these Metallic Tractors,
> He can raise dead malefactors;
> And is reanimating daily,
> Rogues that were hung *once*, at Old Bailey.

Dr. Perkins seems to have deceived himself along with others. Touting a magnetic cure for yellow fever, he went to New York in 1799 to test it and died there of the disease at age fifty-nine, two years before Marsh's birth.[23] His grandson's name acknowledges Elisha Perkins's memory and a legacy to his daughter Susan March.

Four years before her father's medical martyrdom, black-haired, dark-eyed Susan Perkins had rejected haughty Charles Marsh in favor of the romantic lyricist and lawyer Josias Lyndon Arnold, a Rhode Island emigré to St. Johnsbury, Vermont. Susan coped easily with frontier hardships in the northern wilderness, but within a year her husband fell ill and died. Likewise widowed, Charles Marsh renewed his suit, and this time she accepted him.

Of Susan Perkins Arnold Marsh's long and useful life little is known; George himself rarely wrote of her. "Her beauty was of the queenly type," a contemporary reminisced, "Juno and Venus in one."[24] Mental and moral training she left to Charles, but, unusually demonstrative for a New England mother of her day, she taught her children not to be chary of expressing affection.[25] Sympathetic warmth, common sense, and good humor stand out in her dealings with her husband, her children, and—crucially for her son George, as will be seen—her grandchildren.

The good people of Woodstock have less incentive than others to yearn for heaven," proclaimed Vermont Senator Jacob Collamer in the mid-1800s. The Woodstock of Marsh's boyhood was already an uncommonly attractive village. In 1800 its forty-five frame houses held two hundred and fifty inhabitants, with two thousand more in the surrounding township, the fourth most populous in Vermont. Scores of small farms were scattered over the neighboring hillsides, for the migration that later emptied the hills to swell the villages and lowlands had hardly begun. "The first roads ran along the ridges," where settlers also built, Marsh later wrote, "because there only was the earth dry enough to allow of their construction." But low-lying land cleared by logging had since become drier, enabling highways and habitations to move "from the bleak hills to the sheltered valleys," in Marsh's eyes "one of the most agreeable among the many improvements" evident in Vermont.[26]

Turn-of-the-century Woodstock epitomized Timothy Dwight's paean praising the Connecticut Valley. Unlike Hudson Valley towns "where nothing but mercantile and mechanical business is done [and] beauty of situation disregarded," Woodstock was "a place, not where trade compels, but where happiness invites to settle, [and] the beauty of the scenery, scarcely found in the same degree elsewhere, becomes a source of pride as well as of enjoyment." Early on, Woodstock was known as "an elegant little place."[27]

Crossing his father's bridge to broad, tree-shaded Elm Street, young George first passed the small schoolhouse, then his father's law office, Charles Dana's dry goods store, John Carleton's brick shop and saddlery, and Amos Cutler's cobbler's shop. Past the rickety county jail the road branched northeastward downstream along the Quechee, and southward upriver along the Green. Around this open oval stood the village's chief buildings: the grand new courthouse overlooking the river, Benjamin Swan's pearlash factory at the upper end, and the premises of Isaiah Carpenter, bon vivant and bass violist, from whose printing press emerged the *Northern Memento*, Woodstock's weekly paper.[28]

Opposite the courthouse was the two-story Eagle Hotel, ample enough to lodge half the state legislature when it met in Woodstock in 1807. A rendezvous for farmers, merchants, and the courthouse crowd, the Eagle dispensed salt cod and gingerbread with Woodstock's famous cider brandy, gin sling, maple rum, potato whisky, punch, toddy, and eggnog. Distilling was a major industry in turn-of-the-century Vermont; liquor flowed freely. Other types of license also throve; Woodstock was infamous as a place "where the greatest indecorum between the sexes, is habitually practiced and countenanced." But George's morals were not sullied by association with the young bloods at the Eagle or the night owls at Samuel Chandler's countinghouse. The Marshes not only kept a strict Sabbath; they adhered to Sabbath sobriety all week.[29] The lawyers George met at his father's house and office little resembled the unlettered boors Jeremiah Mason had complained of: many were college men versed in English letters, their courtroom prose modeled on Addison, Swift, Steele, and Johnson.

George had no lack of companions. Young Danas, Swans, and Churchills abounded, and a score of cousins lived nearby at Quechee and Hartford. Elder brother Lyndon was an inseparable playmate. A favorite and gifted storyteller was the future sculptor Hiram Powers, grandson of Dr. Thomas Powers; Marsh and Powers remained lifelong friends.

Often-visited friends were the Paines, forty miles north in Williamstown. Ex-senator and four decades a federal judge, Elijah Paine kept daily weather records for Zadock Thompson's pioneering Vermont almanacs, turnpiked a twenty-mile route to Montpelier, and made merino wool (from Portuguese sheep imported in 1810 by William Jarvis south of Woodstock) into Vermont's major industry, through Paine's broadcloth mill in Northfield. By the time George Marsh left Woodstock in 1825, still more forest had been cleared to pasture Vermont's half million sheep, denuding even heavily bouldered slopes. Overstocking quickly eroded the glacially swept thin soils, causing streams to silt up and floods to devastate their shores, in Marsh's own classic account in *Man and Nature*. The multifaceted Paine was one model for George Marsh's own career, unsavory as were Paine's addiction to snuff and his hypocritical temperance preaching.[30] Paine's daughter Caroline would later travel with the Marshes in Turkey and Egypt. A son, future governor Charles Paine, would ruinously enmesh Marsh in railroad speculation.

The picture later drawn by Marsh's wife, Caroline, of George as a timid, gentle, solitary bookworm who preferred to be indoors with his sisters' friends, must be qualified by Marsh's own memories. A half century after their encounter as six-year-olds, Marsh wrote that a St. Johnsbury relative of his mother's "professed to have forgiven, though he had not forgotten, [my] getting mad at him, chasing him over the house, and kicking him when I caught him." A Woodstock companion recalled roaming far and wide with George and Lyndon, "stretching willows for whistles or elders for pop-guns, climbing high rocks, ascending Mt. Tom, losing ourselves in the woods," where moose, wolves, and catamounts were still to be seen. George reconnoitred the Quaking Pogue, a fear-inspiring bog of unknown depth on the far side of Mount Tom.[31]

He learned marksmanship from Revolutionary War veterans, fished in the Quechee, watched wrestling matches, barn-raisings, sheep-shearings, and horse-racing along the river road below his home. Regimental musters after the harvest were festive events; and as the militia marched out from the Common, cider, gingerbread, cakes, and pies filled parade-ground booths in the meadow east of the Marsh place.[32] During the War of 1812, zealots paraded with snare drum and fife. The Marshes, Federalist peace-lovers, kept aloof from some of this frivolity, though Charles Marsh officiated at such state visits as President Monroe's and the aging General Lafayette's. And George recalled Isaac Hull's "Old Ironsides" naval triumph of 1812 over the British as vividly as his mother had delighted, when a child, in feting General Rochambeau's troops after the victory at Yorktown.[33]

*T*he Quechee, "most *rubato* of Vermont rivers,"[34] was the region's chief focus. Like most tributaries to the Connecticut, it is nowhere navigable but was full and swift enough in Marsh's boyhood to power a score of mills. The river was the font of farming as well as of factories. The light-brown, fine sandy loam of the fluvial terrace on which lay the Marsh farm was the most productive and easily tilled soil in eastern Vermont.

From its Green Mountains' source the Quechee alternately dashed over rapids and meandered along fertile intervales, here roaring through a deeply incised gorge, there moving placidly past alder-covered banks. Icebound during winter, the river opened up in April, overflowing the meadows beyond its banks. In summer it shrank to a trickle, and the mills shut down until the fall rains came to turn their wheels again.

Turbulent before white settlers came to the valley, the Quechee's flow was still more erratic by George Marsh's youth. When farmers cleared fields and cut down trees for timber, fuel, and potash, and then for pasturage for sheep, the denuded hillsides could no longer absorb as much rain and snow. Instead of percolating into forest litter and soil, water rushed unchecked into the river; snow melted precipitately in spring thaws; floods were more frequent. And in summer the river might dry up entirely. Even as a youngster Marsh noted these extremes intensifying.

Destructive freshets were now common. Charles Marsh's Quechee bridge had to be rebuilt three times in a decade, and the Royalton coach shook it so severely that passengers were relieved to reach the safety of the Eagle Hotel. In July 1811 a flood breached the bank, destroying Charles Marsh's high stone wall and a sawmill upstream. Despite such losses, Woodstock boasted many flourishing mills. Besides Jabez Bennett's village-center gristmill, there were a flaxseed oil mill, carding machines, a fulling mill, clothier's works and dye house, and a gin distillery and malthouse; in West Woodstock, Moses Bradley manufactured pots, pitchers, and milk pans.

Woodstock's agricultural and woodland produce were carted over the plank turnpike south to Windsor or on the common road east to White River Junction, thence across the Connecticut and south to Hartford, Connecticut, or east to Boston. After 1810, when canals were built around Quechee Falls at Hartland and Olcott's Falls south of Hanover, the Connecticut River offered a cheaper shipping route. Log rafts and flatboats fitted with sails went all the way from Wells River, fifty miles north, down to Hartford. Tea, coffee, salt, spices, molasses, rum, and household furnishings reached rural Vermont and New Hampshire in exchange for farm produce, wool fleece and cloth, timber, potash and pearlash, maple sugar and syrup, and ginseng.[35]

As a farmer, Charles Marsh was closely involved in all of this. Although hired hands did most farm work and road and bridge building, farm matters absorbed him. So did cooking and eating. George's account shows that home held more than the cold comforts of Calvinism and litigation:

That I am addicted to the pleasures of the table I utterly deny, but I confess I am a little critical in roast ham & pork & beans. This however is but a proof of my filial affection. I have to thank my parents both for my taste & my knowledge. . . . My mother . . . considered within herself the nature & capabilities of pork, & the exigencies of the human palate, and she created, evolved out of the depths of her own consciousness, the splendid result—roast ham. . . . Well, of course my father could not be otherwise than a lover of baked pork & beans. What good man is? Some there may be who never tasted them, having been cursed with a birth out of N[ew] E[ngland]. Others there can be none. . . . But as all wise men know, there are varieties of beans, some good, many indifferent, more bad. My father emigrated from Connecticut. His first crop (all the seed he carried from C) was cut off by a late frost. He tried the neighbours, sent hither & thither, bought Shaker seed beans, but all to no purpose. A baking bean he couldn't get. . . . In this extremity what did he do? Did he turn Jew, or Mohammedan, & forswear pork? Did he profess himself a Pythagorean & renounce beans? Not a bit of it. He just sat down & invented a new bean for himself. . . . My father's bean—it is a bush bean, of course; everybody knows that—is a small white bean, of regular shape & proportions, nearly cylindrical, with hemispherical ends, skin as thin as Mrs. ——'s cuticle, & flesh when baked as soft as her hand. No damned crust, no globular segregation into indigestible pellets, but a carnation to the eye, homogeneous to the touch, ambrosial to the palate.[36]

Though never a devotee of fleshpots, George Marsh wore no hair shirt and did not abstain from creature comforts. "You used to think," recalled an elder brother, "that Father in his occasional remarks at family prayers, dwelt [too] largely upon the 'blessings of poverty', which you could not appreciate."[37] Later travels among the poor of the Old World persuaded Marsh that such supposed "blessings" were simply humbug.

George was a habitually serious, precociously adult-looking child. At the age of five or six he began Latin and Greek, tutored by his eldest brother. He owed much to Charles Jr., "who excited my curiosity about books, when I was not much more than an infant, and who kindled my love of knowledge to a passion."[38] The passion endured throughout life.

His mother tried to pry him from his books, but the boy read everything he could get hold of. One favorite was Abraham Rees's *Cyclopaedia; or, Universal Dictionary of Arts, Sciences, and Literature*, whose huge volumes were almost too heavy to lift. George enjoyed astounding his father with the scope of material he absorbed. But poring over the encyclopedia's pages in poor light strained his sight so severely that when he was seven or eight he almost went blind. For many months any light caused intense pain, and for four years he could hardly read at all.

But Marsh gained something from this ordeal. Having to be read to by others, he developed awesome powers of memory; and reading aloud became a lifelong habit. What he later termed "an excessive, almost preternatural, sensibility or sensitivity to sound" gave him an acute ear for foreign languages and a love of music; he could always identify by tone the type and number of instruments in a band, and when convalescing would compose music in his head.[39]

He also gained a love of nature that never ceased to sustain him. When he left his darkened room, his vision was at first so blurred that he could scarcely make out familiar landmarks; but as his sight improved, the boy explored landscapes with the same zeal he had studied Rees's encyclopedia.

Marsh was fortunate to be thus driven at an age when his experience of nature was direct, intimate, naive, and vivid. He grew up a practical-minded Yankee who prided himself on devotion to utility, yet he was equally a romantic, attracted to Thoreau and to the Transcendentalist mystic poet Jones Very, whose lyrics accorded with Marsh's "sense of the delight of life in close contact with nature." Very's sonnet "Nature"—"The bubbling brook doth leap when I come by / Because my feet find measure with its call"—echoed Marsh's own Woodstock childhood. Marsh was "forest-born," he reminisced; "the bubbling brook, the trees, the flowers, the wild animals were to me persons, not things." As a lonely boy he had "sympathized with those *beings*, as I have never done since with the *general* society of men, too many of whom would find it hard to make out as good a claim to *personality* as a respectable oak." Marsh took it amiss when friends failed to share his joy in nature; he felt Rufus Choate's "want of sympathy with trees and shrubs and rivers and rocks and mountains and plains" a decided character defect. A similar "total want of sympathy with nature" Marsh felt disqualified John Ruskin as a "guide to any knowledge, physical or moral."[40]

Young Marsh would have been false to his father's training had he neglected the science of nature. He recalled jolting along ridgetop roads in a two-wheeled chaise, sitting "on a little stool between my father's knees," when he was four or five. "To my mind the whole earth lay spread out before me. My father pointed out the most striking trees as we passed them, and told me how to distinguish their varieties. I do not think I ever afterward failed to know one forest-tree from another." His schooling in geography began when his father "called my attention to the general configuration of the surface; pointed out the direction of the different ranges of hills; told me how the water gathered on them and ran down their sides. . . . He stopped his horse on the top of a steep hill, bade me notice how the water there flowed in different directions, and told me that such a point was called a *water-shed*."[41] Marsh never forgot the form of the land or the forces that shaped it. But the awareness of watersheds so crucial to his *Man and Nature* has more to do with a

later, topographically opposite sense of the term—not as a line dividing drainage areas but as the whole gathering ground of a river system,
whose waters must be controlled and conserved.[42]

George's formal schooling was sporadic and negligible. Latin and
Greek with his brother, geography and morality with his father, and his
own encyclopedic reading taught far more than intermittent attendance at the common school on Elm Street. Even after he regained his
sight, his eyes remained weak and his hearing impaired, and he was often too ill to go to school. At other times, sickness shut down the school
altogether. An 1811 Woodstock epidemic of "spotted fever" (cerebro-
spinal meningitis) led Charles Marsh to send George away to school in
Royalton, fifteen miles north. At Woodstock in 1814, interim teacher
John Powers Richardson (Hiram Powers's cousin) recalled Marsh among
other boys on the left; on the right were "the red-cheeked Marsh girls,
the round-faced Warrens, the pale-visaged Holtons."[43]

Religious instruction was stern but sporadic. Woodstock's Congregational meeting house was, like many, little more than a temporary
base for itinerant preachers. In 1810 Charles Marsh persuaded Walter
Chapin, a young Middlebury College graduate, to fill the Woodstock
pastorate. But Chapin's missions to convert the heathen required frequent jaunts all over New England. When in Woodstock, the amiable
minister dwelt more on the forms than the spirit of religion. Adherence
to doctrine was primary; without rigid principle, Chapin was fond of
saying, "morality was but the 'ghost of departed virtue.'"[44] Such sermons inoculated Marsh against later proselyte fervor in revivals where
many classmates succumbed to piety. But Chapin cannot be blamed for
the doubts that precluded Marsh from becoming a full Congregational
communicant. Facile faith in the triumph of truth, *magna est veritas et prae-
valabit*, always struck Marsh as dubious. "Even when a boy," he later recalled, "I used to say 'Satan is mighty and will prevail.'"[45]

At twelve George was set to go to North Yarmouth, Maine, as a
pupil of the Reverend Francis Brown, later president of Dartmouth.
"Master George is very impatient for the time I shall send him to your
care," wrote Charles Marsh. He added that "the poor boy . . . reads more

hours every day, besides going to a common school, than anyone in the family."[46] But George's health precluded his going.

Bent on his son getting a dose of conservative orthodoxy, Charles Marsh in 1816 sent him to Phillips Academy at Andover, Massachusetts. George boarded in town with three future ministers and two others who did not survive their schooling. Some ninety students suffered prison life at the country's first prep school. The curriculum was mainly Latin and Greek, which George already knew, and religion and morals, in which he was not remiss. Besides daily prayers, the boys recited on Saturday a ten-page lesson from John Mason's *Self-Knowledge . . . and the Way to Attain It* (1745), spent all of Sunday at church and Bible classes in the unheated Meeting House, and on Monday abstracted Sunday's sermons.

The headmaster was John Adams, a noted disciplinarian. "He was very religious," summed up the cultivated Josiah Quincy, who had endured Adams's rule a few years earlier, "but had no literary tastes." Adams aimed to secure faith, not to instill zeal for learning. Frequent revivals converted scores of youths. "There will be a prayer-meeting," Adams would thunder as the regular service ended; "those who wish to lie down in everlasting burning may go; the rest may stay."[47] George stayed only a few months at Andover, but Adams immunized him for life against religious authority. "The sons of schismatics," Marsh later exhorted, "let us not dishonor our parentage by anathematizing schism among ourselves."[48] Blind obedience begot only abject conformity.

That George Marsh should go to Dartmouth was foreordained. Brother Charles had graduated; brother Lyndon and cousin James Marsh were there. For a Marsh to desert Dartmouth in its hour of peril was unthinkable. A year before George came, the trustees had deposed John Wheelock, son of the founder, and offered Francis Brown the presidency. "Should you disappoint us," Charles Marsh wrote to Brown in the name of the Board, "we shall be thrown into a state of absolute despair, and the College must sink . . . into a seminary of Socinianism." Terming Marsh and his fellow "aristocratic" trustees bigots and persecutors, Wheelock persuaded New Hampshire's ruling Democrats to amend Dartmouth's charter, disempowering the existing Board. Charles Marsh's men held fast against the "mob tyranny [of] agrarians, infidels, Jacobins, sans-culottes." George Marsh reached Hanover in the fall of 1816 to find his college battling for its very buildings against a rival state university.[49]

Before Marsh graduated, the U.S. Supreme Court—moved by Daniel Webster's legendary plea that though Dartmouth was "a small college, . . . yet there are those that love it"—famously vindicated the trustees. But the students derived small advantage from the decision; the curriculum remained as narrow and humdrum as before. Dartmouth studies were largely a continuation of those at Andover, and George Marsh got little pleasure or profit from them. What he learned of value he picked up independently. He spoke of his college days later in life only to chide himself for having wasted time.

At fifteen, Marsh was three to five years younger than most of his thirty-three classmates. Diffident, shy, studious to excess, Marsh had no close companions in his class; he was intimate only with his cousin James, a senior, and the future lawyer and statesman Rufus Choate, a sophomore, his intellectual equals. Caring neither for sports nor social affairs, Marsh had little in common with the rest. They in turn found him aloof. They admired his learning and his dry, quiet wit but did not warm to this solemn, bespectacled young man who "was indifferent to all external objects, save some book," a classmate recalled, "and then he placed the book very near his eyes."[50]

Few students shared Daniel Webster's love of Dartmouth. Undergraduates complained then, as ever since, of Hanover's paucity of attractions. It was a small town; some sixty white houses stood around the Green, with three-storied Dartmouth Hall on one side. Marsh lodged in the village his first three years, in Dartmouth Hall as a senior. Tuition was $21 a year, plus incidental expenses: a $2 library fee, 25 cents for a copy of the rules, fines for their infraction—amounting to 60 cents for Marsh's entire four years. Such sums his father could easily spare, with pocket money besides; Marsh was one of the "richer" students.[51]

For most classmates, college was more formidable in both cost and effort. Although frugality and plain dress were customary, many had to eke out expenses by teaching school. The trustees may have been "aristocrats"; the students were not. Some came via preparatory schools, but most had passed formative years on New Hampshire and Vermont farms. In contrast to men from Harvard and Yale, Dartmouth graduates were said to be rough-and-ready fellows of little learning or culture. This reputation was on the whole merited; discipline was strict, but scholastic demands were low. However lazy or incompetent, a student who paid his bills and kept out of trouble usually got his degree. But poverty and illness took heavy tolls—one in three of Marsh's class failed to graduate.

Students assembled in the chapel daily at five in the morning, or "as early as the President could see to read the Bible"; neither light nor heat was provided. The chill may have sped President Brown's premature demise. After chapel, they dispersed to recitations, each year's students in one room. Breakfast followed, then a period of study—or slumber—then a second recitation; after midday dinner, study again, and an afternoon class at three or four. Evening prayers were at six, "or as late as the President was able to see." Sunday was a day of enforced rest; students went to morning and evening chapel, but save for meals were otherwise confined to their rooms.[52]

In the few hours of recreation, students played football on the Green and swam in the Connecticut River; in the winter there was ice skating. Gambling was forbidden, but even Marsh sometimes took a hand at whist. Like Yale president Timothy Dwight, he learned a lesson from this frivolous indulgence. Asked fifty years later to join a game, Marsh said: "No, I believe not. I did too much of that in my college days, and I have never taken a card in my hand since." Dwight bemoaned his card playing as a moral lapse; Marsh merely regretted the waste of time.[53]

There were few pleasant ways to spend time at Dartmouth. Two literary societies, the United Fraternity and the Social Friends, engaged in debates; membership in one or the other was mandatory (Marsh was assigned to the Social Friends). Phi Beta Kappa, to which Marsh belonged, a theological group, to which he did not, and a sporadic Handel and Haydn Society were the only other organizations.

The college year began in September and ended the next August, with a seven-week midwinter vacation. Commencement Day was a grand occasion; parades, fireworks, refreshment booths, and sideshows around the Green gave it the aspect of a county fair. The trustees met, some celebrity gave an address, and all endured hours of student oratory.

The narrow, rigid curriculum was entirely compulsory. Classics and mathematics dominated the first three years; seniors got doses of metaphysics, theology, and political law out of Jonathan Edwards's *Freedom of the Will* (1754), John Locke's *Essay on Human Understanding* (1690), Dugald Stewart's *Elements of Philosophy* (1792), and William Paley's *Evidences of Christianity* (1794). Of natural science there was virtually none; like other colleges of the day, Dartmouth held most science subversive of religion and government. Not until 1836 did chemistry, mineralogy, and geology enter the curriculum. The little astronomy that Dartmouth thought safe to impart in Marsh's time was taught by Ebenezer Adams, a lusty-voiced ignoramus who relied mainly on what Marsh told him; indeed, Marsh was "not corrected," recalled a classmate, "for any mistake or fault" in his college years.[54]

Bishop Paley's *Natural Theology* (1802), which Marsh read in his junior year, was a canonical American college text, much admired for buttressing religious faith with purportedly scientific evidence.[55] One Paley doctrine was geological catastrophism, the view that the earth had actually undergone all the violent upheavals delineated in the Bible. Another Paley premise was that only a purposeful Creator could have made the wonderful forms of nature; a third was that God made everything in nature for man's use. Marsh found Paley intellectually vacuous and spiritually abhorrent.

Schooling in languages was little better than in science. Marsh's Greek and Latin far surpassed the curriculum; "when he left college," remembered a classmate, "he read the Greek poets and historians with as much ease as an ordinary man would read a newspaper." But not all classical teaching at Dartmouth was second-rate. "President Brown hears us in Horace," wrote Choate, "and it is our own fault if we do not make progress." No modern languages were taught, but Marsh mastered the Romance tongues—Spanish and Portuguese as well as French and Italian—by himself. With Choate and James Marsh, in what Choate recalled nostalgically as a "magic circle," George Marsh met regularly to explore classical and contemporary European literature.

They followed the lead of the German philosopher Johann Gottfried von Herder, who ascribed to each nation and epoch a unique soul and mind to be gleaned from specific native texts.[56]

Pedagogy like subject matter at Dartmouth left much to be desired. Actual instruction was minimal. Recitations tested what students, called on in alphabetical order, had memorized. But if classwork was perfunctory, the three-man faculty—Brown, Adams, and Roswell Shurtleff, who taught theology, moral philosophy, political economy, and mathematics—were not wholly to blame; they were during Marsh's first two years shut out of the college library and their own classrooms by the state of New Hampshire. Never well paid, they were now not paid at all, and if the trustees lost their case they would have no jobs.

The condition of the college was, as Choate said, "critical in the extreme." During midwinter in Marsh's first year the state university preempted all the college buildings. Undaunted, President Brown held classes in the Rowley Assembly Rooms over Stewart's hat shop. "It was a pleasing tho solemn sight," reported the *Dartmouth Gazette*, "to see the students, who before had been accustomed . . . to flock to the chapel at the welcome sound of the bell, now punctually flocking to this retreat of persecuted innocence."[57]

Meanwhile faculty and trustees suffered wretched months. New Hampshire's Supreme Court ruled against the trustees in November 1817, and Charles Marsh made plans to transfer George and other students to Middlebury if the federal appeal failed. But in February 1819, Chief Justice John Marshall read his historic decision, disallowing New Hampshire's interference: "The college is a private . . . institution, unconnected with the government [and] the charter of such an institution is plainly a contract"; contracts were inviolable.[58] The rival state university disbanded, and Hanoverians celebrated with cannonades.

George Marsh, himself so intimately involved, later termed the Dartmouth College decision "vitally important to the cause of education." The controversy had "excited a sympathy between two vocations before thought antagonistic—the academic and the forensic,— . . . with favorable results to both of them." But later still, when corporations cited Marshall's ruling to claim immunity from legislative control, Marsh assailed this reading of the sanctity of contracts as "an old legal superstition." He now lauded efforts to curb corporate power spear-headed by his nephew, Senator George F. Edmunds, that culminated in the Sherman Antitrust Act of 1890.[59]

Marsh's senior year was uneventful. As excitement over the case ebbed, morale declined, discipline deteriorated. Students were repeatedly fined for absence and "inattention"; Rufus Choate, now faculty secretary, had to warn that any senior who "habitually neglect[s] any of the exercises [shall] be refused examination with his class."[60] But most squeaked through to Commencement on August 20, 1820.

As the band played, students paraded to the meeting house under Colonel Amos Brewster, Grafton County high sheriff, whose stentorian voice, pompous swagger, cocked hat, sword, gold lace, and sash were indispensable accoutrements. The audience settled to an orgy of speeches: "The Decline of Eloquence," "The Moral and Religious Character of the First Settlers of New England," "The Expulsion of the Moors from Granada" (poem), "Piety Essential to the Highest Enjoyments of Taste." Hours later came Marsh's Herder-based valedictory: "The Characteristick Traits of Modern Genius, as Exemplified in the Literature of the North and of the South of Europe." Compared with others, Marsh was termed plain, direct, unornamented.[61]

I hate boys, hate tuition, hate forms.

Marsh to Spencer Baird, April 25, 1859

Just two weeks after graduation, Marsh returned to Norwich, Vermont, across the Connecticut River from Hanover, to begin teaching. The salary was small, but the title grand for a youth of nineteen: George Perkins Marsh, A.B., Professor of the Greek and Latin Languages at the American Literary, Scientific and Military Academy.

Norwich Academy was emphatically experimental. Its eccentric and ebullient founder, Captain Alden Partridge, had been dismissed as superintendent of the U.S. Military Academy at West Point in 1817 and cashiered for insubordination (noting the captain's "zeal and perseverance," President Monroe commuted the sentence). Partridge advocated military education for all. "A systematic knowledge of fortifications and tactics" would equip young men to understand history, "a large portion

of which is made up of descriptions of battles and sieges," and thus to
defend their country. He also promised to keep young men from evil
habits by eliminating their leisure time; at Norwich, drill and exercise
occupied "those hours . . . generally passed by students in idleness, or
devoted to useless amusements."[62]

For otherwise misspent vacation times, Partridge mounted "excursions," marching his hundred cadets all over northern New England.
The captain was an experienced surveyor, and on these trips took altitudes of many peaks. Marsh went along in the summer of 1821, learned
to use the Englefield barometer in the Green Mountains and on Mount
Washington, and gained familiarity with local land forms.[63]

But if the summer jaunt appealed to Marsh, nothing at the Academy did. The four-story brick barracks, surrounded by a high fence
and guardhouses, were depressingly gloomy, the blue-coated, high-
collared cadets depressingly stupid; Marsh lacked the patience and skill
needed to pound the classics into their heads. Unwilling to suffer fools
gladly, Marsh preferred self-education to teaching others. Less misan-
thropic later in life, he generously aided bright young protégés, but a
teaching career never attracted him. Bored by his classes and by the
self-important Partridge, he spent what time he could in the Dartmouth library, reading works in German and Scandinavian languages
until late at night. This regimen brought Marsh's eyesight to an end
with the school year; he quit his job and sought an oculist.

His eye trouble, the same he had suffered as a child, proved obdurate; for several years he could read very little. During futile winters in New York, Philadelphia, and Providence doctors blistered and
cupped him but failed to mend his vision. Marsh returned to Woodstock weak, thin, and discouraged. There he prepared for the law by
being read to at home and hearing cases in court.

Marsh's horizon at this time was bounded by family. Often in pain
and scarcely able to see, cared for by his mother or a cousin, he countered misery with mordant repartee and multilingual puns. In better
hours he talked law with his father and brother Lyndon, now practicing in Woodstock; played with the new baby, Charles; and joked with
a visiting Perkins cousin. She showed him a landscape she had just
painted. "How do you get off that bridge, cousin?" he asked her, smiling. "Sure enough the bridge ended at one extremity square up against

the old castle. . . . After that 'cousin Sarah's bridge' was not seldom referred to when some hideous mistake occurred."[64]

Unable to work at anything substantial, Marsh spent solitary days roaming the countryside. After long abstinence from print, his eyes gradually improved. Later Marsh looked back on this period, then seemingly sad and barren, as one of growth and maturation, enhancing his powers of memory and capacity for reflection.

*F*our years passed quietly in this way, enlivened for Marsh by few contacts beyond Woodstock. He did take on one unusual task—a state-mandated inquiry on deaf-mute education. Family links with Governor Cornelius P. Van Ness prompted the legislature in 1823 to ask Marsh what Vermont should do for these unfortunates. The issue was then topical: Philadelphia educator Thomas Hopkins Gallaudet, having studied sign-language training under the pioneers Charles-Michel de l'Épée and Roche-Ambroise Sicard in Paris, had returned to America with their star deaf-mute pupil, Laurent Clerc. In six years Gallaudet and Clerc had revolutionized deaf schooling through American Sign Language, a variant of the French lexical system.[65]

Marsh visited Gallaudet and Clerc's American School for the Deaf at Hartford, Connecticut, inspected asylums in New York City and Philadelphia, and corresponded with teachers at Danville, Kentucky, and Canajoharie, New York. Rather than recommending a separate asylum for Vermont's few dozen deaf-mutes of educable age (then teenagers), he advised that the state allocate $150 a year for each to go to Connecticut. In 1826 nineteen Green Mountain deaf-mutes began five years' schooling in Hartford, a transfer scheme that lasted several decades.[66]

Marsh's detailed report far transcended the practical issue at hand: it explored the mental world and future prospects of those who could neither hear nor speak. Common stereotypes then ranked deaf-mutes as cretins unable not only to communicate but to contemplate, remember, or envision; their perceptions were therefore thought to be limited to isolated and momentary sensations. With language providing the only means "of reflection and the comparison of ideas," those without it were indeed pitiable, wrote Marsh, elaborating on Abbé Sicard's seminal *Cours d'instruction d'un sourd-muet de naissance*: "The deaf and dumb, in a state of nature, possess no means of receiving ideas from others." Thus all their ideas "must be original, and derived only from personal observation." Their impressions were transient, their sensations fugitive. They could not "have the satisfaction of reflecting on the past" or anticipating future enjoyment. "Inferior to the savage, who possesses the means of communication with his fellows [and] enjoys the advantages of society," the deaf-mute was "but a moving machine, possessing little to elevate him above the brutes."

But this hapless view was quite false, insisted Marsh. Most deaf-mutes *had* language, as any observer could see. They conversed together in signs all the time. Formal sign-language training gave them ready communion with the hearing; manual alphabets, Marsh noted, conveyed words four times faster than they could be written. Indeed, sign language was more universal, more notably human, than were spoken tongues mutually unintelligible since the fall of Babel. "If you bring together two uneducated but intelligent deaf-mutes from different countries," Marsh later added, "they will at once comprehend most of each other's signs, and converse with freedom."[67]

Marsh felt sign language superior in other ways, too. "The deaf-mute learns nothing by rote, but every thing by analysis; [schooled deaf-mutes] not only acquire the precise meaning of words, but become good grammarians and logical reasoners." In sum, their minds were "precisely like" those of others. Hence "to educate a deaf mute is to render a new member to society, to give life to that which was before dead, and almost to convert matter into mind."[68] Like most educators, until Horace Mann and S. F. B. Morse regressively enthroned oral training a half century later, Marsh thought teaching deaf-mutes to speak was a time-wasting detriment to their linguistic progress. Marsh's interest was life-long; in 1870 he supplied Gallaudet's son, Edward Miner Gallaudet, with data on deaf-mute education in Europe.[69]

This is not the place to argue the merits of sign language, which were as patent to Marsh as they are to modern educational reformers. What is extraordinary is that young Marsh, already a linguist to whom spoken and written words were meat and drink, esteemed signs not merely as a precondition of or adjunct to articulate speech, but often as preferable to it. In southern Europe, he later noted, "telegraphic communications by hands, face, feet, the whole person, in short, are every-where kept up, as qualifications of animated oral discourse." Thus a for-eigner, "who understands no language but that addressed to the ear, loses much of the point of the lively conversation around him." And the language of gesture may be an advantage to a speaker who cannot be heard by a tumultuous crowd, or in "despotic countries [where] every man knows that he is constantly surrounded by spies, and it is therefore safer to express himself by gestures [which] cannot be so easily re-corded or repeated, even when understood." It was said that "the fa-mous conspiracy of the Sicilian Vespers [1282] was organized wholly by *facial* signs, not even the hand being employed."[70]

Late in Marsh's life, sign language led him to speculate on intrin-sic forms of thought. Whether mental images were audible or visual seemed to depend, he surmised, on whether someone was at a given time more habitually exposed to spoken or to written words. But words, whether written or spoken, were intrinsically less veracious than invol-untary *facial* gestures. "So much more truth-telling than words, in fact are these self-speaking muscles," that Marsh commended the aphorism "that language was given us to enable us to conceal our thoughts."[71]

Happier in lofty erudition than in petty legal details, Marsh yet persevered in preparing for the bar. A committee of three, including fu-ture Vermont Senator Jacob Collamer, examined and admitted him as an attorney of the Windsor County Court in September 1825.[72] Soon after, Marsh set out for Burlington, on the other side of Vermont, which became his home for most of the next thirty-five years.

2. Burlington: Blunders and Bereavements

THE EIGHTEEN YEARS FROM MARSH'S ARRIVAL IN Burlington to his election to Congress in 1843 were full of adversity. Sampling a dozen careers, Marsh was no nearer knowing what to do with his life than at the start; indeed, only after eighteen years more, at the age of sixty, was his course clear before him. Meanwhile he vacillated: law and politics he found dull and degrading, scholarship ill-paying. Business for its own sake never suited Marsh; he hated dealing with men whose sole concern was money. Forays into finance, manufacturing, railroad and land speculation invariably proved disastrous, miring him further in callings he abhorred. Pursued by some evil genius, he shifted course time and again. In law Marsh was crippled by the untimely death of his first partner, and enmeshed in calamity by his second. In politics he made two false local starts before finding his way on the national stage; in business all went wrong wherever Marsh took a hand. His family life was mutilated by tragedy.

Only scholarly pursuits gave Marsh solace and let him sometimes forget his terrible losses. In these tragic years he began to build his great library, to study art, languages, and history, and to lay the foundations for the chief works—linguistic, historical, environmental—of his later life.

"B. F. Bailey & G. P. Marsh, attornies at law, Have formed a co-partnership under the firm of BAILEY & MARSH, and will attend to the business of their profession at the office hitherto occupied by B. F. Bailey, in Pearl-Street," ran a notice in the Burlington *Northern Sentinel* on October 28, 1825. The partnership was successful from the outset. Handsome, popular, brilliant Ben Bailey—at twenty-nine the leading attorney in Chittenden County and Burlington's representative in the state Assembly—brought in most of the work, and his breezy wit charmed juries. Standoffish young Marsh did the office chores. Bailey & Marsh's chief business was civil law. The firm argued fifty to sixty bank-ruptcy, trespass, estate claims, foreclosure, and damage-suit cases a year before county, state supreme, and federal circuit courts. Most of them bored Marsh thoroughly.

Partnership and family connections early pushed him into public affairs. Town leaders relied on his literary flair to pen their legislative petitions. In 1829 he gave Burlington's Independence Day oration, an Anglophobic accolade to American democracy and stability that eschewed party politics. As a town selectman in 1831–32, Marsh's most noteworthy act was getting Burlingtonians inoculated against smallpox at public expense.[1]

Politics took up much of the partners' time and energy. Bailey was a perennial candidate for office. He and Marsh were protégés of Governor Cornelius P. Van Ness, long Vermont's political strong man; but in 1828 Van Ness, a onetime crony of Martin Van Buren, fell from local favor by supporting Andrew Jackson for president—a switch embraced by Bailey but not by Marsh. Jackson was locally unpopular because Vermont sheep raisers relied on the high protective tariff on wool, which they rightly feared he opposed.

Bailey ran for Congress in 1830. His contest against Federalist incumbent Benjamin Swift and Anti-Mason candidate Heman Allen consumed two years and eleven ballots. Marsh's political debut as Bailey's campaign manager was made miserable both by having to disclaim his

Jacksonism and by the rise of Anti-Masonic clamor. United only on hatred of Vermont's "Masonic aristocracy," the oligarchy that had long run Green Mountain affairs, Anti-Masons captured the state government for five years and carried Vermont for William Wirt in the 1832 presidential election—the only state the Anti-Masons won.

As proprietor with Bailey of the Burlington *Northern Sentinel*, one of the town's two weeklies, Marsh heaped insults on Bailey's opponents, while the Burlington *Free Press* assailed Bailey & Marsh's connection with Van Ness and the Democrats. Anti-Masons lambasted Bailey as a lewd loafer who swaggered about taverns (though professing temperance reform), Marsh as a stuck-up patrician trading on his illustrious grandfather's name. After the fifth inconclusive ballot, Marsh blamed his opponents for calling eight thousand voters away from work at so busy a time; the *Free Press* rejoined that only one "nurtured in the lap of a purse-proud insolent overbearing aristocracy [would] thus wantonly insult and revile a large portion of the community." The man who had disrupted the district by this long campaign was Marsh, who "prostituted himself soul and body to pander for the inordinate and shameless lusts of his partner." "Petty cabal of intriguers!" jeered Marsh; "Base calumniator!" retorted the *Free Press*.[2] But in May 1832, Ben Bailey suddenly died of measles; Heman Allen was at last elected. Marsh looked for a new law partner and a new political alliance.

The Marshes had never been Masons, but animus against Vermont's "elite conspiracy" ran so high that anyone of "good" family was fair game for Anti-Mason assault. Hence the *Free Press* assailed "Squire Marsh" as a man "whose veins are bursting with the insolence of aristocracy and who has been taught from his cradle, that family and wealth give those who inherit them, an indubitable right to be supercilious."[3] Why Anti-Masonry now swept Vermont throws light on the feuds that embroiled Marsh. Freemasonry had flourished among early Green Mountain patriots more than anywhere else. By the 1820s, Vermont's tight-knit legislative and legal oligarchy included so many Masons that some charged the order with subverting justice. Anti-Masonic fears were inflamed by the 1826 Morgan affair—the abduction and presumed murder by upstate New York Masons of William Morgan, an apostate who had published an exposé of Masonry. This was prime campaign fodder: as the politico Thurlow Weed said of an unidentified body fished out of the Niagara River, "It is a good enough Morgan till after the election."[4]

Hard on Morgan's heels, Vermont Anti-Masons forged their own outrage—an alleged conspiracy by Vermont State Prison officials, all Masons, to fake the death of an inmate, Joseph Burnham. Burnham had been jailed in 1826 for raping a fourteen-year-old apprenticed orphan girl in his house at Pomfret, near Woodstock. Savaged by prison guards (who taunted him for needing two women to help him rape a child), Burnham died in jail. But the odium of his crime made him an ideal target for Anti-Masons, who spread the rumor that he was not dead but had been sprung from jail by Masons, and had been spotted in New York City. No matter that the New York "Burnham" proved an impostor, that his corpse was exhumed in Montpelier, that Masons denied he had ever been one of them; these were all dismissed as Masonic "lies." Linking Vermont's most notorious criminal to the state's Masonic elite gained Anti-Masons mass support in the November 1829 elections, paving the way for statewide victory in the 1830s.[5]

As a scion of Woodstock's "evil" elite, Marsh became a made-to-order Anti-Mason scapegoat in Burlington. This unjust taint by association with a secret society sharpened his abhorrence of secrecy in politics and religion. "There are few secrets which do not cover a wrong," Marsh later observed, "perhaps none which does not involve a lie." As for his own views, Marsh echoed his father's jibe that "Masonry was the silliest thing in the world except anti-masonry."[6]

Only odd or perverse people go to law.

Marsh to Lucy Wislizenus, Rome, February 14, 1875

When Marsh's political career died with Ben Bailey, his law practice also collapsed; the legal work he undertook fell off sharply. Before 1832, Marsh argued half a dozen cases annually before the Vermont Supreme Court, but he appeared only three times in the next two years and not at all after 1835.

In 1833 Marsh took on a new partner: his younger sister Sarah's husband Wyllys Lyman, who had studied law in Woodstock. Theirs

was more a business than a law firm; Marsh's court appearances increasingly concerned his own rather than clients' troubles. Ever more
averse to legal work, he began to close down his practice and planned
a trip to Europe. Prior to 1835 he regularly served as justice of the
peace and in other bar offices, but now he relinquished all such posts.
His final legal task was to be County Commissioner of Bankruptcy in
1842. The following year, the last before Marsh went to Congress, Lyman & Marsh handled just five cases on the county court docket and
one in the supreme court.[7]

What made Marsh quit? For one thing, his clients disgusted him.
He moaned that law forced him into "constant association with low,
ignorant, and even depraved men . . . against which I have, for several
years, enjoyed almost no counteracting influences."[8] These wretches
not only consumed Marsh's time, they corrupted his character. But his
main complaint was that legal effort seemed as apt to pervert as to promote justice. Brought up to abhor casuistry, he could not abide defending a guilty man. Of a "wealthy lawyer [said to have] acquired his
great riches by his practice," Marsh riposted, "Yes, by his practices."
Learning late in life that his young friend Samuel Dana Horton's father
intended him to study for the bar, Marsh objected, "he doesn't really
mean to make a *working* lawyer of you, does he? . . .'Tis the next best
thing to doing nothing."[9]

On occasion Marsh moderated his aversion; in one way law
seemed "the best of professions." Requiring constant cerebral agility, it
was "the calling in which men . . . retain their mental faculties the
longest." Marsh enjoyed some legal issues as abstruse puzzles. And
however grim, law had its comic side: "the strange people [lawyers]
come in contact with," he reminisced, "the strange language they hear,
furnish countless occasions for mirth."[10] A grudging tribute to his clients! As in much else, Marsh had limited sympathy with the mundane
concerns of lesser, looser men.

Yet Marsh's legal reputation was considerable. Although he was
too unworldly to be shrewdly pragmatic, clients found him diligent and
thorough. Impressed by Marsh's "strong will . . . obdurate firmness . . .
fearless and effective" exposures of evil, one colleague judged that "as
an advocate and debater, he ranked with the foremost of his age." A
more judicious appraiser termed Marsh "not exactly fitted for country

law, as he knew too many precedents," while his addiction to "labori-
ousness" and regard for logic "could never adapt itself to a country
court." Marsh himself later concluded that he had "lacked the special
gifts of the profession, & [was] but an indifferent practitioner."[11]

Given Marsh's tragic personal life after his first years in Burlington, no occupation could have contented him. Yet fortune initially
smiled. He found congenial folk at the University of Vermont and built
up a circle of close friends. And not long after his arrival he fell in love
with the vivacious Harriet Buell. When they married in April 1828
Marsh was twenty-seven, Harriet twenty-one, a cheerful foil to her reserved and often solemn husband. Harriet's father, Colonel Ozias Buell,
was a pillar of Burlington's commercial community, a founder of the
Congregational Church, and treasurer of the University of Vermont.
Buell's farm machinery and dry goods business allowed him to be generous; Marsh left Sion Howard's hostelry and moved with his wife into
a house Buell gave them at 12 Maiden Lane (now North Union), exchanged a few years later for a wisteria-covered cottage at Pearl and
Willard streets, near the university.

Within a year a son, Charles Buell, added to Marsh's married happiness. It was not to last. Even before the child was born, Dr. Benjamin
Lincoln warned Marsh that Harriet had a serious heart condition. Constantly worried, Marsh did not tell Harriet, who appeared well. Young
Charles, fair, sturdy, "extremely gentle & affectionate," seemed to his father "of very good promise, intellectually." Marsh cultivated that promise. By the time Charles was four and a half "he read very well. This I
had taught him, by carefully printing, with a pen, letters, words, & little
stories, in parchment books." Marsh was a shrewd pedagogue: "Children are most interested in that which is done expressly for themselves,
& I have no doubt that the manufacture of the books under his own eye,
& for his use . . . stimulate[d] his zeal to master their contents."[12]

Pregnant again in 1832, Harriet was unwell before and after the
birth of a second son, George Ozias. By the following winter Marsh
could no longer conceal from her that she was dangerously ill. Doctors
in New York and Philadelphia shared his anxiety, but Harriet was more
concerned over young Charles, who had caught scarlet fever. They

rushed back to Burlington to find him apparently past the crisis. But the trip taxed Harriet's strength; failing fast, she died August 16, 1833. Exhausted and grief-stricken, Marsh now cared for Charles, who had relapsed. Deafness ensued; to his son's questions Marsh wrote out answers on a slate. Eleven days after his mother, Charles too was dead.

The double blow devastated Marsh. He sent baby George Ozias to his own mother in Woodstock and went to stay with his sister Sarah and her husband Wyllys Lyman. "It was well for me," he wrote thirty years later, that business "drove me into a constant succession of severe labors, of a very engrossing character. This only saved me from madness, for I was never alone . . . for more than a year, without bursting into a fit of uncontrollable grief. Every night was one long wail of the deepest sorrow, for even my dreams were full of death." For several years he lived alone, taking meals at a hotel and sleeping in his desolate house. His expression grew graver than ever, his bearing more somber. Proffered solace he rejected; friends "almost drove me mad by their ill-judged attempts to console me. They told me 'my wife and child had been taken away for *my* good.' The idea that the young mother had been snatched away from her helpless children, and our boy from the light of life, for the *good* of such a one as I, implied to my mind a horrible injustice. . . . Such an interpretation . . . was to me blasphemous, and I could not bear it." [13]

The Burlington that Marsh came to was a fast-growing town of 1,650, doubling by 1830, up to 4,300 by 1840, and 7,600 by 1850. In Ira and Ethan Allen's day a few "savagely raw and shabby" settlements, sundered by a steep ravine, had been interspersed among gravelly farms and wasteland. But the grandeur of the site redressed its drawbacks, making Burlington what Henry James later termed "the most truly charming of . . . New England country towns." From the ridgetop university the viewer's eye swept westward over the town to Lake Champlain 250 feet below, and across that wide expanse to the azure Adirondacks. Far to the east was the other rim of Burlington's vista: the jagged peaks of the Green Mountains from Mount Mansfield to Camel's Hump. In contrast to Woodstock's enfolding hills, which had preserved many rural patterns of Marsh's early childhood, the ranges bounding

Burlington embraced wide horizons with good natural highways to the world beyond.

"Delightfully free, noble, and open,"[14] Lake Champlain's waterway ties to New York City, Quebec, and the West earned for Burlington the name of Queen City. Blessed with the state's finest port and harbor, Burlington had for decades sent Vermont timber to Quebec, but now shipped Canadian lumber to New York via the new Whitehall-Albany canal that had linked Lake Champlain with the Hudson in 1823. Canal boats and schooners thronged Burlington's wharves and mooring sites. Above the docks, sixteen dry goods stores, seven blacksmiths, six taverns, five mantua-makers, four tailors, five joiners, three masons, three saddlers, two barbers, two watchmakers and jewelers, and a tobacconist catered to Chittenden County. Two banks, two meeting houses, a courthouse, and a stone jail filled out the institutional scene.[15]

Burlington's seventeen attorneys trebled to fifty within twenty years. Not all were so ignoble as might be supposed from Marsh's anti-legal strictures. Indeed, several lawyers became close friends, notably cultivated, witty John Norton Pomeroy. Besides Wyllys Lyman, Marsh saw much of James W. Hickok (a family connection of Marsh's wife), Heman Allen, Alvah Foote, Albert G. Whittemore, Asahel Peck, David Read, Charles Kasson, and Jacob Maeck, the gadfly of the county bar.

But it was to academe that Marsh turned in leisure hours. The University of Vermont throve under the guiding genius of James Marsh, George's Quechee cousin and Dartmouth companion. "The son of an intelligent farmer," George later wrote of James, agriculture "continued to have so strong a charm for him that . . . he entertained the project of retiring permanently to his father's farm."[16] Dogged by poor health, he instead went to teach at Andover Theological Seminary, and at length to Burlington in 1826. Tall but frail, his high-pitched voice often barely audible, James Marsh recast New England college teaching as profoundly as he transformed New England philosophy. He liberalized admissions, abolished petty disciplinary rules, curtailed formal instruction and textbook memorizing, encouraged wide latitude of thought. Meanwhile James Marsh's trustees filled college coffers with hard-won Vermont dollars and restored buildings wrecked by fire in 1825. Within a few years the university gained national repute.

The Marsh cousins spurred each other on. They revived their

Dartmouth literary club, devoting weekly evenings to Greek, German, and English philosophy. James had already embarked on what Rufus Choate termed "that ocean of German theology and metaphysics." German literature and abstract theory were little known in America. In the late 1820s, Kant and Coleridge, as glossed by James Marsh, became the philosophic linchpins of American Transcendentalism under Emerson's guidance.

With his Massachusetts disciples James Marsh did not see eye to eye. Though an idealist and mystic, he was also a tough-minded Vermonter who saw no prospect of perfecting humanity. He mocked Brook Farm reformers who hoped "to redeem the world by a sort of dilettante process, to purge off its grossness, to make a poetical paradise in which hard work shall become easy, dirty things clean, the selfish liberal, and the churl a churl no longer." And he thought their scholarship as shoddy as their morality. George Marsh shared his cousin's principles and prejudices, and lamented when James's death in 1842 bereft him of a friend of "philosophical genius and culture." [17]

In Vermont's medical faculty was another cousin, James's brother Leonard Marsh, termed by George Marsh "the profoundest thinker, by many degrees, in my knowledge." Shyness hid Leonard's talents—he was "like the iron steamer *Great Britain* in being built in a dock with gates so small, that it can't be got out, except in fragments." [18] But the fragments were explosive; searing invective suffuses Leonard's abolitionist *A Bake-Pan for the Dough-Faces* and *Apocatastasis; or, Progress Backwards* (both 1854), sardonic ripostes to John Henry Hopkins, Vermont's Episcopal bishop. Hopkins was a near neighbor whose high-church, pro-slavery views George Marsh vehemently opposed. Yet he found this polymath ecclesiastic engaging. Hopkins was an accomplished musician, painter, and practical architect, for whose journal, *The Churchman*, Marsh wrote occasional verse.

Among other academic comrades Marsh enjoyed botanical outings with popular science professor George Wyllys Benedict, later editor of the *Burlington Free Press* and, as state senator, promoter of telegraph links with Boston. With Dr. Benjamin Lincoln, his family physician, Marsh discussed music and mathematics. The conservative Reverend John Wheeler, James Marsh's presidential successor, shared George Marsh's interest in Champlain Valley fruit growing. Another companion was the eclectic Zadock Thompson, geologist, statistician, historian, compiler of textbooks, gazetteers, almanacs, and weather records, Episcopal minister, and editor of the *Iris*, a "Literary and Miscellaneous" periodical. In his spare time Thompson ran a seminary for young ladies; he and Marsh together founded the Burlington Lyceum.

This was the Burlington circle in which Marsh took intellectual refuge, and which held him in high esteem. Terming Vermont "as famous for her Marshes as her mounts," and recalling George Marsh's accomplishments in law, history, and literature, President Wheeler judged him "possessed of more talents . . . than any man I ever knew."[19]

M arsh possessed more than talents during his early years in Burlington. From his wife and father-in-law (who died in 1832) he inherited considerable property. Five years later he had real estate valued at $7,500 and $10,000 in cash and stocks; by 1843 his land was worth over $16,000.[20]

Allied by family and fortune to most Burlington enterprise, Marsh strongly backed facilities for extending credit. Ozias Buell had been a director of the Bank of the United States Burlington branch, established in 1830. Four years later, President Jackson killed the United States Bank as a privileged monopoly. Speaking for local business interests, Marsh censured the government's withdrawal of deposits as "impolitic and unjust."[21] Together with ensuing credit restrictions, the withdrawal doomed the Burlington branch.

As that bank closed, Marsh, Lyman, and the John H. Peck mercantile firm moved to recharter it as a private bank. "Business of every kind grows with the facilities for its transaction," noted Marsh's 1834 petition to the state legislature; in five years the branch bank had expanded local capital fivefold, partly by providing short-term loans, which the "older merchants" of the stodgy Bank of Burlington refused to do. Burlington needed another bank "with a capital sufficiently large to . . . furnish those means of exchange, discount, & negotiation of bankable paper, which the business of the country requires."[22] The state assented; in 1836 Marsh and his associates took over the U.S. branch bank assets for $142,000, payable in four years. But the panic of 1837 made these assets, like those of many banks, worthless. The long unpaid debt sowed the seeds of Marsh's subsequent calamities and, in the end, his virtual bankruptcy.

Prime among Marsh's real estate was the large Castle Farm at Shelburne Bay south of Burlington, with a flock of six hundred merino sheep. Sheep were much in demand by farmers. Losses caused by the wheat midge, combined with high federal import tariffs (1824 and 1828) on wool and woolens, had revived the Vermont vogue for merinos; from half a million sheep in 1824, Vermont's flocks numbered more than a million by 1836, 1.7 million in 1840.[23] Shelburne Bay's heavy, well-drained soils made superb pasture, and Marsh's purebred flock was among the state's finest. A keen stockbreeder, Marsh drove out Spear Street to Castle Farm most summer days.

Another Marsh property was a narrow strip north of town along the Winooski River, including its 38-foot lower falls. Owning raw wool and water power, Marsh decided to merge the two. Why ship wool to Boston when it could be made into broadcloth on the Winooski? Joined by brothers-in-law Lyman and Hickok and others, Marsh began mill-building in autumn 1835.

The Winooski (formerly the Onion) is Vermont's most useful though nowhere navigable river. Narrow, rocky, often deeply incised, it gouges out the best route, from Montpelier to Burlington, across the Green Mountains. All Vermont's main east-west arteries—turnpike, railway, modern highway—have used its valley. At the lower falls Ethan and Ira Allen had dammed the river and erected a forge and furnace, destroyed and rebuilt in 1790, but again wrecked by flood in 1830.

Marsh's Burlington Mill Company, the first river enterprise since then, was promptly chartered (Marsh was on the Legislative Council). Financing the factory and laying a road to the falls proved more formidable tasks. Marsh planned and oversaw the roadwork, "often lifting heavy logs and stones with his own hands," reported the local press, "to quicken the zeal of his Irish laborers." The panic of 1837 slowed work, but the locust-shaded, four-yard-wide turnpike (today's Winooski Avenue) opened a year later, fostering trade to St. Albans and the north. The *Free Press* praised Marsh and Lyman's civic virtues (the road had cost far more than their $1,500 reimbursement); "few men among us have contributed more liberally . . . to promote the substantial interests of the town." The *Sentinel* envisaged the day "when the *loom, shuttle,* and *mechanical arts* will arrest the now waste of waters of [the Winooski], and, like another Lowell, a busy and active population spring into existence."[24]

The seven-story mill held sixteen sets of looms, dye, stapling, and drying houses, machine shop, packing house, and twenty tenements, all steam heated. But just as production began, disasters piled up. A fire badly damaged the mill in 1838; in spring 1839 an ice pack burst through to batter the factory. But fire and ice were minor irritants next to Congress's tariff cuts. In 1839 the price of wool sank so low that farmers let sheep perish rather than feed them through the winter. Cheap foreign woolens forced manufacturers like Marsh to sell cloth for less than the raw wool had cost them; the Winooski factory lost five thousand dollars a week. The 1842 tariff gave partial relief, but that of 1846, leaving domestic woolens unprotected while levying 30 percent duty on raw wool, put Marsh out of business. Like mill owners over much of New England, he sold out at a ruinous loss.[25] Sheep raisers soon felt the pinch, which sharply reduced land values around Burlington; as in the 1820s, thousands of Vermonters emigrated westward. That path, as will be seen, Marsh was twice to sample—and reject.

Banker, farmer, manufacturer, esteemed public figure, Marsh had regained political redemption after the Bailey debacle. He was a prime choice for the 1835 statewide Whig slate for the Supreme Legislative Council, Vermont's upper chamber. The Anti-Masonic crusade had forced most Masonic lodges to disband; moderate Anti-Masons now made common cause with the new Whig party against Jackson's "executive tyranny" and "fiscal madness." A coalition of Anti-Masons and Whigs put Marsh, who ran well ahead of his slate, on the Council in October 1835.

Their first task was to choose a state governor. Since neither the Anti-Mason incumbent nor his Whig challenger had a majority, the election was thrown into the legislature for the fifth consecutive year. For five weeks Marsh and his Council confreres descended twice or thrice daily from their third-floor room in Montpelier's State House, "hardly less inaccessible than Camel's Hump," one councilor grumbled, to the Representatives' hall, a chamber indelibly darkened by tobacco juice despite an abundance of spittoons.[26] After casting a few futile ballots, the Council trudged wearily upstairs again to do such business as it could without a governor. Following sixty-three inconclusive votes, Marsh, as chairman of the Judiciary Committee, broke the logjam by decreeing that the state constitution permitted them to appoint lieutenant governor Silas A. Jenison, a Whig moderate, as acting governor.[27]

The site of these deliberations was claustrophobic. A small village strung out along the upper Winooski, Montpelier remained state capital only owing to rivalries among larger, less centrally located towns. Montpelier was extraordinarily isolated; "turn the eye in whatever direction," commented a visitor, "and mountains . . . obstruct the view. You almost feel shut up from the world, imprisoned." Legislators traveling weary miles along mountain roads to reach Montpelier agreed on its notorious unfitness as the capital. "If the State House had not been here," declared Marsh's nephew George F. Edmunds, when Speaker twenty years later, "no man who was not fit for a place in the insane asylum, would believe that its location in Montpelier could be thought of."[28]

Marsh lodged at the Pavilion, largest of Montpelier's four hotels, kept by "Prince of Landlords" Mahlon Cottrill, who encouraged boarders to make the most of their brief furloughs from farm and family. While enacting laws urged by hometown temperance zealots, legislators enjoyed their liquor in Montpelier, far from constituents' surveillance.

Owing to the election imbroglio, the 1835 session endured a record thirty-five days, yet accomplished even less than its usual slim quota of work. Marsh's Legislative Council dealt mostly with private bills: charter requests for new railroads, canals, manufactures, schools, insurance companies, roads, and banks; quarrels over town boundaries; petitions for state manifestos on slavery, tariffs, the national bank; appeals from prisoners for pardons and from the poor for charity; pleas for the preservation of fish in Lake Memphremagog and smaller ponds.

Debt imprisonment was of greatest statewide import. The abolition of this hoary penalty, which jailed four thousand Vermonters a year, had been defeated at every session since 1820, Vermont lagging behind the rest of New England. Calls for reform were loudest in Burlington, where in one year thirteen men were locked up for other crimes—and 487 for debt. "The voice of the people is against the law," exclaimed the *Northern Sentinel*, "why then is it not repealed? The answer is this—a few cold-hearted, miserly men control the elections of the freemen of Vermont." Marsh was an early if lukewarm reform advocate, arguing that "imprisonment for debt can be justified only on strong presumption of fraud." The law was ameliorated in 1830 and 1834, exempting household goods from seizure, ending jail for women, and freeing debtors who surrendered property.[29]

But reformers now pressed for full repeal, and a bill to abolish debt imprisonment in toto passed the Assembly in 1835. For Marsh this went too far. Total abolition "would be injurious," he warned the Council, "by increasing the difficulty of obtaining credit." There was no "satisfactory substitute for the security to the creditor which justice requires, and the arrest and imprisonment of the person supply."[30] Marsh was persuasive; although the Council's youngest member, he was its guiding force. Suspending the Assembly's bill—a rare event—the Council stymied reform.

This defiance, hard on the heels of the failure to elect a governor, sped the Legislative Council's own demise. The Septennial Council of Censors of 1834–35 had, as before, proposed a truly bicameral legislature; the election and debt-reform imbroglios now gained this measure popular support. "The Council suspended three bills at the late session," complained a Whig paper. "Why not give us a Senate, and do away with this *half way* business of legislation?" And so they did: the ensuing Constitutional Convention voted by more than two to one to replace the Council with a democratically chosen Senate.[31]

Thus ended Marsh's second foray into politics. In shelving debtors' relief, he effectively eliminated his own office.

All these enterprises left Marsh deeply unfulfilled. In his discontent he devised a trip abroad, then thought of quitting Vermont for warmer or wealthier climes. But an excursion to the South in the winter of 1836–37 proved unrewarding, its people uncongenial, its landscapes uninviting. The following year Marsh ventured west as far as the Falls of St. Anthony (later Minneapolis) on the Mississippi, then a lumber frontier. In bustling Chicago his friend William Butler Ogden, now mayor, who in two years had made a fortune out of real estate and railroads, offered Marsh a partnership. The fertility of the prairies and the pace of progress there amazed the Vermonter. But the monotony of Midwestern scenery and the thinness of culture persuaded him to stay home and make the best of New England's unique virtues. Only one blind to nature's beauty "would exchange . . . the unrivalled landscapes unfolded from our every hill," as Marsh told Vermont farmers a decade later, "for the mirey sloughs, the puny groves, the slimy streams, which alone diversify the dead uniformity of Wisconsin and Illinois!"[32]

Back in Burlington in January 1838, Marsh joined more in social life, exchanging solitary reading for evenings with friends. He soon rejoiced that he had not moved west, having met slim, dark-eyed Caroline Crane, the twenty-one-year-old daughter of a Massachusetts mariner turned farmer. Caroline lived with an elder brother, the Reverend Silas A. Crane, assistant to Bishop Hopkins. For three years she had taught in Silas's Episcopal boarding school for twenty girls—"the one matchless *pearl* in the centre of a circle of *emeralds*," in an admirer's phrase. That Marsh had not before met Caroline in so small a town (the Crane school on College Street was scarcely three blocks from his home) shows what a recluse he had become since Harriet's death.[33]

Caroline's suitor, now thirty-seven, looked more a farmer than a lawyer or scholar. Slender in youth, Marsh had filled out; road building and hill climbing kept him fit. Marsh's "firm step and erect bearing" impressed Caroline, but "his habitual expression was grave; the firm-set mouth might even be called stern; and his earnest grey eyes always seemed to look through the object they were resting upon."[34]

This intensely serious man did not immediately charm her. The next winter Caroline went to New York to teach in a school run by kin of Silas's wife—a dour milieu termed "severe and puritanical" by Henry J. Raymond, later founder-editor of the *New York Times*, who on Marsh's advice briefly taught there. Determined to marry Caroline, Marsh visited often and penned ardent, amusing, and doleful letters insisting he could not do without her: "No man was ever so constituted as to require the society of a sympathizing friend more imperiously than I. . . . I not only feel the necessity of communicating my thoughts and feelings, but it seems to me I cannot even *see* sightworthy objects alone."[35] Within a few months he won her over. On September 25, 1839, Silas Crane married them in New York, at six o'clock in the morning, an hour reputedly chosen to ensure privacy.[36]

Caroline was sixteen years her husband's junior, but neither the age gap nor Marsh's awesome erudition made her passive or subordinate. Admiring his work and sharing his intellectual and aesthetic interests, she kept up her own literary pursuits and contacts. Her influence on Marsh was more social than cerebral, though they both would have denied the distinction. Marriage made Marsh, never gregarious, more genial, relaxed, receptive. From Caroline he learned to expect less of others, and to enjoy them more. He even came to expect a little less of himself. Although Caroline, like Marsh, was brought up in Christian orthodoxy, her faith was more evangelical and pragmatic. Her lenient charity tempered Marsh's moral austerity. He paid her benign influence continual tribute; she magnified his enthusiasms, minimized his misfortunes, enhanced his pleasure in all new experience, even taught him a little patience. When Marsh once inveighed against printers' errors, she gently rebuked him for "grumbling at everything short of perfection and even that is not quite good enough."[37]

At first all went happily for them. Marsh had brought back his son George, now seven, from his grandparents in Woodstock, and moved two blocks to a gambrel-roofed, heavily shaded house on Church and Pearl streets. A new fireproof brick shed nearby housed his 5,000-volume library and art portfolios, "one of the most glorious places" a belletrist visitor had ever seen. Save for the books and engravings that lined the walls, Caroline found the furnishings plain, even bare; while acceding to his spartan taste, she added a maid to his two servants. This dwelling remained the Marshes' home, barring spells in Washington and New York and the Middle East, until they left for Italy twenty-two years later.[38]

Marsh rarely broke his set routine. An early riser, he was in his study by five, winter and summer. His desk was piled high with books and papers in five or six languages; he turned from one to another, seldom dwelling more than an hour at a time on any subject. He breakfasted at eight, then went to his office. There time was precious. Clients were urged, sometimes not gently, to come to the point; Marsh would rise impatiently when he felt the subject had been dealt with. (Any importunate caller who came to his house after hours met too cool a reception to repeat the visit.) A nap was imperative after the midday meal at home. Marsh returned to the office about two, spent a few hours there, and devoted the rest of the day to playing with his son and reading aloud with Caroline.

But distress again shattered this agreeable regime. Never strong, young George fell seriously ill, and the Marshes had few unbroken nights. The early 1840s brought grievous family losses. Marsh's older sister's daughter, Mary Burnell, succumbed to tuberculosis in Woodstock; his beloved sister Sarah Lyman died after a long and painful illness; then his younger brother Joseph and his cousin James Marsh perished. Soon after, Caroline was beset by maladies not diagnosed for decades; unable to walk more than a few steps without pain, from 1842 she could read for only a few minutes a day. Tracing a map of a proposed Vermont railroad route, Marsh again damaged his eyes. Mounting business troubles and woolen factory losses made impossible a long-cherished plan to visit Europe.[39]

While Marsh's material prospects seemed ever bleaker, he turned with increasing relief to scholarly matters.

3. Puritans, Vikings, Goths

"MY EARLY INTELLECTUAL ADVANTAGES," MARSH complained when he was thirty-eight, "were of the commonest description." More keenly than most self-made American scholars, he regretted the cultural lacunae of his formative years. And since then things had worsened. "Between weakness of sight, business cares, and domestic sorrows, I have had absolutely no time for regular study or thought since I left college." Hence, he told Caroline as they were about to marry, "I have lost much of the precision and accuracy, and . . . the general knowledge, of my youth; my powers of observation have been weakened by disease, my enthusiasm cooled, and my spirit sobered by labors, disappointments, and griefs."[1]

This was an unduly dismal résumé. Marsh's schooling had indeed been narrow, but he took pains to enlarge his skills and enrich his studies. In observing nature he did not lose old talents but gained new insights; "sight is a faculty, seeing is an art," he often remarked, and in the art of seeing he schooled himself well. Finally, consider Marsh's complaint that he lacked time for study. What would this man have done with *more* time, when with so little of it he encompassed art and architecture, measured mountains, learned twenty languages, compiled an Icelandic grammar, translated German verse, Danish law, and Swedish belles lettres, and gave recondite talks on topics too manifold to classify? To discuss all of his interests would take an encyclopedia; a handful of Marsh's avocations must suffice to suggest their range.

Tools and materials had fascinated Marsh since childhood. Wood, glass, iron, stone, and brass were a pleasure to handle and fashion; only half in jest did he say that his father "spoiled a good tinker when he made a bad *scholard* of me." Fortunately Marsh never put his mechanical skills to an economic test. His episodic inventions and mathematical instrument designs were pretty much limited to his own not always appreciative household. That every man should be his own architect often proved a costly ideal, Marsh confessed: "Persons have ridiculed some of my highest flights of architectural genius as clumsy, awkward, disjointed; and a carpenter, though in my pay, ungrateful varlet! once told me that every dollar of money I expended on my old den added a new blemish."[2] But when others executed Marsh's plans, the results were better. Many a monument and public building—notably Vermont's State House and the Washington Monument—owe much to his expertise and judgment.

The graphic arts engrossed Marsh even more than mechanic skills. Collecting prints, etchings, and engravings was therapeutic. "I fear you will think I am becoming *pazzo per l'arte* [mad about art]," he wrote Caroline; "but when you consider from what lower professional cares and business perplexities it partially withdraws me, you will rejoice that I am now making art, in its higher manifestations, my principal hobby."[3]

It was also expensive. Besides purchasing books on painting, wood engraving, and printing, he was spending thousands on works by Titian and Guido, Reynolds and Rubens. Marsh had to ask himself "whether a strong passion for art, without the means of often gratifying it, is a blessing or a curse." He shielded his pictures from clumsy Burlingtonians, resolving after a few mishaps that "visitors who see with their fingers' ends . . . shall cultivate their sense of *touch* at the expense of my engravings no longer."[4]

Like most connoisseurs of the time, Marsh couched aesthetic tastes
in moral terms. Physiognomy reflected character, but actual human features were imperfect: "human expression must be ennobled, because
man is depraved, . . . we need a purer expression than belongs to mortality." The role of the portrait painter was to exemplify how people
ought to look, especially in capturing those "fleeting changes of linea-
ment, which give moral expression." Hence Marsh admired the bank-
note engraver Mosely Isaac Danforth's idealizing portraiture, and his
old friend Hiram Powers's sentimental neoclassical sculpture; Powers's
The Greek Slave was "the most beautiful object I have ever looked upon."[5]

In landscape painting, by contrast, Marsh extolled realism: because nature lacked ideal forms to strive toward, the artist should simply aim to delineate it with faithful precision. In Marsh's estimation
most landscapists failed in this because they were untrained in botany,
geology, geography, or even in rudimentary observation.[6] Moreover,
many painters were misled by Ruskin's "mistaken" theory that "the
works of nature are admirable only as the poor life of man has illustrated
them, and consequently that the face of creation is an unworthy blank"
in most of the New World. Ruskin seemed to suggest that "wanting ancient memories, American landscape can have no present beauty, and
that which God has created cannot acquire picturesque significance . . .
till man has consecrated it by his doubtful virtues, his follies, or his
crimes!"[7] This error Marsh ascribed to Old World myopia; "eyes familiar only with scenery of so artificial a character . . . must feel the want
of traces of human life" in America. To the contrary, rejoined Marsh,
landscapes anywhere "need not the hand of man, or any memorials of
his virtues or his vices," to be admired in their own right.[8]

Yet "behind Marsh's admiration for nature," as Roger Stein concludes, "was his belief man, not nature, was the repository of human
values." Landscape painting could not be an ideal art because, Marsh
held, "no landscape is a whole, or even a complete part of an organic
whole."

Every landscape is merely the fragmentary contingent resultant of unrelated
forces successive in time, discordant in action, and tending to no common
aim. . . . Landscape painting, then, is but the portraiture of inanimate nature,
and as a moral teacher it can but repeat her lessons. . . . This falls short of the

dignity of historical painting, [just as] unconscious Nature is beneath the rank of divinely endowed man.[9]

As will be seen, it was to man's divine powers and purposes that Marsh later ascribed humanity's unique impacts on the fabric of nature — impacts often as awesomely malignant as benign.

As a congressman in Washington in the 1840s, Marsh became an *intime* of many young artists. His home was a rendezvous for G. P. A. Healy, Thomas Crawford, Eastman Johnson, and Charles Lanman. Johnson and Healy did portraits of their host; Marsh jested that Healy's "portrait of myself (the greater the subject the greater the work) is his *magnum opus*." He praised Healy's semitransparency, "which all untanned (I don't mean un-*sun*-tanned, but un-*oak-bark*-tanned) human skins more or less possess," and his skill "in seizing & portraying the best characteristic expression of his sitter."[10]

Marsh's collection of etchings and drawings was by then one of the finest in the country; no other, he judged, matched his "in *historical value and interest*." Buying it from the impecunious Marsh for the Smithsonian Institution in 1849, the librarian Charles Coffin Jewett lauded the "educated eye . . . cultivated taste [and] earnest study of the history of art" that had formed it. In his day, held one obituarist, Marsh was widely considered "the only authority in this country on the art and history of engraving."[11]

In hindsight, Marsh's taste seems largely derivative — academic classicism, Romantic mysticism, New World chauvinism — and moralistic. To write that "the fathomless depths of sadness revealed in the eye of Murillo's Spanish beggar-girl [offered] a profounder and more touching moral lesson than can be learned from the most brilliant portrait of inanimate nature that Turner ever colored" is sentimental pathos.[12] That Marsh was so highly regarded a connoisseur reflects the dearth of art criticism and the sway of tradition in America. Marsh's penchant in this realm lay in expressing fairly conventional views on a wide range of topics with forceful clarity.

Marsh justified his delight in art, as in many avocations, in terms of utility and patriotism. As he saw it, appreciation of fine art spurred social and technological progress, and vice versa. Technology diffused art more widely, helping to vanquish ignorance, backwardness, and tyranny.[13] His own mastery of technical crafts and the mechanics of re-production bolstered his self-image as a practical-minded patriot.

An acute ear, a superb memory, and intense phonetic drill en-abled Marsh to speak a foreign language almost as soon as he could read it; he took pride in precise pronunciation. For self-teaching he re-duced syllables to morphemes. "Many of the sounds generally sup-posed to be simple may be resolved into yet simpler elements," he explained. Once these were familiar, "phonology may to a certain ex-tent be taught by books." Only later did he realize that sounds specific to other tongues were not only inimitable "but at first absolutely in-audible" to adults who had not grown up hearing them.[14] Having picked up Romance languages in his college years, Marsh turned to German, Dutch, Danish, and Swedish while at Norwich Academy in 1820. Eye ailments and legal work slowed his studies, but in the next two decades he acquired most Nordic tongues and thousands of Scan-dinavian texts.

Desperate to occupy his mind after the loss of his first wife and son, Marsh begged the Danish antiquary Carl Christian Rafn for help with Scandinavian studies. As a lawyer, Marsh "often had occasion to trace principles to a northern origin. Strongly interested in the history and character of the Scandinavian people," he had set out to master their languages and literature. But obstacles proved "insuperable": he could get no Icelandic, Danish, or Swedish books in America, and lacked access to those abroad. Would Rafn "excuse an obscure stranger" and oblige Marsh with suggestions?[15]

Eager to spread Scandinavian studies, Rafn deluged the American with books and periodicals. At the start Marsh was buying fifty to a hundred a month, "but *proh dolor!* (Anglicé—alas for my dollars!) the prices of some of them are such as astound even me, accustomed as I am to be fleeced in this way." He tried narrowing down to Iceland and ancient Scandinavia, but soon again was buying promiscuously. By 1849 he had most editions of the Eddas and sagas and all extant works on Icelandic history, law, and travel; on Scandinavia his collection was little inferior. A century later, an Icelandic specialist judged it "in the day of its owner, in a class by itself in America, and it still remains [at the University of Vermont] one of the most significant collections of its kind."[16]

Marsh's fascination with the north had as many roots as the legendary tree Yggdrasil. Some grew out of reading German philosophy with James Marsh. Others came from his interest in legal and linguistic origins. Ethnocentric pride played its part: the fancied likeness of Scandinavian and New England landscapes, and of Goth and Puritan virtues, turned Marsh toward Nordic lore. Other Americans increasingly shared his ardor. A summer in Sweden, Walter Scott, and Icelandic sagas steeped Longfellow in Nordic themes and meters. Scandinavia inspired the poet James Gates Percival; the blacksmith Elihu Burritt translated Old Norse at the American Antiquarian Society; Emerson deified the visionary Swedenborg. With these men, Marsh transformed Scandinavia in American eyes from an "Ultima Thule of ice, snow, semibarbaric folk, and militarism" into a heroic land of fortitude, faith, and freedom.[17]

Above all, Viking Atlantic voyages riveted American attention; reputed Viking traces in New England thrilled Anglo-Saxonists who much preferred Icelandic to Italian origins. The Fall River skeleton, Dighton Rock inscriptions, and Newport stone tower (all later shown to be of Indian or colonial origin) persuaded many that Vikings had settled along the New England coast. Marsh himself believed that Rafn had "indubitably established" Northmen's presence in Massachusetts around 1000, but was too skeptical an observer to accept the Dighton rock inscriptions as "Scandinavian."[18]

Danish scholars led by Rasmus Christian Rask interwove these philo-Nordic strands. Rask had rescued ancient manuscripts, reanimated Viking virtues from the Eddas and sagas, shown the vital import of Old Icelandic for comparative philology. In 1825 he and Rafn founded the Danish Royal Society of Northern Antiquaries, whose first major project was *Antiquitates Americanae*, a corpus of sagas, codices, and other data on the Iceland-Greenland-Vinland voyages.

Impressed by Marsh's zeal, Rafn in 1834 made him their American secretary and fund-raiser, aiming thereby to resume printing *Antiquitates Americanae*, held up for some years. Marsh circulated the Society's reports, sought subscribers, contributed $500 of his own, and roused interest in runic stones and other presumed Viking artifacts. By 1835 Rafn so relied on Marsh that "we chiefly build our hopes of a favorable result on your exertions."[19]

Published in 1837, *Antiquitates Americanae* fared better than Marsh had dared expect. It was highly praised by such scholars as George Folsom, Edward Everett, and Henry Rowe Schoolcraft. Everyone was talking about Scandinavia, and Marsh was delighted to assure Rafn that "the *Antiqs. Amer.* have done more to create this interest than any other means." Overjoyed, Rafn urged Americans to rename their country "Vinland." The mania for Viking origins led James Russell Lowell to quip that there were three types of runic inscriptions: those that could be understood by the Royal Society of Northern Antiquaries and Professor Rafn, those comprehensible only to Rafn, and those that no one could understand; the last were the most valuable, since they could be read to say whatever was wanted.[20] Thus spurred, the heroic Viking myth long outlived Marsh, becoming a major strand of fin-de-siècle American filial piety.[21]

*T*he renown of *Antiquitates* encouraged Marsh to press ahead with his own related work. The *Compendious Grammar of the Old-Northern or Icelandic Language*, largely compiled from his translations of three Rask texts, was the first ever made in English. He had proposed the idea to Rafn as early as 1834, when the Danish scholar was helping Marsh to master Icelandic pronunciation. The task was soon done, but Marsh's frequent absences from Burlington and the lack of type for Runic characters held up publication. Now, appending hasty revisions, Marsh had 300 copies printed early in 1838.

His *Grammar's* aim was twofold: to "facilitate access to [Icelandic] literary treasures" and to spur interest in the origins of English. As the closest surviving cognate of Anglo-Saxon, Icelandic could be of immense benefit to English etymology. And in its own right, this copious, flexible, forceful tongue had no rival for "spirited delineations of character, and faithful and lively pictures of events, among nations in a rude state of society." Marsh's piquant additions appear on almost every page. For instance, he notes that the English equivalent *span-new* of Icelandic *spán-nýr* (spánn-chip), though obsolete in England, remained common in New England (as is still the case). He speculates on why ancient Icelandic lacked strong or sweeping terms of praise or censure. He likens Icelandic *út* and *utan* (out and in), vis-à-vis Norway and Denmark, to the *out* and *home* of English North American colonists. Some of Marsh's original ideas on Icelandic inflections and syntax are still current.[22]

Other Nordic work followed. The Swedish scholar Carl David Arfwedson, American consul in Stockholm, one of Marsh's regular suppliers, begged translations from him. Touching on divers aspects of Scandinavian life, Marsh dealt with current Swedish trade, the history of toll rights in the Danish Sound, and biographies of early Norwegian patriots.[23] His renditions were painstaking: expertise in many languages led Marsh to consider most translations deplorable. Of a polyglot edition of Gray's "Elegy" he judged that "every particularly fine thought is spoiled in all the translations. Even the German has managed to sink the original." Save for Shakespeare rendered into German, Marsh had "hardly seen a good translation of any English poet."[24]

Translation *into* English seemed to him more feasible. Its "piebald and Babylonish composition" allowed English to express the thought and feeling of most other tongues; it was flexible enough "to imitate the emasculated delicacy of the Italian, the flippant sentimentality and colloquial ease of the French, the stiff and unbending majesty of the Spanish, and even the Protean variety of the German and the Greek." But pitfalls were legion. One was anachronism: Marsh felt translations must reflect the style and manner of the original. Early foreign texts should be rendered not into modern English but into "a dialect used in England in [a comparable] period of . . . cultivation and polish."[25] But few translators knew enough history and culture to select the right period, let alone to achieve fidelity of mood and thought.

Marsh's own translations reflect his precepts. His snippets from Norse sagas and from Christian Molbech's memoir of the altarpiece painter Pehr Hörberg are quaintly archaic. So is his version of Olof Rudbeck's *Atlantica*, a seventeenth-century encomium on Sweden as the cradle of mankind. Marsh's parodic rendition of Olof Siljestrom's epitaph ridicules Rudbeck's "unsupported hypotheses . . . maintain[ing] the antiquity and glory of Sweden, distorted and garbled" citations, and extravagant claims "that all the learning of the Egyptians was borrowed from the sages of ancient Scandinavia."[26]

Fidelity to original texts proved more awkward in Marsh's sorties into verse. Terming Longfellow's version of Matthias Claudius's "Rhein-weinlied" tame and unfaithful, Marsh offered his own translation of the German drinking ballad:

> The Rhine! the Rhine! leaf, tendril, grape, there flourish!
>   Shower blessings on the Rhine!
> His rugged banks they overhang, and nourish
>   Us with this genial wine.

While cleaving to Claudius's burlesque intent, Marsh admitted his was "a more literal, if not a more finished version" than Longfellow's. Less clumsy are Marsh's renditions of Friedrich von Matthisson's "Fairies" and (below) "Gnomes":

> Elf, night-mare, goblin-sprite,
> Through caves of pitchy night,
> We glare with emerald eye,
> And smallest mote descry.
> There we nectareous naphtha drink,
> With vitriol blue our visage prink,
> Then down on puff-ball pillow sink.
>
> No music know but Satan's choir,
> Tickling the ear with discords dire.
> Such are the Gnomes, if you inquire.[27]

But Marsh's command of vocabulary and German idiom were more conspicuous than his poetic talent; perhaps he was wise to confine his efforts to second-rate originals.

With Caroline, Marsh translated the Swedish Gothicist poet Esaias Tegnér, notably his long narrative *Axel*, again aiming to improve on Longfellow. Marsh's version of Tegnér's "The River" accurately conveys its meaning. But where Tegnér is taut and concise, Marsh is stilted, wordy, prosaic.[28] He understood the original better than did Longfellow, but was not poet enough to put it into English verse.

Yet these translations along with *Antiquitates Americanae* gained Marsh repute on both sides of the Atlantic. "Are you acquainted with Mr. Marsh?" a Danish caller asked Longfellow. "He is the most eminent

Scandinavian scholar I have met with in America." Told that Marsh was "the most learned man in the United States," Scandinavians made Burlington their mecca. Marsh and Rafn corresponded up to the latter's death in 1864. Although "business and many deep sorrows have now for several years swallowed up my time," Marsh's last letter—in impeccable Danish—assured Rafn that "I have never entirely given up my love for Scandinavian literature." But he now "leafed rather than read" the Scandinavian books he still bought. And he resigned himself to picturing the North from afar. "A journey to Scandinavia," he ruefully told the American Minister to Sweden in 1875, "had always been one of my strongest desires," but he never had the time or the money for the trip.[29]

Though Marsh did no more Nordic work after 1851, his efforts and counsel inspired many successors, notably Willard Fiske at Cornell. Recognized as "the pioneer American scholar in that field," Marsh had a lasting influence on Scandinavian studies.[30] But Marsh himself came to slough off the racial and environmental premises that had initially spurred his own scholarly zeal.

Marsh justified his Northern studies as morally worthy. Antiquarian pleasure in Icelandic and Old Norse was not enough; he felt a need to claim the inherent superiority of Nordic (or Gothic) languages and peoples. And in ascribing the same virtues to his fellow New Englanders, Marsh linked them by descent, in a Nonconformist, racialist harangue:

The intellectual character of our Puritan forefathers is that derived by inheritance from our remote Gothic ancestry, restored by its own inherent elasticity to its primitive proportions, upon the removal of the shackles and burdens, which the spiritual and intellectual tyranny of Rome had for centuries imposed upon it. . . . The Goths . . . are the noblest branch of the Caucasian race. We are their children. It was the spirit of the Goth, that guided the May Flower across the trackless ocean; the blood of the Goth, that flowed at Bunker's Hill.[31]

In *The Goths in New-England* (1843) and in his *Address* at the New England Society of New York City (1844) Marsh ascribed to the Goths and their descendants virtues surpassing even the historian E. A. Freeman's famed Anglo-Saxon encomia. Much as nativists a half century later sought the sources of democracy among German forest tribes, Marsh traced New England institutions to Gothic origins. Such notions were neither wholly new nor exclusive to New England: a half century previously Jefferson, who similarly judged most English history since the Norman Conquest an unhappy aberration, had urged that the Great Seal of the United States depict "Hengist and Horsa, the Saxon chiefs from whom we claim the honor of being descended, and whose political principles . . . we have assumed."[32] But it was Marsh who put flesh on the bare bones of Anglo-Saxonism. His Nordic myth embraced a congeries of themes: an idealized pagan past, racial determinism, militant anti-Catholicism, nativist exclusion of aliens, praise of rural virtues, fear of city vices. These were then common prejudices, and many (especially Whigs) heralded Marsh as a spokesman.[33]

It was to quicken American patriotism that Marsh exalted Gothic roots. He felt the want of an "intelligent national pride," owing to the lack of any "well-defined and consistent American character," any venerated common heritage. In the wake of Napoleonic incursions, Europeans were striving to forge popular, autonomous ethnic identities, based on Johann Gottfried von Herder's view of each nation as a uniquely sacred organic unity. "The phlegmatic Northman, the ardent son of the fervid South, the philosophic German, the mercurial Frenchman, and the semi-oriental Sclavonian" had each awakened national zeal by exploring their ancestries, purifying their vernacular tongues, and celebrating their folk legacies. Marsh urged his countrymen likewise to study and celebrate their roots.

Love of country came, first and foremost, from knowing its history; Americans would prize their institutions better by studying their origins. As things stood, Americans were too busy making history to study it, let alone preserve its traces. Like Tocqueville, whose *Democracy in America* Marsh had recently read, he warned that it might soon be too late to recover that past. Americans wrongly dismissed their own history as brief: assuming it began only with the Revolution, they forgot that "we were a nation a century before we became an empire." Moreover, "antiquity is relative. It depends not upon mere lapse of years; . . . to the true American, the hoariest antiquity has no memories more venerable than the landing of the Pilgrims." Rather than postponing American history until after that of Greece and Rome, England or

France, schools should immerse pupils in the American past from their primary years.[34]

They should above all study and emulate the Puritans. And latter-day Puritans must take pride in being Goths, scions of hardy free men whose purity, strength, and tribal democracy had rejuvenated the decadent Roman world. If not pure Goth by birth, the Puritans had clung to the Gothic spirit. In *The Goths in New-England* Marsh penned a series of antitheses, pitting Protestant, democratic, pious, hard-working Goths against Catholic, despotic, sensuous, lazy Romans.

*T*o explain the sterling character of his Goths, Marsh relied on an environmental determinism of ancient Greek origin. "Soft countries breed soft men," Herodotus's Cyrus warned the Persians, who thus "chose rather to dwell in a churlish land, and exercise leadership, than to cultivate plains, and be the slaves of others." Marsh likewise had chosen the rough hills of Vermont over western prairies and southern savannas. And he had nearer authority than Herodotus at hand. His Washington archivist crony Peter Force had recently reissued John White's eloquent *Planters Plea*, the Puritan leader's retort to men tempted to quit harsh New England for presumed ease in the West Indies:

The overflowing of riches [is] enemie to labour, sobriety, justice, love and magnanimity: and the nurse of pride, wantonnesse, and contention. . . . If men desire to have a people degenerate speedily . . . let them seeke a rich soile, that brings in much with little labour; but if they desire that Piety and godlinesse should prosper, let them choose a Country such as this . . . which may yield sufficiency with hard labour and industry.[35]

Not mere necessity, but stern necessity, was the font of virtue.

Marsh also echoed Montesquieu's view that "barrenness of earth renders men industrious, sober, inured to hardship, courageous, and fit for war."[36] Goths had prospered because nature was so hard on them. Warmth bred stagnation; in "the sunny climes of Southern Europe" where lazy folk dwelt outdoors, averred Marsh, "the charms of domestic life are scarcely known." Wintry days in "the frozen North" allowed no such indolence; the domestic hearth was the sole refuge. "Secure from the tempest that howls without, the father and brother here rest from their weary tasks; here the family circle is gathered around the evening meal, and lighter labor, cheered, not interrupted, by social intercourse, is resumed." This was an idyll of Marsh's own early years.

Here the child grows up under the ever watchful eye of the parent . . . lisping infancy is taught the rudiments of sacred and profane knowledge, and the older pupil is encouraged to con over by the evening taper, the lessons of the day, and seek from the father or a more advanced brother, a solution of the problems, which juvenile industry has found too hard to master.

Reverence for "home" and "woman," family and schooling, affection and cooperation: everything civilized stemmed from Gothic domesticity. The outdoors had its role: the grandeur and solitude of Northern nature disposed the Goth to pensive patience, spiritual reflection, above all self-reliance. "The necessity of waging a perpetual war with a sterile soil and an angry sky" instilled fortitude and "a fixed habit of untiring industry."

Thus environment shaped lasting social traits. "These hereditary propensities our ancestors shared in common with all the descendants of the Gothic stock," asserted Marsh. Landscape ensured their endurance. "So long as the great features of nature are unchanged, so long as the same mountains and plains and stormy shores shall be exposed to the same fierce extremes of cold and heat, so long will the character of New-England be conspicuous for the traits which now distinguish it."[37]

Marsh tracked his Gothic forebears from Scandinavia down through the vicissitudes of English history. The familiar legend is the crux of Whig history: fifth- and sixth-century Angles, Saxons, and Jutes brought to Britain the seeds of democracy, planting Gothic virtues too deeply for the Norman Conquest to uproot. The English tongue likewise survived the "mass of alien words, that monkish superstition, Gallo-Norman oppression, scholastic pedantry, and the caprice of fashion, have engrafted upon it."[38] And in the seventeenth century—the age of Cromwell, Milton, the King James Bible—"Goths" defied "Roman" and "Norman" rulers to purge England of alien tyranny.

This epoch also gave birth to New England. Marsh's Founding Fathers came from "the class most deeply tinctured with the moral and intellectual traits of their Northern ancestry." And because they "were 'harried out of the land,' before that [noble Gothic] character had become enervated, or its lofty energies spent," they soon shed "the last remnant and most offensive peculiarity of the Roman spirit, religious intolerance." Marsh palliated the Pilgrims' "transient" proscription of outsiders as "rather an error in fact than in principle."[39]

Over time, Marsh divested himself of Gothicism, shed myths of racially innate virtue, and came to crave the sun and shun the cold. But his anti-Catholicism became not less but more intense; four decades later he would devote his last book to an exposé of supposed papal iniquities.

A tall old codger . . . came to Burlington, and gave himself out for a prophet. . . . He selected me for his interpreter, and gathered his flock in a large room. . . . He placed me in a chair where I could see both his audience and him, and commenced his revelation, but stopped suddenly, came up to me, put his hands under my knees, and said, "you solemnly swear to be my deacon forever." So said I, "I shall swear to no such thing!!" At this, he flew into a violent passion, and said, "Then I turn you out of my church and anathematize you!" "Well," said I, "you old scoundrel, turn me out then." He then began to curse, whereupon I took up the puddling stick my mother had to mix hasty pudding with, and gave him a pair of as sound blows on the ear as you shall see of a summer's day.

Marsh dream, 1844[40]

As his dream suggests, Marsh detested authoritarian faith. Above all, recrudescent Catholicism alarmed him. He blamed the Church for the political reaction that had engulfed Europe since Waterloo and feared lest the Holy Alliance "rebuild not only what Napoleon, but even what Luther overthrew." To Marsh, the Oxford Movement and the influence of Pusey, Keble, and Newman in the Anglican Church presaged its Romanization. "England is Greek by birth, Roman by adoption," in his antipodal figure. Her "moral grandeur" and intellect "she owes to the Greek mother, while her grasping ambition, her . . . exclusive selfishness, are due to the Roman nurse."[41]

Worse yet, Catholicism was winning adherents in America. Episcopalians like Bishop Hopkins of Vermont adopted High Church liturgy, and several prominent Vermonters, a daughter of Ethan Allen among them, had become Catholic converts. Catholicism was even gaining civic respect. In 1847, three years after Marsh's New England Society speech, New York's militant Bishop John J. Hughes was given a seat of honor as the Society saluted "Pius IX, Pope of Rome." "What strange changes!" wrote shocked ex-mayor Philip Hone. "The sons of Pilgrims toasting the old lady, whom their fathers complimented with the titles of 'whore of Babylon' [and] 'red harlot'!" Marsh had cause to worry that "even with us . . . the evil leaven is at work."[42]

In America, as in England, many ignorant of medieval history had begun to admire things medieval. Over the next few decades Marsh deplored a "revival of ecclesiasticism in religion . . . followed by a like revival of mediaeval taste in art, and by the unearthing of multitudes of half-forgotten popular superstitions, which any person of ordinary intelligence would have been ashamed to own half a century ago." It became "fashionable for Protestant gentry to attend . . . semi-popish places of worship, to build new churches after Middle-age models of most ungraceful, clumsy, and barbarous styles of architecture, [and] to talk of . . . the devotional feeling of the builders and carvers and painters of the 'Ages of Faith.'"[43]

"It is from England," Marsh charged, "that this poison mainly distils," but he like many saw the Irish as its chief purveyors. It was against Irish Catholics that Know-Nothing and other nativist movements chiefly thundered. As famine multiplied immigrant numbers—by 1851 there were over a million Irish in the United States, half the foreign-born—hostility erupted not only in tirades like Marsh's but in acts of violence. When St. Mary's Roman Catholic Church in Burlington was torched in 1838, Catholic immigrant laborers there—French Canadian and Irish—numbered two thousand. Though (or perhaps because) many of his own workers were Irish, Marsh's anti-Irish animus persisted. He "trust[ed] the intelligence and the virtue of the Africans," urging suffrage for ex-slaves after the Civil War, "sooner than that of the Hibernians, whom I have long looked upon as the most pestilent, truly barbarian element in our social life."[44]

Like other Yankees who equated civic virtue with rural scenes, Marsh assailed the mushrooming cities that drew most immigrants. What an indigestible mess simmered in the American melting pot! "Every foreign heresy or folly in religion and in government finds a congenial soil in that corrupted [urban] mass of outlandish renegades and adventurers." Marsh feared lest "the infusion of extraneous elements will cause a fermentation, which can only end in their own violent expulsion, or the corruption of the whole mass." These latter-day Roman legions might soon swamp Puritan America. As early as 1835 Marsh had backed a petition to Vermont's General Assembly to prohibit monasteries and nunneries, as breeding grounds for enslaving minds, destroying liberty, and fomenting despotism. In Congress a decade later Marsh sought, in vain, to lengthen citizenship residence requirements from seven to twenty-one years.[45]

Marsh's Gothic encomiums got mixed reviews. One critic deplored his harshness toward Romans and city dwellers; another rebuked him for "sheer rationalism in religion, and sheer Jacobinism in politics." He also derided Marsh's climatic theories. "The moral superiority of the Gothic race Mr. Marsh attributes to . . . a bad climate and a bad soil." If inclement weather improved people, why were the natives of Kamchatka and Labrador uncivilized? "Marsh's explanation would be . . . that whenever bad weather fails of developing Puritanism, it is not the kind of bad weather which his theory requires."[46]

The most trenchant critique came from Marsh's good friend the Reverend George Allen, a classicist disciple of James Marsh's who later turned Catholic. "You have started," Allen charged Marsh, "from a petty sectarian prejudice—such as one picks up in a little village— such as the half-educated son of a narrow-minded Calvinistic deacon might be expected to entertain," and had been blinded by bias ever since. "You credit the Puritans with your own modern Rationalism, and you charge upon Episcopacy what is not just even of Popery." It was absurd for Marsh to glorify the Puritans, the most illiberal of sects. "If you were compelled to live with the *real* Puritans, George Marsh, you would be driven to hang yourself—to escape being burned by them." George might have heeded his friend Choate's admonition to James Marsh, that "very much narrowness of mind and very great soundness of faith do sometimes go together."[47]

*I*n Marsh's Gothicist rant, Enlightenment rationalism warred with romanticist idealism. Self-contradiction was rife: in the same paragraph he praised the Industrial Revolution and condemned cities, vaunted present comforts and mourned past virtues. He relied on distinctions between *reason* and *understanding* invented by Greek philosophers, given piety by seventeenth-century Cambridge Platonists (Ralph Cudworth, Henry More, Joseph Glanvill), sophisticated by Kant and Coleridge, and Americanized by James Marsh and Emerson. *Understanding* yielded the facts of material existence by means of sensory impressions; *reason* supplied inspiration beyond mundane experience. The meaning of life could not be discerned nor religion justified by worldly existence, as Locke mistakenly claimed, but only through reason's spiritual intuition. "Reason focused on the unity of phenomena, Understanding on their differences."[48]

Yet "in all these insistent declarations of unity," notes Laura Walls of James Marsh and Emerson, "we are aware of nothing so much as a cascade of dualisms." Although monistic in theory, Marsh's metaphysics were dualist in practice. He sought a unified view of the world, but Calvinism made him dialectical and dichotomous; he could not help seeing things in fiercely partisan terms of good and evil. So he set mind over matter, preferred spirit to the senses, opposed the ideal "indwelling, life-giving principle" to superficial form, favored human over bestial, organic over inorganic, truth above utility. These were the values of the Cambridge Platonists Marsh found "most pious and learned"—a necessary juxtaposition. Just as Marsh's ethics were grounded in metaphysics, so scholarly morality buttressed his credos in politics, religion, science, and art.

Hence the rhetoric of Marsh's Gothic-Roman antithesis. "The Goth is characterized by the reason, the Roman, by the understanding; the one by imagination, the other by fancy; the former aspires to the spiritual, the latter is prone to the sensuous." In art and literature, according to Marsh, the realist Roman excites admiration by outward form and sensuous display; the Goth "pursues the development of a principle, the expression of a thought, the realization of an ideal."[49] Puritan-Catholic contrasts fitted the same schema: the Puritan focused on religion's inner meanings, the Catholic on its outer trappings; the Puritan upheld spiritual freedom and welcomed discussion, individuality, diversity; the Catholic prescribed ritual conformity and bowed to priestly authority.

Marsh relied on reason but exalted the supernatural. He spurned the "sensuous philosophy of Locke, and its wretched corollary, the self-ish morality of Paley," long dominant in American pedagogy. Locke's rationalism neither warmed the heart nor roused the spirit. Still worse was Paley's quid-pro-quo doctrine, "which solves all questions of duty by a calculation of the balance of profit and loss":

> Whatever, Lord, we lend to Thee
> Repaid a thousandfold will be;
> Then gladly will we give to Thee . . .

This was "degrading and demoralizing in ethics [and] fatal in religion."[50]

Transcendent intuition was all very well; revivalist fervor and "animal sympathy" were not. Marsh castigated the itinerant preacher Jedidiah Burchard, erstwhile thespian circus rider, whose charismatic conquests relied on hellfire and mass hysteria. Few of this "great im-poster's [converts] kept anywhere within the bounds of decency," railed Marsh's father after a Woodstock revival meeting at which Burchard had hectored Marsh's brother Lyndon:

"Will you give your heart to God?" "I don't know." "Why! I am astonished that a man of your standing—the son of one of the first lawyers in Vermont—a man of college education, and the cashier of the Bank of Woodstock—I am as-tonished that you *don't*. Well, Mr. Marsh, will you do it?" . . . If the answer was in the *negative* [Burchard] would generally reply, "Then you will be damned," or "Go to hell."[51]

At Burlington in December 1835, Burchard "saved" over two hun-dred, college students among them. The Marsh cousins, "the wise ones of Burlington," branded him a "public nuisance" whose shallow and tran-sitory conversions undermined true faith. To the Marshes, Burchard's revivals seemed obscene perversions of the spiritual humility and in-tellectual rigor of true evangelists such as Bunyan and Henry More. Burchard's "unfailing success," charged George Marsh, came from prey-ing on the feelings of the ignorant and the unschooled; it was emotion "devoid of reflection." But Burchard scoffed at the Marshes' insistence on "giving people time *to think* . . . I tell you, *people won't think. They are too dull and lazy to think!* They want *excitement* . . . sinners are converted by be-ing excited." In the ferment of Vermont's episodic social turmoils this was patent.[52]

After the Burchard tornado, the Marshes faced a new storm in the *patriote* crisis of 1837, when political unrest in Quebec fanned religious and ethnic nativism. Chronic discontent among Lower Canada's French majority, aggravated now by economic misery, exploded into open rebellion when Britain spurned pleas for a measure of self-rule. On No-vember 23, a French *patriote* band routed British infantry at the Riche-lieu River. Subsequently repulsed, *patriotes* fled south into New York and Vermont, enlisting support—and arms—for cross-border forays. Vexed by the imposition of martial law in Quebec, a hindrance to both legal trade and lucrative smuggling, bellicose Vermonters gave vent to ever endemic Anglophobia. Rumors spread that British troops had "wantonly invaded our territory and murdered our citizens," arousing ardent sympathy for "just such a cause as our fathers were engaged in."[53]

Appalled by *patriote* rabble-rousing, Marsh led Burlington's elite to petition Vermont's Governor Silas Jenison to proclaim strict neutrality, for the sake alike of commerce with Canada and peace with Britain. Americans ought not "dignify every case of resistance to an established government with the name of liberty," warned Marsh. "Though it may often be generous, it is not always just to adopt the quarrel of the weaker party"; in this instance it was clearly wrong. Pressured also by U.S. Secretary of State John Forsyth (and indebted to Marsh for his 1836 governorship), Jenison swiftly affirmed his state's pacifism and en-joined Vermonters to shun Canada's "intestine broils."[54]

Marsh's anti-*patriote* stand reflected both propertied prudence and ancestral propriety. "The blessings of order and law are certain—the benefits of revolution are always beforehand doubtful," he admonished; precisely the same views voiced by the orthodox establishment five years later in Rhode Island's Dorr Rebellion. And Marsh felt America's Revolutionary forebears grossly defamed by being likened to Quebec's Catholic peasants, "nine tenths of whom can neither read nor write, who certainly have no strength of religious or even moral principle," as James Marsh put it, and who were "utterly incapable of self-government."

Some in Burlington excoriated the Marshes for craven Anglo-philia. Their petition, sneered the *Free Press*, should have gone not to the governor of Vermont but to the queen of England, "for surely never were more loyal or royal sentiments offered at the foot of the throne."[55] At least the monarchy was untainted by Catholicism. Royalist the Marshes were not; Anglo-Saxonist and anti-Catholic they most certainly were.

The quasi-mystical idealism Marsh expressed in the 1840s proved increasingly at odds with his developing insights into human impacts on nature. When both "reason" and "understanding" persuaded him that he had mistaken the causal connections between society and environment, he amended his views of causality. By the time Social Darwinism and national chauvinism had made racial and environmental determinism attractive to other scholars, Marsh had dropped them from his philosophy. His later work is the better for largely disavowing at least some of his youthful prejudices.

4. Congress and the Smithsonian

MARSH'S ACCOLADE TO GOTHIC NEW ENGLAND coincided with his advent in Washington as a Vermont congressman. He had reentered state politics in 1840 to back William Henry Harrison as Whig candidate for president. The Whig party reflected Vermont antislavery and tariff protection concerns; Marsh helped draft his state's Whig reform platform. The Anti-Masons had run their course; Jacksonian Democrats had crippled Vermont by scuttling the United States Bank and the woolen tariff. As Whig county keynoter, Marsh assailed the dozen years of "an imbecile & intriguing administration" that left the country depressed, industry prostrated, and the "political rights & sacred barriers of the constitution trodden underfoot."[1] Harrison swept Vermont by almost two to one, a wider margin than in any other state, and paved Marsh's way to Congress.

Reapportionment after the 1840 census cut Vermont's numbers in the House from five to four. A new congressional district centered in Burlington merged parts of two old districts: one had been the seat of scholarly Augustus Young, who now retired; the other of William Slade, popular statewide but as a Middlebury man unwelcome to the new district's northern voters. In Marsh, even his old Burlington opponents found an ideal spokesman for local interests. Twelve years back the *Free Press* had stigmatized "Squire Marsh" as "a man whose veins are bursting with the insolence of aristocracy." The same journal now enthused that Marsh's "age, in the very bloom and vigor of life and the palmy state of his intellectual usefulness . . . his high personal character, his unimpeachable integrity, and the acknowledged purity of his life," made him the perfect candidate. On the third ballot, in June 1843, the Whigs nominated Marsh over Slade, shortly to become governor.[2]

Marsh vowed to foster Vermont interests—economic "freedom" from Britain, sound currency, a protective tariff—but refused "to sacrifice the general good" of the country to promote regional matters. Nor would he be bound by instructions from constituents. All were equal under law, but it did not follow that all should *make* law; few were "both *competent* and *prepared* to decide upon all questions." Elect me, said Marsh, and trust my judgment. The Democrats denounced him as "high-toned" and "aristocratic," and some in his own party termed him too tepid against slavery. But Marsh, campaigning strenuously all summer, "avowed himself to be every inch a whig, a whig protectionist, a whig abolitionist." Running well ahead of his ticket in the September election, he beat Democrat John Smith of St. Albans by 1,600 votes.[3]

But Marsh was in no fit state to celebrate. In the midst of the fray Caroline had an apparent stroke, and soon after young George came down with typhoid; both required full-time nursing. Worn out by these cares, Marsh himself fell ill. Three days by steamer down Lake Champlain and the Hudson and by rail from New York brought the ailing congressman, his son, wife, and her sister Lucy to Washington in early December. The capital was cold and damp; muddy slush paved the way to the Exchange Hotel on C Street, where Marsh took a suite of small, unkempt rooms. "Weak and lame, sadly dispirited about my family, I ought honestly to be in bed, but I keep about, & am trying to learn my trade, as well as I can." Yet, Marsh fretted in the tone of hapless gloom that recurrently beset him, "every step I have taken has been met by some disappointment that has thwarted every calculation I had made."[4]

John Quincy Adams judged the 28th Congress "the most perverse and worthless . . . that ever disgraced this confederacy."[5] His appraisal was too harsh. Adams himself was a potent member, and men of stature abounded: from his own state, Rufus Choate and historian-editor John G. Palfrey; poet-lawyer James Dixon of Connecticut and jurist Levi Woodbury of New Hampshire, future Secretary of State Hamilton Fish of New York, Benjamin Tappan of Ohio, James Buchanan and litterateur-diplomat Joseph R. Ingersoll of Pennsylvania; and from the South, future Confederate vice president Alexander Stephens, Henry Foote, William Cabell Rives, and Thomas Clingman. But Congress's performance did not match its members' prowess. Little could have been expected: the House was Democratic, the Senate Whig; John Tyler, the Virginian who became president on Harrison's early demise, Whig by election but Democrat by sympathy, was at odds with both.

In the House minority party, Marsh's role was often frankly obstructionist. He was heartened by a Democratic schism between the Van Buren and Calhoun factions, allaying Whig fears that annexation of Texas might be imminent. Marsh mastered congressional procedure, met fellow congressmen, sampled Washington social life, and looked for a decent place to live.

His Vermont House confreres were all Whigs but one, and all were new to Washington. Tall, beak-nosed, magisterial Jacob Collamer, an incarnation of Yankee common sense, became Vermont's chief dispenser of political patronage. Marsh had long known Collamer in his adopted Woodstock, but there was little love between them. Collamer had no interest in Marsh's science, art, and literature; Marsh scorned Collamer's stag parties and crass opportunism. "He is an able man," Collamer's colleague Solomon Foot said of him, "but so selfish & so disgustingly conceited, as greatly to impair his influence."[6] Representative Foot himself, a modest Rutland schoolteacher who rose to political eminence by selfless good works, was Marsh's generous lifelong friend. With future governor Paul Dillingham and with Vermont's senators, ardent abolitionist William Upham of Montpelier and talented but bibulous Samuel S. Phelps of Middlebury, Marsh had less contact — wisely in the last case, for he was soon queried by Vermont Whigs about Phelps's conduct. Was it true that the senator indulged in "intoxicating beverages, or used garrulous, vituperative, vulgar, or profane language"? If so, was the cause "a defect of physical, or moral, power, an aberration of intellect, or a servitude to passion and feeling induced by habits of improper indulgence"? Marsh's response is missing, but Phelps weathered these well-founded charges.[7]

Marsh's closest Capitol Hill associates were Rufus Choate and, from 1845, Robert C. Winthrop of Massachusetts. Choate he had esteemed since Dartmouth; Winthrop appealed as a fellow scholar. Personable, eloquent, wide-ranging, Winthrop's moderation on slavery endeared him to Southern Whigs. Early allies, Marsh and Winthrop at length diverged; Winthrop became a leading Cotton Whig friendly to Southern interests, Marsh an uncompromising Republican. But they kept up their intimacy, exchanging views on art, literature, history, and politics for more than thirty years.

With John Quincy Adams, Marsh was on easy terms. He backed Adams's antislavery crusade for the right of petition, and committee work brought together the Vermonter and the pungent old ex-president. Marsh saw a good deal of Daniel Webster, but despite Dartmouth family bonds was never intimate with the "Mighty Daniel," a notorious deadbeat. Of a mutual crony, Vermont-born ex-Congressman Edward Curtis, Marsh later wrote, "I believe he was the only friend Webster had, who was wise enough to keep his money from his friend." With Henry Clay Marsh's relations were likewise attenuated. He shrank from heroes and hero worship; the adulation showered on famed figures repelled him. When Clay came to Burlington prior to the 1840 campaign, Marsh sourly noted the "great shouting, throwing up of caps, and blowing of brazen instruments," and promised to save Clay "one shake by keeping out of his reach."[8]

In his first Washington winter Marsh went often to presidential levees, teas, dinners, even balls. If the Exchange Hotel held no charms, Washington offered a wider social circle than Burlington. The Vermonter's command of languages gave him entrée to the diplomatic corps; his interest in science and technology led to friendships with James M. Gilliss of the Naval Observatory, Alexander Dallas Bache of the Coast and Geodetic Survey, and Joseph G. Totten of the Corps of Engineers.

The physical as well as the social climate agreed with Marsh. He regained health in the mild Washington winter, enjoyed long spring walks, relished the roses and strawberries of luxuriant early summer. All in all, Washington pleased him. True, the odious slave mart operated almost within sight of the Capitol; but the city was no longer the abode of vice that Charles Marsh had witnessed in Congress; "no disgusting sights of beggars or prostitutes [now] met the eye."[9] Much of Washington was still raw, ugly, squalid, but Marsh was grateful for the warmth of the sun.

Not all was sunshine and roses. A congressman had to hustle for his $8 daily pay, especially a Vermonter whose constituents ever fretted about what free traders were up to. Although Marsh had insisted he would take no "instructions," he needed to keep an eye on next summer's election. "My correspondents are very numerous," he groaned; "the day that I do not receive ten letters requiring replies is an easy one."[10] A rheumatic arm made writing an added misery.

Myriad miscellaneous matters engrossed his time. Committee meetings were at nine in the morning; the House met weekdays and many Saturdays at eleven or twelve. Marsh ate nothing after breakfast until evening, making for a long, fatiguing day. He found "confinement to the House tedious" and irksome. "It is indeed a disorderly body," agreed Collamer. "It is difficult to hear & very few men or subjects will engage attention. Most of the members are engaged in reading, writing letters, franking documents or in private conversation, not generally in whispers."[11] Ladies crowded the galleries to hear gentlemen hurl abuse at one another. But Marsh was punctilious in attendance. Even at the session's end, when Congress met nights as well as days, he seldom missed roll call. Many were less scrupulous; it was often difficult to get a quorum.

Hard of hearing, Marsh took little interest and small part in most House debates. He wrote his official letters, then relaxed. "It became . . . almost a standing pleasantry among some of the frequenters of the galleries, to predict when J. Q. Adams and G. P. Marsh" would doze off, Caroline related; "it was amusing to see them both quietly alert until they saw who was to have the floor for the next hour, and then . . . dropping at once into a profound sleep, broken simultaneously by the sound of the Speaker's hammer."[12] Daily naps enabled Marsh to attend social functions past midnight yet rise at half past four in the morning.

The most crucial House issues involved slavery. A notorious gag rule empowered the Speaker to table unread any petition bearing on abolition. At first a lone voice, John Quincy Adams ceaselessly attacked the rule as unconstitutional; by and by others joined him. In February 1844 even some Southerners voted, in vain, to repeal it. Their stand made Marsh sanguine about freedom. The once preponderant South now elected only two-fifths of the House; shorn of coercive power, its "better elements" seemed ripe for reform. "Slavery is on the defensive," Marsh judged. "Slaveholders find themselves overborne by the force of public opinion through the civilized world." Southerners assured Marsh that emancipation in the border states was imminent.

Hence he at first saw eye to eye with moderates like Georgia Senator John M. Berrien, styled "the American Cicero" for his orator's skills, and North Carolina's Thomas L. Clingman, who had come North to put their case. Marsh felt Yankee extremism counterproductive. When Vermonters grew impatient with Marsh's silence in Congress, he explained that he could not satisfy himself or his constituents "without giving offense in quarters which it would now be bad policy to stir up the animals." Why vex them needlessly when "they are preparing (with the exception perhaps of the extreme South) . . . to abolish slavery"? The Whigs "did not wish to provoke those who are already giving up." [13]

Nor was Marsh at first alarmed by Secretary of State Calhoun's aim to annex Texas, which had broken away from Mexico and become an independent republic in 1836; against what many feared would extend slavery Marsh expected Southern Whigs to hold firm. Initially he proved right; in April 1844 the Senate voted down President Tyler's annexation act by two to one. The future seemed safe; the leading presidential contenders, Clay and Van Buren, both opposed annexing Texas. But Van Buren's stand doomed his candidacy; Democrats instead nominated annexationist James K. Polk. Animus over Texas and Mexico embroiled Congress from this time forth.

The tariff was as hotly debated as Texas and the gag rule. In 1842 the Whigs had barely managed to raise duties on materials used in Northern manufactures—wool, glass, iron. But the high cost of protected goods hurt Southern planters; the Democrats now sought to reduce imposts. With the tariff, as with Texas, Marsh at first expected a Whig victory, but his optimism soon waned. As a known moderate, he hoped to sway the decision. In his maiden speech on April 30, 1844, he rebuked in quiet, even tones the low-tariff Northern Democrats who

put party power over "their own solemn convictions of . . . their country's good." Marsh waxed sardonic. Free traders professed constitutional qualms on protection; yet the same men who would now deny
Congress the power to tax had themselves scuttled sacred rights of
petition! "Gentlemen who swallowed those camels will [n]ever be
strangled by so small a gnat." Marsh scouted their scruples as frivolous:

The constitutional colic is, indeed, a grievous complaint, oftentimes an excru-
ciatingly painful disease, but, happily, it is never mortal. Gentlemen are frequently attacked by it, they sicken, they suffer . . . but die never. In the long
rows of our departed predecessors, in yonder cemetery, you find the monuments of those who have fallen a prey to death in all its varied shapes. Gout,
apoplexy, consumption, fever, and even the hand of violence, each hath its victims, but constitutional scruples, none.[14]

Marsh went on to explain why existing duties must be kept. Fixed
prices were essential to Northern industry. "Myself unhappily a manufacturer, I know too well the indispensable necessity" of estimating
profit and loss in advance. A small reduction of duties, which "shall
scarcely save a penny to any individual consumer, may work utter
ruin to the manufacturing capitalist, and the hundreds who depend
upon him." Any who thought Northern millers affluent ought to heed
his Winooski fiasco. "Scarcely a woolen factory in New England,"
Marsh told the House, "has not lost a sum equal to its entire capital,
since 1837."

To wintry Vermont protection was vital. "We of the extreme
North . . . contend with physical difficulties to which the more favored
South and West are strangers. Our territory is mountainous—our soil
rugged and comparatively unthankful. . . . Our climate is of even fearful severity." The 1842 tariff did not enrich Vermont, but at least it kept
mills rolling. Fear lest Congress remove this slim protection "has already produced a panic, whose influence upon the price of our only
staple will cost the wool growers of Vermont not less than half a million. . . . Prostrate our manufactures, deprive us of this one resource,
and you plunge us into absolute, hopeless, irretrievable ruin."[15] The author of *The Goths in New-England* sought to shelter his Puritan constituents against undue hardship.

Vermonters, delighted "that our mountains can breed a lion," lauded Marsh's "rare rich racy style." For once heeding the Green Mountain lion, the House narrowly (105 to 99) shelved tariff reduction. (The respite was fleeting; two years later, deaf to Marsh's indictment of free trade as an "unholy combination between our own Government and the capitalists of Europe," Tyler's tariff repeal wrecked Marsh's woolen enterprise.) [16] But as of June 1844, Congress postponed both Texas and tariff until the election provided a clear mandate.

It was high time Marsh was home. The presidential campaign was on, and Marsh had his own fences to mend, first as an unsuccessful senatorial candidate, then for reelection to the House. Initially he had "no fear, or next to none," that Henry Clay would lose, but as Clay wavered on Texas and slavery, Marsh despaired. "Mr. Clay is infatuated," he reported after seeing him late in October; "he believes he shall carry *every* state but *two!* If there were a few weeks instead of days between this and the election, I verily think he would manage to *lose* every state but *two*." [17] Marsh's own campaign was more effectual. The Whigs swept Vermont in September, Marsh topping the ticket with double his previous majority. Though Vermont held firm for Clay, Polk won the presidency by a narrow popular but large electoral margin. Apprehensive, Caroline Marsh reported the Washington scene:

An hour after midnight, a yell of triumph, protracted, hideous, demoniac, rang out from one end of the city to the other. . . . Early the next morning [November 10] Mr. Marsh called a friend to the window and pointed out a huge flag floating over a distant quarter of the town. "Do you know what that flag is waving over?" he asked, with an excitement of manner very rare with him. "There is the *Slave Market* of Washington! and that flag means *Texas*; and *Texas* means *civil war*, before we have done with it." [18]

Texas was on every mind. The Democrats made the election a mandate for annexation, and Marsh felt prospects "gloomy in the highest degree." Aware the decision depended "upon no debate within doors, but upon appliances without," he yet hoped by plain speaking to change a few minds. [19] What use was Texas? he asked the House. Would it augment Northern manufacturers' markets, as Southerners promised? Hardly. Texas was poor and thinly populated. Would acquisition "extend the area of freedom," as some claimed, by removing blacks from the existing South? Texas was surely freer now than it would be with slaves, and any reduction of Southern slave surpluses would be temporary at best. Such "reasons" were to Marsh mere ruses: the real goal of annexation was political—Texas was wanted, perhaps carved up into as many as five or six new states, to restore Southern hegemony in Congress.

Southerners should not be so anxious. The North had no power, moderate Northerners no desire, to curtail slavery in the South. It was not to destroy slavery but just to stop it spreading that Marsh spurned Texas. "We of the North repudiate and disclaim all purpose of impairing the moral or legal right of the South, yet we are resolved that the extension and perpetuation of an evil which is its misfortune should not be our fault." It was wrong to "stigmatize the people of the North . . . as abolitionists, fanatics, incendiaries . . . with a reckless hatred of the South." Slaveholders' chimerical fears might, Marsh warned, foment dissolution of the Union and civil war, which would be "hailed by the despotic governments of Europe . . . as the signal of a richer partition than the dismemberment of Poland."[20]

Deaf to Marsh's moderation, five days later the House approved annexation by 120 to 98. As Choate had warned him, Texan congressmen sat next to Marsh in 1846.[21]

Marsh found committee work more constructive than speeches to closed minds. The Naval Affairs Committee, his first assignment, dealt with pension claims, disciplinary complaints, dockyard management, supplies, and naval architecture. "All mechanical matters in my committees are referred to me," Marsh was pleased to note; but he found "hosts of projectors" constantly at his heels. He paid scant heed to lobbyists but "listened patiently" to would-be inventors, to whom he "pointed out the fatal flaw, [yet seldom] persuaded the disappointed, half-demented victim of false hopes . . . to give up his project."[22] Technical matters aroused Marsh's curiosity and enlarged his contacts in scientific Washington.

In a typical case, two Navy machinists devised a rope-making machine that produced superior rope at half the former cost. The Navy denied them compensation, holding that anything its employees made was government property. But Marsh ruled that the men had been paid only for their labor, not "for the exercise of mechanical ingenuity." They were "employed, not as *inventors*, but as labourers—not to *contrive*, but to *build* machinery." His committee awarded them $500—a sum Marsh reckoned the machine saved the Navy annually.[23]

Simple sailors fared less well with Marsh. He wanted flogging continued as standard shipboard punishment, but Congress abolished it despite him. He backed a temperance crusade to ban the Navy liquor ration. The Navy itself opposed the ban, fearing mass discontent and rum-smuggling; recruits would balk at a teetotal service, veterans be seduced into foreign navies. But for Marsh "the advantages to health, discipline, efficiency of the crew, important saving in stowage and freight, removal of temptation from apprentices and other young persons on board ship, and, above all, the moral bearing of the question" outweighed Navy arguments. However, the House let Marsh's bill die. Even temperance backers were of two minds, as Horace Greeley related. Voting to abolish grog, one congressman said to another: "That was a glorious vote we have just taken." "Yes, glorious," said the other; "let us go and have a drink on the strength of it." Not until the Civil War did a New England-led Congress commute the Navy liquor ration, at a miserly five cents a day.[24]

The Joint Library of Congress committee was a Marsh assignment heralding his major role in the new Smithsonian Institution. Congress's then minuscule stock of books would be lost to view in today's mammoth collection. Congressmen were not book-minded; Marsh had to pass up a great European bargain because he could not get "so large an appropriation for the purchase of *any* library, let its character be what it might."[25] But his committee did persuade Congress to double the annual library budget from $2,500 to $5,000.

The Library Committee's major task was to dispose of materials (rocks, fossils, plants, and animals from Antarctica, the South Pacific, and the Americas) collected by Commodore Charles Wilkes's 1838–42 exploring expedition, the nation's greatest since Lewis and Clark. Under the aegis of the National Institute for the Promotion of Science, a private agency formed in 1840 by Secretary of War Joel R. Poinsett, the Wilkes Expedition material was merged on arrival with other moldering hoards—odds and ends from Lewis and Clark, Lewis Cass, Henry Rowe Schoolcraft, Stephen Long, and the dauntless Zebulon Pike. In 1842 Wilkes's able naturalist, Charles Pickering, began to classify and catalog all this; in 1843 the National Institute asked Congress to defray his costs. The request precipitated a fight. Library Committee chairman Senator Benjamin Tappan of Ohio accused the Institute of usurping his committee's functions.

Marsh arrived to find the outraged Tappan on one side, on the other Institute scientists, backed in the Library Committee by Rufus Choate. A third bloc wanted the government to stay out of science entirely. Like Choate, Marsh felt only the National Institute could do the job. On behalf of the Library Committee, he urged in June 1844 that the specimens be declared federal property under the Institute's continued care. Marsh envisioned a great government museum that would give "character, consistency, and unity to national science." By thus promoting the "liberal arts," America would become "a leading example of the benefits of free and popular institutions."[26]

Despite the efforts of Marsh and John Quincy Adams in the House, and of Choate and Woodbury in the Senate, Congress terminated the Institute's trusteeship, and the collections lay dormant until acquired by the Smithsonian Institution in 1857. In time, as Marsh had dreamed, they formed the nucleus of the United States National Museum. Marsh and other Institute friends meanwhile turned their attention to the new Smithsonian, already riven by party feuds and personal ambitions. In this new agency the line between "genteel" scholars and "practical" frontiersmen would be sharply drawn.

A laboratory is a charnel house—chemical decomposition begins with death, and experiments are but the dry bones of science. . . . Without a library . . . all these are but a masqued pageant, and the demonstrator is a harlequin.

Marsh, *Speech on the Smithsonian* (1846)

Marsh spent the summer recess of 1845 trying to recoup his personal fortunes. Nothing went well: the Winooski woolen works continued to lose money; the Lyman & Marsh firm borrowed heavily to cover ill-fated investments. Marsh's one cheer was finding "some remarkably fine engravings, which I could not resist."²⁷ Culture was to him an essential, not an extravagance; in this realm the nation, too, deserved the best. An opportune moment for deciding what was best arose, soon after Marsh returned to Washington in December, in congressional hearings on the Smithsonian Institution.

Denied adequate homage at home by his illegitimate birth, the English chemist James Smithson had left his large fortune "to found at Washington . . . an establishment for the increase and diffusion of knowledge among men." A decade after Smithson's death in 1829, $508,318.46 was deposited in the U.S. Treasury—to the dismay of many legislators. Some mistrusted Smithson's motives, suspecting a British plot to regain New World hegemony. Others felt the foreign gift demeaning. Even those grateful for Smithson's bequest wondered what to do with it, so vague was the stated purpose. John Quincy Adams wanted an astronomical observatory; others backed a national university, agricultural schools, popular lectures, educational pamphlets, nautical almanacs, essay prizes, basic or applied research.²⁸

Out of the welter of Smithsonian schemes two chief camps emerged by 1844, one "practical," the other "intellectual." The former —mainly western Democrats like Senator Benjamin Tappan of Ohio and Representative Robert Dale Owen of Indiana—favored agricultural training, experimental science, and other projects of immediate public use. The scholars, notably Adams, Choate, and Marsh, preferred either a museum or a national research library. "Nothing is more imperiously demanded by all great American interests," Marsh asserted, "than enlarged, multiplied, & diversified collections of books."²⁹

Skirmishing between scholars and apostles of utility was fierce in the Joint Library Committee, to which Congress first referred the Smithsonian issue. Here were leaders of both factions, the militantly practical Tappan and the erudite Choate and Marsh. Tappan had just routed the National Institute; now he won the first Smithsonian skirmish. Why create another library, he asked, when the Library Committee was already buying all the books worth having? The committee majority backed his scheme for agricultural experiments and training. But Tappan's bill failed in the Senate, while a Choate countermeasure to devote Smithsonian income to book-buying lost in the House.

As a new member Marsh had little to do with the Smithsonian, but in the 29th Congress he was appointed, with John Quincy Adams, to the seven-man Smithsonian select committee, chaired by Owen. A redoubtable minority duo, Adams and Marsh secured a compromise bill after weeks of argument: the library would get $10,000 (half the income), the rest going to schools, an agricultural experiment station, and popular science tracts. But in the House, Owen again moved to halve the library funds. What need had Americans to emulate the "vast and bloated book-gatherings . . . of European monarchies?"[30]

Salvaging immortal texts from philistine disdain, Marsh refuted Owen in his best congressional speech. His main argument was pragmatic. A keen advocate of utility, Marsh knew the value of experimental research. But true science "must be drawn from deeper sources than the crucible and the retort." Too few realized that America's vaunted technology depended on basic research; excessive zeal for practicality was itself impractical. Marsh cited astronomy, chemistry, music, and optics to show how applied science derived from pure science. And "higher knowledge" not only catalyzed technical progress, it "serves to humanize, to refine, to elevate, to make men more deeply wise, better, less thoughtful of material interests, and more regardful of eternal truths."

Marsh countered Owen's lowbrow argument with an appeal to national pride. Statesmen ought to be learned; most of America's founding fathers "spent some part of their lives in scholastic retirement." Neither virtue nor freedom could survive ignorance; far from being "intuitive or even instinctive," as Owen claimed, American liberties had "necessitated the labor of successive generations of philosophers and statesmen." Marsh also invoked Anglophobia: a great library would be "the most effective means of releasing us from the slavish deference, which, in spite of our loud and vaporing protestations of independence, we habitually pay to English precedents and authorities."[31]

Owen had claimed Americans needed few books: "shall we grudge to Europe her antiquarian lore, her cumbrous folios, the chaff of learned dullness that cumbers her old library shelves?"[32] We should indeed begrudge this, retorted Marsh: just "because a newer, or better, or truer book, upon a given subject, now exists, it does not necessarily follow that the older and inferior is to be rejected. It may contain important truths or interesting views that later . . . authors have overlooked—it may embody curious anecdotes of forgotten times—it may be valuable as an illustration of the history of opinion, or as a model of composition." Just as no one could hope to master all knowledge in any field, so "every good book supposes and implies the previous existence of numerous other good books." As Thoreau would say, "decayed literature makes the richest of soils."

American libraries were pitiably small. Göttingen University's library was six times the size of America's largest collection. With only 40,000 volumes, the Library of Congress was "miserably deficient" in every field; of the million and a half German books printed it had fewer than a hundred. Of 20,000 books on American history, "every one of which it would be highly desirable to possess," the Library of Congress had fewer than one-tenth. An annual outlay of $10,000 would need over a century to build a library on the order of Göttingen's. Marsh urged spending the next thirty years' Smithsonian income mostly on books, the rest on a museum and an art gallery.[33]

Marsh's was "one of the best speeches ever delivered in the House," thought John Quincy Adams; the *New Englander* wished him a long life in Congress "to present [to] our western members, what many of them have never seen, the spectacle of a living scholar." But Marsh converted few unlettered Westerners to book-buying. Isaac Morse of Louisiana held Smithson more "practical" than Marsh supposed; admiration of the steam engine, the cotton gin, and the telegraph had spurred the Britisher to house his bequest in America. Such inventions, other Westerners chimed in, were far more useful than shelves of musty books.[34]

Galvanized by rumors that Owen's schools scheme was a "diabolical plot to spread Fourierism" (communal utopias such as Brook Farm), House scholars again rebutted. Adams decried Owen's plan to train normal-school teachers, termed "farcical" by future President Andrew Johnson of Tennessee, himself unschooled. Johnson scoffed at the idea of a young man "educated at the Smithsonian Institution and brought up in all the extravagance, folly, aristocracy, and corruption of Washington go[ing] out into the country to teach the little boys and girls to read and write!" Few would stay teachers: "ninety-nine out of a hundred . . . would hang about a law office—get a license—become a pack of drones."[35]

Owen's scheme had by now alienated most of the House. Section after section of his bill was struck out; finally William J. Hough of New York offered a substitute, with riders added by Marsh, "to direct the appropriation entirely to the purposes of a library." One Marsh amendment did away with lecturers; a second raised the library budget to $25,000; a third funded the publication of scientific treatises. The House passed the Hough-Marsh bill on April 29, 1846, by 81 to 76.[36]

Despite the adoption of Marsh's plan, he was not one of the three House members put on the Smithsonian's fifteen-man Board of Regents. Instead, Speaker John W. Davis selected his fellow Hoosier and Marsh's chief adversary, Robert Dale Owen, along with New Yorker William J. Hough and Alabaman Henry W. Hilliard. Marsh's legislative victory seemed doomed to administrative defeat; for although Senator Choate was a regent, Owenites on the Board outnumbered library backers. Prodded by Bache, dominant in federal science circles through family connections, Owen got the regents to agree that the Secretary (full-time director) should be "a man capable of advancing science . . . by original research." This was a blow to Choate and Marsh, who had hoped to appoint Charles Coffin Jewett, librarian at Brown University. The Board instead elected the eminent Princeton physicist Joseph Henry; Jewett became Assistant Secretary and librarian.[37]

Manipulated by Bache, however, Owen had undone his own cause. Bache urged his close friend Henry to accept the Smithsonian post so as to "redeem Washington. Save this great National Institution from the hands of charlatans." By charlatans Bache meant not the library men but the Owenites, whom he had used to elect Henry. Like Bache, Henry thought Owen's grandiose building, popular education, and cheap instructive tracts ridiculous. The austere Henry and the ebullient Owen had nothing in common. As Owen saw "all vestiges of interest in the common man" vanish, he lost influence on the Board; when the Indianan left Congress in December 1847, Winthrop (now House Speaker) made Marsh regent in his place.[38]

Marsh's appointment followed a compromise between the library men and Secretary Henry. Henry was not against the library in principle, but he feared the Smithsonian's entire income would not suffice to buy one of the size Choate and Marsh wanted, let alone pay for cataloguing, binding, and upkeep. He interpreted the enabling act as permitting, not mandating, $25,000 a year for books, and threatened to resign unless he were given a free hand. He and Bache got Choate to agree that half the annual income would go to the library and museum, the rest to research, publications, and lectures.

Marsh at first opposed the compromise "as a departure from the spirit if not the letter of the law," and warned of a congressional exposé. But talks with Henry persuaded Marsh it was politic to accept the equal split; Congress was unlikely to award the library more. Though service on the Board of Regents convinced Marsh that "it must be very difficult for a man of common sense . . . to agree with Prof. H[enry] in anything," Marsh cultivated him for his own purposes, and later attested that "our action was entirely harmonious during my continuance on the Board."[39]

Like Owen, though, Marsh was outmaneuvered: the library compromise was to take effect only when the building was completed. Bache and regents' chairman Joseph G. Totten spread construction costs over four or five years, meanwhile paring other expenditures to the bone. The only books bought were "such valuable works of reference as . . . may be required." Although opposed by Marsh and Choate, and never openly declared, retrenchment became actual Smithsonian policy. Nothing like the allocated sum ever was spent on the library. In 1847 book purchases were only $546, the following year still less, despite exceptional bargains Marsh located in Europe. The peak of splurging came in 1849 and 1850, when the Smithsonian spent $6,100 on books and pictures, half of it to buy Marsh's own collection of prints and engravings.[40]

Marsh made one last effort to restrain Henry and Bache. In December 1848 he backed Andrew Johnson in urging a congressional inquiry "as a most wholesome and necessary check" on Smithsonian operations. "The Board of Regents," added Marsh, "ought to have been long since made acquainted with its direct responsibilities to the power which had created it."[41] But the motion failed, and Smithsonian controversy simmered down for some years. (The issue of congressional responsibility for—or interference with—the Smithsonian resurfaced 150 years later in the quite different context of the Enola Gay exhibit.[42])

Marsh never realized his dream of a great national library. But he went on to play a prominent role in shaping the Institution's research and publications programs.

Although internal squabbles and external self-defense cost the regents time and energy, they achieved much of moment. Marsh's partners on the Board were men of wide experience in science and public life. Secretary Joseph Henry, discoverer of self-induction and a pioneer in electromagnetism, though often dilatory, proved a high-minded and honest if acerbic chief. The most powerful regent over two decades was A. D. Bache, virtual czar of American science. A surveyor of everything from tornados and magnetism to harbors and coral reefs, Bache directed the Bureau of Weights and Measures as well as the Coast Survey. Many regarded Henry as a rubber stamp for Bache, whose imperturbable good humor and tact usually gained his objectives. Bache's own mentor, Corps of Engineers chief General J. G. Totten, was a polymath enthusiast in every realm from conchology to crystallography—a catholicity congenial to Marsh.[43] Richard Rush, veteran diplomat and cabinet officer, had been influential in securing the Smithson bequest. William W. Seaton, co-editor of the Washington *National Intelligencer* and sometime mayor of Washington, chaired the regents' Executive Committee. The national capital had no group more distinguished than the Smithsonian trustees.

Minutes of one executive meeting show the tenor of their concerns. Present August 10, 1848, were Henry, Seaton, Totten, Marsh, and Jefferson Davis. They declined G. Nye's collection of paintings but would buy Skinner's *Farmer's Library and Journal of Agriculture*. They priced the first volume of "Smithsonian Contributions to Knowledge" at 35 cents. For this series Marsh recommended a bibliography of pre-1700 Americana to be compiled by Henry Stevens, Jr., a Vermont-born, London-based bookdealer (the work was stillborn, but Stevens acted for decades as Smithsonian and Library of Congress purchasing agent). Philadelphia brewer Dr. Robert Hare, inventor of an oxyhydrogen blowpipe (for welding), had given the Institution enough chemical apparatus to "fill a canal boat." Other gifts were a paper-holding press, from Virginia diplomat Nicholas P. Trist, and a bronze bust of the

Danish classical sculptor Bertel Thorwaldsen. Henry spoke of James P. Espy's new scheme for reporting weather data by telegraph (the start of scientific forecasting); Smithsonian instruments were being sent to weather observers in Santa Fe and California. Marsh noted an offer "to sell to the Institution a collection of shoes, illustrative of the changes of form in this article of dress. The purchase was declined; the articles would be accepted as a gift."[44]

Among these minutiae were germs of major enterprises. By 1852 more than two hundred weather stations all over the country were supplying temperature, pressure, precipitation, and wind data to the Smithsonian. Under Henry and the Swiss-born geographer Arnold Guyot, its meteorological seed flowered into the U.S. Weather Bureau in 1869. Other Smithsonian programs gave rise to its own Bureau of American Ethnology and to the U.S. Geological Survey.

The Smithsonian also pioneered in the making of scholarly books; the high cost of engraving maps and drawings and of reproducing plates had deterred many from publishing their research. The handsomely produced "Smithsonian Contributions to Knowledge" brought the work of American scientists, explorers, and naturalists to a wide public; many expeditions, notably to the trans-Mississippi West, were under Smithsonian aegis.

Marsh himself chose the first "Contributions" volume, Ephraim G. Squier and Edwin H. Davis's *Ancient Monuments of the Mississippi Valley*. Squier, a young teacher and journalist who had examined the massive and then mysterious Indian mounds of the Midwest, showed Marsh his work in 1846. Marsh persuaded Joseph Henry to print it. Sparing no effort, Marsh read proof, raised money, calmed Squier's royalty concerns, and puffed the book. "It is fortunate for the cause of American ethnology," wrote Marsh, that this "first systematic attempt at its elucidation" was of such quality.[45]

In private, Marsh was more critical of the pugnacious, self-centered, touchy Squier and was mortified to be thanked in the preface for "his sound and critical judgment." With no pretensions to archaeology, Marsh understood field techniques better than Squier, who had failed to note either site locations or floristic contexts. What was the character "of the forest growth on and near the works," asked Marsh, "as compared with that in localities not likely to have been inhabited?"

After all, "the primitive forest is probably different from that now growing on grounds once cleared." Squier replied casually that so far as he
recalled, the vegetation was the same both near and remote from the
ruins. Marsh regretted that so important a fact—"*the* most important
with regard to their chronology"—had been neglected.[46] Marsh's stress
on plant succession as a clue to settlement history was uniquely preco-
cious. So too was his plea, discussed in the next chapter, for conserving and recording site strata. And Marsh would glean fruitful lessons
from the Indian mounds for his *Man and Nature*.[47]

Published in 1848, *Ancient Monuments* was well received by American scholars. Until its appearance, the mounds had been widely believed to be of pre-Indian, even of Welsh or Israelite, origin. Squier and
Davis helped to squelch such fantasies.[48] Marsh subsequently aided
Squier's appointment as U.S. chargé d'affaires in Central America,
where he negotiated for an abortive Nicaraguan canal while carrying
on further archaeological work.

M̲arsh's main Smithsonian legacy was the choice of his own protégé, the naturalist Spencer F. Baird, as Henry's assistant. As curator of
the museum, then as Henry's successor, Baird built the Smithsonian's
great collections, set up exchange networks the world over, and sent
out the geologists, botanists, and zoologists who first systematically explored America.

When Marsh met him in 1847, the twenty-four-year-old Baird, a
keen ornithologist (Audubon had wanted him for a western trek), was
teaching at Dickinson College, Pennsylvania. He had just married the
daughter of General Sylvester Churchill of Woodstock, Vermont. Already fond of chatty Mary Churchill, Marsh took an instant liking to
Baird. Among other aspirants for the Smithsonian museum post was geologist David Dale Owen, Robert's brother. But by 1848 Representative
Owen had left Washington, and Marsh had more influence. He gained
support from other regents and found Henry well disposed toward
Baird. But the Secretary did not want a curator appointed until the
museum was ready to house natural history collections, perhaps five
years on. Marsh had to deal warily with Henry, surmising that the Sec-

retary thought him "strong in punctuation, & not wholly ignorant of Low Dutch, but in 'science' a dummy." He urged Baird to come to Washington and plead his own case, "*provided* that you will . . . lay aside a little of your modesty, and swagger enough to make a proper impression." Marsh advised Baird to cultivate Bache as "more prompt" to act than Henry.[49]

Meanwhile, Marsh set about training Baird. He gave him the run of his library, tutored him in German, and stimulated his study of other tongues. Samples of Marsh's advice show what Baird was up against: "You'll have no trouble with Danish. . . . Study the language *per se*, and the analogies will come fast enough to embarrass you, without being sought. Dutch can be learned by a Danish & German scholar in a month." Baird must read Ludwig Tieck's *Kaiser Tonelli* and the Flemish language devotee Hendrik Conscience's *Eenige Bladzyden uit her Bok der Natuer.*[50] Study, study, study.

Baird soon put this hard-won knowledge to use. The New York bookdealer and publisher Charles Rudolph Garrigue asked Marsh to translate and edit the famed *Brockhaus Bilder-Atlas zum Konversations-Lexikon;* Marsh recommended Baird in his place. Baird would get a dollar a page, editorial experience, and an international reputation. "How many pages can you do in a day?" asked Marsh, tormenting him with some tale of his own speed; Baird's life became "regular almost to monotony." When the four-volume *Iconographic Encyclopaedia* appeared in 1852, even the perfectionist Marsh rejoiced.[51] The work helped to form Baird's judgment and enlarge his horizons, brought him in touch with eminent scholars, and made him after Agassiz the best-known scientist in America.

In 1850 Baird told Marsh of "the consummation of what you, more than anyone else, have put into train"; Henry had appointed him Assistant Secretary to the Smithsonian, in charge of museum collections and natural history. Baird stayed thirty-eight years, the last ten as Secretary. The Smithsonian continued to profit from Marsh's sagacity and aid, above all as Baird's dazzling lifelong correspondent, to whom Baird confided his plans and hopes as to no one else. The relationship revealed in their letters is joyously warm, Marsh writing to "My dear Boy," Baird addressing Marsh as "My dear father" from "your affectionate Son."[52]

The Smithsonian story typifies the shaping role that Marsh played again and again. He was not a great statesman; aloofness and aversion deterred him from seeking political prominence. He was not a great scientist; omnivorous curiosity and restless practicality precluded any single-minded dedication. It was in the borderlands linking science and the public weal that Marsh made his lasting contributions. He applied science to life, not with the disinterested precision of an engineer, but with the goals and means of a humanist. Marsh was a superb promoter of knowledge. The Smithsonian—its aims, its activities, its personnel—in fair measure stemmed from Marsh's efforts as an impresario of ideas.

5. American History
from the Ground Up

MARSH'S SMITHSONIAN COMMITMENTS WENT HAND in hand with an enrichment of his social life after 1845. No longer were the Marshes transients in Washington; they were acclimatized veterans. After two years of boarding, Marsh rented a small house on the south side of F Street between 19th and 20th streets, midway between the Potomac and the White House. Here he, Caroline, and her sister Lucy lived until 1849. The new home badly needed repairs and furnishings, but once things were fixed and bought they "never for an hour regretted trying the experiment of housekeeping," wrote Caroline, and had "never been so happy in Washington before." Marsh termed these years among the happiest of his life.[1]

Marsh found his daily walk along elm-shaded Pennsylvania Avenue to the Capitol invigorating, the two-mile distance enough to deter the office-seekers and lobbyists who had previously stolen his time. Life in Washington's west end was relatively calm and quiet. Finding congenial neighbors among nearby officials and foreign diplomats, Marsh went less often to political soirees. The scholarly Prussian Baron Frederick von Gerolt was a Marsh favorite; others included the Russian Minister, Count Alexander de Bodisco, literary Calderon de la Barca of Spain, Colonel Beaulieu of Belgium, Steen Bille of Denmark, and the Austrian chargé, Chevalier Johann Georg Hülsemann.

Marsh's closest new friend was Colonel James Bucknall Estcourt, sent from Britain to settle the Maine-Canada boundary dispute that had festered since 1837. Cultured, sensitive, self-effacing, Estcourt was for Marsh the Anglo-Saxon par excellence. Yet this friendship made Marsh not less but more Anglophobic. That England boasted such luminaries as Estcourt made it all the worse that haughty, ignorant aristocrats ran British foreign affairs. In Washington the presence of Estcourt made more glaring, in Marsh's eyes, the defects of dissolute Henry Stephen Fox and stupid Richard Pakenham, Fox's successor as British Minister in 1844.

The Estcourts shared the Marshes' passion for most things German: the two couples spent evenings reading aloud from German poets, philosophers, and natural scientists who had "done more to extend the bounds of modern knowledge than the united labors of the rest of the Christian world," in Marsh's opinion; "every enlightened student . . . readily confess[ed the] infinite superiority" of German literature. Marsh feasted on the philology of Jacob Grimm, the geography of Humboldt and Ritter, the classical history of Barthold Georg Niebuhr; for relaxation he turned to Ludwig Tieck's romantic comedies, Jean Paul's perfervid fictions, poetry of every era from the *Nibelungenlied* to the latest ballad. Sundays he attended the Evangelical Lutheran Church, whose Göttingen-trained pastor, earnest young Adolf Biewend, often came and read to Caroline or joined the Estcourts, von Gerolt, and others at Marsh kaffeeklatsches.[2]

When revolution ignited Europe in 1848, most of Marsh's diplomatic *intimes* left Washington, some to defend their regimes, others to join the rebels. Political refugees in turn fled to the United States, and many uprooted savants came within Marsh's compass. In sympathy with liberal uprisings, Marsh denounced reactionary sovereigns' "flagitious betrayal of human liberty" and spoke on the rebels' behalf in the House Foreign Affairs committee. He also succored them in defeat. The Danish historian Adolphus Louis Köppen stayed a month at F Street, an exhausting guest "by reason of the multitude of his talks." The Prussian

"Lischke, deplorably given to the shooting of little innocent . . . birds, impaling of insects, disembowelling of fish, and pickling of crustaceans," Marsh put in touch with Baird "to exchange bloody trophies." The most frequent visitor was the Thuringian naturalist/physician Frederick Wislizenus, an explorer of New Mexico whom Marsh described as "very full of prickly pears, burrs, and cacti."[3] Wislizenus became Caroline Marsh's doctor, her sister Lucy's suitor and then husband.

Marsh was a fine host, despite the constraints of his budget and household. Dining one Christmas Eve, Jacob Collamer found the quarters cramped: "half-a-dozen gentlemen [were] as many as his rooms will accommodate at one time." But young Donald Grant Mitchell, whose comic reveries gained him fame as "Ik Marvel," termed Marsh's "cosey, modest [home] one of my pleasantest eating places," an oasis in the arid tedium of Washington society. Marsh was "a thorough scholar [with] a splendid library and fine old engravings," Caroline engaging, accomplished, and exceptionally pretty.[4]

Caroline's continuing ill-health made it far easier to entertain at home than to go out. Among F Street visitors Mitchell noted Robert Winthrop, talking "of Texas and Houston"; Rufus Choate, chatting "of old days at Dartmouth"; Matthew Fontaine Maury, the Virginian director of the new Naval Observatory, who was charting ocean currents and sea lanes; the portrait painter G. P. A. Healy, fresh from successes in the courts of Europe; and the young lawyer Charles D. Drake, future Missouri senator. Marsh's strictures against intoxicants in the Navy and the halls of Congress were waived at these gatherings. Drinking his excellent wine, his guests toasted their generous host, "the stout master flanked by a modest Bocksbeutel, expressing his old Teuton love for the modest juices of the Stein-wein."[5]

Save for occasional headaches and neuralgia, Marsh's health was good. Sedentary life had added inches to his girth—he was now over two hundred pounds—and his face had filled out. He wore his dark brown hair long, brushed back over a high forehead; a massive brow, spectacles, firm lips, and determined chin gave him an aura of force and forthrightness.

Marsh's intense reserve did not put strangers at ease. Shy young George F. Edmunds, the future Vermont senator who married Marsh's niece Susan Lyman, was at first "greatly awed" by this "very dignified man with great glasses, and surrounded with an army of books." But Marsh's "gentleness and kindness" soon relieved Edmunds, who left "with a sense that my own importance had increased from the interview." Marsh made his presence felt unobtrusively. He never swore, seldom raised his voice, yet exerted a pervasive quiet authority. Only children, whom Marsh treated with mature gravity, were not inhibited by him; some neighbor's youngster often joined him hand in hand along F Street.[6]

Caroline, though happy in Washington, was more than ever an invalid. She could not read, found the briefest walk torture, and endured a host of other ailments. Despite doctors' care, cold showers, iron preparations, aconitine (an alkaloid sedative), sea bathing, outdoor trials, and devoted nursing by her husband and sister, she recurrently relapsed, losing weight alarmingly. Marsh feared an incurable spinal illness. In the evenings he read aloud to her or Lucy acted as amanuensis. Although in pain much of the time, Caroline kept her spirits up and hid her maladies from visitors. The painter Charles Lanman vividly recalled Marsh carrying her in his arms into a Washington drawing-room.[7]

After the first winter, young George was no longer with them. Schooling in Washington was poor, and Marsh feared the influence of slave society on an impressionable twelve-year-old with "the faults of an only child, and a delicate one at that." He felt George needed old New England virtues, firm handling "in the family of some clergyman." The unhappy boy was shifted from place to place. For two years he boarded in Burlington with his aunt and uncle Hickok, who found him disobedient, stubborn, hostile. Marsh found a school at Newton Centre, Massachusetts, run by a Herr Dr. Siedhof, who could teach George good German. "While your organs of speech are flexible, and your ear delicate," ran a typical letter to his son, "I hope you will practice the analysis of sounds." George must make the most of this opportunity, as financial setbacks "may very probably put it out of my power ever to afford you hereafter advantages so great as you now enjoy."[8]

Underwhelmed by Siedhof, George thought his father "imposed upon." Latin and Greek were perfunctory, arithmetic not given, "and all the English branches are miserably taught, now how shall I ever enter college from a school where there is such teaching." Classrooms were crowded and unventilated, meals meager and miserable; "yesterday the meat was so, excuse the term, *rotten*, that you could smell it." Other parents agreed that George had cause for complaint; Siedhof scanted the needs of hungry American boys, and "Mrs. Siedhof has been *very unfortunate* in her cook and help generally." And Siedhof was "capricious, irritable, & furious in temper."[9]

Worse followed. Unaware the school forbade the boys to spend money, Marsh told George to buy *"three pair of thin pantaloons, & two thin coats, & in procuring them, I hope you will consult economy;* & particularly that you will not have them made so tight as to expose them to wear out too fast, & above all, that you will have them made as gentlemen, and not as loafers, wear them." When George bought these clothes, Siedhof taxed him with breaking rules; Mrs. Siedhof "got me a little mad & I uncautiously let out that you had written to my grandmother that if my clothes had not been poorly washed I should not want any more shirts, so . . . she went off in a rage, and . . . he informed me I could ornament his institution but . . . eight days longer."

Marsh was furious both with Siedhof, who had earlier said he was *"vollkommen zufrieden"* (wholly satisfied) with George, and with his "insolent and improper" son's "impertinence and disobedience."[10] George protested he had *"not* abused the confidence you 'imprudently' reposed in me but that I have only done what was due to *your* own dignity & mine"; his father should at least "'forgive & forget' what was caused only by my thoughtlessness." But Marsh was having too much trouble finding another school for his son—one "where he can't *help* learning, if there be such"—to forgive him. Finally he sent George, now a slight, dark-haired, defiant fifteen-year-old, to Kimball Union Academy in Meriden, New Hampshire. There George's Greek and mathematics improved, but he remained, wrote the headmaster, "a little too independent in his bearing—indulges in the foolish habit of smoking—does not love to be constant at church."[11]

Bright and high-spirited, George was argumentative and outspoken. He did not hesitate to chivvy his father. When a tenant who owed Marsh money turned up in Burlington after a long absence, George wrote: "Hadn't you better pitch into him for some of that rent?" Marsh was too hard on his son. More anxious than affectionate, he was quick to find fault, prone to rebuke lack of tact as disrespect. Back in Woodstock, George Ozias's grandmother pitied his loneliness; he longed to be with his parents in Washington but sensed he was not wanted there. Susan Marsh also pitied her son, who with so many other cares was at such a loss with his own son.[12]

Wherever I go, I find the mudpiles better worth study than the superstructure of the social edifice.

Marsh to William Tecumseh Sherman, February 27, 1872

Marsh's summers away from Washington were by and large agreeable. He enjoyed jaunts in both the Green and the White Mountains, raised building and equipment funds as a governor of the now indigent University of Vermont, and entertained the Bairds and the Estcourts on Vermont visits. To Baird he touted Burlington, but Marsh found the "best aspects" of American society in Boston. In 1847 he relished days there with Choate, with his uncle Jeremiah Mason, and with the historian George Ticknor, "one of the most thoroughly well-educated and well-informed men I have ever met," recalled Marsh, "though not *deeply* learned after the German model." He hobnobbed with Charles Sumner, William H. Prescott, George Hillard, and John Lothrop Motley. Sumner, ten years Marsh's junior, a scholar-politician of his own type, became a lifelong friend. The botanist Asa Gray struck Marsh as "a man of great learning, of a high order of intellect, and of a truly noble and generous character."[13] He enjoyed conversing with the newly arrived Swiss zoologist and geologist Louis Agassiz, whose mix of scientific exuberance and expressive piety made him lionized in Cambridge society.

Putting scientists and statesmen together sometimes had bizarre results. To one Marsh dinner in Boston came Agassiz, Choate, and Baird, who brought a frog to show Agassiz. Caroline Marsh tells the tale:

The box containing the treasure was produced, and the animal gently lifted from his bedding. "Oh, Spencer!" exclaimed his newly married wife, snatching up the delicate web that lay at the bottom of the box, "this is my wedding-handkerchief, for which I have searched so long! How could you take it for such a purpose?" "I was looking for something very soft, and did not notice that it was a handkerchief," said the husband. . . . Mr. Choate, whose extremely nervous organization is well known, had been much disturbed by the unsightly aspect of the strange animal, and perhaps even more, by a small snake that had been brought out at the same time, and was "dragging its slow length along" from one side of the table to the other. . . . Quietly Choate [asked Caroline], "May I beg you to intercede with Baird for the re-incarceration of these enormities of nature?" Then, seeing Mr. Agassiz absorbed in contemplating the singular little frog, he added, quickly, "No, no; that gentleman is happy, I see—I shall do better to retire," and he was gone.[14]

Sumner invited Marsh to give Harvard's Phi Beta Kappa address in August 1847. "One who speaks beyond an hour and a quarter, speaks at a venture," warned Sumner, but added cordially, "*you* can speak as long as you wish." Marsh's lecture *Human Knowledge* proved rambling and abstruse, "elaborate and instructive," thought Harvard President Edward Everett, but "dull." Yet it was lauded by Choate, Sumner, Mason, and Ticknor, and the Boston press claimed "every sentence . . . an essay [by] one of the ripest scholars of the age"; nay-sayers were just unused to hearing "valuable truths . . . in such rapid succession."[15]

After a week in Cambridge Marsh spoke at Union College, Schenectady, New York, on the unique nature and needs of American history. Thence he returned to Vermont to judge oxen, swine, and maple sugar at a Rutland County Agricultural Fair in September. At Rutland he gave a third talk, commending progress in farming but cautioning against heedless depletion of resources.[16]

"*If I can find time,*" Marsh told bibliographer John Russell Bartlett, he would translate the Danish geographer Joakim Frederik Schouw's *Skandinaviens Natur og Folk* (1845), "combating the common notion that national character is influenced by physical causes." Schouw led Marsh to reconsider his own earlier environmentalism and spurred his nascent insights into mankind's potent agency in reshaping the earth.[17] At this time Marsh was also writing on the sands of the sea, the proper proportions of pyramids, and the explorations of Japetus Steenstrup and Johann Jakob von Tschudi. How he found the leisure to prepare the three major lectures he delivered that summer is hard to imagine. Like the 1846 Smithsonian speech, these essays presage his greatest historical and environmental insights.

Common to all three 1847 essays is the theme of utility. *Human*

Knowledge elaborated on Marsh's Smithsonian argument that applied science throve only when backed by theoretical science; mere empirical knowledge was useless without analytic coherence. Why science knew and predicted the remote better than the close at hand was the puzzle that engaged him.

The astronomer can compute with unerring precision the courses of the stars in all past and all coming time, but no seer can tell whether favorable or adverse breezes will impel the ship that this day ventures forth upon the uncertain deep. [Man was] at home among the stars, a stranger upon earth; unable . . . to weigh a scruple, yet calculating the specific gravity of Herschel; knowing not the ingredients which enter into his own nourishment, yet decomposing the sunbeam.

The familiar aspects of nature nearest in scale and tempo to everyday human existence were the most variable and mutable, the least uniform and explicable by universal laws. Realms of practical utility, Marsh concludes, elude understanding based only on sensory data and logical analysis; their elucidation requires historical, social, and spiritual reflection as well.[18] His empirically based skepticism later, as will be seen, led Marsh to doubt the sweeping theories of Max Müller in philology and Charles Darwin in evolution.

The American Historical School, his Union College address, stressed the social utility of understanding the national past. In *Goths in New-England* Marsh had lamented the excessive optimism and mobility that deprived Americans of a proper "reverence for antiquity." He now called for new historical insights derived from uniquely American events. Europeans dwelt on the pomp and circumstance of armies and aristocrats — drum-and-trumpet histories of little meaning to citizens of a democratic republic. History *for* the people must be *about* the people, not just about their rulers; barren political and military chronicles should give way to all-embracing annals of everyday life.

How could one open new windows on the American past? Marsh staked out a remarkably prescient terrain:

To know what have been the fortunes of the mass, their opinions, their characters, their leading impulses, their ruling hopes and fears, their arts and industry and commerce, we must see them at their daily occupations in the field,

the workshop and the market; . . . invade the privacy of their firesides and un-
veil the secrets of their domestic economy; we must live and suffer and toil with
them . . . to determine both what and why they were.

Everyman's history would emerge from such mundane sources as
municipal proceedings, records of trade, education, living standards,
health and welfare, leisure pursuits, ephemeral literature, "the private
biographies of the humble as well as the great." Such data, which Old
World chroniclers had "scorned to record, as beneath the[ir] dignity,"
would reveal more of "the true history of man than the annals of ages
of warfare, or the alternate rise and fall of rival dynasties."[19]

Above all, American history ought to encompass things along with
words, not just archives and genealogies but mundane tangible relics.
Artifacts were more vivid and memorable than written texts; "the pic-
tured tombs of Thebes are fraught with richer lore than ever flowed
from the pen of Herodotus, and an hour of buried Pompeii is worth
more than a lifetime devoted to the pages of Livy." Americans needed
palpable mementos of their forebears' circumstances. "Let us build monuments to the past," urged Marsh's Whig contemporary Judge Job Durfee of Rhode Island. "Thus shall the past be made to stand out in
monumental history, that may be seen by the eye and touched by the
hand!"[20] In contrast, Marsh cherished the past's unintended vestiges
more than the present's self-conscious memorials. Abandoned tools and
other humdrum relics would reveal to Americans their own actual daily
history.

Our forebears' natural world was, Marsh feared, beyond recovery.
But what they had fashioned from the products of nature could yet be
found. As befit a Smithsonian trustee, Marsh called for "a complete collection of their agricultural and mechanical implements . . . furniture
and domestic utensils." A display of such ancestral bygones would impart "a lively conception of their mode of life."[21] Here was the first proposal for a museum of American life. Anticipating Artur Hazelius's
Skansen in Sweden by half a century, Marsh's phrasing and purpose
presage today's folk museums. Workaday things of ordinary folk; the
humdrum and the earthy; a vivid sense of past life roused by seeing and
handling artifacts; an emphasis on environmental context—all are fore-
shadowed here. And decades later Marsh added relics of wild nature to

the roster of features he would steward for future instruction and delight. He might even have welcomed Dorothy Canfield Fisher's suggestion, a century after his own, to turn all Vermont "into a National Park of a new kind—keep it just as it is, with Vermonters managing just as they do—so the rest of the country could come in to see how their grandparents lived."[22]

A few Europeans, inspired by Herder's visionary folk nationalism, had begun to envision such a social history, and the French historian Jules Michelet was beginning to put it into practice. But none had conceived the comprehensive study Marsh now called for. In America, most historians of Marsh's era left ordinary people out of their chronicles save as mythic founders or anonymous victims. Only George Bancroft fused national purpose with democratic historical vistas, and Bancroft's chief concern in limning the past was to glorify the present.

Like others of his time, Marsh sought a purposive history that would illumine mankind's transcendent aims. But he never avoided mean or inconvenient truths. And his quest for data—most notably in *Man and Nature*—was accordingly far more comprehensive than that of academic historians, who long neglected the social and the mundane. It took an amateur like Edward Eggleston to plead, half a century after Marsh, for "the history of culture, the real history of men and women." Only in the next century would Finley Peter Dunne's "Mr. Dooley" demand a "real" history: "If any wan comes along with a history iv Greece or Rome that'll show me th' people fightin', gettin' dhrunk, makin' love, gettin' married, owin' the grocery man an' bein' without hard-coal, I'll believe they was a Greece or Rome, but not befure." Not until eighty years after Marsh's *American Historical School* did such precepts inspire the "New History." Only now is history from the bottom up de rigueur.[23] In infusing history with archaeology, geography, and folklife, Marsh was more than a century ahead of his time.

Marsh's vision of progress called for the same kind of populist awareness of past technology as Henry Ford's a century later. His museum of artifacts aimed to open young eyes as no books could do. Hefting their ancestors' farm tools, children would learn to fathom change and to esteem the legacy bequeathed them. "Observe the father . . . at his heavy toil; in felling the woods, & breaking up their virgin earth, how rude his implements, how slow his conquest of an untamed soil.

Compare his clumsy plow, his ill-forged axe & his heavy hoe, with the light, well-balanced & neatly-finished tools of his descendants." These axes and plows and hoes were, for Marsh, prima facie proof of *improvement*.

Such progress was social, too, for "with all this toil . . . in the now smiling fields & verdant hills of our Eastern States, . . . your own fathers had to be ever ready with loaded musket to repel the lurking savage."[24] Marsh saw the conquest of wild nature and primitive indigene promoting art as well as agriculture. This vision of progress, now wholly out of fashion, was then a pervasive, indeed mandatory, concomitant of American faith in the future.

Yet despite all technology's wonders, the continuance of such progress seemed to Marsh far from certain. The unintended consequences of settlement and conquest showed fearsome pitfalls ahead. To guard against these in good time, Americans must become more mindful of their past, through traces inscribed on the ground as well as in archival memories.

Besides treasuring their own forebears' tools and records, Marsh urged Americans to conserve the relics of other cultures. He faulted Squier and Davis's book on Mississippi Valley earthworks for failing to show "the *precautions* to be observed on opening mounds, & making other excavations, with reference to the injury likely to be done both to the mounds & their contents, by carelessness, & upon the preservation of earth structures." He deplored "the lamentable destruction of so many memorials [of Aztec and Inca civilizations] by the ignorance and bigotry of the so-called Christian barbarians who conquered them." In the Old World as in America, Marsh censured treasure hunters and academic excavators alike. "Foreign collectors and archaeologists," he observed at Luxor and Esna in upper Egypt, "have done almost as much to deface the monuments as the barbarians."[25]

It was not only to husband prehistory but to sustain material progress that Marsh stressed the need for environmental foresight. Showing how past misuse of land and water, forests and soils had reduced present resources, and urging environmental reform as essential to future well-being, Marsh's Rutland address sketched a program for environmental management that presaged some of the most important insights of his *Man and Nature*.[26]

Marsh's domestic pleasures in Washington were overshadowed by public miseries. The winter of 1845 sealed Polk's annexation of Texas, and in May 1846 the United States went to war with Mexico. Conflict with England loomed in the Northwest over expansionist demands of "Fifty-four forty or fight." Party and sectional discord marked the first session of the 29th Congress, bitter acrimony the second. Marsh drew away both from Southern Whigs and Northern Democrats. Before the 30th Congress assembled, General Winfield Scott took Mexico City; the war was won. But a fiercer battle erupted over the fruits of victory. When Winthrop was at last chosen Speaker in December, Whigs ruled the House for the first time since Marsh had come to Washington.

Northern Whigs wanted peace without territorial expansion. In and out of Congress Marsh stressed the cost of the war and the worthlessness of Mexican lands. Earlier he had promoted an impassioned antiwar pamphlet by Albert Gallatin; he now fed Gallatin data exposing the administration's financial sins and got enough copies of the new pamphlet printed to bring it "within the reach of every person in the U.S."[27] Public fury, he hoped, might force Congress to reject the $18,500,000 loan sought by Polk.

Marsh issued his own blast against the war loan in February 1848. Polk meant to pay for the Mexican conflict by taxing mainly Northern imports: tea and coffee, gold and silver. "How much tea and coffee are consumed by the three or four millions of Southern slaves? . . . How many of them consult gold and silver watches, to know their hours of labor?" Northern principles and purses were alike despoiled. Deluded Southerners truly expected Texas to enrich them, but nothing could excuse their allies in the North:

He who would write the blackest page in American history must ferret out the secret . . . intrigues, by which the Texas Revolution was fomented; . . . depict how the hopes of Texan stock-jobbers fell and rose as this or that Northern Democratic member exhibited tokens of rebellion, or meekly gave in his adhesion to the slavish policy of his party . . . explain how that contemptible faction, that so long swung . . . between the law of conscience and the dictate of party, alternately betraying each, was at length fixed; and, in fine, tell what votes were extorted by craven fear, and what purchased by damnable corruption.

Having sacked Mexican cities and slaughtered Mexican civilians, the administration now proposed to dismember Mexico—an act not only immoral but unprofitable. Marsh would refuse Mexico as a gift. So alien an accretion would cost America more in sectional strife than it would gain in resources, for the North was bound to prohibit slavery there, the South to insist on it. "Why conquer or buy provinces which will be but an apple of discord?"[28]

Marsh arrested "the profound attention" of the House[29] but did not stem Manifest Destiny. The Senate ratified Polk's Mexican treaty, thereby adding California and New Mexico to the United States. Meanwhile old John Quincy Adams succumbed at his seat in the House. In the spread of slavery and the loss of freedom's champion, Marsh foresaw tragedy for the Union.

The Northwest was a further source of strife. Free Soilers seeking a quid pro quo for Texas backed expansionist calls for Fifty-four forty. But Marsh liked Oregon as little as Mexico, and was relieved when Polk compromised the boundary at the 49th parallel. In August Marsh called for an explicit ban on slavery in the Northwest. Southerners held that slavery could never pay there: nature forbade it. So why insist on "a barren, disputed right [it] never intended to exercise?" The reason, Marsh charged, was that slaveowners knew otherwise; history showed climate no barrier to slavery. Common in ancient Greece, Rome, even Britain, slavery was "everywhere profitable under the management of a prudent master," notably in newly settled lands where free labor was scarce. Unless banned, slavery was bound to filter west; already, Marsh heard, there were slaves in Oregon.[30]

The Vermonter spurned the whole expansionist credo. Why annex Oregon, California, and New Mexico? Even were they flowing with milk and honey, laden with gold and silver, rich in wheat and cotton, Marsh would not want them. Nations should grow by organic accretion, not by instant conquest. Hasty gobbling up of huge chunks of land dissipated national energies, forfeiting ripe fruition of the old for heedless gutting of the new domain.

Marsh's ideal nation—a land inhabited by like-minded citizens of similar culture—echoed Herder's 1794 dictum: "The most natural state is *one* people with *one* national character." To be French or German, Czech or Polish was to have inherited the essence of some primordial organism. The viable identity of a sovereign state or ethnic entity inhered in the soil and soul of a people defined by descent.[31]

As in *Goths in New-England*, Marsh lauded racial, linguistic, and cultural homogeneity. The contented nation, however small, was one united in character. "The citizens of the little republic of San Marino, and of the duchy of Tuscany, are as happy and as prosperous as if they were annexed to the kingdom of Sardinia, or even enjoyed the paternal discipline of the gentle Metternich," said Marsh. Nothing was gained by increased size, and much was lost. The cost of expansion was loss of liberty. The larger a country, the harder to defend—the more need for a standing army, centralized power, executive patronage. Americans must heed the fate of the Roman Empire: "the soldiery raised to protect the frontier may supersede your electoral colleges, and impose upon you a Dictator."

Since size as such had no merit, America was best off with *no* new accretions. Marsh doubted the wisdom of venturing even to the Mississippi; the original thirteen states met all national needs. The United States had trebled in size since 1803; and every new state was a slave state. The Old South would come to regret competing with Texan sugar and cotton; New York State farmers would suffer when western wheat captured their markets; Boston would lose its China trade to San Francisco. Adding the West to the United States was geographically absurd; separated by sterile deserts and impassable mountains, "they can never have a common interest with us." The Far West and Southwest were moreover inhabited by Spaniards and Indians "of habits, opinions, and characters incapable of sympathy or assimilation with our own, . . . unfitted for self-government, and unprepared to appreciate, sustain, or enjoy free institutions." The West was "unsuited to the genius of the people of the United States" and bound to weaken the nation; Marsh would sever "the unnatural connection between us and these remote regions."[32]

New England partiality buttressed Marsh's ethnic bias. His 1837 trip had turned him against everything west of the Hudson. His anti-Mexican animus came from General Sylvester Churchill ("no sensible or northern man" would want to live there), Lieutenant William H. Emory, and Frederick Wislizenus, who held all northern Mexico worthless for farming. Marsh supposed New Mexico and California still

1. *Charles Marsh, by J. A. J. Wilcox, portrait in Windsor County Courthouse,*
Woodstock, Vermont

2. *Woodstock from Mount Tom, 1859, sketch by Henry P. Moore*

3. *Burlington, Vermont, overlooking Lake Champlain, lithograph by T. Wood, 1846*

4. *The Woodstock green, early nineteenth century,*
from Abby Maria Hemenway's Vermont

5. *Woodstock courthouse and schoolhouse, 1849*

6. *View from Marsh house, looking east, ca. 1865*

7. *Quechee Valley, Woodstock, 1888*

8. *Marsh house as remodeled by Frederick Billings, 1870*

9. *Marsh-Billings house, after 1886*

10. *George P. Marsh, by G. P. A. Healy, 1844, portrait at Dartmouth College*

11. *Caroline Crane Marsh, pencil drawing, ca. 1840*

12. *G. P. Marsh, Caroline, and her sister Lucy Crane, daguerrotype,
Washington, D.C., ca. 1845*

worse; "the best informed explorers" told him they were mainly "unsuited to agriculture, unable to sustain a dense population, adapted only to the lowest form of semi-civilized life—the pastoral state." Not even Oregon was much good, Marsh surmised.[33]

Marsh was not alone in his low estimate of the arid West, soon to be stereotyped as the Great American Desert. Yet it is strange that a man so enamored of technology, so conscious of man's power over nature as his Rutland speech showed, should scout transcontinental transport as chimerical. And his dim view of western resources proved as myopic as his notion of the difficulty of getting there. At the peak of the gold rush a few years later, Marsh took a more sanguine view of California and Oregon, at least "as a good drain for some classes of our population whom we can well spare."[34] But only Old World engineering feats at last persuaded Marsh the American West might be made fruitful. He now and then envisaged striking gains in the conquest of nature, but his accounts of human folly were always more graphic. He was a born critic, not a booster.

Could Vermont Whigs stomach a Louisiana slaveholder and Mexican War general as their presidential candidate? Zachary Taylor's nomination in 1848 put Marsh in trouble at home. While Congress dragged on through the summer, Marsh noted "symptoms of rebellion" in Vermont;[35] two days before he reached Burlington the Free-Soil Convention nominated Martin Van Buren, threatening substantial inroads in Whig strongholds. Antislavery Vermonters were tired of being told to be patient; the Free-Soil Party offered an outlet for their abolitionist fervor.

Like most Vermont Whigs, Marsh was a Free-Soiler at heart; but party loyalty, political realism, and mistrust of Van Buren kept him in Taylor's camp. Yet he feared that the Southern general's ticket was "so unpalatable to my constituents that my support of it may put an end to my political life." In 1846 his margin of victory had dropped to a paltry 900. Fortunately for Marsh, his opposition was now divided between a Democrat, Burlington lawyer Asahel Peck, and Free-Soiler Stephen S. Keyes, "who was a Texas proslavery man," scoffed Marsh, "until he was nominated for Congress by the Van Burenites." In the September poll

Marsh led but fell short of the required majority. "I am not *Free-soil* enough," he found, "and many of my old friends are going for a Loco" (Locofocos were populist reform Democrats). He shuddered at the prospect of a prolonged electoral battle like the Bailey-Allen marathon of 1830–32. "To have three-fourths of the state called out again in January," Marsh wrote Collamer, "would be a serious inconvenience, & I am moreover persuaded that the sooner this question is decided, the better are our chances of success."[36] As two other districts were similarly stymied, Collamer and Marsh induced the Whig-run state legislature to allow election by plurality.

Marsh redoubled his campaign efforts, speaking "sometimes seven hours in a day, and shall speak more, yea, I shall reason with them continually until Nov. 7th." But Taylor made little effort to conciliate Northern Whigs, and Free-Soil orators were "active, unscrupulous, and numerous"; Marsh's opponents accused him of being *pro*-slavery. The practical issue, Marsh stressed again and again, was not slavery but its extension, which Taylor would resist. Southern Whigs opposing Texan annexation had shown "generosity, an incorruptible firmness, and . . . political virtue." Many of them felt slavery both inexpedient and morally wrong; "no man can pass a winter at Washington on terms of familiar intercourse with Southerners without finding that . . . it is in general only political demagogues who seriously defend the institution."[37] Time, calm, and Whig moderation would bring emancipation. Once again Marsh was too sanguine.

Marsh's defense of the Whigs was less convincing than his assaults on their opponents. Here he was on firm ground, the outraged Puritan scourging the wicked, "a great master of strong and condensed expression," recalled a supporter. His fiercest invective hammered Democrats turned Free-Soil: "They profess reformation without repentance, glory in their shame, and with hands still . . . red with the blood of an unholy war, they proclaim themselves the chosen apostles of human liberty. Out upon the bald hypocrisy of these whited sepulchres! The sacred cause of freedom needs no such allies as these."[38]

Election Day set Marsh's fears at rest: Taylor comfortably carried Vermont; as in New York, Free-Soil fervor split the opposition, costing the Democrats the election. Marsh ran well ahead of his ticket with an absolute majority of 1,554. Bone-weary but happy, he looked for a return to "political virtue."[39]

Delayed three weeks by snowstorms and other mishaps, Marsh found Washington in November balmy, roses still in bloom. His "great crony" archivist Peter Force "had saved several bunches of catawba grapes for me and they were still hanging on the vine until I plucked them." But his euphoria did not outlast his roses and grapes. Now on the House Foreign Relations Committee as well as the Smithsonian, Marsh had never been busier. "There isn't any Christmas vacation for us poor dogs now a days," he sighed. "Our mill goes, whether there is any water or no. We . . . work like horses."[40]

With Taylor's inauguration still two months off, by New Year's the capital was already thronged with office-seekers. Marsh himself was one of them. "'Some of my friends,'" he wrote, "think it possible I may be sent abroad in some diplomatic capacity. . . . I have had a puff in a Boston newspaper recommending me as Minister at Berlin on the ground that I 'can speak German like a brick.'"[41] Contact with European diplomats had revived Marsh's urge to visit the Old World, and he yearned for a climate that might help Caroline. But the Winooski mill and other failures had reduced him "from affluence to comparative poverty"; only a diplomatic post would permit him to go abroad. "My principal reason for seeking a foreign mission," he confessed, "has been the hope that the little that could be spared from an outfit would enable me to save my library and engravings" from having to be sold. Ideally he wanted "a commission to visit some terra incognita where my outgoings shall be so much less than my incomings, that I shall wax rich." Rafn in Copenhagen heard rumors that Marsh was to be Minister to Denmark, but this was a minor post with a meager salary.[42] Marsh's talents and leanings made Frankfurt, seat of the German national assembly, and Berlin, the Prussian capital, the places on which he set his chief hopes.

Vermont Whigs and Democrats alike touted Marsh's claims, and even the Burlington *Sentinel* supported him as "an act of simple justice— to which the uniform courtesy of Mr. Marsh . . . to his political opponents, adds obligation." The national press agreed that Marsh's scholarship, language skills, "modest and winning manners," gentlemanly polish, and influence in Congress entitled him to a post abroad. "Gen. Taylor might ransack universal Whiggery with a lantern," held the Boston *Courier*, "without coming across a man better fitted" for Berlin.[43]

But that hope was shattered by outgoing President Polk's last-minute appointment of Indiana Senator Edward A. Hannegan as envoy to Prussia. Notorious for want of tact and sobriety, Hannegan was an expansionist Democrat "of whose vulgarity and profligacy nothing but an actual acquaintance with him can give any notion," fumed Marsh. His account of Hannegan's ratification gives a vivid picture of Capitol Hill capers—and also a more candid appraisal of Southern Whigs, who forced Hannegan's appointment, than Marsh had conveyed to his constituents.

During the night Hannegan sent twelve baskets of champagne to the Senate chamber. The refractory senators were plied with this and other liquors, and as I know from personal observation very many, and as I believe more than half of the senators, were in a state of fearful intoxication. [Some later] spent several hours in drinking, dancing, shouting, like fiends, embracing each other, and testifying in every other brutal manner conceivable their joy at this great triumph. . . . The President was justly indignant at this outrage, and would have revoked the appointment, but was overruled by the same influences, which brought it about, and I learn that [Kentucky Governor John J.] Crittenden has *insisted* that Hannegan shall not be recalled.[44]

Another blow to Marsh was Taylor's choice of Jacob Collamer as postmaster-general; little Vermont would hardly be given two major appointments. Marsh could now expect only a second- or third-rate post; Chile was one such prospect. Losing hope, he reviewed other political prospects: he might head the Patent Office; he might take Phelps's place in the Senate; if he stayed in the House, he might become Speaker. Winthrop's willingness to accommodate slaveholders had cost him Northern support; a caucus of Whigs and Free-Soilers was ready to push Marsh for the Speakership.[45]

In the end, Marsh got a wholly unexpected plum through Colonel William Wallace Smith Bliss, Taylor's chief of staff in the Mexican War and now the president's son-in-law and private secretary. Within a few weeks, Marsh and young "Perfect" Bliss had found mutual interests in languages, philosophy, and mathematics. Their friendship gave Marsh easy access to Zachary Taylor, whom he found "honest, upright, and genial," partial to high tariffs, to freedom in the territories—and to Marsh as an envoy. The president would be delighted to give him a post. There was one obstacle: the Whig majority in the House was minuscule; they could not afford to lose a seat. Could Marsh guarantee that a Whig would be elected in his place? Vermonters assured the president that were Marsh rewarded, his district would be "safe beyond peradventure by more than 800."[46]

Armed with this promise, on May 29, 1849, Taylor named Marsh Minister to Turkey. Green Mountain Whigs were elated; the president had honored Vermont "AS SHE WAS NEVER HONORED." Marsh was pleasantly surprised; the post was "not that which I would have selected nor that where I should expect to be most useful," but the exotic East appealed to his fancy, the climate of the Bosporus would benefit Caroline, and there would be time for travel; the administrative duties were said to be "very light," and the decentralized nature of the far-flung Ottoman Empire led Marsh to suppose "I shall be at liberty to be absent from Constantinople a considerable part of the year."[47]

Marsh planned to sail in July, but Caroline, bedridden in New York, was too ill "to travel ten miles, or even five, by coach." He used the enforced delay to ensure that a Whig, James Meacham, would replace him in Congress. Campaigning throughout Vermont's hottest summer in half a century, on August 29 he gave voters parting promises: the Whig administration would protect Vermont industry with a proper tariff, repeal the subtreasury law preempting all gold for government use, provide generous aid for internal improvements, and ban slavery in the West.[48]

Personal matters also had to be seen to. Marsh packed apparatus for observing and collecting flora and fauna to send Baird at the Smithsonian, as well as household equipment for an indefinite stay abroad. The list affords a domestic glimpse of expected needs and lacunae. There were four crates of (530) books, six more containing an oil stove, a dozen mahogany chairs and two tables, cotton and linen sheets, pillow cases, curtains, towels, and napkins; candlesticks, dishes, cut glass saltcellars, silverware, "a dozen Brittania large spoons," a bread knife, toaster, griddle turner, baking tins, and wicker baskets; a fowling piece, small grindstone, brace and bit, files, saw, chisel, copper wire, and carpet nails; a letter clamp and inkstand; and hardest to fathom for a stay in Constantinople, stocks of coffee and cheese.[49]

Travel expenses and debts to settle left Marsh desperately short of cash. As a last resort he sold part of his library to friends, and to the Smithsonian his collection of engravings and prints, for $3,000, storing many of his own remaining books there as well. But the purchase was of little immediate aid, the parsimonious Joseph Henry delaying full payment for three years.[50] Finally, the Marshes chose those who would accompany them to Constantinople. Besides himself and Caroline, her sister Lucy, and young George, the party comprised Marsh's niece Maria Buell and his old friend Caroline Paine, sister of Vermont's ex-governor Charles Paine.

In Woodstock, where his father had died the preceding January at eighty-four, Marsh said good-bye to his mother, whom he was not to see again. Susan Marsh later dwelt on the difficulties her son had faced and might yet confront. "His trials have been peculiar," she concluded, "& I often feel comforted with a hope that we shall live to see him not only a professor but a possessor of that faith & the love of the saviour of sinners, which will lead to happiness."[51]

Five years in the Levant would bring Marsh much happiness, though not of the kind his mother wished. Old World stimuli made him a consummate professor, not of sacred faith, but of the transcendent secular glories of art and scenery.

6. Constantinople and the Desert

THE MARSH SAILING PACKET, CHOSEN OVER STEAM
for the sake of Caroline's health, took a full month to reach Le Havre,
on October 18, 1849. During the last eleven days head winds buffeted
the ship back and forth in the Channel. Seasick the whole voyage,
Marsh confirmed his view that "ocean, in all its phases, is an uninviting
object." He never got over "the horrors of the sea. I think sailors are
dead souls in Purgatory, perhaps even in a worse place." Before his appetite returned he wondered, "Are people sea-sick in balloons? . . . I'll
go home in one, else."[1]

After this rough passage, the rail trip to Paris was a succession
of delights: terraced foothills framing the fertile valley of the meander-
ing Seine; polychrome crops in long, narrow, unfenced strips; most
striking to Marsh, "the great quantity of woodland, [for] every road,
lane, and every inch of ground not tillable, or which can be spared from
cultivation, is planted with trees as thick as they will grow," and then
closely lopped for firewood. In sum, "everything . . . as unlike America
as can well be imagined."[2]

Weakened by the crossing, Caroline was judged to need a respite in Paris. Lodging lavishly in the rue de Castiglione just off the Tuileries Gardens, Marsh enjoyed the three-week delay. He marveled at monuments and art, combed bookstores and libraries, lingered in the Louvre and the Cabinet des Estampes (prints and engravings) of the Bibliothèque du Roi (now the Bibliothèque Nationale). The seeming absence of vice and misery amazed him: "no drunken men, no brawling, no squalid poverty, and scarcely a beggar," he wrote his brother; "how they can earn their bread I do not know, for one sees few tokens of industry." One clue was the omnipresent soldiery, for "half the men you meet are in uniform, sentries are posted everywhere," seconded by "a large armed police, which is constantly on the alert." Marsh was tempted by Paris prices; he judged "a single man could enjoy all the advantages of the place" for $1,000 a year, "provided he is able to speak the language well enough to make his own bargains." He broke off sightseeing to confer with Turkey's envoy at Paris, Prince Callinaki, on furthering American trade with Black Sea ports.[3]

In mid-November the Marshes resumed their trip south by rail and stagecoach, then by steamer down the Rhone to Avignon, by carriage along the Riviera and across the Apennines. Mediterranean evergreens and fruit trees dazzled New England eyes: ash-green olives, cork oaks, dark cypresses, palms, umbrella pines, citrus groves. That summer's leaves had fallen was no cause for regret; stripped of vines and deciduous foliage the shape of the land emerged more clearly—"our geology and geography [are] the stronger for the winter frosts."[4] Level green plain, blue sea, craggy or snow-capped mountains offered vistas far more vivid than the ubiquitous greenery or brown bleakness of Vermont.

At Florence the sculptor Hiram Powers, Marsh's boyhood friend long resident in Italy, guided him for two weeks among paintings and sculpture. Then on for three weeks in Rome, stunned by the Coliseum, in whose shadow Marsh would later live; seeing St. Peter's "always with renewed admiration, though disgusted with the wretched superstitions of which it is the seat"; thrilled by "wonderful, wonderful pictures, old & new," at the Academy of St. Luke; staggered by Rome's "vast number of great works in marble of the highest excellence." Marsh wrote Powers that "like many other ignorant people, I had been made to believe the ancients couldn't sculpture animals. What an absurdity! Their animals in the Vatican are as miraculous as their gods & men."[5]

Naples provided the macabre spectacle of the monthly dole strewn among beggars in the street, and an awesome eruption of Vesuvius. Despite suffocating fumes and noise "appalling beyond any sound I ever listened to," the volcano lured Marsh back several days running. Climbing within a hundred yards of the source to measure the fifteen-foot-high, cherry-red lava, he sent a detailed account of the flow to the Smithsonian.[6]

A week in the Mediterranean and Aegean aboard the frigate *Mississippi* brought the Marsh entourage through the Dardanelles and the Sea of Marmara, around Seraglio Point into the Bosporus and the Golden Horn. Ahead, the domes and minarets of Constantinople stood out in the morning sun. Frigate and fort gave salute and countersalute; Turkish and American officers came aboard; and the American Minister ceremonially disembarked. It was February 23, 1850, five months since Marsh had left home—and three months after his expected arrival.

For this delay Marsh would later pay dearly, but what with Paris and Florence, Avignon and Rome, pictures and statues, ruins and volcanoes, he had "enjoyed more during the winter, than I thought I could in the rest of my life." As striking as the sights were the sounds: "the great bell of Lyons, the roar of Vesuvius, the echoes of the Golden Horn," which Marsh was hearing while writing to Boston librarian Charles Folsom. "A salute is firing. Every report is repeated, with incredible rapidity, hundreds—I do not exaggerate—of times, & at last comes a concentrated roar, as if the pillars of heaven were tumbling together."[7]

After these European delights, Marsh at first found Turkey grim. The weather worsened; sunshine gave way to snow flurries and cold winds. Ottoman fiat banned Constantinople proper to European residence; the diplomatic corps lived across the Golden Horn in unpicturesque Pera. Pera's cobblestoned alleys rose too steeply from the waterfront for horses to maneuver; as snow fell on mud-laden pavements, the Marshes finished their journey on foot to Dubois's hotel.

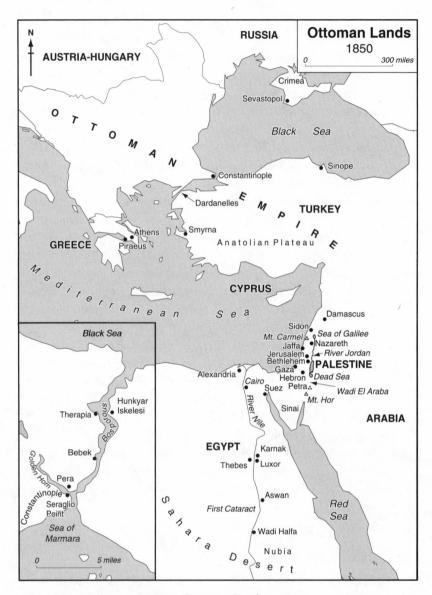

Map 2. Ottoman Lands, 1850

This hostelry proved cold, damp, and sparsely furnished. Their barn of a drawing room had neither stove nor fireplace; after one freezing day Marsh got a large charcoal brazier, around which crouched the seven shivering Americans. Orange peel and incense failed to dissipate the giddy vertigo of the fumes. There were just two sheets among them all; the beds were filthy, the privy broken. "So wretched" did the Marshes find the meals "that we neither did nor could swallow them." Although mulcted for a full month's lodging, after two weeks they moved to Mme. Giuseppina's warmer hotel, backing on to a cemetery of decaying turbaned tombstones. In a nearby parade ground soldiers drilled from daybreak till late at night to the ceaseless monotony of fife and drum.[8]

Marsh had no comfort until late April, when he rented a house at Therapia, a Greek village twelve miles from Constantinople on the European shore of the Bosporus. Perched twenty feet above the landing quay, their three-story abode had ornamented mosaic ceilings, a Turkish bath, a piano, and a terraced garden fragrant with pomegranates, passionflowers, and jasmine. Broad steps lined with crimson dahlias in great stone vases led up to a second terrace of berries and fruit trees, and to a vineyard above. In summer the garden yielded melons and quinces; figs ripened, small but sweet; delicious grapes abounded, though Marsh judged New England's better (he later found "the finest" in the Ionian island Zante).

In Therapia pleasant days passed without incident. Caroline, though ever ill, kept in good spirits. They had an army of servants: a Bulgarian porter apt to embrace Marsh by the knees unless he was quick enough to avert it, a maître d'hôtel who "progresseth continually in the ways of perfection," a cook, a waiter, a gardener, a washerwoman and her maid, a seamstress, a Turkish officer to accompany Marsh and another to carry messages to the Sublime Porte, the "lofty gate" that officially denoted the central offices of the Ottoman government. On horseback and boat trips to Bebek or to the Golden Horn the Marshes stopped now and then for coffee and sherbet with a Turkish host.

June brought a surprise visit from Frederick Adolph Wislizenus. Won by this intrepid suitor who followed her halfway around the world, Lucy Crane married him at the legation, and they left for Germany and home in St. Louis. With Lucy gone, Therapia was quieter than ever. Few visitors came from Constantinople; the Marshes saw only neighboring missionaries and British and French envoys *en vacance* near by. They spent an equable summer and fall listening to the wind sighing through pine and cypress, nightingales singing in their branches, the bleating of goats; and watching the flux of porters shuffling along the quay, donkeys laden with panniers of black bread, effendis with their pipe-bearers. Below, lithe caïques flecked the laurel-fringed Bosporus, darting over its translucent surface to Constantinople or the Black Sea. It was "a good place for a painter," Marsh wrote Powers, but "not for a sculptor. The people are all drapery, the drapery all colour. Form or fold there is none, but you may make a rainbow out of a crowd, and with trees, water, & greenery, get up a landscape. The view of Constantinople & Pera from the water is the finest thing in the world, but within there is not much."[9]

*I*n the narrow bottom of a caïque—rowed by six oarsmen for an extravagant fee—Marsh crouched awkwardly on the way to Teheragan Palace for his first call, March 11, 1850, on his Imperial Majesty, twenty-seven-year-old Sultan Abdülmecid I. It was an imposing spectacle for a Vermonter unused to Ottoman ritual. Marsh presented his credentials in French, congratulated the sultan on the recent birth of twin sons, trusted that Turkish-American relations would remain amicable. The sultan, wearing plume, fez, and diamond agrafe, assured the new Minister of imperial friendship.[10]

For a diplomatic novitiate no place could have been more bewitching than Turkey, long a hotbed of intrigue. When the young sultan came to the throne in 1839, the Ottoman Empire was crippled by corruption and beleaguered by the ambitious Mehemet Ali, viceroy of Egypt. With British and French aid, Abdülmecid had regained suzerainty over Egypt and sought to break the hold of the old aristocracy and of long-entrenched Armenian tax-gatherers. Guided by the influential British envoy, Stratford Canning, the sultan had fostered civil order and religious liberty, safeguarding Christian missions and foreign residents. Yet Turkey on the brink of the Crimean War was still rent by religious faction, economic misery, social turmoil.

Marsh judged the sultan just and generous. Unlike "a long line of cruel predecessors, [he was] almost worshipped . . . throughout his empire . . . excess of liberality indeed seems to be almost his only fault." Too amiable to be effective, however, Abdülmecid was "alone in his empire." Marsh suspected the sultan's conniving advisers, venal nobility, Muslim clergy, Catholic and Greek Orthodox agents were in cahoots with Austria or Russia. "Enormous frauds and peculations" drained the imperial treasury. Armenian bankers kept half the taxes they collected, and were bribed to persuade the sultan to remit levies. Previous sultans had a cure for such evils, "rude indeed, but effectual," noted Marsh. "The Pacha who had grown rich by bribery, or the Armenian who had swindled the government, was strangled or banished, and the treasury replenished by the confiscation of his spoil." The present sultan had renounced this means of redress but "provided no substitute, [so] the government is robbed and cheated with shameless openness, and perfect impunity."[11]

With a government marked "on the one hand by helpless imbecility, and on the other by open and unrebuked corruption," only "mutual distrust and hatred" among racial and religious factions stifled domestic upheaval, and only "commercial and political jealousies . . . among European Powers" staved off external aggression. As Russian invasion loomed in 1852, Marsh warned his State Department that Turkey was "a house of cards; the smallest power in Europe could overthrow it, if the rest would not interfere." With the government virtually bankrupt, "the whole Empire is in a state of complete disorganization, and there is no security for life or property anywhere."[12]

Pera, the diplomats' haunt, for Marsh epitomized Turkish ineptitude. It was "full of villains of every description"; rape, murder, robbery, and vendettas were daily commonplaces. Among the irritants of his "ticklish" post he listed "fleas and scorpions, fellows watching you or following you with an ugly stick or worse; . . . loud screams at midnight from a neighbor's house, that a band of armed robbers was plundering; . . . a report of a gun and then another report that a neighbor's wife had been shot at her window, the assassin expressing much regret at his mistake, having taken her for another lady."[13]

As Christians in a Muslim land, the foreign envoys in this "wretched place" were socially close-knit. Marsh found the diplomatic corps superior to the circle he had known in Washington, and his command of languages made his entrée easy. He had most to do, and many views in common, with the powerful Stratford Canning, long dominant in Turkey. He was on good terms at the start with the Russian envoy Titoff, who shared his interest in ancient Scandinavia; the French ambassador General Aupick and his consul general, Eugène Poujade; Count Portales, Prussian Minister and friend of Marsh's cynosure Alexander von Humboldt; Count de Souza of Spain; Baron Romualdo Tecco of Sardinia; and Austrian internuncio Count Stürmer.

These aristocrats shared traditions of bribery, spying, and secrecy that made Marsh aware "European diplomacy is very nearly as corrupt as it was in the good old times." Their common aim, no less in "liberal" France and England than in despotic Russia and Austria, was to avoid armed conflict while sustaining the existing order. "Beautifully balanced Europe is now," was Marsh's mordant comment, "with Russia giving law to almost the entire continent, and nothing to neutralize her influence there but crazy France!" [14]

In this diplomatic assemblage no one ranked lower than the American Minister; at an official dinner Marsh's place was seventy-second. He was not even a plenipotentiary but a mere Minister; every minor European power had an envoy of higher grade. Again and again Marsh reminded the State Department that his low status degraded the United States in Turkish eyes.

The ill repute of previous American envoys to the Porte was an added vexation. Marsh's immediate predecessor, the bellicose Maryland politician Dabney Smith Carr, knew no French, the mandatory tongue. Carr "was almost never seen in society," Marsh related, "unless in that of some drunken Englishman, & had no better diplomatic position than his horse." Carr's recall had removed "a load of infamy . . . from the shoulders of the Christian community here." Not yet recalled was George A. Porter, consul at Constantinople, "an idle, ignorant, weak, and dissolute person," reported Marsh, and "a discredit to the country that employed him." [15] Other American consulates were as badly staffed; J. Hosford Smith at Beirut was of "doubtful competence [and] little ability," and Edward S. Offley at Smyrna a self-important crank who gave Marsh endless trouble. Many could not cope with the simplest commercial or diplomatic matters; the consular files show Marsh continually bailing them out. American foreign service and prestige demanded drastic reform.

Marsh also hoped to expand Turkish-American trade. England and France, as Turkey's self-appointed guardians against Russia and Austria, took the lion's share of Turkish commerce. But Britain's new free-trade policy had opened Black Sea ports to American vessels. American machinery and hardware and British goods bought with American cotton might be exchanged for Levantine commodities. Turkish duties on American goods were very low. If the United States would reciprocate by reducing tariffs on olive oil, wine, silk, raw wool, and dyestuffs, Marsh envisaged "a very great and profitable trade with the East."[16] Eager to lessen the Porte's dependence on England, Foreign Minister Aali Pasha was ready to meet Marsh halfway.

The first fruit of Turkish-American accord was an 1850 visit to the United States by naval officer Emin Effendi, better known as Amin Bey. Marsh had urged the Porte to send "a genuine Turk," rather than one educated in Europe who "would have acquired prejudices in favor of England or France." Amin Bey was to report back on American ships, docks, and Navy yards and to gauge what American goods might be profitably imported.[17]

Amin Bey was escorted to America by John P. Brown, Marsh's zealous if unworldly dragoman, or translator. At first all went well. "Emin Bey proves to be an intelligent fellow," Brown wrote Marsh from shipboard, "out of whom I hope to make something useful." Feted in New York, the Turk moved on in a glow of good feeling to Washington in September 1850, where President Fillmore and Secretary of State Webster lionized him, and Congress allotted $10,000 for his expenses. Brown was "delighted at the success of our little scheme." But Americans began to suspect they had been too generous. Rumor spread that Turkey had not accredited Amin Bey; Congress had been duped by an impostor. Despite denials by the State Department, which "never for a moment . . . doubted" the official character of the mission,"[18] the press, particularly Horace Greeley's *New York Tribune*, accused Marsh and Brown of foisting a spurious envoy on a gullible host.

Public affection for Amin Bey cooled rapidly; by winter the trip turned into disaster. Weary of dragging the Turk around the country, Brown no longer sang Amin's praises but spoke of him as gross, ill-mannered, ignorant, obstinate. Ottoman reaction to the mission was mixed. Having viewed Amin's official "brilliant reception [as] a new proof of the friendship of the American government for the Sublime Porte," the sultan was "surprised and incensed" by later press attacks.[19] The practical results were in any case minimal; commerce between the two countries remained negligible.

Marsh also tried to promote American trade with Persia. The Persian chargé d'affaires at Constantinople agreed to a treaty of commerce, and Marsh's draft gained the United States useful concessions. All was done in secrecy to prevent interference by France, which at the time monopolized Persian trade. The U.S. Senate ratified the treaty in 1852. But to Marsh's angry dismay, advance news had leaked to the press, and France pressured the shah of Persia to reject the treaty.[20] This was not the only time the State Department was careless, not to say unscrupulous, with Marsh's confidential despatches.

As Marsh had hoped, his regular diplomatic duties were not terribly onerous. But he had many other unanticipated tasks. The most agreeable was lending a hand to the dozen Protestant missionaries domiciled at Bebek, near Therapia. Brought up "to form a fit estimate of your personal characters," Marsh told them, and "to feel a strong interest in the[ir] success," he took their part in frequent brushes with the Sublime Porte. Thus permission to build a mill and bakery at Bebek was withdrawn when the bakers' guild objected. But "it was not found wise," a missionary recalled, "to make a direct promise to Mr. Marsh, and then attempt to evade it"; Marsh got the Grand Vizier to retract "without any harsh words, and without any Oriental *lubrication*."[21]

More troublesome were extraterritorial conflicts. Since the sixteenth century, "Capitulations" or commercial treaties had permitted Franks (Europeans) to conduct economic, social, and religious life free of Turkish control. Exempt from local taxes and immune from the laws of the land, residents of such enclaves were subject only to their own community rule. As most commerce was in European hands, and European practices diverged markedly from Turkish codes, such privileges and immunities were highly prized by foreign residents and deeply resented by Turks.

The American Minister was required both to protect American citizens in Turkey and to adjudicate their disputes with others. But how could Marsh defend American interests and at the same time act as an impartial magistrate? Myriad exceptions to general rules complicated matters. Thus American diplomatic authority covered criminal but not civil cases. Americans and Turks could sue each other, but if a Turk won a civil suit and the American refused to pay, nothing could be done; American consuls could not enforce Turkish decisions. In reprisal, Turks often refused to honor just American claims. Marsh asked that his jurisdiction be extended to such civil cases, but the Senate Foreign Affairs Committee vetoed this proposal.[22]

Marsh's forensic tasks were both arduous and tedious. For example, an English firm sued the dragoman of U.S. Consul Offley in Smyrna; on jurisdictional grounds Offley defied the British suit, which was then appealed to Marsh. In another case, Armenian protégés of the British consul in Jerusalem alleged that America's vice-consul at Jaffa, Jacob Serapion Murad, an ethnic Armenian, owed them $100,000; they asked Marsh to dismiss Murad from his post to enable them to sue him in Turkish courts. Adjudicating such quarrels consumed countless hours.[23]

Disparate American and Ottoman legal codes aggravated matters. Marsh was loath to enforce Turkish prison sentences for nonpayment of debt; he felt uneasy about "exercis[ing] so broad a power, which the American government does not exercise in its own territory." If he now and then strained a point on behalf of Americans, Marsh was on the whole scrupulously impartial. He had "no wish to shield American citizens" from the just consequences of their misbehavior, he assured the Porte. Abuse of privilege Marsh would not tolerate. When Offley tried to leave Turkey without giving surety for his considerable debts, Marsh admonished the would-be decamper: "I cannot suffer the interests of our citizens in Turkey to be hazarded by any attempt to deprive the Turkish tribunals of the jurisdiction which belongs to them." Unless Offley satisfied his creditors, Marsh would "notify the Porte that neither

you nor your interests are any longer under the protection of this Legation."[24] The consul stayed put.

All in all, extraterritorial privileges seemed to Marsh pernicious.
"The whole practice of foreign protections," he advised the Secretary
of State, "is a mischievous anomaly." It made Turks reluctant to engage
in business that might subject them to foreign jurisdiction, and inflamed local hostilities. The entire fabric of privileges and immunities
buttressed European faith in their own superiority—an ethnocentric
arrogance that Marsh saw as self-defeating. Not until 1914, though, did
Turkey expunge extraterritoriality for good.[25]

*The poor Hungarians the martyrs of so saint a cause are about to be cast out
helpless, with no other choise than to shot themselfs, to perish by hunger or to
gain a miserable life by unlawful dishonest means.*

Lajos Kossuth to Marsh, January 1, 1851[26]

Most of the thousands who solicited aid from the American
Legation were foreigners: political refugees, escaped felons, Constantinople flotsam and jetsam. Their numbers were augmented by those
who fled Italy, Poland, and Hungary after the failed 1848 revolutions.
These exiles circulated from legation to legation, begging for protection, employment, money, food and clothing.

Christian legations in Turkey traditionally looked after Europeans
without envoys of their own at the Porte. In this spirit Marsh issued
passports to exiles from Hungary, Poland, and the Papal States. But as
more sought help, the problem grew out of hand. Some were true refugees, but Marsh termed the majority "men of depraved and desperate
characters who ask for protection only as an immunity for crime." Many
held American passports issued blank in Rome and hawked as mer-
chandise. Dabney Carr had sold passports to non-Americans; some
consuls still did so. Such abuses made it hard for Turks to punish and Europeans to extradite their own nationals. Marsh finally refused to shield
any except bona fide American citizens, but he continued to be plagued
by "Americans" who displayed "passports," demanded immunity, extorted "loans," and often had to be deported at legation expense.[27]

One group of refugees, however, Marsh helped to the limit of his powers. This was Lajos Kossuth's band of 1,200 who had fled Hungary in August 1849. Austria demanded their extradition, but the sultan was happy to harbor soldiers as anti-Russian as they were anti-Habsburg. Supported by Britain, Turkey refused to yield them up, agreeing only to hold them for a year in far-off Anatolia. Commending the Porte, Marsh on his own initiative offered Kossuth and some followers asylum in the United States. Americans had lauded Hungary's freedom fighters; Austria complained of Secretary of State Webster's hostility; a breach loomed. Directing Marsh to do what he could for Kossuth, the State Department backed his offer of asylum.[28]

In March 1850, Marsh asked Turkey to let Kossuth and his chief aides go to the United States. But Austria, anxious to stop Kossuth from enlisting English and American support, threatened reprisals if Turkey released the Hungarians before the agreed term. In frequent touch with Marsh, Kossuth detailed the neglect and mistreatment endured by his men. Could they not be brought to the United States and given land?[29] Marsh held out little hope his government would defray transport, but thought Congress might offer land once the refugees reached America. Meanwhile he assured the impatient Kossuth that he had interceded time and again with the Porte; he would not cease to advance "that great cause which has excited so strong a sympathy in my own breast, as well as in that of the entire body of my countrymen."

By July, foreign minister Aali Pasha, now eager to rid Turkey of the refugees, offered passage for two hundred men as far as England if the United States would take them the rest of the way. Alas, Marsh lacked "authority to accept this liberal proposal." Meanwhile he needled his government to uphold America's reputation for aiding the persecuted. By winter their plight was desperate. American press sympathy, unfulfilled promises from Washington, and Marsh's own conscience compelled him "to go much beyond my means in supplying the wants of these suffering outcasts"; they could hardly be let die at the legation door. Webster acted at last in February 1851. Commending Marsh for having succored so many, the Secretary of State offered passage to Kossuth, who boarded the *Mississippi* with his family and fifty comrades on September 10.[30]

But difficulties with the Magyar patriot were far from over. Well aware that Kossuth meant to violate his promise to go directly to America, Marsh warned Captain Long and Commodore Morgan to ensure the Hungarian had no chance to leave the ship. But on learning no stopovers were scheduled, Kossuth protested he was "still a prisoner." When the *Mississippi* put in for supplies at La Spezia, he harangued crowds gathered at the harbor and begged to be let go. The desperate Morgan finally acceded. Kossuth toured France and England in triumph before reaching America in December 1851. Captain Long did not "wish to hear the name of Kossuth ever again," for rumors spread of the Hungarian's "mistreatment" aboard the *Mississippi;* American papers condemned Long's brutality. Marsh came to Long's defense, telling Raymond of the *New York Times* that Kossuth was lying: "It was notorious, from the moment he went on board the ship, that he did not mean to proceed to America in her."[31]

Others came to share Marsh's view that Kossuth's liberation might be more trouble than it was worth. As the fiery Hungarian went from ovation to ovation, preaching revolution against several regimes along with the Habsburg, American leaders grew increasingly uneasy. Kossuth's eloquence did not move them to meddle in European affairs; "many of the politicians," Markoe at the State Department told Marsh, "have been merely coquetting with him." Kossuth now fell out of favor, some weary of his exhortations and appeals, others put off by his arrogance and rudeness. In the end his American visit contributed more to nativist xenophobia than to the cause of liberty.[32]

Marsh regretted that Kossuth had failed to bolster American support for European freedom, but he was not surprised. He found Kossuth personally intolerable. To the Hungarian's selfishness and egotism Marsh now added ingratitude. Kossuth had never "had the decency to utter a word of thankfulness for the unceasing exertions of every member of this Legation in behalf of himself & his associates, or even for the bread which hundreds of them have eaten at the cost of the ill-furnished pockets of myself & my secretaries."[33]

The ingrate foreigner provoked Marsh's contempt. That the State Department treated him with stingy meanness was worse. Diplomatic discretion did not permit him to condemn his own government, but its petty bureaucrats drew his caustic barbs.

November cold drove the Marshes from Therapia back to Pera. But they could find no house fit to live in. None had fireplaces, stoves, or chimneys, though Pera was colder than New York; nothing was available for less than a full year, and rents were exorbitant. Food was equally costly; most had to be imported. Meat was wretched, fruit inferior; Marsh reviled even Smyrna's famed casaba melons.

The price of shelter, food, and fuel made Pera "the most extravagant city in the world." Marsh lived less well on his $6,000 salary than he had as a $2,500 congressman. "I never lived in so poorly a furnished house (we have not a foot of carpeting), never fared so badly, never enjoyed so few luxuries or even comforts as here," he wrote. "I have given no entertainments of any sort . . . yet my household expenditures have exceeded my salary." Ministers were meant to keep large retinues; every Turkish official expected a present on each visit; impecunious Hungarians always needed handouts. Such plaints peppered Marsh's dispatches.[34]

For his country's sake, not just his own, Marsh asked for more money and a higher rank. "The Turks are a rude people," he explained, "and measure the consequence of foreigners by the style they live in." The Austrian and British ambassadors' salary was ten times his; most envoys had winter palaces in Pera and summer places on the Bosporus. "I *know* that if I had the means of living like a gentleman," the pauperized American Minister told Baird, "if I could now and then ask a Turk to dinner, if I could make him a present of a rocking chair, or some other trifle . . . I could accomplish more for the extension of our commerce . . . in *one year*, than my 'competent' predecessors have done . . . to this hour."[35]

Marsh's pleas did not fall on wholly deaf ears. Colonel Bliss convinced the president, and Marsh's promotion to Minister Plenipotentiary was imminent, when he was stunned by news of Zachary Taylor's death in July 1850. This "deep calamity" elevated to the presidency Millard Fillmore, whom Marsh despised, and doomed his raise. Rapport with the new Secretary of State, Daniel Webster, availed him little. Marsh's friend Francis Markoe, head of the State Department's diplomatic bureau, termed Webster "crazier than ever for the Presidency [and] would not recommend an unpopular measure to save the soul of his best friend." Webster did condone "some augmentation" of Marsh's salary but was "very gruff and ill-natured" about the matter, wrote Solomon Foot; "his disparaging remarks in reference to yourself, led me to suppose, you had in some way given the old fellow some offense."[36]

Instrumental in Marsh's setback were Greeley's attacks on him in the *New York Tribune*. "Greeley," declared Marsh, "is an old, and so far as I know, unprovoked enemy of mine, and no new evidence of his malice toward me can surprise me." The two had met in Burlington in 1848; Greeley briefly joined Marsh in the House that December. The crusading editor flayed fellow congressmen for wasting time and money, tried to cut their salaries, and savaged their travel allowance claims.[37] Marsh was one of the few who dared oppose Greeley's mail-route mileage bill as petty and degrading. Like his father, Marsh scorned niggling congressional economies. He customarily opted to defray diplomatic extras, White House refurbishment, landscaping of public grounds— amenities that Greeley thought sheer ostentation. The "best-hated man" in the House, Greeley had earned Marsh's particular dislike:

One day, when a vote was to be taken on a question in which Greeley felt an interest, but apparently Marsh did not, the zealous new member, noticing that the latter made no sign, pulled him by the sleeve and said; "Why don't you vote? Why don't you vote? This is a question on which you ought to vote."

Marsh shrank from such physical contact and resented being told what to do. He made some cold and cutting reply. The offended Greeley "never after missed an opportunity of manifesting strong hostility toward" Marsh, who returned his animus in full measure. "The more I know of him," wrote Marsh when the killjoy Greeley spoiled a Washington social occasion, "the more I believe he is a great knave. A man who can't laugh must be."[38]

Marsh's request for a raise was just what Greeley expected from "a man of fortune, an advocate of High Salaries generally, and a deadly foe to retrenchment." Who cared, asked Greeley, if an American Minister's "parade and glitter of ostentatious display" did not measure up to a Russian ambassador's? The mission to Turkey had been "competently filled" by Porter and Carr "without any complaint as to salary," noted the *Tribune*. "If Mr. Marsh is sick of it, let him drop off and some good man will be found to take his place for the $6000."[39]

Greeley's "malicious falsehood" angered Marsh. In fact, Porter and Carr had repeatedly complained of their salaries, and Constantinople costs had since doubled. Truth meant nothing to Greeley; "he thought his assertion . . . would be believed, & that it would injure me."[40] And it did injure Marsh. His own press allies were outshouted by Greeley's *Tribune*.

Marsh saw one solution to his financial vexations. By wintering in the Turkish provinces, he could avoid the ruinous expense of taking a house at Pera. Marsh figured that they could "spend three or four months travelling in Egypt, Arabia, and Syria, with every convenience, for less money than we can stay here." Caroline's health was worse than ever; Marsh wanted to get her away from the cold, wet Bosporus. Moreover, he assured the State Department in September 1850, it would further American interests for him to see more of the Ottoman realm. Convinced that "escape from the miserable inconveniences of life at Constantinople" was essential, Marsh had already set sail for Alexandria when his *congé*, with Webster's "hearty sanction," came in January 1851.[41]

The Austrian steamer *Schildt* tossed in heavy Mediterranean swells, and Marsh was again sick during the whole week's voyage. "I hate the sea," he wrote Caroline Estcourt, "and would be well content to pay my share of the cost of filling it up altogether." Alexandria, modernized since 1820 by Mehemet Ali, was attractively exotic. Marsh conferred there with Said Pasha, Mehemet's son and future viceroy of Egypt. At Cairo he met Viceroy Abbas Pasha. Far from finding Abbas "foolish and brutal," as he had been warned in Constantinople, Marsh was impressed by his discernment and candor. The conflicting interests of Egyptians, of the Sublime Porte, and of European powers left Abbas's position precarious. As Turkish viceroy, Abbas had to administer the sultan's religious and social reforms, highly unpopular in Egypt, and hence potentially risky to Coptic Christians. If he defied the sultan, Abbas asked Marsh, would England help Egypt? Fresh from talks with Canning, Marsh thought Abbas could count on British sympathy.[42]

This was strange counsel to come from the American Minister to Turkey, given Marsh's habitual disparagement of British policy. A year's contact with Canning had converted Marsh to a qualified admiration of Britain's imperial role in this part of the world. "All the great interests of humanity, our own included," he believed, "will be best advanced by an extension of British sway in the Levant"; he hoped Egypt and Turkey would become English protectorates.[43] One reason was Marsh's alarm at the threat of Russian hegemony, against which Britain seemed the staunchest bulwark.

Early one January morning the Marsh party sailed up the Nile from Cairo. It was a carefree expedition, George Ozias jocular, Caroline Paine "clean daft" in the study of Italian. Evenings they were regaled by four young Italian traveling companions, notably the poet-historian Giuseppe Regaldi, whose fame as improviser Marsh found well merited:

"I will communicate to one of the company a subject, you shall give me two unconnected verses from any Italian poet, and I will improvise two stanzas in the same metre, introducing one of the two verses into each." I gave him one verse out of Tasso, and another from Ariosto, and he instantly produced two stanzas, in *ottava rima*, introducing very felicitously the verses I gave him.[44]

On deck by day Marsh took in the Nile scene. Date palms, sycamores, and mimosas fringed the turbid river. Rice, wheat, sugarcane, and cotton dominated the narrow valley, with here and there a village of dried mud huts. Everywhere bullocks drew water for irrigation in chains of buckets; Marsh heard the creaking of these *norias* from early dawn till late at night. Waterfowl—herons, cranes, geese, ducks, cormorants, pelicans—festooned the shores, and near Aswan Marsh glimpsed crocodiles.

Above the First Cataract to Wadi Halfa the scene surpassed Marsh's expectations. The Nubian desert reached almost to the river, leaving a mere thread of arable land. The days grew warmer, the air soft and mild, with a steady north breeze. Their broad, shallow-bottomed, lateen-sailed vessel contained cabins, hammocks, a fireplace, and a caged ostrich; their cook concocted magnificent meals from dainties bought in Cairo bazaars and from sheep and poultry garnered en route. "We have a good boat," exulted Marsh, "most excellent servants, dragoman and crew, and I have no where enjoyed so many of the pleasures, alloyed with so few of the ordinary discomforts, of travelling, as in this visit to the Nile."[45]

Neither the bites of Nubian flies nor piercing Nubian cries for baksheesh muted Marsh's euphoria, as he told

how Gumbo, King of Nubia, gave Caroline an ostrich, & how, when I gave him in return five candles, a box of mecca dates, & another of guava jelly, he declared, that he had seen in his day a great many Pachas, Beys, Mudirs, Effendis, Lords, Members of Parliament, travelling captains, gentlemen, & Frenchmen, but not any so magnificent & munificent as this Elchi [ambassador]; how the Reis of the Cataracts, tho otherwise a good Musselman, on our passage down that famous chute, drank, solely as a libation for Caroline's sake, a bottle of gin, & then apologized for not drinking more, because he had lost a son only 8 days before; how the German, the regular chaplain of the Cataracts, prayed fervently for the safety of the boat whenever the water was smooth, & bawled, cursed, & swore, at all the ticklish passages.[46]

Returning down the Nile they tarried twelve days at Thebes in a house of "unburnt Pharonic bricks, mummy cases, & dead men's bones," then in an "incredible mud hovel . . . on the roof of a temple at Luxor." At Karnak, carrying Caroline through the temple, Marsh sprained an ankle. He hobbled into Cairo at the end of April still incapacitated. But a one-eyed, shriveled Arab healer put Marsh's badly swollen foot and ankle in warm water, dipped his own fingers in olive oil, rubbed, pressed, and pulled; Marsh put on his shoe and walked away without pain.[47] This miraculous cure tempted Marsh to venture into the desert, hoping that warm sand and dry air would benefit Caroline. They left Cairo on May 7, 1850, with a twenty-six-camel caravan guided by the doughty Sheik Hussein of Aqaba and his sizable retinue.

Like the ancient Israelites, the Marshes took forty days to reach the Holy Land, traversing almost the same route—east to Suez, down the eastern shore of the Gulf, over the central ridge of Sinai, through precipitous wadis and over barren plateaus to Aqaba, thence across Arabia Petraea to the red ruins of Petra, up Mount Hor, and along the Wadi El Araba to the Dead Sea and the groves of Hebron. Here they exchanged camels for mules and went on into Jerusalem. The trip was formidable, the heat intense, but all went well: "any forty days of *stage* travelling in the United States," Marsh wrote his mother, "would involve more of fatigue, danger, and discomfort of all sorts, than this trip has done."[48]

Marsh was captivated by the camels. Surefooted, dependable, not choosy as to food, they carried heavy loads over great distances without needing water. One could bestride a camel "sidewise or backwards," wrote Marsh, "with legs crossed or dangling, and arms folded or akimbo, with no fear that your beast will kick up or stumble and pitch you over his head, or rear and throw you over his tail, or shy out from under you at the sight of an old woman or the bow of a country school-boy, or take the bit in his teeth and run to Quoddy with you."[49]

Given his dread of the sea, Marsh's fondness for camels, with their rolling gait, evil disposition, and pungent smell, seems perverse. But the affection was real and, he claimed, reciprocated. He encountered one camel "in a great rage, making threatening sounds, blowing a sort of bladder out of its mouth, snapping at other people, but didn't bite *me*." To a novice, with breath jerked out of his body, limbs aching, and throat parched, the camel's jolting stride might seem wearisome, but a few days' practice, Marsh promised, made it relaxing. Indeed, the camel's back was so comfortable that robbers lurked, waiting for travelers to fall asleep on their steeds. Arabs ground grain and baked bread while in motion. Others claimed to sew, knit, darn, draw, even peer through telescopes; "by resolving myself into a set of animated gimbals," Marsh boasted, "I contrived to take and record compass bearings."[50]

Their party left camp by two on moonlit mornings to avoid riding in the heat of the day. On the way, Marsh scrutinized the desert landscape, now bounded by the narrow defile of a dry wadi, now a broad stony plateau with clumps of dwarf acacia and stunted tamarisk. Conversation was sparse, for camels were tempted by nearby thistles, which precluded riding abreast and enforced solitary silence.

At the end of the day the Marshes threw themselves exhausted on the sand. While men unloaded and pitched tents, Hussein's servant proffered small cups of coffee spiced with cloves, with the sheik's compliments. This delectable refreshment was, to be sure, "a portion of what [Hussein] daily robbed from our stores for his own use!" Then came dinner and journal-writing, and the party settled for the night, grateful for cool breezes after the scorching day. Evening silence was broken only by camels crunching their beans and by muted Bedouin quarrels over the division of spoils. Sleep normally came soon; but

Marsh recalled one fearful night with "no water—thermometer at 110°, air deathly still, and camels *very* near."[51]

After several days at Hebron, the party went on to Bethlehem and Jerusalem. Marsh swam in the Dead Sea but did not find "the extraordinary buoyancy, stickiness, and other odd qualities" he expected.[52] The way north to Nazareth was long and tiring. Fever had weakened Caroline at Sinai and Bethlehem, young George had been sick in Jerusalem, and Marsh himself now fell ill. A few hours after reaching Nazareth he was unconscious. An old Spanish priest at the convent bled him—with the result that Marsh seemed to stop breathing. The anxious priest squirted liquid ammonia in his face, scalding but reviving his patient.

After a week's recuperation at Mount Carmel they set out for Damascus. At Magdala on the Sea of Galilee both Marsh and Caroline were again desperately ill. Despite the urging of their guide, who feared attacks by robbers, they lay helpless in the birthplace of Mary Magdalene, a mud village in a low, marshy plain, recalled by Marsh as "one of the most horrible places imaginable."[53] They were saved by the advent of an Arabic-speaking Welsh missionary physician, John Bowen. With Arab help, Bowen built stretchers to move the invalids twelve miles to high, cool Safed. There in an old olive grove Bowen and the Nazareth priest, aided by Caroline Paine and Maria Buell, nursed the Marshes back to health. Nine days later, with the help of American consul Ibraham Nakhli (Nachleys) at Sidon, they were under way once more, Marsh on a litter, enfeebled but enthralled by the wild grandeur of southwestern Lebanon.[54] Twelve days more brought them to Beirut; they reached Constantinople August 31, 1851, after an absence of eight months.

Rumor had spread back home that they had all perished in the desert; one friend "had very little expectation of ever looking upon *her* [Caroline's] sweet face again."[55] She was in fact still gravely ill, but a British embassy physician, young Humphrey Sandwith, gradually brought her out of danger. Marsh's own reserves were low after his bout with fever. He came down with erysipelas, then jaundice from a liver complaint; acutely ill, he did not recover until spring.

None of these afflictions dimmed Marsh's exploring ardor. He had marveled at Palestine even more than Egypt and looked forward to further trips. "My anticipations of pleasure and instruction from travel have been much more than realized," he wrote his mother. He had experienced "great enjoyment, great sufferings, and great mercies."[56]

Marsh's travels were enhanced by exceptional powers of observation and breadth of interest. Every aspect of nature and culture absorbed him; landforms, climate, flora, and fauna drew his attention along with relics of antiquity and living cultures. He approached all facets of existence, animate and inanimate, as an intelligent amateur; he ever disclaimed scientific expertise. But this lack enriched his accounts more than it flawed them.

His letters and journals reveal remarkable talent in fusing myriad aspects of landscape into descriptive unity—a talent rare among those trained in narrow specialties, that would become rarer still, as what was called natural history gave way to increasingly narrower disciplines. His most vivid and meaningful depictions are those least analytic, least learned, least encumbered with the technical jargon of science. To Marsh, as to others in his day, "science" had come to mean systematic structuring of measurable data; the sort of synthesis he forged was not "scientific" but belletristic.

Marsh's exhortations on "the extreme importance of keeping a most full and minute record of every observation and every noteworthy occurrence" are gems of common sense, his journals and letters models of their kind. He recorded carefully and industriously. Claims that travelers did better without the aid of diaries were "all moonshine," he cautioned; "you remember no whit the worse, and you observe vastly better, for the practice of full, clear, and accurate description." Later recall brought back only a small part of what one saw "where all—nature, art, man—is new." Journals served to "refresh and revive the fading pictures" that could never be replicated.

Let no excuse of lassitude, no impatience of the inconveniences of writing on your knee in the open air, with insects buzzing about your ears, and the wind scattering your papers and sanding your page before it is filled [deter the diarist]. Trust nothing to the memory. Make no vague entries, such as "fine scenery after sunrise," "remarkable rock far off to the right," "singular appearance in the sky this morning," and so forth, foolishly imagining that you will remember the details, and have the energy to write them out tomorrow.

Today's energy must suffice; "tomorrow will bring with it new observations to record, new inconveniences to surmount, new weariness to combat."[57]

Lack of systematic training did not deter Marsh from conscientiously recording geographical basics. More than most in an age when measurement was a sine qua non for investigating nature, Marsh was equipped by early training and by inclination for recording data. He took compass bearings, measured altitudes, registered temperatures and rainfall, estimated wind direction and force, computed stream widths and rates of flow, and chronicled rarer events: volcanic eruptions, earthquakes, lunar eclipses, northern lights, mirages, typhoons, landslides. Two of Marsh's data-laden letters to Joseph Henry appeared in the *American Journal of Science*.[58] He enlisted missionaries as volunteers in the Smithsonian weather survey, getting meteorological instruments sent from Washington to Bebek and Anatolia.

No less demanding of faithful accuracy than the surveyor Henry Thoreau, Marsh incessantly lamented that his own instruments were too cheaply and carelessly made, too poorly scaled, for precise measurements. A famous Paris firm had "shamefully cheated" him with defective barometers; a level and compass Wislizenus sent from Germany proved unreliable; a thermometer from the renowned firm of Troughton & Sims had "so large an air bubble in the tube as to be quite worthless"; a pair of terrestrial and celestial globes by Malby of London were "rude enough to disgrace an apprentice of two months' experience." Because "instrument making has become a manufacture instead of an art, an *unknown* person cannot order an instrument from any shop in Europe . . . without the *certainty* of being cheated." While Marsh grieved that European artisans had retrogressed, the American thermometers and barometers Baird sent him were no better.[59]

Marsh collected flora and fauna as assiduously as he took temperatures. In Nubia he gathered some eighty plant species for Wislizenus, but they arrived in poor condition because they had been imperfectly dried. The animals he shipped Baird at the Smithsonian at first proved a disaster. Casks leaked, spirits evaporated, specimens spoiled; time and again Marsh's labors went for naught. Finally he got Baird to send out spirits, casks, and strong, wide-mouthed glass vessels with good corks. When these came Marsh could better cope with what Baird's biographers term "the most unusual instructions ever levied on a U.S. Minister." Baird would not "give much for a wild ostrich, but will give a bottle of first rate scuppernong wine . . . for his skeleton. And . . . I must have a camel's head, at least." But Marsh had invited this. "What volant, natant, or repent thing, of those that fly, swim, or creep, over in or around the Bosporus, do you most desire in pickle?" he had asked. "Name [it] and it shall be sent you."[60]

Baird asked especially for fish. Marsh sent twenty species from the Bosporus, along with crabs, tree frogs, and lizards; a "charming variety," conceded Baird, told by Marsh of "humpbacked fishes, wrynecked fishes, knock-kneed fishes, bowlegged fishes." From the Nile Marsh sent asps, horned vipers, "pelicans' heads with the sacs and the parasitic animals that inhabit them"; beetles, ostrich heads, lizards, scorpions, bats, frogs, and toads. Baird craved crocodiles, but Marsh "could get neither eggs nor young. It is a dangerous diversion," he added, "to look for the nest of this bird." In the desert, leopards lurked about the camp "and uttered the fearfullest roar I ever heard." A jackal broke into Hussein's hen coop and killed forty-five chickens, "with the blood of which he was so puffed up that he couldn't get out." Marsh ordered him killed and prepared for Baird, "but he escaped, treacherously let off no doubt by the Arabs, who are very reluctant to kill any animals, except for food."[61]

Back in Constantinople, Marsh gathered grasshoppers, coleoptera, mantises, and other insects for the Pennsylvania entomologist S. S. Haldemann. At Hunkyar Iskelesi on the Bosporus he found a new salamander, naming it *Salamandrosus Maribus* in honor of Mary Baird. Nor did Marsh neglect birds; he promised Baird eggs "enough for an omelet" and Haldemann a few to hatch. But he hesitated to pilfer a stork's nest, fearing "a bullet through me in spite of diplomatic privileges were I to climb the chimney to rob it."[62] He did persuade his niece Maria to plunder the nest of the rare *Jeb*. Marsh could not send the camels, aurochs, or dinosaur skeletons Baird craved, but to that end put the Smithsonian in touch with the grand duke of Tuscany, the emperor of Russia, and several German zoos.

Nothing gave Marsh more joy or so roused his mirth as these collecting excursions. Later he reveled in Sicilian sarcophagi at Agrigento, proposing to send the bones of Empedocles for dissection, so that Baird might "see whether the Greek philosopher and the American savant be of one species." Marsh's fossil finds culminated in "fine osteological collections in [Italian] convents and churches of a bicipitous variety of the human family." This was a race of self-duplicating saints; John the Baptist was "cited as having . . . no less than three heads," but Marsh knew not "whether these heads were successive or contemporaneous."[63]

Marsh also advanced American science in mundane ways. On museum visits he set up Smithsonian swaps of duplicate specimens and journals. From such beginnings emerged Baird's worldwide system of exchanges, which was to make the Smithsonian's National Museum the greatest American repository. Marsh's Old World recipients were well pleased too. "Last year Mr. Marsh promised me the seeds of some rare plants," said Queen Amalie of Greece, amazed that "this year he actually brought them!"[64]

Man's works enthralled Marsh as much as nature's. It was thrilling to see the human past in abundant evidence. And many aspects of life—clothing, houses, farm tools and processes—seemed to him "just what they were in the time of the patriarchs and prophets." Marsh had read Wilkinson, Martineau, and Burckhardt and knew of the "startling conclusions" of Lepsius and Bunsen about the antiquity of upper Nile relics. He scrutinized the form and hues of Pharaonic architecture and sculpture. Like zoology, archaeology was so young a science that even an amateur might find something new; in a ravine near Petra, Marsh found numerous Greek and Sinaitic inscriptions and, still more exciting, a 300-foot tunnel cut through rock, designed to protect ancient Petra against flash floods.[65]

The antiquity sensed by Marsh pervaded the entire Old World scene. "Not the pyramids and temples and tombs only—but the very earth . . . the meadows levelled and the hills rounded, not as with us [Americans] by the action of mere natural forces, but by the assiduous husbandry of hundreds of generations . . . have a hoary and ancient aspect that seems to belong rather to an effete and worn-out planet than to . . . the thousand fresh existences of the new world."[66] It was here Marsh first saw that men had everywhere left their mark; soon he realized how far that touch had transformed nature. The mangled forests and disrupted rivers of New England had already shown him the immediate impact of human improvidence; the deserts of the Levant revealed the ultimate effects of similar processes when long continued. Although thousands of years of history had intervened, the connection was to Marsh abundantly clear and heavy with portent.

7. Missionary Miseries, Mediterranean Jaunts

DURING MARSH'S ABSENCE IN EGYPT AND PALESTINE, a furor arose at his Constantinople legation. Francis Dainese, acting U.S. consul there, had early impressed Marsh as astute and well informed. In April 1850 he recommended Dainese, born of Lombard parents in Pera, be named consul in place of the "useless" George Porter, long absent in Washington. Though the legation dragoman, John P. Brown, warned that Dainese was "notoriously fraudulent" and did "not command the respect of scarcely one respectable person in Pera," Marsh at the time took no heed. As late as September, he was having gifts of cheese and engravings sent from Vermont to Dainese "as some return for the thousand things he has done for us all."[1]

When he went to Egypt in January 1851, Marsh left as chargé d'affaires in Constantinople Henry A. Homes, a scholarly, Turkish-speaking American missionary who had been sixteen years in the Levant. Homes shortly accused Dainese of taking bribes for condoning illegal passports and spiriting goods through customs. Still absent, Porter in July dismissed Dainese as acting consul, naming Homes in his place. But Dainese refused to surrender the consulate and the legation archives to this "proselytizing priest," as he termed Homes. Aided by the Austrian envoy, Homes and Brown had to break into their own consulate to recover the American legation's seals.

Shorn of diplomatic immunity, Dainese fled to Syria to evade his creditors, and thence by January 1852 to America. In Washington he assailed Brown and "Mr. ex-Reverend Homes," who had burglarized his office, "trampled the American flag, & affixed the seals—the double-headed eagle of Austria—to the doors and windows of the American consulate!!!" Had Marsh been on duty, Dainese told Secretary of State William L. Marcy, he "might have prevented the perpetration of these outrages by his subordinates," but the American envoy was "almost constantly absent from his post."[2] Dainese also fed the press data on the Amin Bey "fraud" and other matters "discreditable" to Marsh and Brown.

Back in Constantinople in September 1851, Marsh found all save Dainese's creditors relieved he was gone; diplomats had voiced "astonishment that such a swindler should ever have been in our government employ." Rescinding his "inadvisedly given" endorsement, Marsh now advised the State Department that Dainese's appointment would be "prejudicial to the public interests." But Marsh was thousands of miles from Washington, Dainese on the spot; deaf to his envoy's warnings, President Fillmore in March 1852 appointed Dainese consul at Constantinople. "We have all been suddenly taken aback," Marsh wrote Markoe; "D's reputation here is as bad as possible, & I cannot account for his success in obtaining the recommendations which led me into the unfortunate mistake of [initially] advising his appointment." (The official notice failed to reach Marsh, who learned of the appointment only through the press and hence did not honor it.) Urging revocation, Marsh detailed Dainese's infamies, forwarding a private note of protest from Turkish Minister of Foreign Affairs Aali Pasha. Never operative, the appointment was revoked in October 1852, but Dainese remained in Washington to foment trouble for Marsh.[3]

The Dainese imbroglio all but made Marsh persona non grata at the Porte. Gaining access to Aali Pasha's confidential memo, Dainese threatened to sue the Turkish foreign minister for libel. Not knowing whether Dainese had stolen his dispatch or been shown it by the State Department, the mortified Marsh had to leave Aali "at liberty to suspect" that Dainese knew of his note through "an indiscretion of my own."[4]

Marsh was embarrassed time and again by such disclosures. Then, as since, Americans mistook openness for honesty and supposed privacy incompatible with democracy. American diplomacy suffered as a result. "Important intelligence has often been withheld from our government," Marsh noted to Markoe, "because it is notorious that there is no safety in making confidential communications to it." These betrayals embittered him. "Perhaps it is best as it is," he concluded sardonically. "The contrary rule [i.e., privacy] would sometimes deprive officials of the pleasures of blabbing a secret, and then, great & good men might sometimes be exposed to the great & severe trial of their manhood, independence, and moral courage, if they were expected to . . . refuse to answer impertinent questions." Marsh's reproaches were "as just as they are severe," replied Markoe to a similar earlier charge. But then what could one expect? "I have been nearly 20 years in this [State] Dept.," he added dolefully, "and I despair of ever seeing anything done in it as it ought to be."[5]

What with Dainese afflictions, Kossuth vexations, money worries, and ill health, the winter and spring of 1851–52 were hard on Marsh. "I have no news to tell you," he wrote a friend in Florence. "Men grow dull and oblivious in this sleepy capital, where not even rumour, save in the shape of a Pera lie, even penetrates." At Therapia—where they had returned with enough stoves to make the villa habitable—he took up his travel diaries and began writing "some loose babble about the Desert." His prolixity distressed him; "the moment I begin to treat any particular point, it swells up like a bladder, and I am fearful I shall make a volume on a grain of sand." Finally he sent off an essay on camels, "the silliest thing I ever wrote"—but he enjoyed gestating what was to grow into his next book.[6] He now looked forward, alas mistakenly, to a summer of repose.

What a pity it is that Yankee Missionaries, the world over, are so very Kantankerous . . . ! If they were only round, plump, & jolly, instead of being lean, lank, flabby, & emaciated, I'm sure they would give infinitely less trouble.

Captain Louis M. Goldsborough to Marsh, Piraeus, June 5, 1853

Marsh's dream of an easy Bosporus summer was undone by State Department orders to go to Athens to sort out the tangled woes of Reverend Jonas King. Regret at leaving Therapia was lessened by the hope that this assignment might gain him the additional portfolio to Greece, which had won independence from the Ottoman Empire in 1829. The USS *San Jacinto* was put at Marsh's disposal; he reached Athens on August 1, 1852.

Jonas King had been in Athens twenty-five years, on a relief mission during the Greek revolt against Turkey, then with the American Board of Foreign Missions and as acting U.S. consul. He was embroiled with Greece on two counts. One was a property dispute. In 1829 King had bought a two-acre plot near the Acropolis. City officials warned him the land would one day be needed for public use; and meanwhile King could neither build on it nor sell it. Now the Greek state moved to confiscate the land for what King held was a derisory fraction of its value. The second case was a charge of proselytism. For years King had freely held Protestant services in his home, but in 1850 he was tried, convicted of "heterodox preaching," and sentenced to prison, then banishment.

Marsh's tasks were to gauge what King's property was worth and to determine if the missionary's trial had been "fair and legal." Since Greece had curbed no other religious sects for preaching, King's treatment "looks much like persecution," Webster advised Marsh; if so, the American government would protect the missionary.[7]

Athens was agitated over the affair. Local papers accused the United States of trying to "terrorize" Greece by sending Marsh in an armed frigate; *Jeune Hellas* grieved that the Greek negotiator was Foreign Minister Andronikos Paikos, "a stupid fellow, who will manage the . . . business wretchedly." Marsh's inquiry proved tough and tedious. The trial reports were in chaos, and he had to translate the crabbed calligraphy of "accursed Greek" documents—some "totally illegible, and decipherable only inspiration-wise"—leaving him "purblind as a mole." Every obstacle was put in Marsh's way: witnesses were intransigent, lawyers reticent, judges hostile. Jonas King himself was a nuisance— quarrelsome, plaintive, forgetful of facts and dates, testy over delays, intolerant of compromise. The reedy-voiced New England missionary had alienated most Americans and English in Athens. Even those who thought him wronged were loath to aid him, for he was a perpetual irritant to Greek relations; they would be glad to be rid of him.[8]

After sweltering three weeks in midsummer Athens, Marsh sailed via Ithaca and Corfu to Trieste, where the *San Jacinto* underwent repairs, to write his reports. Monuments to Marsh's painstaking industry, these "most unequivocally" showed King in the right. The missionary's property claim had been treated with "slavish injustice and bad faith"; Greece ought "to release his land from the injunction laid upon it, or to pay him its fair value."[9]

On the religious issue, Marsh concluded that King had been prosecuted simply to coerce him to abandon or compromise his property claim. Irate "that a government could stoop to . . . such base measures to evade a pecuniary liability," Marsh wrote that "the ministry was acting in concert with a debased and fanatic priesthood throughout the whole affair." The trial had been conducted "unfairly and illegally . . . with a gross departure from the spirit of the law." Witnesses perjured themselves, judges turned prosecutors, the Greek Orthodox Synod circulated inflammatory placards and recruited hostile crowds at the court. The United States was duty-bound to intervene on King's behalf.[10]

While awaiting instructions, Marsh took off on a European jaunt. He might "be censured for this deviation," but felt "entitled to some indulgence" after his "severe labors." Webster should let Marsh "rest my old blind eyes and shattered brain."[11] He had lots of time—for which in the end he paid dear. Webster died in October 1852, and not until February 1853 did his successor Edward Everett (immersed in such concerns as the Peru guano trade, fishing rights in the Bay of Fundy, and broils with Britain over Cuba and Central America) find time to commend Marsh's promptness and assiduity and order him back to Athens.

Everett concurred that King's trial had been unfair, the decision "unjust and oppressive." But since the *forms* of law had been observed, he thought redress would be hard to get; and an intemperate reaction might endanger other American missionaries. It was enough for Marsh to convey the president's "decided opinion that . . . the sentence of banishment ought immediately to be revoked." On King's property claim Everett's support was no stronger. If Greece refused to pay what Marsh thought just, he was to request arbitration. Marsh should "avoid the tone or language of menace, but let the government of Greece perceive that the President is quite in earnest."[12]

This craven response outraged Marsh. American prestige required that the president peremptorily "*demand* a remission of the sentence of banishment." Marsh feared "a disastrous effect upon our national character and influence abroad . . . if the weak and unprincipled government of Greece be allowed to enforce the illegal and oppressive sentence of its corrupt tribunals against an American citizen."[13]

Back in Athens in April 1853, Marsh reported "evasion and procrastination . . . the cardinal features of Greek diplomacy." Foreign Minister Paikos, "insolent on paper, & creeping, fawning, sycophantic, timid and false in conversation," kept delaying in the hope that the Russian-Turkish crisis would force Marsh's return to Constantinople or "that the approach of the sickly season will drive both me and the ships of the squadron from Athens." Not until June did he have Paikos's answer, "framed in the tone of lachrymose sensibility and injured innocence, which the diplomatists of Greece invariably assume, whenever a claim is preferred against her."[14] Arguing that the king of Greece had no power to revoke court judgments, Paikos declined on his majesty's behalf to remit the sentence of banishment.

Marsh received this refusal with "severe displeasure." But he had no time for further negotiations, for in July the threat of war called him back to Constantinople. However, Jonas King was never banished; a year later Greece lifted the sentence of exile. In this the missionary saw "the hand of God," though conceding that American aid, and above all Marsh's pressure on the Greek government, had assisted Divine Providence.[15]

The land negotiations were equally tendentious. Paikos first contended that the city of Athens, not the Greek state, was accountable. He then offered to compromise, paying for the land at 1835 values. Marsh proposed instead that King get twelve and a half drachmas an acre, about one-fourth of the original valuation. Sir Thomas Wyse, Britain's capable and influential Minister to Greece, offered to arbitrate; but Marsh was reluctant to involve a mediator. Thus the matter was still unresolved in September, when Marsh was replaced as Minister to Turkey. Confident of success, he would even then have returned to Athens to finish the job, had the State Department desired. Instead, he briefed his successor on the King case, advising Carroll Spence to "shut his ears, hold his tongue, and look savage." He had "the devil to deal with, but the devil is thoroughly frightened, and if Mr. Spence fails of reaping the fruit of my labors, it will be his own fault." The case was finally settled substantially on the terms Marsh proposed, enhancing his fame, wrote Markoe, as "an acute controversialist, a sound lawyer, and . . . a perfect Grecian." [16]

But the King affair left Marsh with an abiding distaste for "Hellas without Hellenes." The country had been despoiled: "the Romans first, then the Byzantines, & finally Lord Elgin and other northern barbarians have so completely stripped Greece, that there is little of art, but the ruins of her architecture left." Marsh dismissed Greece as "a poor country on the north shore of the Mediterranean formerly inhabited by *Graeculi esurientes* [hungry Greeks], now by vermin and Bulgarians." (Bosporus Greeks had previously struck him as "horrid wretches . . . regularly drunk on every fête day, and are then the noisiest brutes in the world.") [17]

In a dispatch to Washington, Marsh traced Greek decline from the 1832 accession of the young Bavarian King Otto, first spineless, then despotic. Westernized philhellenes of the 1820s had given way to profiteering immigrants from Balkan Turkey, bent on gleaning spoils from Russia's imminent dismemberment of the Ottoman Empire. Worst, to Marsh, was the xenophobic Greek Orthodox Church's "miserable fanaticism, which proscribes alike the Catholic and the Protestant, and aims to found a political state upon . . . sectarian bigotry." Under their self-indulgent priesthood, Greeks seemed to him "the falsest and most bigoted of people." [18] After his trials with Jonas King, it is a wonder he found anything to choose between fat priests and lean missionaries.

During the eight months' delay on the King cases after September 1852, Marsh toured central Europe with as much rapture as he had Egypt and the Levant. But intermingled with work on his Greek reports, these new sights and sounds, people and ideas piled up too fast to assimilate. Unlike his orderly Levantine journals, Marsh's European letters were desultory, irregular, hastily scrawled, and kaleidoscopic.

From Trieste the Marshes with Maria Buell and Caroline Paine went northeast through the weird and desolate sinkholed Karst to Lai-bach (Ljubljana), thence over the Julian Alps via Graz to the Austrian spa Bad Gleichenberg. East of Graz, Marsh found a mountain scene "surprisingly like Vermont. The hills are lower, but the trees, & wild as well as cultivated smaller plants seem almost identical with those that grew on Mount Tom, & on the fields and meadows about your father's house and mine," he wrote Hiram Powers. And "if our soil were as neatly cultivated, and the forests as carefully protected, and as judi-ciously and economically managed," he told his brother Charles, "there would be nothing to distinguish the two countries." The comparison became a harbinger of Marsh's *Man and Nature*.[19]

While Marsh completed his second King report, Caroline bathed in Bad Gleichenberg's hot springs and dosed with spa waters. After-noons brought walks or drives followed by tea and talk with the sister-in-law and niece of the British Minister at Athens. Caroline thought the Wyse ladies clever and vivacious: "Tho they are Roman Catholics, we find they sympathize with us Americans in practically all things," and thanks to their Irish birth they lacked "all British stiffness." Marsh found them a bit giddy and was relieved when they parted, "inasmuch as I have told about all I know, though the rest, for aught I see, talk as glibly as ever." For her part, Winifrede Wyse attributed to Marsh "a grand talent *pour le silence*."[20] He took refuge from feminine chatter in the company of Baron Joseph von Hammer-Purgstall, famed Orientalist, at nearby Hainfeld Castle.

In mid-October they went on to Vienna, where Marsh acquired reading glasses and was generally patched up "so that I should run a spell longer." Guided by Hammer-Purgstall, Marsh enjoyed the "cheer-ful, lively" Viennese, "remarkable for devotion & profanity"; so glorious seemed Gothic St. Stephen's that worship in it might almost "justify an apostasy to Popery." With his niece Maria he then made a whirlwind tour of Bohemia, Saxony, Bavaria, and the Tyrol. In Nürnberg, Marsh renewed his admiration for "all that belongs to the middle ages, except their infernal spiritual & temporal tyranny." He was entranced by Rem-brandts and Rubenses in Munich, by the stunning collections of Dres-den's Grune Gewölbe, and by "three striking avalanches" at Königsee. Back in Vienna in December, he found the doctors had failed his wife. "What do you think they prescribed for Mrs. M's eyes and limbs? A pair of spectacles and a stick! You should have seen her with London-made glasses and crooked staff, essaying to read large print . . . and to walk like Mother Goose!"[21] She still could not see to read, and even a short walk brought severe pain, diarrhea, and fever. She remained hauntingly fragile.

Nevertheless, the Marshes went on in December to northern Italy in "fog, fog, fog! and cold enough to keep you shivering, and just at the freezing point, day & night."[22] In Florence early in 1853, Marsh learned of his mother's death; she had long been in poor health, but the bereavement was severe. Grieving, Marsh took solitary walks in eroded Tuscan hills.

Florence also offered civilized distractions—art galleries and the company of American sculptors and painters: Hiram Powers, Miner Kellogg, Eastman Johnson. Powers took Marsh to meet the Brownings. Afterward Marsh saw them often and was deeply moved by Elizabeth Barrett Browning's intense candor. She in turn was attracted to Caroline; as a fellow sufferer, Mrs. Browning lauded the cheerful courage of this "great invalid and almost blind." In evenings over tea and strawberries the Marshes and the Brownings talked of poetry and the radical Proudhon, of the reformers of '48 in Europe and Harriet Beecher Stowe in America, of Dumas fils' newly dramatized *La Dame aux Camélias* and the actress Fanny Kemble. They also discussed spiritualism, a fixation of Mrs. Browning's, unswayed by the skeptical Marshes' scientific citations of Humboldt and Faraday—"I consider facts to be too strong for either of them," she wrote Caroline. The Brownings promised to join them in Constantinople in the spring, but Marsh's protracted tasks in Athens precluded the visit. Elizabeth Barrett Browning's death just after they returned in 1861 immensely deprived the Marshes—"we had always named her almost first when we talked of the pleasures of living in Italy."[23]

At the end of March 1853, when State Department orders came at last, the Marshes waited two weeks in Naples—with "the same strange display of abundance & beggary, rags & finery"—for the laggard *Cumberland*. Commodore Stringham's gingerbread and pickled oysters must have warded off the Minister's seasickness; the delicacies were consumed by the time the *Cumberland* reached Athens, though Lieutenant H. A. Wise wished some had been kept "to fatten the missionaries a bit." "The Piraeus certainly is a horrid hole," exclaimed Captain Louis Goldsborough, explorer and future Civil War naval leader assigned to Marsh, "and I shall be rejoiced to get out of it."[24] Goldsborough, long Marsh's close friend, waited stoically while Marsh and Paikos bickered, until Marsh grew too uneasy about Turkey to delay any longer. When the *Cumberland* steamed into the Golden Horn July 5, 1853, he had been out of Constantinople three weeks short of a year—a hiatus of which his foes would make fell use.

In Constantinople Marsh was at once plunged into a predicament as trying and touchy as the one he had left in Athens. Like Jonas King, Martin Koszta invoked American succor, and like King he became a burden to his benefactor. One of the exiled Hungarians who had gone with Kossuth to America, Koszta had registered his intent to become an American citizen. But eighteen months later he returned to Turkey, ostensibly to survey trade prospects for a New York firm. On June 21, 1853, a three-man gang seized Koszta on the Smyrna waterfront and threw him into the harbor; he was fished out by a waiting dinghy from the Austrian brig *Huszar*, where he was put in irons by Captain von Schwartz, charged with fomenting anti-Habsburg sedition.

The American consul at Smyrna, Edward S. Offley, complained of the kidnapping to Austrian consul-general Peter von Veckbecker. John P. Brown, U.S. chargé d'affaires following Homes's return to America, at once protested to Austria's internuncio in Constantinople, Baron Karl Ludwig von Bruck. Repudiating Austria's claimed legal right to expel Koszta from Turkey, the Americans demanded his release as an intending citizen. In reply, Bruck asserted that the Hungarian was still a Habsburg subject, had come to Smyrna to incite revolution, and would be extradited for trial in Vienna.

Koszta's deportation was forestalled by Captain D. N. Ingraham, a sharp-tongued South Carolinian whose corvette, the USS *St. Louis*, happened to dock in Smyrna the day after Koszta was seized. Ingraham loaded his twenty guns and brought his ship alongside the *Huszar*, preventing the Austrian brig from transferring Koszta to an Austria-bound steamer. On July 2 Ingraham sent von Schwartz an ultimatum: if Koszta were not released by 4 P.M., he would *"take him out of the vessel."* Other Austrian ships readied for the fray, while American merchantmen in the harbor swung the *St. Louis* around to gain effective fire. At the last hour the Austrians backed down, agreeing to release Koszta into the "neutral" custody of French consul-general Pichon pending orders from Vienna and Washington. Koszta was taken from the *Huszar* and his chains struck off; "as he touched shore," in Caroline Marsh's perfervid account, "the shout from the assembled thousands was a tremendous 'Viva la republica Americana,' with groans and 'a bassos' for Austria."[25]

"Now, you gentlemen of the pen must uphold my act," Captain Ingraham wrote Marsh, just back in Constantinople. Marsh strongly backed Ingraham and Brown; Koszta's arrest had been an act "of illegal and private violence." Marsh hoped that Ingraham's forthright action would curb Austrian persecution of other refugees, giving "new force to the hourly increasing respect with which the American government is regarded" in Europe and the Levant. Over Austrian protest, Secretary of State Marcy upheld Marsh, who made ready to return Koszta to America. To this the humiliated Bruck finally agreed, reserving Austria's "right" to imprison Koszta "the moment he is again surprised on Ottoman territory."[26]

Just as the issue seemed resolved, Marsh was mortified to have the wretched Koszta spurn the terms of his release. Previously eager to leave at once, he refused to embark. "Koszta now suddenly became quite cock-a-hoop, forgot his fears and his 'gratitude' together," as Marsh put it, and "boldly declared, that he would return to Turkey" whenever he liked. Offley chimed in: by ceding Austria the right to reimprison Koszta, Marsh and Brown had disgraced America; he (Offley) forbade Koszta's removal from the French Consulate. In vain Marsh argued that he had given up no right, that Austria's stipulation was no part of the actual agreement. "As the head of the Legation, and as a professional lawyer," Marsh could not "be governed by the [legal] opinions of Mr. Koszta," he told Offley. And to Marcy, Marsh huffed that he could not keep "a position of respectability or usefulness, if ignorant and presumptuous subordinate officials" were allowed to defy him "for the sake of giving themselves a factitious importance, or of gratifying a private malice."

Threatening to rescind all protection, Marsh at last got Koszta to embark, the Hungarian maintaining to the last the right of "going and coming wherever my business demands." In fact, his affairs were more convoluted and his motives for delay more devious than Marsh knew. Koszta was mainly reluctant to quit Smyrna because his local mistress had spirited a Parisian portrait painter into the French consular premises to prolong his stay. The flattered Koszta was bent on having a finished portrait, of which he could send admirers inscribed copies.[27]

The Koszta affair was more consequential than these petty shenanigans suggest. It set precedents for invoking domiciliary intention to justify American protection, and for American shows of force against foreign powers. Countering every Hülsemann retort, Secretary of State Marcy won plaudits even from Britain's Lord Palmerston for standing up to Austrian bullying. And Captain Ingraham, later a Confederate naval leader, was lionized as a national hero on his return; it was said that had he wished he might have become president.[28]

No less significantly, the Koszta affair aggravated ill will that helped fuel the Crimean War. Bruck, discredited and indignant, sought to avenge his own and his country's outraged honor not on America but on the hapless Turks, for failing to imprison Koszta in the first place. Bruck forced the sultan to dismiss foreign minister Aali Pasha for "negligence" at Smyrna and launched a vendetta against refugee subversion in Ottoman lands. Having been chosen as internuncio to convince the Turks Austria was their one true friend, Bruck had instead forfeited Habsburg prestige and hardened Turkish resolve to resist Austria's ally, Russia. And Bruck's heavy-handed threats to Turkey led France and Britain to mistrust Austrian aims, weakening hopes of concerted suasion against Russia.[29]

The exit of the ingrate Koszta, whose "folly, impertinences and obstinacy" had all but nullified the efforts made on his behalf, did not end Marsh's worries. "I suppose," prophesied Brown, "that both yourself & me will appear in the U.S. [as] Austrian Agents, as in the affair of Dainese!"[30] In fact, Marsh's enemies now combined against him. Dainese drafted Koszta's statement of grievances and Greeley published it; the whole farrago was shown to President Franklin Pierce.

Marsh was no passive scapegoat. He excoriated Koszta as he had Kossuth in the New York and Boston press. A friend remarked that "going among the Turks had not improved [Marsh's] humanity much— else you had not flayed alive those two very notorious *soi-disant* Americans right in the face of the multitude who had been invited to admire them."[31] Not for much longer would Marsh be out of harm's way; the Democratic victory made his dismissal imminent, and late in 1853 President Pierce appointed his successor.

An ancient dame, who sat on a stone at the receipt of customs, put me this question, "Where do you live?" It puzzled me confoundedly, & does yet. Indeed, where do I really live?

Marsh to Lucy Wislizenus, Milan, January 1, 1853

Marsh felt at home nowhere. After four stimulating years abroad, the prospect of settling down again in Vermont was unenticing. The ending of his Ministry was only one of several misfortunes. Both the post to Greece and an expected salary increase had been denied him. "Cruelly used," Marsh surmised that "an enemy has done this." Yet though meager, his pay as envoy had been better than nothing. As the 1852 election neared, Marsh had been dismayed by the outlook of "being turned out and going home." Financial disaster had overtaken him in Vermont; he wanted to put off "the mortifications & vexations that inevitably await me." He made no plans. "I do not know what I can do at home," he grumbled. "I cannot return to the law, after so long an interval, and I see no other opening."[32]

For a time Marsh hoped that the undesirableness of his post might delay his recall. He half feigned apostasy to Pierce's party. "My wife turned Democrat two years ago because the custom house people at Boston charged duty on some gewgaws she sent home from Egypt. She has been a savage free-trader ever since." Marsh himself felt "mightily inclined to Democracy," he told Markoe; another year or two in the Old World would make him "a desperate Radical." He had "clear given up Tariff (since my factory failed) and most Whig devilry."[33]

Washington friends bolstered Marsh's hope that he might be kept on in Constantinople. Some surmised regional affinity with New Hampshire-born Pierce. Painting Pierce's portrait, G. P. A. Healy found the president "a most unaffected agreeable man, so much so, that my wife cannot believe he is a real Democrat." Pierce told Healy that Marsh would be allowed to remain "as long as he [the president] can keep off those who are besieging him for office, he thinks very highly of your excellency." [34]

But Marsh's enemies were telling the president that the Minister took long vacations at government expense, that Austria had outwitted him in the Koszta affair, and that when most needed in Constantinople he was wasting time in Athens on a worthless missionary! Marsh's former chargé Homes found the president hostile. "I am surprised," Pierce said sternly, "that Mr. Marsh should remain away from Constantinople at such an important crisis"; he heeded no retort that the Minister was in Athens on State Department orders. "I think I can hear the whiz of the axe as it comes down on my neck," Marsh mused. "I pity Gen. Pierce. Don't he wish the Whig office-holders had but one neck so that he could decapitate us all at once?" [35]

While Marsh awaited the end of his Turkish mission, the Ottoman Empire itself seemed about to expire. After many months of fruitless negotiation and intrigue, 40,000 Russian troops invaded European Turkey in July 1853. The Crimean War had begun.

The immediate provocation was a dispute over the Holy Places in Palestine, to which Tsar Nicholas I demanded open Greek Orthodox access, along with the virtual autonomy, under his appointed patriarchs, of Orthodox Christians throughout the Ottoman Empire. The conflict's other roots lay deep in European power politics: Russia's age-old desire to control the Straits, Austria's to repress imperial sedition and to expand Habsburg domains against its perennial Muslim enemy, and England's to prop up Turkey as a client state and security buffer to British India.

Marsh's sympathies were largely pro-Turk. Detesting tsarist despotism, he feared that Turkish collapse might lead to Russian suzerainty throughout Europe. Constantinople in Russian hands presaged "absolutism in politics and obscurantism in religion": in short, the "destruction of existing European civilization. [As] the most effectual obstacle to the advance of political barbarism," Ottoman integrity was crucial to the survival of freedom. [36]

Many Turks called for instant reprisal against Russia, but the sultan's chief advisers heeded Stratford Canning's cautions. In concert with France, Britain urged Turkey to accede to most Russian religious demands. Fearful lest such concessions enrage Muslims and thus endanger Christians in Turkey, Marsh directed Commodore Stringham's flagship to stand by. He thought British appeasement not only "wrong & shameful" but a tactical blunder. Orthodox Greeks in European Turkey outnumbered Muslims three to one; after gaining them further privileges, Russia "will need no war to conquer Turkey." It would then be too late for Britain and France to stop Russia's entire absorption of the Ottoman Empire. Marsh was unaware, or chose to ignore, that Canning's (now Viscount Stratford de Redcliffe) delay was tactical: he sought to win over the British Cabinet, secure French compliance, neutralize Austria, and gain time for massing Britain's naval forces.[37]

Defying nearby British ships, six Russian warships bombarded the Black Sea port of Sinope on November 30, 1853, sinking the entire Turkish fleet. More than two thousand Turks perished. This "massacre" inflamed British public opinion, but even then Canning held back. Failure to redress Sinope persuaded Marsh that Britain meant "to permit the abject humiliation, if not the final sacrifice, of Turkey." Months before Britain and France entered the war in March 1854, with their troops reaching the Crimea in September, Marsh had left Constantinople. He followed the ensuing butcheries from afar, with dismay. The Allied capture of Sevastopol in September 1855 after a year's siege and the ensuing peace treaty left Russia impoverished, Russian influence in southeastern Europe sunk, but Russian ambitions unabated. Marsh feared that Russia after the close of the Crimean War was, if not as strong, certainly as malevolent, as ever.

He was particularly appalled by pro-Russian sentiment in the United States. American bias was actuated by antipathy toward Islam, Anglophobic sympathy for a supposed underdog, ignorance of tsarist oppression, and, most detestable to Marsh, a community of interest between slavery and serfdom. "Much American sympathy with Russia was simply *bought*," he later recalled, "by bribery of journalists, & by appeals to the South on behalf of a people who held a form of slavery to be lawful."[38]

Once back home, Marsh sought to dispel these American delusions. No Russian could be trusted. Even more than conniving Muscovite diplomats he loathed Russian secret agents disguised as tourists, scientists, "servants in great families, political exiles, foreign correspondents of newspapers, Levantine merchants and adventurers, professional gamblers, *chevaliers d'industrie* [swindlers], and even cripples and beggars." Most of all, Marsh hated Russia for throttling secular and spiritual liberty.

At some odds with views he had privately expressed, he now publicly lauded Turks for religious and filial devotion, piety, sobriety, cleanliness, charity, lawfulness—"never have I seen a community so free from violence and crime as Stamboul"—tolerance of Protestant missions, and mistrust of Roman Catholic and Greek Orthodox idolatry. Greek Orthodox were just the opposite: dwelling in abominable filth, they were habitual drunkards, liars, and thieves. It was Orthodox jealousy of the success of Protestant missions in Turkey, Marsh thought, that had in great measure incited the tsar to embark on war.[39]

Marsh's Russophobia never abated. Twenty years later, again alarmed by rumors of an American-Russian alliance that might provoke an anti-Christian backlash among Turks, Marsh urged the Secretary of State to send a fleet to the Levant. But if "unprincipled Russians seem to be doing their best to justify the horrid charges" of the Turks, Marsh could no longer stomach the latter, either. "Satan's house is divided against itself," he judged. "This is a quarrel in which it is impossible for a fair-minded man who knows the parties and their quarrel, to take sides *with* either."

"For heaven's sake," he had exhorted Markoe as the Crimean War ended, "don't sympathize with the hellish nonsense of Russia & Rome— the Czar & the Pope. I hope yet to see them both dangle on a gibbet. . . . Till then, there is no hope for liberty on the continent." Only some English were now admirable in Marsh's eyes. "Don't be too hard on poor old England," he wrote; "in spite of the misconduct of her *rulers,* the existence of her *people* is the sole source of light & hope to European civilization and Christianity. All else is devilish."[40]

Marsh's grudging Anglophilia was fortified by the notable group he knew in Turkey in 1853: Admiral Sir Richard Saunders Dundas, later head of the Baltic fleet; the Earl of Carlisle, belletrist parliamentary and prison reformer soon to be lord-lieutenant of Ireland; the scholarly Rawlinson brothers, George and Henry, the latter consul at Baghdad; and especially Florence Nightingale's brother-in-law, Sir Harry Verney, pioneer in land reclamation who kept in lively converse with Marsh over three decades. Marsh enjoyed many evenings at Canning's house with these men. The admiration was mutual; Carlisle termed Marsh "one of the best conditioned and fully informed men it is possible to find anywhere."[41]

Marsh got official notice of his "decease" in October 1853. "I kiss the firman and put it to my forehead," he wrote Baird. "'Tis the will of our Lord the Padishah. The Democracy is great." In his final dispatch Marsh starchily chided the Secretary of State for addressing the sultan casually, as "Your Highness" (a term applied to mere viziers) rather than "Your Imperial Majesty," as a "want of respectful observance [toward] a very precise and punctilious people in affairs of etiquette."[42] On Christmas Day the Marshes left Constantinople for good.

He was in no hurry to get back to America. Anticipating nothing but trouble in Burlington, he stretched out the homeward trip as long as possible. After six days of seasickness—"we had died," Marsh alleged, "had the voyage been a day longer"—they reached Malta. There Sir Harry Verney entertained them several days before they went on to Sicily. Marsh felt "grievously light-headed and queasy still," but had seen "nothing fairer in Europe" than Sicilian vistas of jagged peaks, stupendous chasms, and smooth seas. At Syracuse he gathered papyrus reed but lamented that the famous fountain of Arethusa, the metamorphosed wood nymph of Greek legend, was now only a vent for washerwomen. In a ravine near Centuripe he picked up a flint arrowhead "just like those of our Indians," he wrote Markoe; "all the Sicilian antiquaries declared [it] altogether unique." Marsh suspected stone weapons had not been found in Sicily only "because eyes familiar with such objects have not sought for them." He later cited Thoreau's response, "when asked where Indian arrowheads were to be found, 'Everywhere'."[43]

Mount Etna was the crowning delight. Marsh, for whom "every bit of lava is a sacred thing," indulged his "most extravagant passion for volcanoes." Beyond the smoking, lava-filled chasm of the Val de Bove, hundreds of cones dotted the sides of Etna, many rising more than five hundred feet. "What is Vesuvius," marveled Marsh, "after this?"[44] He struck up a friendship with the vulcanologist Carlo Gemmallaro and went without sleep to read Seneca's account of earthquakes in *Quaestiones Naturales*.

Marsh was unexpectedly eminent. "Messina made quite a *Pays de Cocagne* for us, [and] Taormina even brighter glories." He found "since I became an *Ex* I am a more important person than when I was in full bloom as a diplomat *en fonction*." But they lodged in "the same union of splendour and discomfort we have so often seen in Italy. Bed tickings of heavy silk, satin coverlids, very fine linen sheets edged with lace, & an embroidered muslin bedspread, with *one* small washbasin & ewer, of heavy silver, but no towels." The Marshes were "loaded with civilities, and every potentate I meet offers me straightway the half of his kingdom. I should like to accept this of Sicily."[45]

Marsh reached Rome in time for carnival, "a foolish festival truly, but showy." They stayed two happy months, as rewarding and unclouded as any in his life. "There is no time," he exulted, "to see anybody or anything at Rome. One is daft at once, and driven hither and thither by all manner of whirlwinds." He avoided other tourists, as "society and sight-seeing are incompatible." This interdiction did not extend to the Brownings, wintering in Rome, or even to his one-time bête noire Martin Van Buren, whom Marsh saw now and then and found agreeable. The former president talked "pretty coolly of all his enemies old and new." A rebuff by President Pierce had angered Van Buren, but Marsh doubted that "the trials of this present administration keep him awake o' nights."[46] Also in Rome were the Estcourts, with whom the Marshes toured the city, until Estcourt—now a general—left for the Crimean War and his death, a year later, of cholera.

In April the Marshes crossed the Apennines to Ancona, Bologna, Modena, and Milan. Spring was late; at the end of May a thick snowstorm caught them in the Alps below Stelvio Pass, and Caroline had to be carted by wheelbarrow across avalanches blocking the road to Innsbruck. Marsh was fascinated to learn that this Alpine region had not long ago (geologically speaking) been submarine. He wrote from the "Primeval Ocean" to inform Baird that Switzerland had

emerged from the waters and become dry land. Trees have grown, shed their leaves and perished, and been succeeded by new forests, and a vegetable soil has been formed, and subdued by tillage, and men have built here a city called

Bern. . . . All this is recent; the oldest inhabitants inform me that it is not above fifty million years since the highest part of town was upheaved above the sea.

With this geological reminder of Old World antiquity, after a few weeks in England Marsh turned reluctantly homeward. Adding love of the Alps to his fondness for bare desert and verdant Italy, he wondered why "men who have time and money should ever do anything but wander over God's fair earth and gaze at its exceeding loveliness."[47]

8. *Debts and Dromedaries*

AFTER FIVE YEARS ABROAD, MARSH REACHED HOME
August 23, 1854. The Burlington Cornet Band regaled his arrival at the
town hall. Vermonters cheered "his safe return to his native State &
rejoice to know that he returns honored and respected for his most
valuable labors & rich in experience."[1] But Marsh came back rich in
nothing else. While townsmen paid him public homage, in private they
thronged to his door to collect what they could as his importunate
creditors.

There have been Marshes who not only had *money, but actually* kept *it till
they died.*

Marsh to Caroline Estcourt, Burlington, March 31, 1857

Marsh's financial distress was largely due to the misdeeds of the
Vermont Central Railroad, in which he had invested heavily. This railroad was one of several chartered during the 1830s and 1840s to link
landlocked Vermont with Boston markets and Canadian timber. Promoted by Marsh's old friend, ex-Governor Charles Paine, and by the

Burlington firm J. & J.H. Peck & Son, Marsh's partner in banking and milling, the Vermont Central had secured a perpetual tax exemption and a route from White River Junction to Burlington via Northfield, Paine's hometown. Its first train steamed into the Queen City in December 1849.

The Vermont Central's financial structure was shaky from the start. Construction costs were four times their estimates; directors voted themselves enormous salaries and commissions; new stock issues at once depreciated; treasurers absconded. Meanwhile the line deteriorated: sleepers rotted, freshets and fires destroyed bridges and roadbeds, and there was no money for repairs. Competition with the rival Rutland & Burlington line led the Vermont Central to pay a suicidal sum for a link with the Vermont & Canada that bypassed Burlington. By December 1852 Paine and the Pecks had milked the Vermont Central dry, and bondholders took over.

Thus Marsh came back deep in debt for the now worthless stock he had borrowed to buy. He was by no means the only sufferer. Vermont Central's feud with the rival Rutland line and its bad bargain with the Canada line had prostrated Burlington. Boston merchants cut their costs by making an end run around rather than across Lake Champlain. With the lake trade crippled and the Winooski factories dead, half of Burlington's firms—the Pecks among them—soon failed.

Marsh lost real estate as well as cash. The Vermont Central spur line into Burlington had sliced through his most valuable holdings. While Marsh was in Turkey, his brother-in-law and partner Wyllys Lyman had pocketed a large sum to persuade other landowners—including Marsh's other brother-in-law and agent Henry P. Hickok—to sell land preempted by the Vermont Central at a paltry $1,000 an acre, although "all parties well knew that I would not have consented to such a destruction of my property for four times that sum." On his return, Marsh demanded $10,000 redress for this ruinous "fraudulent" sale, but the railroad was already bankrupt. He recovered nothing.

Some $50,000 in debt—a million in modern money—Marsh found that his promissory notes were held by the very men who had done most to ruin him: the Pecks. Marsh would have surrendered all he owned to be released from debt, but the Pecks demanded more; in September 1856 they filed a suit against his future earnings. If "the whole of their monstrous claim [were] established," Marsh would "still be liable to the very villains who have defrauded us in a large sum," for although the bankrupt Peck *company* owed Marsh many thousands, that debt could not be offset against what Marsh owed the Pecks *individually*.[2]

Endless struggles to extricate himself from "the villainy of the Pecks and Lyman's imbecility and knavery" left Marsh utterly dispirited. "The 'old man' is getting 'grumpy,'" his son complained to Caroline, "sighs heavily, won't laugh or joke, & is . . . a very stupid companion." Marsh's only consolation was the grim "hope of seeing J. H. Peck on his way to Sing Sing." Caroline was quite prepared for the worst, and "only wish you may have philosophy and faith enough to see all go without too much pain. You must indeed my dear husband remember that these things . . . assume an apparent magnitude which they really do not possess." It was Marsh's moral duty not to yield to morbid self-pity; "you have too much yet to accomplish in this world . . . to allow yourself to think for a moment that you are likely to be broken down by pecuniary *misfortune* or Peck *malice*."[3] When Marsh asked how she liked being married to a pauper, Caroline chided him for petty worldliness.

Not until April 1860 did Marsh finally settle with "the creditors of the rogues who cheated me. [I have] surrendered all the property I have in the world, . . . and now, at the age of fifty-nine, I begin . . . with a debt of $10,000, in good new notes, and not a shilling to pay it." Two months later Congress awarded him $9,000 for extra expenses abroad he had been claiming for the past five years. Adding $500 author's fees, and a month's advance salary on leaving as envoy to Italy in 1861, enabled Marsh to clear his debt. At last unburdened, though penniless, he faced the future with more cheer than he had felt for a decade.[4]

Suits for federal recompense had soured Marsh no less than brawls with Pecks and the Vermont Central. The sums he sought were small but vital to a virtual bankrupt. But though his claims were just, his enemies were many and virulent, and litigation protracted—the time and effort Marsh devoted to it precluded fruitful work.

Marsh claimed $1,000 a year for extra judicial duties at Constantinople. American envoys elsewhere were paid for such work, but Congress had overlooked those in Turkey. Since his predecessor Dabney

Carr had just been so reimbursed, Marsh foresaw no obstacle. He also asked compensation for his Greek mission. During the year away from Turkey he had to pay rent for the Therapia house and wages to servants there; hence Marsh was "a loser to a considerable amount by the performance of the arduous duties imposed upon him." The Senate foreign relations committee approved these claims, Vermont's Solomon Foot securing an award of $13,500. But Marsh's bête noire Francis Dainese, sniping at Marsh anent Amin Bey, Kossuth, King, Koszta, and himself, roused opposition in the House. "This Levantine renegade," as Marsh termed him, "can forge lies faster than I can answer them. . . . In fact out of the pages of Eugene Sue, I have never heard of so accomplished a villain as this man." Pressing his own claim, Dainese corralled an anti-Marsh faction—acting Secretary of State William Hunter, Archbishop John J. Hughes of New York, and several Catholic congressmen, notably Pennsylvania's Joseph R. Chandler, a backer of papal temporal rule in Italy. These men averred that Marsh had already been paid too much for "trifling" services. And to Marsh's shocked dismay, he was denied reimbursement.[5]

His defeat was a casualty of vendettas between ever more envenomed religious coteries. Protestant alarm at massive Irish and Mexican immigration, more intense than when Marsh was in Congress, was aggravated by labor riots, convent horror tales, much-publicized conversions, and Catholic crusades to ban the Bible in schools and to secure public funding of parochial education. Three million new arrivals between 1846 and 1854 had swelled the foreign-born, mostly Catholic, to 14.5 percent of the American population, the republic's largest alien proportion ever. The meteoric rise of the Know-Nothing Party in the mid-1850s reflected Protestant nativist fears.[6] And Catholic reaction to nativism was unleashed in Congress and in the press.

Opposition to Marsh—he was accused of victimizing Dainese out of anti-Catholic animus—typified the backlash against bias voiced by the likes of Mrs. John P. Brown from the legation in Turkey:

[If] what had been done to Mr. Marsh through the influence of D[ainese] and his papal friends . . . does not prove the necessity of a "Know-Nothing Society" I do not know what would. That society ought to have been organized years ago, & then they would have put a stop to the refuse of other nations emigrating into our once happy country, & preventing foreigners from electing & putting into office their papist friends, who will soon . . . bring us under the dominion of the Pope of Rome.[7]

Marsh, though never a Know-Nothing, shared their qualms about immigration. In Burlington, Irish and Quebec Catholics were more than one-third of the population by 1850; Father Louis de Goesbriand was made first bishop there in 1853. Marsh neither concealed nor palliated his hostility toward Catholics. "Our liberties are in greater danger from the political principles of Catholicism than from any other cause," he warned Vermont's governor-elect, just before Catholics in Congress again readied to defeat Marsh's own claim.[8]

As Congress reconvened in December 1855, Marsh shelved his mishaps in Vermont to attend to those in Washington. Alienated by bankruptcy from Burlington, he was glad to be back in the capital. In Turkey he had been "quite homesick for F Street," feeling that if he could "choose my own habitaculum for life, it would be just in that good neighborhood." He stayed with James Gilliss, then with Spencer Baird, but was longest with David A. Hall, Vermont-born lawyer-agent of the American Board of Foreign Missions. Hall's household was quite disorganized, "a regular *Polnische* [topsy-turvy] *Wirthschaft*, the children allowed to do just as they please, and please to do it very improperly." Though Marsh wished the Halls "would show a little more of their Yankee blood in some of their domestic arrangements," he was made much of by their daughters and nieces. He enjoyed Chinese tea, macaroni, mild flirtation, and fierce battledore and shuttlecock with the "well-favoured & well-read" Charlotte Bostwick (Caroline wondered how Marsh had avoided smashing the bird to bits). The provocative Charlotte told him of her "great dread of being left an old maid"; Marsh supposed "she would not make such a confession to a single man." While Caroline got despondent letters from her husband, Charlotte found his "lurking humor" unquenchable. She long remembered the now solemnly bearded gentleman of fifty-five with "that gray pair of pants, those leather shoes, those domestic-looking silk pocket-handkerchiefs and that invincible, respectable, 'Varmonit' look that rode Juggernaut-like over all the last fashions, defying . . . the Capital's best society."[9]

After Charlotte left, Marsh also quit the Hall menage—"I could

not endure the disorder, confusion, & misgovernment of the children"—
to return to Baird, where he was "wretchedly lonely, but not more so
than elsewhere in Washington." At Capital gatherings Marsh often
encountered his arch-foe Dainese, who "flourishes & is received everywhere, [for] no soul can withstand the prestige of bad English, musta-
chios, and brass." To Marsh's mortification, Dainese sported a Turkish
medal; "the compliment to a man whom every officer of the Porte
knows to be a perjured swindler proves that 'the sick man' is *very* sick
indeed." It was small consolation to learn from the faithful John P.
Brown that "the decoration is of the lowest grade, 'only such as is given
to eunuchs, women of the hareem & . . . common soldiers.'" [10]

Meanwhile Congress played shuttlecock with Marsh's claim. Foot
feared another defeat. The "diabolical Dainese" ridiculed Marsh's petition for eleven months' service in Greece, noting the Minister had
spent but twenty-three days in Athens "and devoted the remainder of
his time, as any gentleman of leisure and taste travelling upon the
money and in the time of Uncle Sam might do, in a minute study of
Italy, Germany, &c." Dainese found a new ally in Senator Richard Brodhead, a Pennsylvania Democrat whom Marsh had antagonized a decade earlier in the House by "charges of plunder [and by] his appeal for
a dismemberment of this Union" (Marsh had wanted to slough off the
West). Brodhead computed that of four years abroad Marsh had spent
only two and a half at work, for which he had been paid a total of
$53,000—"a pretty large sum for a brief trip to Europe." [11]

Marsh rebutted these calumnies as best he could. He had received
not $53,000 for two and a half years but $34,500 for three years and ten
months; all his travels had been ordered or authorized by the State Department; his delay in reaching Constantinople was unavoidable; and
his mission to Athens had saved $100,000 and the freedom of an Ameri-
can citizen.

In the Senate, Foot again championed Marsh. Congress had just
conceded the inadequacy of the new Minister to Turkey's salary by
raising it 50 percent to $9,000. "That he is a man of economical habits,
everyone who knows him will admit," yet Marsh had had to supplement his pay with $5,500 of his own. Though Brodhead now assailed
Marsh as "the leader of the niggerites in his native State," the Senate
again passed Marsh's bill. But in the House new charges surfaced as fast

as Marsh refuted old ones; hope for his Greek claim dwindled. Even the judicial claim, which Marsh thought ironclad, was rejected by the foreign affairs committee, Chairman Alexander Pennington of New Jersey fearing lest similar payments to every diplomat in Turkey back to 1830 exert "a giant's grasp upon the public treasury."[12]

As winter and spring gave way to oppressive summer, Marsh's spirits drooped. "I cannot tell you," he wrote Caroline, "how I long for rest, & opportunity to do my own work & think my own thoughts." He would leave Washington, but Dainese was "concocting some new villainy against me, & I must stay & watch him." The new villainy was a "proof" that Marsh had spent his expense allowance on "private" presents to the sultan, "private" use of horses, and the like, and was asking five or six dollars a day for board in Greece, although the Hotel d'Orient in Athens cost "only $2 per day, *everything included except the more costly liquors.*" Dainese having "completely poison[ed] the Committee of Foreign Affairs," Marsh's Greek claim now stood no chance.[13] On August 12 he left Washington for New York and Burlington to contend with Pecks and local politics.

The two years since Marsh's return to America had been "years of almost unmixed bitterness—and the next promises to be like unto them." Marsh saw himself permanently ruined, feeling these setbacks had paralyzed his mental faculties and wrecked his chances of being useful. Penury prevented him even from bringing Caroline to Philadelphia for medical treatment. "My claim, my unhappy claim! Well, I tell her sometimes she deserves no better, for marrying a beggarman." It was no solace that she credited him with "the courage and patience of a martyr."[14]

The Marsh-Dainese feud went on with ceaseless acrimony. Dainese persuaded allies in Constantinople to swear that Marsh had misused funds; Marsh got the Turkish government to testify to Dainese's mendacity. Charges and countercharges filled the press; session after session saw Marsh claims beaten. "If I cannot carry a case with so much of merit as yours I will be compelled to regard my membership of the House a failure," wrote Justin Morrill, a new Vermont congressman, in 1857; but it fell six votes short of the needed two-thirds. A 328-page congressional encomium to Dainese's consular virtues reprinted his indictment of Marsh as a "skulking assassin of private character [who]

disseminates *calumny* in high quarters under the seal of privacy." In rebuttal, Marsh abused the House for giving more credence to the "unsupported statement [of] a foreign perjured swindler" than to "the
solemn declaration of . . . the minister and his two secretaries." But slander found ready ears, and Dainese's congressional allies were numerous
as ever. Dainese "can defeat y'r bill," warned Francis Markoe in 1858, "&
you can defeat his. Will you combine?" Morrill likewise urged that
Marsh and Dainese bury the hatchet. But Marsh felt he could "only lose
by any coupling of my name with that of so depraved a scoundrel." He
wanted to win "without . . . touching pitch which cannot fail to defile."[15] So he lost.

Not until 1860 did Marsh get another chance. This time he and
Dainese agreed not to oppose each other. In April Marsh's claim bill
passed the committee, and Markoe at State enlisted his brother-in-law,
Maryland railroad magnate George Wurtz Hughes, a newly elected
House Democrat. Hughes "disarmed the hostility & won the cooperation of his own party," called up Marsh's bill ahead of schedule, and got
it past the House unopposed. Foot rushed the bill through committee
and secured its Senate passage on June 12. Marsh at last salvaged $9,000
for his Greek mission.[16]

This was small return for six years of misery. The protracted battle
permanently embittered Marsh. Henceforth he habitually saw himself
as persecuted, vilified, wronged, with enemies everywhere. His 1850s
struggles cost him faith in justice and the power of truth—not that he
had ever had much. "Even when a boy," he wrote Caroline, "I used to
say that 'Satan is mighty and will prevail.'" Now he had proof of it. Indeed, the Dainese affair resurfaced even after Marsh's death, when his
disgruntled former consular aide Eugene Schuyler, later a diplomatic
notable, reiterated the calumny that Marsh had been paid a huge sum
for a few weeks' service in Athens on behalf of a missionary's unjustified
cause.[17]

Yet his personal devils remain shadowy figures. It might reveal
much about Marsh to know why his partner Lyman deceived and
cheated him, and why Dainese hated him so much he spent six years
trying to destroy his reputation. Marsh viewed his enemies as malevolent conspirators, but the facts hardly bear him out. His Vermont associates were victims of their own cupidity; Dainese, whose work Marsh

had at first commended, resented the Minister's volte-face. But we know little of these men's motives or of their later histories. All of his enemies save Peck, who was jailed, appear to have escaped lasting ill consequences. But against the Vermont Central Railroad, as will be seen in the next chapter, Marsh mounted a reformer's revenge.

The parlous state of his property made money Marsh's overriding concern. A single criterion now determined all he did: would it pay? The search for a Marsh family fortune in England, begun by a distant Midwestern kinsman and pushed by "one of my swampy namesakes in Canada," intrigued him, but the cost of finding hypothetical Marsh millions was great, the hope of reward minute. How could he settle his debts? Should he teach, write, return to politics, go back to law, reenter commerce? Hiram Powers suggested that Marsh start a newspaper similar to *The Times* of London, a venture Marsh scorned as both costly and corrupt. "Could an *honest* paper be supported in any great city of America or Europe? [Only] venality & want of principle gives currency to the New York Times, the Tribune or the Herald, . . . the pioneer in knavery now left behind by the greater shamelessness of its more youthful rivals."[18]

Writing *for* newspapers was a prospect proffered by his friend Henry J. Raymond. While Marsh was in Turkey, Raymond, then launching the *New York Times*, offered him "as much lucre as you may deem a fair offset" for occasional letters. After Marsh's return, Raymond made an explicit bid for several "strong, clear, ABLE editorials upon topics of current interest" every week. "You are so thoroughly furnished with all the needed knowledge upon all topics of discussion, & you write with such fluency, that an hour a day would probably be all the time you would need to give to it—for half the strength of an editorial lies in its being *off-hand* & extempore." Raymond wanted "an exact transcript of such remarks upon any topic as you would make *in conversation* . . . & the more sharply & strongly worded the better." Marsh's response to Raymond has not surfaced; but his later brief pieces for *The Nation* struck James Russell Lowell as well "worth reading for [their] matter [yet] too palpably parts of a book. He does not get under way quite rapidly enough for a newspaper."[19] Indeed, Marsh's printed work never matched the punch of his breezy letters.

If the world was his oyster, as Raymond implied, the pearls in it proved minuscule. Marsh turned down almost every offer that came his way because none paid enough. Meanwhile he spent his energies on tasks that rewarded him still worse, waiting for something better to turn up. "*Kommt Zeit, kommt Rath,*" he hoped; "a door will be opened in time."[20]

One open door was the law. Francis Markoe obligingly hinted to John A. Rockwell, a Connecticut congressional crony of Marsh's who sought a Washington partner, that Rockwell & Marsh "would form a firm so respectable & strong as to carry all before you!" But Marsh vetoed it. Long away from the law, he would find it hard to brush up enough for Supreme Court practice; certainly he had not pushed his own claim with any success. "I have failed once, from undertaking too many things at once." He had moreover lost faith in himself as a partner, for "I bring ill fortune to all who are in any way connected with me."[21] Besides, the prospect of legal work was as distasteful as ever.

Marsh gave more serious thought to a post at Harvard. The chair of history had been empty since Jared Sparks vacated it for the presidency in 1849. Marsh's historical studies were well known at Harvard; in 1855 George Ticknor and C. C. Felton, professor of Greek soon to be Harvard's new president, proposed and the Corporation unanimously agreed on Marsh. He would find "a noble sphere of usefulness, and very agreeable social relations in the University." Marsh's experiences "in the actual scenes of history, added to your comprehensive studies, seem to point you out," urged Felton, "as the person who, of all the country, can make the most of it."[22] That the "scenes of history" meant the Old World was widely taken for granted, even by the author of *The American Historical School*.

With great sorrow Marsh declined this post. He had longed for such a career—though it was "rather literary dissipation & scientific dilettantism than scholarly *labour*, that I have dreamed of as a release from the uncongenial cares & associations belonging to the life of a country lawyer." He loved the study of history and felt Cambridge society "an irresistible temptation." But fighting lawsuits and pressing claims took more time than a professor should spare, and duty obliged him "to look for the occupation which promises the largest pecuniary rewards."[23] Harvard paid only $2,000 a year. Although Marsh keenly regretted forgoing the Harvard position, his historical forte, more social than political, probably flowered to better effect outside academe.

He turned down many other academic posts. In 1859 the University of Pennsylvania invited Marsh as provost, on Baird's assurance that he was no German mystic but "the stiffest kind of old up and down Congregationalist." Marsh was not tempted. "I hate boys, hate tuition, hate forms," he wrote bluntly; "the objections . . . are infinite." His view of schoolboys had changed little in the four decades since his days at Norwich Academy. But he resumed an unpaid role as University of Vermont trustee. For their now "sadly deficient" library (compared with the 1830s) Marsh drew up a purchase plan, acquired books from the bibliographer Henry Stevens, Sr., and procured a new building.[24]

The world of science also beckoned. Washington friends envisaged Marsh replacing Joseph Henry at the Smithsonian. Baird reported Henry in "apparently irreconcilable warfare" with Assistant Secretary C. C. Jewett, who continued to champion Marsh's dream of a large library. Jewett, Baird, and others hoped that Henry would resign; he "would lead a far happier life," averred J. M. Gilliss, "with equal if not greater usefulness to the world in his old pursuits." If Henry left, would Marsh allow his name to be put forward? Terming the Smithsonian "an Augean stable—a disgrace to the country," Markoe looked to Marsh to clean it up. "Many of our leading men," he assured Marsh, "agree with me that you are the fittest man in the U.S. to be at its head."[25]

But Henry stubbornly stayed on. Bache was the real power, Gilliss told Marsh: "he winds Henry about his finger and in fact controls the Smithsonian as well as the Coast Survey." Backed by Bache, Henry dismissed the stunned Jewett in July 1854, igniting a public fracas. Choate resigned from the Board in protest; others defended Henry. Berating both camps, Horace Greeley called for a congressional overhaul of the Smithsonian. Meanwhile, Henry channeled more funds to research, abandoning the library altogether. The Smithsonian's 40,000 volumes went to the Library of Congress, which thus became the largest in the country, as Marsh had urged in the early 1840s.[26]

Marsh kept out of the new Smithsonian controversy, largely for Baird's sake, though in private he vilified Henry's administrative conduct. "In all matters of business," he had earlier confided in the antiquarian John Russell Bartlett, "Professor H[enry] is as imbecile a person as I ever met, & a man more utterly unfit for his place could hardly be found." But now Henry invited him to lecture at the Institution, assuring Marsh he was not trying to "purchase his silence"; the $134 Marsh got for four Smithsonian lectures could hardly be construed as a bribe. Indeed, Marsh judged that Henry had breached the library-research compromise "injudiciously & what is worse *unfairly*," a view many shared. Marsh did prepare a lengthy résumé of the Institution's history for the House Select Committee on the Smithsonian. But his restrained and dispassionate comments left Henry grateful for Marsh's "fairness."[27] Marsh's relations with all the Smithsonian staff remained cordial. Although many wished to see Marsh at its head, it was fortunate for his future achievements and well-being that his ambitions lay in other directions.

Some Washington friends hoped Marsh might return as a senator. In January 1853 Vermont Senator William Upham died suddenly of smallpox; to prevent incoming President Pierce from foisting an interim Democrat appointee, S. S. Phelps was hurriedly called back to finish Upham's term, but was unseated by the Senate in March 1854. Phelps's "Cotton Whig" stance and "sour, saucy, & doggedly aristocratic" demeanor made him anathema to numerous Vermonters, wrote Justin Morrill. Would Marsh contest the seat? Many statesmen in Vermont and Washington would "be rejoiced to see you in the Senate."[28] But Marsh lived on the same western side of the Green Mountains as new Senator Solomon Foot. This was a fatal handicap in Vermont. Ever since the conflict between Joseph Marsh and the Allen brothers, east and west each had its own senator; the "mountain rule" meant that no eastern Vermonter held that seat until direct primaries led to the election of Ralph E. Flanders in 1946. "Why shall I not advertise a place for you at Woodstock," urged the ever helpful Gilliss, "or some other place on the right side of the mountain?"[29] But Marsh doubted that the change-of-residence ruse would work. He did accept the Whig nomination for Vermont's Council of Censors, chosen every seven years to revise the state constitution, but was edged out by a Know-Nothing campaign plank promise to abolish the Council of Censors entirely.

The Whig Party was, in fact, badly eroded by the growth of Know-

Nothing and other radical groups, and at length irrevocably split by the Kansas-Nebraska Act of 1854, repealing the Missouri Compromise that had outlawed slavery in both territories. Most Vermont Whigs joined the new Republican Party, which won control of the state legislature in 1854, and in 1856 Marsh enlisted heartily for that party's presidential candidate, the explorer John C. Frémont. Animus against Fillmore strengthened Marsh's Republican sympathies. "I have never liked Mr. F, & have a crow to pick with him when I get home," he had written in 1853. Fillmore as vice president had opposed Marsh's appointment to Turkey, and as president had treated him "basely." But James Buchanan, the Democratic candidate, seemed to Marsh even worse. To avert a Democratic victory, he pressed conservative Boston friends to back Frémont. Otherwise, he warned them, Frémont if elected would pick all his advisers from a "dangerous radical cabal."[30] But Marsh made no converts among Massachusetts Cotton Whigs anxious to conciliate Southern slaveholders, and he never quite forgave Winthrop (who backed Fillmore's Know-Nothing American Party) and Choate (who supported Buchanan) their "apostasy." Marsh did not stop to think that his own increasingly abolitionist stance had changed more than theirs.

Marsh again turned down congressional bids. He refused the seat left by James Meacham, his 1849 successor in the House, who died during the 1856 campaign. He rejected Solomon Foot's offer of his Senate seat, hesitant to bind himself for life "to such slavery as a senatorship now is, and that with so meagre a compensation." No sooner had Marsh said no than Congress raised its own pay. Now he would "gladly accept," but rightly surmised that the salary increase would "remove Mr. Foot's wish to resign." This deeply embarrassed Foot, who felt "largely indebted for the place I hold to the generous support of your [Marsh's] friends." Were Frémont elected, Foot promised to get Marsh "a first class foreign mission"; should the Republicans lose, he was "ready to resign my place in the Senate" on Marsh's behalf. But Marsh could not accept this from his old friend, even though party leaders urged him to run as a candidate who could win "without struggle or opposition."[31] Marsh's parlous financial plight had clearly not dented his political viability in Vermont.

Marsh's personal stake in the election was, indeed, another diplomatic post. In July he had "a long and satisfactory private interview" with Frémont; in August he hit the hustings with all his energy, for three months devoting himself entirely to the campaign. But though Vermont voted four to one for Frémont, the national election went to the Democrats. "Well, exit Pierce, enter Buchanan!" Marsh wrote in despair. "Which is the poorer creature of the two will soon be known to those who are able to dive low enough for such investigations."[32]

Marsh never again ran for elective office. In 1858 he turned down the Republican nomination—tantamount to election—for governor; he "could not live on the salary" without social mortification. Small wonder! The Vermont governor was paid just $750 a year, less than in any other state.[33]

Judging from his catalog of refusals, one might suppose that Marsh had truly confined himself to the narrow path of pecuniary duty. But he was an unregenerate dabbler. "What *will* you undertake next?" wondered Gilliss, acknowledging some maple seeds Marsh had sent him. "Arboriculture, mathematical Instrument making, Architecture—three professions mentioned in one letter! If you live much longer you will be obliged to *invent* trades, for you will have exhausted the present category."

Marsh had always enjoyed working with his hands, and liked to pretend that "when my friends took old iron & files & hammers away from me, & put books in the place of them, they did me & the world a mischief."[34] Fed up with the crude measuring and surveying devices he had used abroad, Marsh "Yankee-like set to work to invent" better. Why not engrave glass rulers on their underside? He bought a fine diamond point and went to work etching and coloring glass. By the summer of 1855 he had constructed three instruments: a plate glass protractor leveled on one edge, with a metallic or ivory tongue and engraved parallel lines on the under surface; a similar ruler; and a circular protractor engraved on a square plate. He was also making a dioptic vernier so as to more accurately measure distances on paper.

Marsh took pride in being "quite a dextrous glass worker," drilling through half-inch plates "holes less than one fortieth of an inch in diameter, and (in one *painful* instance) even polishing the same." He boasted of his instruments' beauty, "certainty & celerity of adjustment,

[and unmatched] precision." Hoping they might "sell if but as an ornament," Marsh showed his "divers cunning inventions" to his friend
John H. Alexander, surveyor and engineer. "Are these instruments *new*?"
Marsh asked. "Are they good for anything? . . . How can I turn them to
account?"[35]

Alexander was discouraging. Marsh's use of transparency was
thirty-five years old; "an obscure wretch," grumbled Marsh, "has had
the meanness to anticipate it." And glass instruments generally were no
longer popular; they broke too easily. Only the dioptic vernier, if Marsh
could finish it, would be a real advance; the others were not worth
pushing commercially. And so it turned out. Marsh could get no pat-
ents; his work had gone for naught but the thrill of creation. "Tho' the
rude accoucheurs of the Patent Office suffocated the birth," Alexander
comforted him, he had at least had "the sweet spasm of conception &
the pleasant pangs of parturition." Alexander himself was preparing
a dictionary of the Delaware Indian language, an enterprise Marsh
termed "worse than my Icelandic foolery."[36] Of his own many careers,
that as instrument-maker was the briefest.

Marsh never, to be sure, put his main hopes for solvency in glass
etching. Marble quarrying in Burlington and lecturing away from home
occupied much of the time he could spare from litigation. It took some
years of setbacks to convince him that riches lay along neither road.

Fine marble outcroppings had been found at Mallett's Bay, six
miles north of Burlington, by a Marsh cousin-in-law, David Read. With
Marsh, Read incorporated the Winooski Marble Company in 1855, for
"quarrying marble, slate, & other stones & minerals in the state, and
manufacturing & selling same."[37] At this time marble used in America
was almost all imported, chiefly from Italy. Marsh's was one of the earliest native ventures.

Operations had already begun. "Quarrying, which is now the great
affair of the nation," wrote Marsh in autumn 1855, "goes on well. I go
down every day." Despite bruises and sprains he enjoyed scrambling
through rocks and dense brush, finding new colors and varieties, checking on progress. Other veins of marble found near Burlington were no
threat, for they were "inferior in quality, and in advantages of quarry-

ing and shipping."[38] The Mallett's Bay strata were nearly horizontal, with beds of marble up to six feet thick directly at the surface. The marble was extremely hard, which made it difficult to work, but resistant to corrosion and capable of taking a high polish. Marsh superintended cutting and sawing, laboriously done with iron saws, sand, and water. The blocks were then carted to the bay and carried off by steamer.

By the end of September 1855, Marsh and Read had sold two boat-loads (more than 1,200 cubic feet) in New York, and Captain Montgomery C. Meigs, construction chief at the federal Capitol, ordered 400 cubic feet from Mallett's Bay. "Everybody is running marble mad," Marsh exulted; "I hope we may take advantage of the excitement, and sell out at a fair profit." But demand fell off as fast as it had grown. The high cost of cutting and polishing Winooski marble led buyers to prefer softer foreign stone for ornamental work; though Marsh got an order for a dozen Capitol fireplaces, Meigs found his marble "too hard and brittle to work well" for mantelpieces and too costly for plain surfaces.[39]

Far from becoming a bonanza, Mallett's Bay never paid its way. Marsh took a slab with him to Italy and reported that "none of the Italian variegated marbles have anything of the strength and freedom from flaws of the Winooski." Nothing at the Florence marble exposition of 1861 came up to the Winooski brilliant red. But Marsh's then quarter share in the Mallett's Bay quarry fetched a paltry $5,000 in 1869.[40]

I shall hang my "lecturing" on the same peg with my other failures and follies. It must be a long peg and a strong peg to hold them all.

Marsh to Caroline Marsh, February 22, 1857

Lecturing was the last string to Marsh's bow. In this he catered to American ardor for exotica of all kinds. Bidden by countless civic groups, Marsh puffed up for popular consumption his experiences in the Levant. He launched his speaking career in Burlington in September 1854, went on to Boston, New York, and Washington, visited upstate New York in January 1855, and made extended tours the next two winters, addressing middling audiences in scores of dispiriting venues.

Marsh was less adept at lecturing than at quarrying. His tone was dry, level, humdrum; the culture circuit evoked none of the pungency that made him a fiery campaigner. Marsh himself was bored—how could he stay fresh while delineating "The Environs of Constantinople" for the tenth time? He was chagrined that his auditors showed more appetite for odd facts about Bedouins than for discourses on Nature. The Boston press, to be sure, spoke of "large and appreciative audiences" ("that is," Marsh translated, "an invisibly small one"), and commended his "statesman-like views." But even a paper that praised him as an "easy and vivid narrator" noted the need for utter silence, as the slightest sound drowned out Marsh's voice. His quiet manner, rapid utterance, compact language, and "the absence of all personal adventure," concluded his wife, were "not likely to draw large audiences."[41]

The disenchantment was mutual. If Marsh's hearers did not take to him, he found them, their towns, the whole lecture routine repugnant. Traveling in Europe had been the keenest pleasure; in America Marsh found it utter torment. Trains were crowded, dirty, delayed, perilous, thronged with addicts of "tobacco, solid, fluid, and fumiform"; roads were rutted seas of mud or dust; the countryside was mainly flat and dreary; baggage was lost or damaged; mail failed to arrive; accommodations were meager, dark, dirty, often a "miserable old-fashioned country tavern, where I cannot find even a room with a fire, to say nothing of other comforts."[42] Exposure, exhaustion, cold, hunger, thirst were Marsh's routine lot.

His laments make better reading than his lectures. On a chilly trip to New York he was detained two and a half hours "in a miserable hole," captive audience to "a discussion between a Know-nothing and a half drunken German." Near Elmira, Marsh's train overtook one that had derailed, "and our engine finished the mischief by running into the train, breaking two cars and carrying away its own smoke pipe, cropping one man's ear, and otherwise maltreating sundry good citizens." But his main gripe was low fees. Too often he gave a "thin lecture to a thin audience" for a thin reward. Since money was his only object, it was dismaying not to cover expenses. One "very large audience" of 1,200 "to their shame, paid me only $30"; his average return was $50.[43] Perhaps he was too diffident to demand more; other circuit riders fared better.

Self-imposed constraints on content further limited Marsh's appeal. There was so much he dared not mention! With his claim under beady-eyed congressional scrutiny, he needed to seem a single-minded public servant, a man with no eye for anything but his job, a paragon who never took time off. As he cautioned Caroline Paine, then also recounting their travels in the Levant (in her *Tent and Harem*, 1859), "any notice of any *civilities, kindnesses* or *favors of any sort* shewn our party by persons of any rank . . . might very seriously *injure me.*"[44] Marsh's own circumspect talks were purposely impersonal.

Yet even these travelogues were at times provocative. Eager to enhance American prestige abroad, he cited foreign views about the United States to urge domestic reform. Old World attitudes sharply conflicted: while reactionary autocrats loathed America as an exemplar of freedom, common folk in Europe looked to it as "the great source from which universal liberty is to flow," and much admired Americans: "An American passport is worth more to a traveller than a British dukedom." But America's reputation for defending its own citizens and for championing freedom abroad was, Marsh felt, undermined by home events. The image of liberal America was corroded by bellicose expansionism, apologia for slavery, and a spoils system that bred envoys infamous for "ignorance, vulgarity, and vice." Marsh had earlier foretold that America was "to be disgraced and misrepresented at three of the most important points on the continent [Berlin, Frankfurt, Constantinople], by abandoned, ignorant, and lowbred clowns."[45]

Besides reformed diplomacy, improved husbandry was Marsh's favorite lecture topic. "You don't know how scientific I'm getting," he wrote Baird of his 1855 Hartford talk, "The Study of Nature," on which he enlarged a year later in New Hampshire. In these harbingers of *Man and Nature*, Marsh contrasted Old with New World climate, terrain, housing, public works, railroads, crops, and farm techniques. He urged better practices on both sides of the Atlantic, but above all at home. While remaining "the architects of your own fortunes," Americans could learn much from Europe, he told New Hampshire farmers. They should harness science to agriculture, domesticate exotic species, and stem excessive runoff and soil erosion by conserving forests.[46]

In comparing landscapes Marsh relied on his own firsthand depictions. "Exceeding smoothness of surface" struck him as arable Europe's most striking aspect; centuries of cultivation had "obliterated the minor irregularities, . . . reduced the sharpness of the angles, removed the smaller rocks, filled the dried up water courses, and thus given the whole landscape a rolling outline [usually un]broken by hedge, fence, or other artificial enclosure." Also marked was "the absence of any thing [like] an American's idea of a forest"; lopped and trimmed like hop-poles, the trees bordering Continental roads and streams looked wholly unlike "the umbrageous oaks and elms and beeches" of America (and of England).[47]

From comparing landscapes Marsh went on to recommend reforms in husbandry. While America relied mainly on annual crops such as wheat, maize, and cotton, much of Europe was planted to perennials—grapes for wine, mulberries for silk, olives, cork oaks—that needed far less labor. All these might be cultivated in the United States, though Marsh admitted they were less suited to New World climates. On the other hand, the "ruinous effects" of citrus blight and grape and silk-worm disease abroad showed the folly of exclusive reliance on a single crop. Americans should shun European trends toward monoculture.

Efficiency and labor economy Marsh felt uniquely American vir-tues. In his view they stemmed from a healthier diet, better tools, "superior intelligence and ingenuity," and "voluntary contract" employ-ment set by the worker's "own agreement, and not by the pleasure of an arbitrary lord, or by external circumstances [coercing] half-remuner-ated labor or starvation." Praising the longer American hoe, heavier ax, dexterous rake and pitchfork, scythes sharpened not by hammering on stones but by grinding or whetting, Marsh found European (notably Austrian) peasants "so ignorant of mechanical powers, that I have seen four men exert for a long time their utmost strength in trying to shove a stone over the rail of a wagon body on a pair of skids, when one man, by raising the lower end of one skid with each hand, would have tilted the stone into the wagon with entire facility."

On the other hand, European public works were far superior—canals "familiar with the laws of hydraulics"; railroads smoothly tracked, with paved or sodded embankments and well-staffed cars, stone and masonry bridges and viaducts of "the best possible workmanship"; best of all, wide, macadamized, moderately graded common roads, well cambered and tunneled. What explained these virtues? European states aimed at general public convenience, and esteemed public works as proud national icons. Their construction was also a safety valve against rural unrest. "Providing employment for the poor, and thus sustaining them at the expense of the rich, . . . and at the same time . . . attaching the laboring classes to the power that gives them bread," was laudable national policy.[48] Europe's public works programs showed rare awareness of needs to mitigate appalling economic inequalities. Enhancing popular attachment to communal patrimonies was especially commendable because it fostered public stewardship.

Lasting attachment to local landscapes distinguished rooted Europeans from restless Americans. In a Woodstock Fourth of July oration in 1857, Marsh contrasted Old World sites where "every rock has its name, every landscape its history" as the intimately familiar locus of lengthy occupance, with American scenes where, "on the contrary, all things, the works of nature or the works of man, the woodland and the meadow, the river and the highway, the church and the cottage are in a condition of perpetual fluctuation and change, and the lapse of half a century so transforms the features of every landscape, that it is no longer recognizable" to anyone absent even for a few brief years.

With New World history, as with nature, the past was left to perish with few regrets. While Europeans preserved their historical monuments, most American revolutionary landmarks had already vanished, and "many an ambitious improver looks with an evil eye upon what he considers but a useless encumbrance to the convenience of trade. . . . Physical improvement and a constitutional want of respect for personal and hereditary relics combine to promote the destruction or change of all the material memorials" of previous epochs. Writing to an English friend, Marsh wondered "whether the *newness* of everything in America strikes a European as powerfully as the antiquity of the Eastern continent does us."[49]

Pleading for conservation, Marsh invoked Old World losses to open American eyes to threats to land as well as to landmarks. A dismaying saga of environmental depletion was at last impelling Europeans to curb erosion. Because similar New World damage was relatively recent and limited, Americans did not realize that they too were "already beginning to suffer from the washing away of vegetable soil from our steeper fields, from the drying up of the abundant springs which once watered our hill pastures, and from the increased violence of our spring and autumnal freshets." Old World history revealed the ensuing disasters; but "only in countries that have been laid bare . . . for generations [could] the extent of the devastation thus produced be comprehended." Alerted to Old World ruination, Americans might heed present European efforts to stem erosion: forbidding tree felling where avalanches or landslides threatened, restrictions against grazing, mandating reforestation and stream control. Thanks to millions of acres annually replanted, Marsh forecast that "Europe will be better supplied with wood in the next century than it is in this."[50]

The Old World offered bright new prospects as well as cautionary lessons for remolding nature in America. "Almost every plant which we grow as food for men, except Indian corn and the potato, and all the [domestic] animals," he reminded New Hampshire farmers, "have been introduced into this continent in the space of three centuries." Fruitful transfers were bound to continue; "our fields are destined to be enriched and enlivened by plants and animals, until now quite strange to us."[51] One such quite novel proposal shortly met with dramatic success: Marsh's campaign to bring camels to the American West.

Like Franklin and Jefferson as envoys abroad, Marsh was keen to furnish America with Old World plants and animals. His idea of importing camels surpassed his predecessors' wildest fancies. As early as 1846 Marsh had told Congress that "the alpaca, the Thibet goat, and perhaps even the camel" could prove useful in the United States; dromedaries "might thrive on the sands of the South or the great prairies of the Southwest." Home from the Levant, he refloated the idea in a January 1855 talk at the Smithsonian. The arid Southwest supported neither horses nor mules. But camels from Saharan and Arabian deserts could surely find ample water, acacia, and mesquite in far less barren America. Lauding their hardiness, Marsh envisioned camels adapting to New World locales with "ultimate success."[52]

In America the camel might be both beast of burden and weapon of war. Marsh had once scoffed at "subduing the Comanches . . . and other Rocky Mt. Bedouins, with corps of dromedary dragoons." But Egypt and Palestine persuaded him camels could aid military "economy, celerity, and efficiency" by sustaining supply routes and hauling guns and matériel over steep slopes. Marsh had seen "few more imposing spectacles than a body of armed men, advancing under the quick pace of the trained dromedary"; such a fearsome sight "would strike with a salutary terror the . . . savage tribes upon our border." He recommended not Egyptian but Algerian camels "accustomed to much severer winters, a moister climate, and a rougher country." French military use in North Africa offered leads for choosing which breeds and gear to import.[53]

Marsh was not America's only camel enthusiast. For three years Secretary of War Jefferson Davis and others had sought a camel trial. But it was Marsh's lecture that spurred Congress in March 1855 to send a Navy ship with Major Henry C. Wayne and Lieutenant David Dixon Porter to North Africa and the Near East with $30,000 to buy camels. In May 1856, thirty-three camels were debarked at Camp Verde, Texas, and forty-one more in 1857.

Transport and acclimation proved easy, but other problems ensued. Soldiers disliked the smelly animals and got seasick atop them; horses went wild with fright. Some camels were slain by Army men, others stolen by Indians. With further advice from Marsh, the rest got on well enough. Surveying a route to the Colorado River in 1857, Lieutenant E. F. Beale found the "noble & useful brute" surefooted and docile. Secretary of War John Floyd hailed "the entire adaptation of camels to military operations on the plains" and hoped to buy a thousand more, but the Civil War intervened. Confederate troops captured some camels; the rest were used in Nevada silver mines or sold to circuses. Of their descendants, the last died in Los Angeles' Griffith Park Zoo in 1934.[54]

A modern authority on camels charges Marsh with praising their virtues inordinately while minimizing obstacles to adaptation. But for all his optimistic advocacy, Marsh had never dreamed of camels in a *settled* West. "So essentially nomade indeed is the camel," he noted, "that the Arab himself dismisses him as soon as he acquires a fixed habitation."[55]

Marsh meanwhile turned public interest in the venture to personal advantage. It took him only ten days to enlarge his Smithsonian lecture into a book; the literary *Camel* came out just as the Army's beasts reached Texas. The book covered species and breeds, anatomy and diet, speed and gait, modes of training, harness and burdens, and geographical range. Owing much to his crony Baron von Hammer-Purgstall, to the German geographer Carl Ritter, and to the French engineer Jean Luc

Sébastien Carbuccia, Marsh knew his own to be "the fullest account of the animal" in English.[56]

Critics lauded The Camel's "extensive scholarship and acquaintance with many tongues, for which the writer has so enviable a fame." But only an English reviewer realized that the "racy, piquant [extracts] from the unpublished journal of a traveller" were Marsh's own. The author thought it "odd that nobody here had the wit to perceive that I was quoting myself."[57] There were too few such asides as Marsh's comparison of the camel with the ostrich (in Arabic known as "camel-bird"):

Apart from the number of legs, there is not much to distinguish the profiles of the bird and the beast. The ostrich himself seems to have a dim consciousness that he is a sort of equivocal middle-thing between a volatile and a quadruped, for when they said to him, fly! he answered, "I cannot, I am a camel;" when they said, carry! he replied, "I cannot, I am a bird."

And Marsh's book omitted the apt explanation his "Desert" essay had supplied: "Every body knows that husband and wife come at last to look wonderfully alike; . . . if such effects are produced in the human, by 5, 10, or 20 years of cohabitation, why should not the ostrich and the camel, who have been near neighbors and familiar gossips for as many thousands, contract an equally strong resemblance?"[58]

More drollery might have sold more copies. In July Marsh mourned that The Camel had passed almost unnoticed; in October his publisher, Gould & Lincoln, pronounced it "very near a total failure," in the end remaindering half the 3,000 print run to Ohio school libraries. They did better with J. G. Biernatzki's The Hallig; or, The Sheepfold in the Waters, Caroline's translation of an evangelical pastor's romance set on a small Danish island in the North Sea, to which Marsh added a memoir of Biernatzki. But Camel and Hallig together earned the Marshes only $125. Terming Gould & Lincoln "very stupid or very stingy, probably both," Marsh decided to carry his "next grist to another mill." But intended volumes on Lapland and on Oriental topics were shelved, then abandoned, Marsh having concluded that "nothing sells that is solid."[59]

Fall and winter 1856 marked the nadir of Marsh's career. The marble quarry was moribund, The Camel redundant; Dainese had downed his Greek claim again; creditors importuned as new debts surfaced daily.

"Everything goes wrong with me," Marsh lamented; "I am *d'une humeur de dogue*" (in a huff). Then came the long, hard campaign—and the Republican defeat. Marsh tried to see Frémont in New York, "but got entangled in that vile vulgar labyrinth of squares and places about 8th & 9th streets"; his feet were so blistered he could not walk for days. Back in Burlington, marble mishaps and new Peck charges brought on a headache that drove Marsh "nearly crazy." He spent a bitterly cold Christmas in Woodstock preparing a western trip to combine recuperation with remuneration. "If ever an old vessel wanted mending," he wrote dejectedly, "it is I."[60] On January 29, 1857, he set forth on his longest— and thankfully last—lecture tour.

This minute and faithful record of my travels . . . is composed with no view of magnifying my perils and mine adventures, but solely to the intent that a curious and enlightened posterity may have authentic information on the modes and facilities of locomotion practiced and enjoyed in this century.

Marsh to Caroline Marsh, Davenport, Iowa, February 18, 1857

Veteran traveler in the Levant, Marsh viewed his five-week Midwest expedition of 1857 as the worst journey he ever made—a fitting climax to this chapter of fiascos. His Niagara lodging of January 31st "was so filthy . . . that even the cataract could not wash it clean." By Cleveland he had already repented his venture. Leaving Cincinnati at 5:00 A.M. Marsh *"failed to connect* at Richmond, the train having been purposely delayed for some rascal reason growing out of a Railroad quarrel, and *so lost* Lafayette. We were put into an extra train at Richmond, with orders to keep *behind* a freight train," and reached Indianapolis, sixty miles away, six hours later.

Marsh was then to start for Chicago, "but while I was expecting the train *inside*, it went by on the *out*, carrying my trunks with it . . . so that my Chicago appointment is lost also. . . . The natives I meet in the cars don't know so much as a yellow dog. I am much inclined to return directly home." The next day Marsh's train "met a freight train that had run off and broken down the tracks. We then walked a quarter of a mile, in Indiana mud, and sat eight hours in the cars, waiting for a messenger to go nine miles on foot, and get a locomotive to supply ours, which broke down just after we got started again." He reached Chicago after twenty-five hours, "making four nights out of seven in the cars, and to make it pleasanter, I had nothing to eat from 7 yesterday morning until 1½ P.M. today, all our stoppages being in deep cuts, where there was no provender for man or beast. . . . The only pleasant thing I have enjoyed thus far is the anticipation, that a woman in the Nursery Car where I was, who promised each of her six children a sound whipping, will be as good as her word."[61]

In Chicago Marsh took solace in seeing his old friends William B. Ogden, now one of the Midwest's wealthiest men, and the portrait painter G. P. A. Healy, newly resident there under Ogden's patronage. About the city itself Marsh was caustic. Although the inhabitants had done "everything possible to combat the natural disadvantages of the locality," it was not enough. Marsh compared Chicago unfavorably with the Levantine swamp where he and Caroline had been so ill: "only Migdol has no city, which is one advantage, and has mountains in sight, which is another. As for the inhabitants . . . I think them somewhat alike, though the stray Bedouins are more *picturesque* than Chicago travellers." But he was stuck there, for "the prairies, railroad tracks and all, were under water—[a] state of things as exceptional as the rains in Styria—an exception that occurs every year."[62]

Chicago to Galena on the Mississippi, 150 miles away, took Marsh's train twenty-nine hours:

We were obliged to cross on the ice, or temporary foot bridges, or ferries, five streams where the bridges had been carried away by the freshet, or broken down by trains; were two hours in freight cars without light, fire, or seats . . . and at one time stood on the ice two hours waiting for a ferry boat to take over 200 passengers, at 8 or 9 a trip. The weather has been excessively cold, and many of the passengers froze their ears, and some their fingers and feet, while on the ice. On the whole, it was the most laborious and disagreeable journey I ever made.

Marsh neither ate nor slept during the entire trip. Only "virtuous indignation at my own stupidity, in coming out here," kept him going.

He reached St. Louis at three in the morning on February 15, after a week with "*two nights in bed*, [the] rest in cars or at depots, *travel*

600 miles, *receipts* $65, which does not leave a large balance in my favor." With his sister- and brother-in-law Lucy and Fred Wislizenus he talked nonstop for two days, "like some radiant comet lighting up for a space our Stygian gloom." Also in St. Louis were Marsh's old legal friend Charles D. Drake; Adolf Biewend, the Lutheran pastor of Washington days, now at Concordia College; and Orientalist Gustav Seyffarth from Leipzig, more refined "than most learned Deutschers I have seen." Marsh was made Fellow of the St. Louis Archaeological Society, "an easy & honourable position, duties nix, emoluments ditto, . . . nothing to pay but my postage. Truly, I am a Hans in Glück!"

But the lecture circuit once more submerged him. Marsh found "the mud of Bloomington [Illinois] deeper and blacker than any East of it, but far behind that of Davenport in profundity":

We reached Mendota at 4 in the morning, waited 4 hours, and were then treated to a ride of 30 miles in a freight car. After this, was a hiatus of a mile, and no conveyance between. I could not hire anyone to carry my trunk, so I passed my hand through the handles, and swung it over my shoulders and commenced to march. I lost my india rubbers at the first plunge, but recovered them, and waded to the other depot. I then went twenty miles on a locomotive, then 35 in a freight car.

But he was too late for his Davenport lecture. In twelve days away from Chicago, Marsh "travelled 1200 miles, and received $115, which, after deducting expense, wear & tear, & loss on wildcat money received in Iowa, won't leave me more than twelve dollars. Still, I have seen a great deal of *perary*." He reckoned were he "to go directly home now, I should reach Burlington with less money than I started, but have I not seen the Great West?"[63]

The return trip was no better. "Another month of squalling babies would extinguish me," he wrote Caroline. The "fine generous people" of Kalamazoo gave Marsh $22 for a lecture, "& even their fair words won't tempt me among them again. In fact I have made up my mind after this trip, to give up the thankless task of enlightening the world, & I shall let wisdom die with me, an' it will."

If the trip put nothing in Marsh's pocket, it added fuel to two of his pet hates—the railroads and the West. The wide empty prairie repelled him; "the impression is striking, but so simple that, once made, it lasts forever, and there is no novelty thereafter" save seasonal changes in the vegetation. Lucy Wislizenus feared that her brother-in-law had taken "such a *disgust* at the entire West that he will never venture out among us again." She was right. "Not all the gophers in Illinois," Marsh wrote Baird, "would tempt me to dwell in this famed West, even were I as desperate a naturalist as you."[64] With all its handicaps, Vermont was not so bad as this.

9. Vermont Public Servant

The changes . . . wrought in the physical geography of Vermont,
within a single generation, are too striking to have escaped the attention
of any observing person, and every middle-aged man who revisits his
birth-place after a few years of absence, looks upon another landscape.

Marsh, *Address . . . Agricultural Soc. Rutland County,* 1847

AFTER HIS GRUELING WESTERN JOURNEY, MARSH
returned to Burlington to be "grievously tormented of Satan incarnate
in Pecks." Family problems taxed him further. His son George, happily
reconciled during their Old World stay and now studying law at Harvard, again came down with typhoid at Woodstock. Near death for a
month, he never fully recovered, became a heavy drinker, and hatched
delusions of parental persecution. His worried father wished George
"well married to some wise & prudent woman, who would govern & direct him." [1]

When not at his son's bedside in July and August 1857, Marsh inspected damage caused by spring floods that had burst the Quechee milldam and filled the channel below with sand (fortunately, the river-banks and streambed did not seriously erode, as he had feared).[2] He also recalculated the altitudes of Mount Tom and other Woodstock hills with his younger brother Charles, using barometric techniques the geographer Arnold Guyot had shown him. (He would revisit Mount Tom for the last time in February 1860, treating Caroline to a dazzling day's ride on an ox-drawn sleigh.)[3] But the 1857 winter's lecture trip and his son's summer crisis had left Marsh exhausted. In November he caught a severe cold; neuralgia recurred, and for two months he was seriously ill. Dr. Leonard Marsh prescribed a change of air; Marsh went for several weeks to Boston and New York. Away from Pecks and Dainese, he regained health and spirits in congenial company.

Just when all Marsh's private projects were foundering, a modest stopgap emerged from a public quarter. In November 1857 he was appointed Vermont Railroad Commissioner—at the grand salary of $1,000 a year. "Among my many blanks I have drawn one small prize," he wrote Baird. "Nothing could come in better time, and I am by no means disposed to look this gift horse in the mouth."[4] Governor Ryland Fletcher had already made Marsh Vermont's Fish Commissioner, State House Commissioner, and head of an Ethan Allen Monument Committee, but those posts were wholly or virtually unpaid.

The fish assignment was the most conceptually fruitful of these odd jobs. Fishing in Vermont was not what it had been when, as a boy, Marsh noted aquatic changes in the Quechee and other Connecticut River tributaries. Vermonters bemoaned both the loss of salmon caused by building dams and the extinction of trout by the voracious pickerel that early settlers had introduced. Sporadic efforts had been made since the 1820s to restock inland fisheries,[5] and every year the state passed acts to safeguard particular lakes and streams, but even when enforced, such protection was not enough to sustain fish numbers. Marsh was now asked to survey replenishment prospects. He appended papers by European experts on fish-breeding and kindred studies from southern New England to his own fifteen-page commissioner's report.

While fish breeding had recently seen laudable advances, Marsh warned that it was by no means a panacea and often had injurious repercussions. For instance, indiscriminate damming of streams for fisheries "seriously impeded the drainage of the soil [by] obstructing and diverting the natural flow of water." Marsh had seen fish ponds at Sicilian monasteries whose construction had converted adjoining land into "barren and pestilential wastes." Careful engineering and water control were essential to curb injurious side effects.

In general, though, Marsh deemed Vermont an auspicious locale for breeding both indigenous and exotic species. For a start he proposed restocking Lake Champlain with the shad, salmon, and trout "which formerly furnished so acceptable a luxury to the rich, and so cheap a nutrient to the poor of Western Vermont, but which now are become almost as . . . extinct as the game that once enlivened our forests."[6]

Marsh blamed this decline on several human agencies. One was blind greed, notably "the improvidence of fishermen in taking [fish] at the spawning season." Industrial growth was another cause: dams athwart rivers obstructed seasonal fish migration, and sawdust and chemical refuse poisoned fish habitats. But the gravest effects Marsh laid to deforestation. The wholesale felling of trees increased runoff, exacerbated fluctuations in stream flow ("spring and autumnal freshets are more violent, the volume of water in the dry season is less"), and depleted insect life, a major food source for fish larvae.

The clearing of woods [and] the removal of many obstacles to the flow of water [induce] more rapid drainage . . . Rain and snow find their way more quickly to the channels of the brooks [which] run with a swifter current; streams formerly clear, gentle, and equable, are now dry or nearly so in the summer, but turbid with mud and swollen to the size of a river after heavy rains or sudden thaws; *torrential* water courses [suffer] great changes in the configuration of their beds.

Inundations swept away fish and their eggs, filled channels with mud, and deranged stream banks, "render[ing] it impossible for fish to return to their breeding place to deposit their spawn." In sum, supposed "human *improvements* have produced an almost total change in all the external conditions of piscatorial life."

Like Marsh's agricultural talks, his fish report showed him alert to the interplay of plant and animal habitats, and alarmed by human impairment of their intricate linkages. But Marsh was no out-and-out protectionist, no sentimental foe to all development. He knew the abuses he deplored could at best be mitigated: "We cannot destroy our dams; . . . we cannot wholly prevent the discharge [of noxious effluent]; we cannot check the violence of our freshets or restore the flow of our brooks in the dry season; and we cannot repeal or modify the laws . . . of nature" governing the food chain. The most restrictive codes could never "restore the ancient abundance of our public fisheries." The most extensive fish-breeding schemes could do so only in part. But without *some* public curbs, Marsh warned that Vermont risked losing not only fish but forests and topsoil.[7]

The legislature lauded Marsh's "noteworthy" report but found it "inexpedient" to act upon it. Relevant action came fifteen years later at the national level, spurred by a U.S. Commissioner of Fish and Fisheries report. It was no coincidence that this commissioner was Marsh's Smithsonian protégé Spencer Baird, who owed not only his insights but his appointment to Marsh, through Marsh's nephew Senator George F. Edmunds. Baird's 1872–73 report credited Marsh with initiating salmon restoration in the United States and restated Marsh's conservation precepts. Baird's memoirist in this domain judges Marsh's fish report "one of the most influential, thoughtful, and prophetic studies ever written on the subject."[8] As in so many environmental realms, Marsh had played a truly pioneering role.

His ichthyological labors netted Marsh just one hundred dollars, much of it splurged on an aquarium, where he watched his own fish "chaw one another up most catawampously." Intellectually the investigation proved more rewarding. It inspired Marsh to ponder anew the complex, sometimes bizarre, often self-defeating effects of human enterprise on the economy of nature. For example, the extirpation of voracious seals and otters by fishermen had multiplied fish populations in many locales. But this "by no means compensated [wasteful man's] own greater destructiveness"; Marsh estimated that losses in netting, curing, and transport exceeded the total numbers of fish consumed as food and used for fertilizer.[9]

Himself an avid angler in youth, Marsh urged restocking not just for food and profit but for recreation and fortitude. He lamented that Vermonters had lost pioneering self-reliance along with their fish and game. In this he presaged Theodore Roosevelt and later devotees of muscular engagement with nature, of hunting as "the moral equivalent of war."[10] "The people of New-England," Marsh charged, had become "more effeminate, and less bold and spirited. We have notoriously less physical hardihood and endurance." Why was this so? Because we "are suffering, both physically and morally, from a too close and absorbing attention to pecuniary interests."

As Marsh saw it, "the courage and self-reliance, the half-military spirit" of early Vermonters (like that of English landed gentry) had been instilled by the dexterity and invigoration of the chase. But these traits had succumbed along with moose, deer, catamount, wolf, lynx, and the vast flocks of pigeons and waterfowl now almost unknown to Vermonters. The constraints of civilization precluded replenishing these birds and beasts. On the land, "we must . . . accept nature in the shorn and crippled condition to which human progress has reduced her." But in the waters, restocking and breeding might revive abundance and promote angling as "an innocent and healthful recreation," nursing qualities of self-reliance Marsh felt essential to the survival of American liberty.[11]

Much of this paean to hunting—the faith in health through strength, the plaudit to a trained militia, the plug for England's hunting-shooting-fishing tradition, the revulsion against "effeminacy"—seems strangely regressive. Indeed, *Man and Nature* only a decade later is laden with censure of sporting slaughter as "most discreditable" cruelty. Yet these views reflected a deeply felt anxiety, common in Marsh's time, that heroic Founding Fathers had given way to unworthy sons in a lesser ignoble era, when character no longer counted and only lucre mattered.[12]

Such views also conformed with Marsh's enduring rural bent, stressed a year earlier (1856) in his New Hampshire talk. For Marsh, the Crimean debacle demonstrated "that the English, in ceasing to be a rural, had also ceased to be a martial nation, and that a manufacturing and a civic [city] people have neither the moral nor the physical qualities which made the British armies so formidable" against Napoleon. Indeed, urban growth at rural expense, "even in our young republic, has proved a token and a near precursor of national decrepitude and decay." City life was profoundly unnatural: "man's true strength of mind and body [are best] evolved, trained and perfected in modes of life akin to that which God prescribed to our first parents when he made them tillers of the ground." [13]

Yet even those who tilled the soil were now apt to be infected by bias born of urbanite ignorance. In rural Vermont, ancestral familiarity with nature had given way "to a mistaken prejudice which often ascribes mischievous propensities to particular birds, quadrupeds and reptiles." In his fish report Marsh spelled out why it was wrong to stigmatize such creatures: they "much more than compensate the little injury they inflict upon the crops" by consuming "vast numbers of noxious insects." Many birds and animals "popularly supposed to be destructive to grass and grain, in fact depend for their sustenance almost wholly upon insect life, and are accordingly useful as protectors, not injurious as destroyers, of the food of man." [14]

Nothing Marsh undertook during this decade was of greater import for his maturing environmental insights than this brief essay, ostensibly on Vermont's fisheries, actually on the whole fabric of multifarious nature.

Marsh's State House chore stemmed from a catastrophe. One January evening in 1857, with furnaces stoked for a meeting of Vermont's constitutional convention, Montpelier's twenty-year-old State House caught fire. Lacking pumps, townsfolk hastily packed the hall with snow, but to no avail; the building was gutted. Nothing survived but the portico and a granite shell. After a long debate on whether to relocate, the legislature opted to remain in Montpelier. Governor Fletcher asked Marsh to head a three-man commission to design a new State House. Marsh felt the $40,000 allocated too small, the two-week deadline for an initial plan too soon, but he was at length persuaded to take on the task. He met daily through March with fellow commissioners Norman Williams, Woodstock merchant and legislator, and John Porter of Hartford. [15]

The plan that emerged was Marsh's. He alone knew anything of architecture; he alone had definite views. "Both Judge Porter and myself desire you to exercise your own taste & judgment," Williams reaffirmed. "We shall adopt what you propose." The legislature had mandated keeping the classical form of Ammi B. Young's previous state house, with its much-loved Doric portico. But Marsh's new design enlarged the earlier structure, too cramped, as he put it, for either convenience or beauty. The building was extended by one bay at each end, the central bulk deepened, a cupola added beneath a heightened dome; it was to be steam heated, gas-lit, and as fireproof as economically feasible: no wood was to be used in the first floor, staircases, or partitions.

For want of time, the commission left details of finish and decor to the virtually self-appointed superintendent, Dr. Thomas E. Powers of Woodstock, ex-Speaker of the Vermont House, and to Marsh's chosen architect, young Thomas W. Silloway of Boston, former assistant to Ammi Young.[16] Marsh worked closely with Silloway in drawing up specifications but overruled several of his decorative proposals. In particular, Marsh insisted that the pediments and portico be left unadorned.

By the time construction halted for the winter in November, the walls, partitions, floorings, and columns were nearly complete. But before work began again in spring 1858, a storm blew up between Powers in Montpelier and Silloway in Boston. Rubbishing the architect's detailed designs as "like Hebrew to me," the overbearing Powers economized by using cheap materials (including fire-damaged stone and brick from the old building), sought to flatten the dome, and aimed to build staircases of wood instead of iron. Alarmed, Silloway pressed to be allowed to supervise construction; Powers told the architect his presence would not be worth the expense of his coming to Montpelier. Given Powers's ignorance of roof trusses and other matters, a subsequent inquiry judged this "an economy so intensified as to become equivalent to extravagance."

In early March the frustrated Silloway resigned. Powers at once hired a more subservient architect, Joseph R. Richards. Aghast, Marsh bade Silloway go to Montpelier to rescind his resignation; Powers paid no heed. By April Marsh, Silloway, Richards, and Powers were all on site at cross-purposes. Silloway's dismay at Powers's follies is detailed in fifty long letters to Marsh. Although Marsh insisted Powers should do "*exactly* what the commissioners desired," Powers scrapped their intended truss work and stairwells and, instead of replacing the bottom drums of the damaged portico columns, simply patched them. The legislature's Buildings Committee chastised Powers but deemed it too late and too costly to reverse most of his errors.

Despite the imbroglio, the building was ready for occupancy in just two years, by October 1859. It had cost less than the $150,000 allocated and subscribed (Montpelier residents bore part of the burden). This was most economical; Ohio's state house of the same vintage took twenty years to build and cost ten times as much.[17] Marsh's vigilance had "rendered the state a good service," Silloway attested; "nothing but devotion to the cause on your part has arrested the evil." In return, Marsh sought with little success to give due credit to Silloway, the "real architect."[18]

Vermont's 1850s State House still stands, perhaps not America's "finest specimen of Greek architecture," as the architect Stanford White wrote, but a worthy example of neoclassicism and a handsome monument to Marsh's informed taste. Happily, Powers failed to scuttle the wooden statue representing *Agriculture* that Marsh commissioned from Larkin Mead to crown the dome, where it remained for eighty years until dry rot forced its replacement by a copy. Likewise by Mead, a marble statue of *Ethan Allen*, shown demanding the surrender of Fort Ticonderoga "in the name of the Great Jehovah and the Continental Congress!" would soon stand on the State House portico. "You see," Marsh crowed to Hiram Powers, "art is progressive in Vermont."[19]

Art was moving ahead in Washington too. Indeed, Marsh felt more strongly about the look of the federal Capitol than he did about the Vermont State House. Under the aegis of Captain Montgomery C. Meigs, supervising engineer of the Capitol extension, the entire building was in the 1850s embellished and "improved." Meigs aimed to rival the Parthenon.

Congressman Marsh in the late 1840s had turned a mordant eye on several painters later to win Capitol commissions, notably Emanuel Leutze, whose "libellous picture of the Iconoclast Puritans" Marsh refused even to view, having "read so many *printed* lies about the Puritans that he did not care to see a *painted* lie." Likewise scorning his friend

G. P. A. Healy's "insipid" historical canvases, Marsh thought "Healy must confine himself to portraits. He will succeed in nothing else."[20] But he liked Robert W. Weir's *Embarcation of the Pilgrims at Delft Haven, Holland* (1843), which was "at first placed in the worst light in the Rotunda," Marsh wrote Lanman in 1847, "but a very judicious exchange was made . . . between Weir's painting, and that poor bald daub of [J. G.] Chapman's [*The Baptism of Pocahontas at Jamestown*, 1840], by which both were great gainers,—Weir's being seen to much better advantage, & Chapman's no longer in danger of being seen at all."[21]

In 1855 Marsh turned his critical gaze on Washington's sculptural landscape. Since "we are decorating the capitol & our other public buildings on a scale of great magnificence," Marsh urged Hiram Powers to come back from Florence and reactivate the dormant commission for his classic figure *America*, which would make him "*the* Government sculptor at once!"

One obstacle to Marsh's hopes for Powers was his prolonged absence from America. As Francis Markoe warned, Powers had been "*out of sight* so long, he is now almost *out of mind*. Half his countrymen believe Powers an idea! a myth!! they think of him as dimly & indistinctly as of Phidias or Lysippus and query whether he didn't make the Venus de Medici as well as the Greek Slave. If he will take the trouble to come on here while Congress is in session, he will carry all before him."[22] A more serious impediment was Powers's presumed antislavery stance, apparent in his enchained *Greek Slave*. Southern sympathizers had cause to suspect Powers's *America* might be decked out with abolitionist subversion. For this reason presidents Pierce and Buchanan delayed finalizing Powers's commission, demanding photographs and testimonials. In fact, Powers had been persuaded by Edward Everett to eliminate the Liberty Cap in *America*'s uplifted left hand. But the sculptor then substituted an even more overt emblem of emancipation, manacles beneath the left foot.[23]

Another problem was Powers's jealousy of Thomas Crawford, whose 19½-foot-tall *Freedom* was to decorate the Capitol dome. Powers felt he deserved to have a no less colossal statue in a prominent place. Marsh demurred. "Heaven save us from Brobdingnagianism in art! The heroic size is large enough, and I would have it in short measure, at that."

Moreover, at a height of 300 feet Crawford's statue would scarcely be visible; "a work of art might as well be in the capitol well as on its highest pinnacle."[24]

For much of what *was* prominent in Washington Marsh had a cosmopolitan connoisseur's contempt. Horatio Greenough's *Rescue* group for the Capitol Rotunda portico pleased Marsh "even less than when I saw it in an unfinished state in G's studio, but [it had] the advantage of making [Luigi] Persico's [facing] abomination yet more abominable." Blatant expansionist emblems, Persico's *Discovery of America* (1844) and Greenough's *The Rescue* (1853) became favored backdrops for antebellum presidential inaugurals. (These sculptures, deeply offensive to Indians, are now consigned to oblivion in storage.) Randolph Rogers's bronze *Columbus Doors* (installed at the Capitol in 1863) were meant by Meigs to emulate Lorenzo Ghiberti's at San Giovanni in Florence. But Marsh, familiar with Rogers's work, had been "too long an observer of art, to believe that even he can do as much in 40 days as *Ghiberti* in 40 years."[25]

Facing the White House was Clark Mills's equestrian Andrew Jackson, the first large bronze cast in America. "It was 'creditable' to the artist," Marsh agreed with Edward Everett, but only "considering his utter want of professional training, experience and observation." Mills had never even seen an equestrian statue, and he had "no conception of dignity, grace or proportion," in Marsh's view. "The *horse* has been a good deal complimented as a *spirited* figure," but it was "generally conceded," even by congressmen, that "the rider is abominable." How then had Mills got the commission? "The fact that he was a South Carolinian . . . would have turned the scale in his favor, even if Lysippus had been his competitor." But Marsh had no fear the excrescence would endure: "it will not stand. So large a mass on so slender a base cannot long resist the action of the wind, changes of temperature, oxidation of the iron supports in the legs, . . . & some heavy gale will overthrow it by the time we have a new hero ready for the pedestal."[26] Yet in defiance of Marsh's expertise, Mills's horse and rider still stand a century and a half later in all their naive exuberance.

The decay of commercial morality . . . is to be ascribed more to the influence of joint-stock banks and manufacturing and railway companies . . . than to any other one cause of demoralization. . . . Private corporations . . . may become most dangerous enemies to rational liberty, to the moral interests of the commonwealth, to the purity of legislation and of judicial action, and to the sacredness of private rights.

Marsh, *Man and Nature*, 1864[27]

*F*irst fish, then architecture, now railroads; Marsh delighted in being master of all trades. His exposé of the state's railroads was a minor classic of corporate malfeasance.

Railroad building in the late 1840s and early 1850s had attracted substantial foreign investment and encouraged an agricultural boom, but the extension of credit and the taking of quick profits also had ruinous repercussions. As elsewhere, Vermont's railroads proved a mixed blessing. The Vermont Central–Rutland & Burlington dispute and ensuing bankruptcy led to calls for state intervention. A governor's commission in 1855 found corporate cupidity, fraud, and reckless rivalry rife, and advised that public safety "imperatively require[s] some efficient power of constant inspection and control." A state commissioner should monitor accounts, regulate rates, and notify if "any railroad corporation has exceeded its legal powers or . . . incurred a forfeiture of its franchises."[28]

Railroad interests expressed outrage. The proposed "extraordinary powers [might] be executed with most summary despotism, to the detriment and even destruction" of Vermont's railroads. "Such extreme and odious legislation" would betoken "a deep-seated and inveterate hostility . . . to any further investments in railroads"; capitalists would thenceforth shun Vermont.[29]

Notwithstanding the railroad lobby, the commission came into being, but the reform candidate, Marsh's old crony George W. Benedict, did not get the post. The Vermont Supreme Court instead chose Charles Linsley, a Vermont Central engineer, whose 1856 report found the railroads too impoverished to be forced to improve services. Railroaders were thus dismayed when in 1857 the state legislature canceled the Supreme Court's appointive power and replaced the compliant

Linsley with their inveterate foe, George P. Marsh, who clearly had it in for them. The Vermont Central had ravaged his real estate, bankrupted him, and prostrated Burlington. For years Marsh had traveled about the state finding fault with the railroads, suggesting state control, even ownership.[30] And he was just back from the Midwest, where bleak ordeal inflamed his animus against the corporate iron horse.

Commissioner Marsh was as critical as the railroads feared. He was also extremely thorough. In summer and autumn 1858 he rode every mile of track in Vermont (543 in all) at least twice. He posed the corporations stiff queries: How much were their lawyers paid? Had any officer or employee a personal interest in contracts? How were contract tenders weighed? Why were repairs delayed? Who got passes (railroad officials, convention-goers, clergymen, other "manifest objects of charity")? Were existing stopovers and transfers necessary? Although managers were "dilatory" in replying, by October Marsh had finished. His report might "bring a hornet's nest about my ears,"[31] but he had no mind to draw the stings.

Marsh's exposé of the "sorcery by which [railroad managers] turned corporate misfortune into individual gain" was a harbinger of Charles Francis Adams, Jr.'s 1871 muckraking classic, *Chapters of Erie*. Marsh termed the mishaps of Vermont's railroads the good luck of its citizens. If investors were ruined, the absence of rich stockholders was "eminently favorable to the independence, the impartiality, and the purity" of the state legislature. Vermont's railroad managers had funds sufficient "to serve as a temptation to private fraud and peculation," but not ample enough to corrupt public officials.

In the long run, held Marsh, even stockholders were fortunate that their shares were worth so little. Only too aware "that a still further decline was at all times probable," they would not be fooled into buying again. And their losses were a dire warning to the uninitiated. Lacking any temptation to speculate, Vermont might keep its virtue. "Thus far," Marsh reasoned, "we have escaped the enormous moral, political, and financial evils to which the almost universal corruption of great private corporations has elsewhere given birth."

Even in Vermont, to be sure, shareholders had sometimes got or given cushy posts and lucrative contracts at public cost. But no longer.

The bankrupt railroads were now "waifs and strays . . . subject to ab-solute disposal and control" for the public good. When private fran-chises became "nuisances or abuses," they ceased to be inviolable. Bankrupt lines must not be run for their managers' benefit; it was the legislature's "right and duty" to take them over. To that end, wholesale revision of state railroad law was imperative.

Turning to specifics, Marsh listed myriad grievances. Vermont's railroads were "all imperfect in construction," tracks badly built and un-safe, routes poorly planned—flagrantly defective compared with Euro-pean railways. These defects were habitually concealed. "Engineers and directors of railroads, with long grades above one hundred feet to the mile, have regularly sworn in their annual reports . . . that there were no grades exceeding half that elevation." Freight trains were appallingly slow—why should goods require seventy or eighty hours to reach New York or Boston when passengers got there in ten or twelve? Marsh insisted on a host of bridge and roadbed repairs and safety devices. Pas-senger complaints were equally well founded. Railroads opened ticket offices only five or ten minutes before trains left, yet charged extra for tickets bought on board. Passengers were often carried beyond their stations because conductors failed to announce stops.[32]

Most vexatious were transfers required between lines. Some rail-roads knowingly coerced passengers' choice of route. The most flagrant offender was the Vermont Central; originally chartered between Bur-lington and Windsor, it no longer operated through trains to either terminus. Travelers to and from Burlington had to change at Essex Junc-tion, seven miles northeast, where the Vermont Central connected with its subsidiary, the Vermont & Canada. From this desolate spot a spur line squirmed perilously along steep grades to meet the Rutland & Bur-lington line in Burlington. The Vermont Central aimed to wreck the Rutland by denying it an outlet north and west; the Burlington bypass, as shown in the previous chapter, had proved disastrous to the town and to Marsh personally.

Vermont Central's trustees contended "a change of cars at Essex is deemed necessary for the proper accommodation of the travelling pub-lic." This Marsh termed monstrous. "Whatever advantage may accrue to the Railroads, . . . the 'travelling public' is in no respect 'accommodated' by submitting to [this] annoyance." [33] Indeed, the Vermont Central rigged its schedule to make northbound Rutland line passengers wait nine hours or more. Essex Junction was infamous; E. J. Phelps's "Lay of the Lost Traveller" spoke for all Burlingtonians:

> With saddened face and battered hat
> And eye that told of bleak despair,
> On wooden bench the traveller sat,
> Cursing the fate that brought him there.
> "Nine hours," he cried, "we've lingered here,
> With thoughts intent on distant homes,
> Waiting for that delusive train
> That, always coming, never comes,
> Till weary, worn,
> Distressed, forlorn,
> And paralyzed in every function,
> I hope in Hell
> His soul may dwell
> Who first invented Essex Junction! [34]

Essex Junction dismally failed the charter obligation of a good Burlington connection, Marsh ruled. The Vermont Central must cease to mistreat those who preferred the Rutland or a steamboat route to Burlington. Failing a proper junction with the Rutland & Burlington, the Vermont Central should forfeit its franchise. The legislature concurred, giving the railroad two years to rectify this "gross case of abuse." Thus mandated, the Vermont Central completed a route into Burlington across a ridge of windblown sand in May 1861. But landslides episodically put the spur line out of action. More and more fill was brought in to stabilize the roadbed, but the extension was totally abandoned in spring 1863. [35]

Corporate pressure on Vermont politics remained more intense than even Marsh had supposed. "There has never been a time," wrote an assemblyman, "that rail road influences were brought to bear so directly on members" as after Marsh's 1858 report. The bill embodying Marsh's proposals and enlarging the scope of his powers died in the Committee on Roads, made up of "creatures of the Rail Roads." The railroad lobby staved off his reappointment as commissioner until the third ballot. They then slashed the salary from $1,000 to $500, hoping Marsh would resign.[36]

But fighting corporate interests was for Marsh a labor of love. "My anathema on the scoundrel railroaders," he later wrote, "whom I loathe with all my soul." He exempted no one. "All Trustees, Presidents, Directors, Lessees, Conductors, Engineers, and Stokers of Railways are presumably rogues, and when they are charged with thievery, they can be taken as prima facie guilty. . . . A gallows-car should be attached to every train, and a director hanged at each trip." His second report chastised the Assembly for ditching the urgent reforms mandated by his first. State control would benefit everyone "except peculators and speculators." But the legislature hung back, and reform hopes expired with Marsh's commissionership in 1859. His successor, loyal to the lobby that put him in office, located "a community of interest between the public and the railroad companies" and rejected "faultfinding" as no part of his job.[37]

Despite dread of being overgoverned—a common American shibboleth—Marsh became convinced that all transport and communications should be government owned and operated. He well knew the perils of bureaucracy but argued that "the corruption thus engendered, foul as it is, does not strike so deep as the rottenness of private corporations." The depravity born of government patronage was "less evil than the wide-spread demoralization and the vast amount of private ruin and misery" that resulted from corporate greed and stockjobbing. Securities speculation "converted many mercantile communities into hordes of plunderers," causing each to fear or loathe his neighbor. Unless public responsibility curbed private avarice, all would suffer.[38]

Equally pernicious was the power of trade in shaping state policy. "Everywhere the implacable foe of conscience and honor and generosity," wrote Marsh in 1859, unbridled capitalism was "selfish, grasping, grovelling, unjust, whether under the harshest despotism or the most enlightened democracy, whether battening on the life-blood of colliers and factory-children, as in England, or pampered with sugar and poisoned with cotton, as in Boston and New York." So rapacious were Vermont's corporate interests that during the cold winter of 1857, when the poor in Burlington suffered severely for want of fuel, railroad managers would not transport firewood already felled in the forests lest the new market demand increase the price of locomotive fuel.[39]

The gigantic frauds of the post-Civil War era, when business interests corrupted every branch of government, fortified Marsh's hostility toward corporations in general and railroads in particular. "Joint-stock companies have no souls; their managers . . . no consciences," he declared in *Man and Nature.* "In their public statements falsehood is the rule, truth the exception." Ten years later a niece of Marsh's died in a shipwreck off the Scilly Isles, Britain's Lighthouse Board having failed to provide telegraph service there "because 'it would not pay.'" Marsh concluded that "the great companies of transportation in the United States, in England and in France, are far more powerful than the governments or the people. And they defy all control, but that of the purse."[40]

Marsh's corporate animus was not only consonant with his mandates for conserving resources, but causally related to them. In both arenas he saw private interests endangering public welfare; in both he opted, as a pragmatist, for public control rather than private reformation. Since narrow self-interest so easily thwarted or throttled stewardship, the state must serve as custodian of the common weal.

10. The English Language

It is a rash thing for a Yankee man to attempt to teach English,
but we are a bold people, and I am resolved to venture it.

Marsh to Caroline Estcourt, June 3, 1859

EARLY IN 1858, MARSH ARRANGED TO LECTURE ON
English language and literature the following winter at Columbia College in New York City. Unlike the teaching jobs he turned down, this
was a short-term commitment; the base pay was $1,500, and Columbia
trustee Samuel B. Ruggles encouragingly added that "the course can
hardly fail to yield the Professor a considerable amount in fees."[1]

Over the next months Marsh snatched moments from public tasks
and private perplexities to work up lectures. By October he had reviewed enough English, German, Catalan, and Italian material to keep
him a few jumps ahead of his students. Clearing out his study with
"a holocaust of all the old medicines, old papers, old shoes and other
rubbish I could find,"[2] he settled in a boarding house at 22 University
Place, near Ninth Street, and started teaching.

197

Marsh began with "An Apology for the Study of English," a plea for serious inquiry into what most dismissed as a nursery topic. Addressed "to the many, not to the few," Marsh's lectures were a grab bag of miscellanies on the origins of speech, the uses of etymology, the structure and vocabulary of English compared with other tongues. He dwelt on inflections, the influence of the printing press, rhyme, linguistic corruption, American English, and the English Bible—for him the language's supreme masterpiece.

Most of these topics were then utterly novel. English had no place in any American college curriculum, save at Lafayette, where it was taught by Francis A. March like an exercise in Greek or Latin grammar. The view that English was unworthy of study, though deplored by Marsh's Harvard colleagues Francis Child and James Russell Lowell, persisted throughout the nineteenth century.[3]

Marsh found preparing without sources at hand "making bricks without straw"; he was reduced to "evolving my facts out of the depths of my own consciousness." But lack of documentation helped make his lectures lively. The course went well, though the class was smaller and his income less than Marsh had hoped. The German-born political economist Francis Lieber, also new at Columbia, faced a similar shortfall. "My dear Fellow-Lecturer to many benches and few hearers," Lieber wrote Marsh, "had we not better turn Buddhists at once and lecture, in deep meditation, to ourselves, on ourselves, and by ourselves?" Lieber blamed the meager turnouts on their refusal to entertain; both men "loathed that curse of American oratory . . . pityful things to make the audience laugh." Marsh dropped all thought of staying a second year. "The course does not *pay*," he wrote flatly; "a business which yields no profit, ought to be given up."[4]

The winter was a therapeutic and social if not a financial success. Caroline was better; she could walk a quarter of a mile without pain, and Marsh envisioned waltzing with her "to the admiration of the beholders." New York was intellectually stimulating. He relished the geographer Arnold Guyot and the geologist James Dwight Dana, both at Columbia that winter. Through Raymond of the *Times* Marsh met such litterateurs as William Cullen Bryant, Richard Grant White, and Vincenzo Botta, Italian refugee scholar at New York University. Marsh renewed contact with Caroline Paine, now married in New York, and Charlotte Bostwick, who had found a husband after all, a Brooklyn clergyman. He saw much of the fiery Polish exile Adam Gurowski and the Swedish philologist Maximilian Schele de Vere, who "listened very well to what I said, though a capital talker himself."[5]

More than anyone, the ebullient Francis Lieber made New York congenial. Marsh found Lieber *"so subjective* in conversation, that he doesn't give one a fair chance to understand him," but the two shared a similar wry humor and wide-ranging passions and prejudices, literary and political. Lieber begged Marsh to stay on as Columbia's librarian "so that we might 'Beaumont & Fletcher' it," but Marsh felt the previous librarian, Joseph Green Cogswell, had "set a bad example, worked too hard and too cheap." Marsh craved "small duties & large pay."[6]

To publish his Columbia lectures he spent months tracing sources and checking references. For the ever thorough Marsh, the Boston and New York libraries were dauntingly voluminous. In July 1859 he returned to Burlington "like an escaped convict to his cell," grateful only that he had "too much work to do, to dwell on the *Widerwärtigkeiten* [adversities] of our position there." Besides coping with lectures and litigants, Marsh ground out his second railroad report, some "slovenly and bitter" pieces for Edward Everett Hale's *Christian Examiner*, and two poems for Abby Maria Hemenway's Vermont anthology.

All this kept him "a writing and a writing, with all my might, for fear the printer's devil, now full 300 pp. behind me, will *emporte* [overtake] me." Gone were the carefree days when he and Baird used to "scribble about nothing . . . before we came to be a couple of such sapless dry old sticks as time, trouble, and Satan have made us." Toiling day and night, Marsh did not "exchange a reasonable word with anybody once a month. . . . Whereas I used to be a conversable and . . . witty person, I am grown the dullest old owl in Christendom."[7] As usual in Burlington he missed both the books and the social life of Boston and New York, and "wished the road ran the other way." His Gotham boarding house had been far from dull:

The chairs in the dining room are mounted on casters as fickle as the wheel of fortune, and we had a melancholy exhibition of their instability today, at dinner. A fair young lady who sat opposite me, in bowing an acknowledgement of a compliment from the gentleman who attended her, leaned too far forward, & her chair shot out from under her. She disappeared beneath the table, & I . . . thought only of Mrs. Radcliffe's trap-doors and cellars full of grinning skeletons. She however, recovered her position, made a light dinner, & through the repast, looked red enough without rouge.[8]

Printing was done by October for both Marsh's *Lectures on the English Language* and Caroline's *Wolfe of the Knoll*, poems inspired by her German reading and travels in the Levant, but Scribner's held both books over until spring 1860. Marsh was enraged; the delay "defeats one of the principal motives I had in publishing. . . . If I were to hear tomorrow that the plates of my volume were destroyed, I should be rather glad than sorry." At length he conceded that since the world "has already waited 60 centuries for us, [it] will probably hold out a couple of months yet."[9]

"The appearance of Marsh's book," Lieber exulted, "big, substantial, wholly excellent—together with his wife's poetry—small, aethereal, wholly charming—is one of the finest literary phenomena." In ice-bound Vermont the Marshes were warmed by New York and Boston reviews. *Wolfe of the Knoll* was uniformly praised; even the pompous London *Athenaeum* allowed that this was no ordinary transatlantic drivel, "a third-rate Germanism," but on a par with "our own romantic poetesses."[10]

Marsh, too, fared well. Acclaiming "a man of profound and varied scholarship" and "a philologist neither perversely wrong-headed nor the victim of preconceived theory," reviewers touted his *Lectures* as "the best guide we know of" and "one of the most valuable contributions to American literature" ever made. Though censuring Marsh's "heroic attempt to defend American pronunciation," the English *Critic* lauded "the most perfect philological treatise" on English yet produced. Another English reviewer found Marsh "Teutonic enough in theory, [but] he Latinizes terribly in practice . . . and is carried away in a mass of long and hardly intelligible words."[11] This was perceptive and just; Marsh often forgot his own precepts. Yet whatever its defects, *Lectures* was generally lauded, widely cited, and extensively reprinted (and pirated); it was reissued in 1997 as a seminal text in American linguistics.[12]

Marsh held his *Lectures* overpraised. He professed to be amazed he had written it. Had he suspected eighteen months before that he would be "the author of an actual, printed book in 8vo of 700pp. the 'English Language' would have been one of the last things that would have occurred to me as the probable subject." The lectures were "an accident growing out of another accident."[13] He characteristically dismissed each of his works on completion as an unintended and trifling diversion.

Marsh's English inquiries were less fortuitous than he professed. The subject was much in the air in the late 1850s; the philology of Richard Chevenix Trench, the ethnographic comparisons of R. G. Latham, the grammatical texts of W. C. Fowler, the Sanskrit studies of William Dwight Whitney spurred converging interests in the shape and roots of English. Marsh's lectures, grounded in matchless familiarity with European languages and literatures, at once made him an authority; he was inundated with requests for collaboration. George Ticknor found Marsh's remarks on stylistic revision hugely helpful. George Bancroft acknowledged "a strong desire to write English correctly . . . by the aid of the severe judgments of the master"; Marsh read a draft volume of Bancroft's *History of the United States* with "critical and hypercritical care." This was revenge for Bancroft's corrections of Marsh's own mistakes, notably in attributing "the well of English undefiled" (a reference to Chaucer) to Milton instead of Spenser. "T'wasn't a blunder," confessed Marsh. "It was a crime. I knew Spenser first said it [but] quoted the greater name to give the more weight to the expression."[14]

Thanks to Marsh's imminent release from his creditors, his energy was once more boundless. He hoped to prepare a new edition of *Appleton's Cyclopaedia*. He aided Philadelphia librarian Samuel Allibone's *Critical Dictionary of English Literature* (3 vols., 1858–71). He traded ideas with Whitney and biblical scholar Josiah W. Gibbs at Yale and with the rhetorician Henry Coppée in Philadelphia.

Marsh's major intended venture was the London Philological Society's projected *New English Dictionary*. He welcomed the advent of "a more complete common thesaurus of the English tongue than now exists of any language, living or dead," as vital to joint Anglo-American concerns. As the Society's American secretary, Marsh was to select scholars to survey American and eighteenth-century English literature. The chosen readers were to send Marsh etymologies of all words, phrases, and idioms not in existing concordances.

Marsh was mortified when most he approached treated his invitation with "utter neglect"; he doubted he could find acceptable readers even for American authors, who were then viewed as unworthy of notice by most American scholars. But the untimely death in 1861 of editor-in-chief Herbert Coleridge shelved the whole project until James Murray took it over in 1879.[15] The *New English Dictionary* did however bring Marsh in contact with English philologists and a few cooperative Americans, most notably the art historian and Dante scholar Charles Eliot Norton, who became a close lifelong friend.

Marsh took a leading role in the notorious "War of the Dictionaries" that raged between the reformer Noah Webster and his followers and Joseph E. Worcester, American editor of Samuel Johnson's *Dictionary* (1828). Webster accused Worcester in 1830 of plagiarizing his own 1828 *American Dictionary;* Worcester and others charged Webster with linguistic corruption, William Cullen Bryant vowing to extirpate "every trace of Websterian spelling . . . from the land." The dispute embroiled America's leading literary lights—Worcester's backers largely Harvard-based linguistic conservatives, the Websterians radical Americanizers from Yale. While Marsh was "a philologist only by courtesy," in H. L. Mencken's phrase, it is far from true that "the regularly ordained schoolmasters were all against him." In fact, Marsh's backing was eagerly sought by both camps.[16]

Both Worcester's new *Dictionary of the English Language* and the Merriam brothers' revised and enlarged Webster came out in 1859. In extended reviews Marsh rated both dictionaries "far from perfect," but on most counts he much preferred Worcester. Webster's pronunciations confounded unlike sounds, such as the *a* in *fate* and *fare;* his "quackish" spelling reforms, aimed at forging an American tongue independent of England, Marsh dismissed as "stereotyped cacography." Worcester's vocabulary was fuller, his etymologies more accurate. Webster, ignorant of much English literature and most cognate tongues, had relied on speculative "internal" evidence instead of "reading and excerpting real live English books." Webster's notions of verbal affinity derived mostly from his reading of Scripture, for him the sole source of linguistic truth.[17]

The Merriams took Marsh's criticisms to heart, offering him an editorship (which Marsh had to decline) to rectify these defects. Their next (1864) edition acknowledged Marsh's "valuable suggestions [on] the principles which should be followed" in dictionary-making. Although Marsh again chided the Merriams for having "from sheer obstinacy, retained some of Webster's foolish spellings," he was pleased that "they threw his entire etymology (more absurd even than his orthography) overboard, & set [Karl August Friedrich] Mahn of Berlin to make a new one, which, though not perfect, is the best general etymologicon we have of our tongue."[18]

But by 1860 Marsh was editing another dictionary: he had been asked to revise Hensleigh Wedgwood's *Dictionary of English Etymology* for the American market. Ever obsessed by changing word uses, Marsh accepted with alacrity. Wedgwood's first volume, A through D, had just appeared (1857); this 500-page tome detailed the roots and careers of 2,000 selected words. Although Marsh rated it "the ablest work" to date on English derivations, he felt that Wedgwood, like Webster, depended too much on other dictionaries, not enough on actual literature. Marsh hounded friends for help in correcting and amplifying Wedgwood. Could H. A. Homes look out Turkish words? What about *bosh* (yes) and *derrick* (no)? Would James Russell Lowell check the *Laws of Oléron* (fourteenth-century Bordeaux maritime codes sent to London) for *buoy*? Harvard professor E. A. Sophocles, whose 1860 *Glossary of Later and Byzantine Greek* "hugely delighted" Marsh, confirmed his derivation of *carboy* (a demijohn) from the Persian, and of *qarabah* (wine vessel) from Arabic, rather than Turkish *karà boya*, black dye or vitriol, as Wedgwood had it. Henry Swan Dana of Woodstock, hired by Merriam-Webster to mine the works of Pliny, asked Marsh's help on *oss* (Arabo-Spanish *ojalà)*, *inshallah* (if God please), *stoupla* (stove), *parpine* (wall-binder), and *dromond* (a large medieval ship). With Wedgwood in press, by September the weary Marsh was finding etymologies "poor intellectual food" and regretting "too much time foolishly spent [on] a sour task."[19]

To Wedgwood, Marsh added "historical illustrations of the etymology of some 200 words" from old Scandinavian, Dutch, Catalan, and Hispano-Latin sources "in great part new to English etymology."[20] His capsule sketches of *average, awning, baggage, ballast, canoe, ceiling,* and *cheese* are wide-ranging, erudite nuggets. Pooh-poohing surmises based solely on similarity, Marsh time and again deploys history to reject Wedgwood's conclusions. Thus Wedgwood fancied a hewn beam was called a *balk* from its "resemblance in shape to a balk [unploughed ridge] in a ploughed field"; Marsh spurns this as "an inversion of the natural order of derivation, [since] beams for houses were in use, and must have had a name, long before the plough was invented." Wedgwood thought all cognates of *boot, bet,* and *better* "derived from the cry *bet,* used in setting dogs on their prey"; rejecting this "extraordinary theory," Marsh notes that "proof of the antiquity and wide diffusion of the alleged root [is] wholly wanting."

Frequent citations of a word proved little, for "there is always cause to suspect that one author has followed another without much critical investigation." Voyagers' testimony on language "is to be received with caution, both because of their general looseness and inaccuracy in all philological matters, and because it proves too much." For example, the word *canoe* was reported all the way from Amazonia to Canada, but if Columbus, Vespucci, and others offer "conclusive evidence" that *canoe* is American Indian, "precisely similar evidence, from other navigators of about the same period, will show that it is East-Indian also, which is not believed to be the case."[21]

Reviewers acclaimed the American Wedgwood. "The acute mind of Marsh, the peer of etymologists," had produced "one of the most fascinating books of the time . . . to be read through both for delight and instruction." But the $400 Marsh got for his labors did not cover the cost of his reference materials. He could not afford to carry on with the remaining volumes.

Yet he could not resist continuing. In 1865 Marsh published a twenty-page supplement to volumes 2 (1862) and 3 (1865), and he went on annotating Wedgwood, Mahn, and other etymologists, up to his last hours of life. Wedgwood adopted many of Marsh's corrections, some without credit, but balked at some of the Vermonter's cautions; in tracing *church* (Anglo-Saxon *cyrice*) from the Greek *kyrios* ("the Lord"), which Marsh felt "wants historical probability," Wedgwood thought "it is carrying scruples to an extravagant length to doubt the identity of two words, because we do not know how the Greek name came to be employed instead of the Latin equivalent *dominium.*"[22]

A late inquiry into *battledoor* recalls Marsh's shuttlecock games with Charlotte Bostwick. Refuting Webster, Marsh shows the name of the game had diverged from George Fox's "*Battle Door* for Teachers and Professors" (1650), deriving instead from a double gate made for wider and easier entrance. To those who might scoff at such "trivial observations on a trivial subject," Marsh stressed their high philological import. "The humble radical *bat*, . . . which had remained almost sterile for uncounted centuries . . . now counts a forest of descendants" spread throughout Anglophone and Romance realms. And "existing literary monuments enable us to trace the annals of this family of words with a [rare degree of] certainty and minuteness."[23] Marsh traced the history of words with the same zeal that he would trawl the history of woods.

That wild man up in Vermont is writing seven books, each in seven quarto volumes, all to be finished seven weeks from this date.

Marsh to S. F. Baird, May 10, 1860

Not all Marsh's labors were linguistic. He wrote on Italian affairs, essays germane to his ensuing diplomatic role. He gathered data on environmental change for what became *Man and Nature*. He sketched the life of his cousin James Marsh. He importuned *Burlington Sentinel* editor John Godfrey Saxe, whose "Proud Miss McBride" Marsh thought the funniest poem in English, to publish a "funny book" by Francis Markoe. To Markoe he relayed his own addiction to puns: "When I tried in vain to turn a screw with a shilling piece, I said, 'This won't do it. There's no *purchase* in money.' . . . When my sister-in-law ripped her dress in travelling I said 'Lucy, you rip where you didn't sew!' Pretty good, aren't they?" He contended that English friends unable to pronounce "Wislizenus" (Marsh's brother-in-law) had dubbed him "Herr Weiss-nicht-was." "I never make puns," Marsh pretended while making another. "I hate 'em, but I couldn't help this."[24]

But linguistic work predominated. Marsh was to speak on the origin and history of English at the Lowell Institute in Boston in winter 1860–61. Although "cabbaged, cribbed, confined by erysipelas," he worked fourteen hours a day to begin his lecture series in mid-November. Boston and Cambridge were more enticing than ever; there were Atlantic Club dinners, evenings with Everett and Ticknor, sessions with Bancroft, Lowell, and new friends like Norton and the ballad scholar Francis J. Child. But Marsh's eyes again failed. On December 19 he lectured against doctor's advice, and inflammation set in that night.[25] Holed up for the winter in far-off Burlington, he grumbled at the world and labored with amanuenses to finish Wedgwood and ready his Lowell lectures for the printer.

Marsh's *Origin and History of the English Language and of the Early Literature It Embodies* (1862) was more learned but less readable than his preceding volume. In tracing English from Anglo-Saxon through Elizabethan times, Marsh focused on grammar and word use in *Piers Plowman*, Wycliffe, Chaucer, and John Gower's *Confessio Amantis*. Peppered with Marsh's tenets on translations, dictionaries, and slang, his book built on the new evolutionary philology of Jacob Grimm, Franz Bopp, Rasmus Rask, and Max Müller, whose Indo-European linguistic insights were, Marsh regretted, still little known in the New World. Seen by some as the best extant history of English, *Origin and History* was much praised by Müller himself. But while *Lectures* was reprinted four times in two years, used as a text by Child at Harvard, and pirated in England, the public found *Origin and History* formidably abstruse, with its great gobbets of line-by-line comparisons of biblical and other texts in ancient Greek, Latin Vulgate, Moeso-Gothic, Anglo-Saxon, Wycliffian, and other variants. It was a book for scholars.[26]

In his own day, Marsh gained greatest renown as a philologist, and even now he is apt to be classified as such. More of his published work concerns language than any other topic. So far as he admitted to being a professional in anything, it was in the history of languages and literature. If his environmental work is today rated more pivotal, his linguistic expertise was phenomenal and often innovative. Moreover, the two realms had manifold links; analogies between the evolution of language and the evolution of nature were pervasive even before Darwin's *On the Origin of Species*. And like Darwin, Marsh made fertile use of the parallels between biological and linguistic development—although, as will be seen in Chapter 14 below, his conclusions differed from many of Darwin's. In particular, Darwin relied heavily on Wedgwood (who was his cousin and brother-in-law) for precisely those etymological genealogies and imitative parallels—"bow-wow" and "pooh-pooh" derivations that Marsh often rejected as historically unfounded.[27]

Never was Marsh's scholarship neutral. Partisan invective, often in tangential asides, is at once a weakness and a special charm of his books.

Thus he lamented that printing had cheapened not only the cost of books but also their quality, multiplying half-baked readers and low-brow writers eager to supply popular markets. "None seek the audience 'fit though few,' that contented the ambition of Milton." And popular taste tended to be not just banal but vicious:

The dialect of personal vituperation, the rhetoric of malice . . . the art of damning with faint praise, the sneer of contemptuous irony, the billingsgate of vulgar hate . . . are enough to make the fortune of any sharp, shallow, unprincipled journalist, who is content with the fame and the pelf, which the unscrupulous use of such accomplishments can hardly fail to secure.

Marsh aligned "purity of speech, like personal cleanliness, . . . with purity of thought and rectitude of action." Impure in his eyes were affected usages, Cockneyisms (*directly* instead of *after*), Americanisms (*community* stripped of "the"), vulgarisms (*in our midst, unbeknown to me*), preciosity (voicing the *t* in *often*), and provincialisms (*Ohiuh* for *Ohio*).[28]

Vogue usages Marsh severely rebuked. Like *brilliant* more recently, the word *lovely* was then "the one epithet of commendation in young ladies' seminaries and similar circles, where it . . . is applied indiscriminately to all pleasing material objects, from a piece of plumcake to a Gothic cathedral." Marsh blamed Ruskin for having "adopted this school-girl triviality, and, by the popularity of his writings, has made it almost universal, thereby degrading, vulgarizing, and depriving of its true significance, one of the noblest words in the English language." With officialese Marsh had no patience. "Tell the party that wrote that letter," he answered one from Baird, "that th[is] party . . . don't intend to deal any more with the party that wrote the letter; [this] party *does* know English enough not to call parties any persons except parties that call other people parties."[29]

Marsh saw each language encoding a particular culture, shaping and shaped by all its aspects. Grammar and vocabulary revealed marks of history and character unique to each people. "A foreigner, writing for foreigners, has a totally different set of ideas to express and a totally different mode of similar ideas." An English Goethe could never have written *Faust*, nor a Frenchman Shakespeare's plays. Hence translations were bound to be defective. One should strive for a translation that "transforms the *reader* into the likeness of those for whom the story, the ballad, or the ode, was first said or sung." Thus Marsh applauded Sigurd, apostle of eleventh-century Sweden, who substituted *cold* for *heat* in limning hell's torments.[30] To be frozen was for Swedes more hellish than to be fried.

Marsh still gauged language through an Anglophiliac lens. He lauded English as "the reflection of the waking life of an earnest, active nation, not, like so much of the contemporary expression of Continental genius, a magic mirror showing forth the unsubstantial dreams of an idle, luxurious, and fantastic people." Yet Marsh was no longer the fanatic Gothicist he had been. He liked to "throw stones at some of the current *idola tribus*, & particularly at the theory . . . that the English language, English literature, & English genius owe nothing to the influence of Norman-French speech, poetry, & blood." To the contrary, Marsh now esteemed the English as "the only Gothic tribe ever thoroughly imbued with the Romance culture, and at the same time interfused with southern blood, [with] the best social, moral, and intellectual energies of both families." The amalgamation made English "superior in power" to both Teutonic and Romance literature.[31]

Linguistic merit stemmed from political liberty; linguistic corruption from subjugation. Thus trans-Alpine tyranny had deformed Italian into a "base and hypocritical" tongue, by turn abject and pompous, its strength drained and its meaning perverted by cloying superlatives and diminutives:

A bold and manly and generous and truthful people certainly would not choose to say *umiliare una supplica*, to humiliate a supplication, for, to present a memorial; . . . to speak of taking human life by poison, not as a crime, but simply as a mode of facilitating death, *ajutare la morte*; to employ *pellegrino*, foreign, for admirable; . . . to call every house with a large door, *un palazzo*, a palace . . . ; an alteration in a picture, *un pentimento*, a repentance; a man of honor, *un uomo di garbo*, a well-dressed man; . . . or a message sent by a footman to his tailor, through a scullion, *una ambasciata*, an embassy.[32]

But this degradation was not incurable, and when Italians gained freedom Marsh trusted that "their speech, like themselves, will burst its fetters and become once more as grand and heroic as it is beautiful." Two decades on, Marsh noted many improvements, such as ridding

Italian of the extended circumlocutions of stereotyped greeting and leave-taking phrases formerly deemed essential to politesse.[33]

Marsh shared the common Anglophone preference for English words of Gothic origin over Latinate ones. But he refuted the historian Sharon Turner's canonical rank-order of Germanic over Romance word ratios (Shakespeare and Milton on top, Gibbon and Johnson far down), as based on woefully meager samples.[34]

He had no faith in pedants "who know languages only by vocabularies and paradigms"; wide reading of "real live books" was needed to trace how and understand why meanings changed. "We must know words not as abstract grammatical and logical quantities," he urged, "but as animated and social beings." To Ticknor he confessed his "instinctive dread of such melancholy osteologists as [R. G.] Latham and [W. C.] Fowler," in whose arid tomes he found nothing "but heaps of leached ashes, marrowless bones and empty clam-shells." As Marsh put it in *Origin and History*, "you may feed the human intellect on roots, stems, and endings, as you may keep a horse upon saw-dust; but you must add a little literature in the one case, a little meal in the other."[35]

Worse yet, "the ignorance of grammarians has done much to corrupt our language, the dullness of orthoepists to confuse its pronunciation." Experts who relied on brief phrases detached from textual and historical context shamefully misinterpreted familiar quotations. Thus Cicero's *errare malo cum Platone*, "it is better to be wrong with Plato," was often cited to exemplify excessive deference to authority. But anyone who took the trouble to read *Tusculanae Disputationes*, said Marsh, would see that Cicero and his pupil chose to accept not Plato's *authority*, but simply his *conclusion* that the soul was immortal, "preferring to share with him the beneficent possible error of eternal life, rather than the fearful and pernicious truth, if it were a truth, of his [Epicurean] opponents [who] denied the immortality of the soul." Theory and logic were no substitute for detailed evidence from complete sources, not mere snippets.

Two decades later Marsh complained of his old friend Max Müller's "unsatisfactory speculations" about etymology and language origins, while lauding W. W. Skeat's *Etymological Dictionary of the English Language* (1879–82) as historically sound.[36] Marsh remained committed to what Stephen Alter terms "the contingent and capricious realm of real history" as opposed to the hypothetical family trees posited by both philologists and Darwinian naturalists. He had more in common with later neogrammarians who urged linguists to "forsake the hypotheses-laden atmosphere of the workshop in which Indogermanic root-forms are forged, and come out into the clear light of tangible present-day reality" through the study of living languages.[37]

*H*ow mutable was language? How mutable *should* it be? Such issues engrossed nineteenth-century philologists. Tracing how dialects waxed and waned, Marsh realized that all languages were in flux. Along with the art of archery had vanished a huge related vocabulary, while new ideas, tools, and milieus ever bred new words and syntax. And all word meanings changed over time.

Some such changes were easier to trace than others. Certain words are carapaced by their past meanings: *exorbitant* patently derived from its now obsolete denotation of heavenly bodies that deviated from planetary orbits; the Smyrna peddler's cry of "Americani!" was not a plea to New World visitors but a term touting New World cotton goods once held peerless. In contrast, other current usages revealed no trace of earlier modes of parlance. Marsh instanced punctuation marks, unknown in ancient times, but now essential ocular crutches:

Beginning with air-bladders, we never learn to swim without them. Every parenthesis must have its landmarks, every turn of phrase its finger-post. We think by commas, semi-colons and periods, and the free movements of a Demosthenes or a Thucydides are as unlike the measured, balanced tread of a modern orator or historical narrator, as the flight of an eagle to the lock-step of a prison convict, or to the march of a well-drilled soldier, who can plant his foot only at the tap of a drum.

But only because most were unaware that the past had lacked punctuation did they suppose these marks eternal verities rather than recent contrivances. When anachronistically inserted into ancient texts (as with chapter and verse divisions of Scriptural narratives), modern punctuation often subverted meaning or spoiled rhetorical effect.[38]

The function of linguistic studies was "to teach what is, not what ought to be." But Marsh did not share Latham's view that "whatever is, is right." To see flux was one thing, to embrace it quite another. That change was essential to living tongues did not make it always desirable. Some changes were beneficial: thus the spread of printed books and journals had helped purify many European languages, promoting "one-ness of thought and oneness of speech."[39] National vocabularies were enhanced by finding new as well as by reviving old modes of expression. Marsh applauded neologisms like "outsider" (1844), whose politicized usage he was the first to remark. And he urged abandoning the untenable distinction between *shall* and *will*, a "verbal quibble" he hoped destined soon to vanish, its departure "not a corruption, but a rational improvement." Though then opposed by most authorities, Marsh's prediction on this point, as Mencken noted, proved accurate.[40]

But many losses diminished the force and beauty, many innovations were "repugnant to the genius of a language." Marsh lamented the passing of the Anglo-Saxon inseparable particles *van-*, *be-*, and *for-*. "What a losing bargain we made when we exchanged those beautiful words, *wanhope*, for despair, and *wantrust*, for jealousy or suspicion." Just as peoples might degenerate, so might languages. "To treat all its changes as normal, is to confound things as distinct as health and disease." Marsh exhorted that "corruptions of speech, [like] moral delinquencies, or vulgarisms of manner, . . . be exposed, stigmatized, and corrected."[41]

On balance he thought it best to stem change. Between the passion for novelty and attachment to tradition some compromise was needed, but "love of innovation is more dangerous, because the future is more uncertain than the past, and because the irreverent and thoughtless wantonness of an hour, may destroy that which only the slow or painful labor of years or of centuries can rebuild." And *rapid* linguistic change deprived people of their own literary heritage, rendered unintelligible to them by the too swift metamorphosis of their everyday tongue. Besides, linguistic conservatism accorded with public feeling. "The popular mind shrinks from new words, as from aliens not yet rightfully entitled to a place in our community, while antiquated and half-forgotten native vocables, like trusty friends returning after an absence so long that their features are but dimly remembered, are welcomed with doubled warmth."[42]

English had, after all, improved little since Elizabethan times, and further progress seemed to Marsh unlikely; it made good sense, then, to keep it more or less as it was. Corruptions willfully introduced should be weeded out. Marsh condemned the "newly popular" passive continuing present ("the house *is being built*" in place of "the house *is building*") as a solecism "originating not in the sound common sense of the people, but in the brain of some grammatical pretender. [This] awkward neologism, which neither convenience, intelligibility, nor syntactical congruity demands, . . . ought therefore to be discountenanced," as a misguided "attempt at artificial improvement."[43]

On this point Marsh has been shown wrong; the new passive voice was not artificial or deliberate, it was gradual, unprovoked, and unstoppable, like uses of *to be* and *to have* as auxiliary verbs. So total is the change that we now find it hard to see why the new usage grated on mid-nineteenth-century nerves. In chiding would-be reformers, Marsh failed to realize that he, a would-be conserver, was as powerless to stem language change as reformers were to induce it.[44]

More trenchant were Marsh's views on American usage. Just as he felt that New World democracy demanded a comprehensive history of everyday life, so did American versatility require all-embracing language competence. As jacks of all trades, Americans unlike Europeans needed to master all vocabularies. "Every man is a dabbler . . . in every knowledge. Every man is a divine, a statesman, a physician, and a lawyer to himself." Hence Americans needed "an encyclopaedic training, a wide command over the resources of our native tongue, and . . . a knowledge of all its special nomenclatures."[45]

Marsh was one of the first philologists to compare "American" favorably with "pure" English. Noah Webster aside, most Americans had been "uncompromising advocates of conformity to English precept and example," in Mencken's words; they "combated every indication of a national independence in speech with the utmost vigilance." Marsh was no out-and-out Yankee apologist. As noted above, he thought Webster's spelling reforms foolish and divisive. But while conceding the elegance of educated English writers, Marsh judged common American English "not at all inferior to that of England [in] syntactical accuracy." Unlike many English, for instance, few Americans deployed "different *to*," or used *has been* when they meant *was*. Marsh also upheld many supposed American vulgarisms as historically more valid than pallid English variants. And many supposed American archaisms reflected greater familiarity with English Scripture.[46]

Moreover, Americans *spoke* better. Such interjected hiccups as "you know" and "don't you know" Marsh found far more frequent among English speakers inarticulate in other ways as well:

Cockney affectation has sanctioned the disgusting practice of sputtering out one half of the word, and swallowing the other, [and emasculating] our once manly and sonorous tongue by ... crowding a half dozen syllables into one explosive utterance ... that clipping, crowding, and confusion of syllables, which three centuries ago led Charles V to compare it to the whistling of birds.

To be sure, this "nauseous thickness of articulation" was rooted in a deeper source—the same "piebald and Babylonish composition" that made English uniquely flexible. Germanic tongues accented first syllables; the English extended this habit to polysyllabic Norman-French words, truncating their later syllables—Cholmondeley to *Chumley*, Saint John to *Sinjon*, Cirencester to *Siseter*.[47]

Americans had outgrown this stutter and gave each syllable its due. This Marsh ascribed to three causes: climate, foreign borrowings, and literacy. "Distinct articulation belongs to a dry atmosphere, and a clear sky." Like southern Europeans who stressed word endings, southern Americans brought out all the syllables. Marsh thought this southernism infectious; "many a Northern member of Congress goes to Washington a *dactyl* or a *trochee*, and comes home an *amphibrach* or an *iambus*."[48]

Second, Americans acquired accent and intonation from immigrants "who, in adopting our speech, cannot fail to communicate to it some of the peculiarities of their own." On a train trip Marsh overheard two women conversing "in a language unintelligible to me":

One [had] just arrived from the Scotch Highlands, the other ... had been in the United States some years. The new-comer, though understanding English, did not speak it; the other, without having forgotten her native tongue, had been long enough in America to have lost her fluency in the use of it. ... Each was speaking the language most familiar to her [but with the same] *native intonation* and *accent*. The *tones* and *modulations* of their voices were so exactly alike that I could not distinguish between them.[49]

English in America could not escape Celtic, Germanic, and other intonations.

Third, most Americans read more than most English people. Together with geographical mobility, widespread literacy made American speech strikingly uniform. Exposed so much to books and newspapers, Americans learned to pronounce words less as they heard them and more as they saw them spelled. "From our universal habit of reading, there results not only a greater distinctness of articulation, but a strong tendency to assimilate the spoken to the written language." This was one reason Marsh objected to Webster's spelling reforms; with the loss of -*elling* and -*eller* "our grand-children of the year 1900 might talk, as the Websterians analogically now should do, of 'trāve-lers' and 'drīve-ling'."[50]

Some of this was fanciful. Marsh's climatic argument was based on faulty premises; long American vowels were not more but less diphthonged than standard English; oral assimilation is everywhere the main route to learning words. But the twentieth-century scholar George Krapp found more than "a small grain of truth in [Marsh's] theorizing." Many of Marsh's distinctions are now conventional wisdom. Even in the television age, the relative uniformity of American speech reflects more reliance on the eye, less on the ear, than is the case in Britain.[51]

Yet though partial to many American traits, Marsh deplored any transatlantic divergence of English. He chided Anglophobes "who would imbitter the rivalries of commerce by the jealousies of a discordant dialect" and thereby

cut off the sons of the Pilgrim and the Cavalier from their common inheritance in Chaucer and Spenser, and Bacon and Shakespeare, and Milton and Fuller, by Americanizing, and consequently denaturalizing, the language in which our forefathers have spoken and prayed, and sung, for a thousand years. If we cannot prevent so sad a calamity, let us not voluntarily accelerate it. Let us not, with malice prepense, go about to republicanize our orthography and our syntax, our grammars and our dictionaries, our nursery hymns and our Bibles,

while these still remain our peerless heritage. When Americans excel Bacon and Shakespeare "it will be soon enough to repudiate that community of speech which . . . still makes us one with the people of England."

Marsh's fears of schism ebbed over time. He was later amazed by how fast newly coined phrases crossed the Atlantic in both directions. "Many an American has been astonished to hear from an English acquaintance an expression . . . utterly new to him, and very un-American besides—followed by: 'As you say in America.'" The superiority of English in general was for Marsh the cardinal point—a supremacy he laid to its multifarious sources, its composite syntax, and its easy ability to assimilate new forms. "English is emphatically the language of commerce, of civilization, of social and religious freedom, of progressive intelligence. . . . Beyond any tongue ever used by man, it is of right the cosmopolitan speech."[52]

Marsh held language a legacy to study so as the better to steward it, a heritage to cleave to (and if needed purify) as integral to national and cultural identity. His advocacies and his antipathies strike the modern reader as those of a political moralist, not an impartial scholar. We are apt to forget that in his day most scholars were avowed moralists, who routinely reified languages as living beings, with "little ills which could be cured by appropriate remedies prescribed by good grammarians." Yet partial as he was, an awesome cosmopolite erudition informed Marsh's linguistic insights. "That one living man should have known all the[se] recondite facts" was amazing enough; the wonder for one reviewer was that "in any year of his life, one man can arrange them in their precise places" to so clarify "the study of this grand organized living being, the English language." While Marsh's etymological work is now mainly forgotten, some of his linguistic insights are yet cited as prescient and enlightened.[53] And they were, as will be seen, in many ways inseparable from his ecological insights.

Never had Marsh known Burlington to be so unpleasant and unhealthy as in the winter of 1860–61. Diphtheria was epidemic, and he saw "no house without a very sick patient." His own ill-health kept him from going to Washington in pursuit of a diplomatic post—not that he had recently done much politically to merit such a reward. Fear of jeopardizing his claims bill had prevented him from joining Vermont delegates at the Republican convention in Chicago in May 1860, but he exulted in Lincoln's nomination as "probably the strongest that could have been made." And as a Republican standard-bearer Marsh congratulated Vermont on "the bright prospect of a return" to virtue—Vermont's 76 percent was the highest Lincoln received—and sought service in the new administration.[54]

"George P. Marsh of Vermont and Jay [Edward Joy] Morris are pushing for the same position," their rival, Carl Schurz, told a fellow-Wisconsonite.[55] The post was that of envoy to Italy, previously the Kingdom of Sardinia, at Turin, King Vittorio Emanuele's Piedmontese capital. Others, Marsh's congressional colleague Anson Burlingame among them, also sought this attractive new mission.

Sardinia was the island tail of a national dog now looking to rival the mythic wolf that suckled Romulus and Remus. The unification of Italy under the House of Savoy, long resisted by Austria, France, and the papacy, had brought most of northern Italy into the Piedmontese fold in 1859–60. Garibaldi's conquest of Sicily and Naples, followed by Piedmontese advance into Umbria and the papal Marches, now assured the kingdom the whole peninsula except Venetia, the Roman states, Trieste, and Trentino. The prospect of a reborn Italy was warmly greeted in America, many envisaging a resurgent antiquity, a renewed Renaissance, or a seedbed of European democracy.[56]

In love with Italy, Marsh had in 1856 dreamed of Rome as "a very desirable diplomatic residence," hoping Frémont would appoint a "fit person" in place of "the poor creature who now disgraces us there" (Lewis Cass, Jr., was America's envoy to the Papal States). Tuscany he found more charming still. "I commend Florence to you," he wrote Markoe. "As a residence the city is delightful, & all the advantages very great." Marsh had not visited Piedmont but particularly admired the Piedmontese, of whom he had met many in Turkey. The Kingdom of Sardinia was the only one in Europe "that has made any progress in recent times," he judged in 1851. "Sardinia is waging a noble struggle against the power of Austria, the intrigues of France, and the machinations of the Jesuits. No Europeans have more sympathies with . . . Americans." Further impressed by the constitutional monarchy's progress under Cavour, Marsh thought the new Italy "better prepared for free institutions than any other country on the continent."[57]

Piedmontese triumphs roused Marsh's fervor anew; he longed to smash Austria's "hell-born tyrannies." "My red republicanism, which is always hard to keep down, is at present more rampant than Mrs. Marsh makes my Calvinism," he wrote Lieber. "I am a Goth as well as you, [but] I never felt half the interest in any foreign political question that I do in the *liberazione dell' Italia dai Goti*. I wish I was 30 years younger, and *kugelfest* [bulletproof], & had a *Heckethaler* [magic self-multiplying coin]. I would do fine things for liberty; but old, poor, & above all not shot-proof, really I can't afford it."[58]

Political sentiment heightened Marsh's passion for the people, climate, and art of Italy, and he lobbied strenuously for this new ministerial plum. "From much study of the language, history, and especially the present political relations of Italy," he wrote his nephew George F. Edmunds, who was fast becoming Vermont's most influential Republican, "I could be more useful there than elsewhere." William Cullen Bryant, who had been in line for the post, now withdrew and wrote Lincoln recommending Marsh for his "immense fund of information and . . . great personal merit." Vermont's congressmen vaunted Marsh as "among the first scholars & statesmen in the nation"; Vincenzo Botta knew no one else "who would be so acceptable to both countries";[59] Caroline's nephew Alexander B. Crane (close to Lincoln's crony Judge David Davis of Illinois), Raymond of the *Times*, and many others urged his appointment to Italy.

Of the other contenders, Schurz, who had fled Germany after the failed 1848 revolution, was Marsh's most formidable rival. As February ended and illness still kept Marsh from Washington, he began to despair of "anything but a chance to decline an unwelcome offer." On March 7 the papers reported that Schurz would get the Sardinian post. Though Marsh could "hardly think it possible that Pres't Lincoln will do so absurd a thing," he all but lost hope. "Indeed," he wrote Edmunds, "I should hesitate about accepting Spain, and I would certainly not go to St. Petersburg, Berlin, Paris, or any part of Spanish America." France was too expensive, Prussia and Russia unendurably cold, and "of course any second class mission would be entirely out of the question with me." Solomon Foot repeated his generous offer: "You shall have the mission to Sardinia or my place in the Senate. The minute Carl Schurz is nominated I shall resign my seat and telegraph Governor [Erastus] Fairbanks to give it to you."[60]

Marsh's despair was short-lived. Schurz had been Lincoln's first

choice, but German birth militated against him. "The *national* feeling of
the Germans is so intense that [Marsh had] never known one who did
not more or less sympathize with Austria," the arch-foe of Italian unity.
Liberal as Schurz was, Marsh's liberal Italian friends felt he "could
hardly be acceptable to an Italian government," while conservatives
opposed him as a "revolutionary, socialist follower of Mazzini." On
March 16 Italy's Washington envoy Giuseppe Bertinatti warned Secretary of State W. H. Seward that Schurz would be persona non grata in
Turin. On March 17 Marsh reached the capital, and on March 18 Lincoln appointed him Minister to Italy "because of the intense pressure of
[his] State, and [his] fitness also."[61]

"Beate!" exclaimed Lieber on hearing Marsh had been "excellenci-
fied." "To be in the midst of a forming Italia, and a scholar like you! Now
then for the collection of memoires, by you and Mrs. Marsh—historical and social, instructive, bright and salient memoires!" Except for
Horace Greeley's *Tribune*, which feigned surprise to learn Marsh was
a Republican, the press held him "exactly the best man for the place
. . . one of our few able & experienced diplomatists." Greeley vilified
Marsh as a spendthrift plutocrat, less a present-day democrat than an
eighteenth-century Royalist, a patrician unsympathetic to Italian republicanism. Raymond retorted in the *Times* that the Italian people, if
not Mazzini's hotheads, had no better friend than Marsh.[62]

Delighted to be chosen, Marsh was quite unprepared. "Nobody
thought Mr. Marsh would get it. Nothing but the feeling that [he] was
breaking down under his bookmaking induced us to [try] for this appointment," Caroline wrote her sister, "and even then both he and I felt
so drawn in different directions that we scarcely knew what we wished."
Marsh had books to finish at home, and he was anxious to help his son.
He was also uneasy about leaving with the Union in peril. But facts
"stared us in the face, we were earning the barest living with the hardest work, Mr. Marsh's eyes were failing, and two years had made his
shoulders stoop, and changed him almost to an old man." Aside from
small stipends at Columbia and as Lowell lecturer, Marsh had earned
virtually nothing from three years' literary toil. Health was an even
greater inducement. Marsh confessed he was "fast wearing out," but not
until the day he was appointed did Caroline discover "how strong the

impression was" in Burlington that he "could not long endure his present mode of life."[63]

The decision made, Seward was eager that Marsh reach Turin as
soon as possible. The days passed in a fever of preparation. There were
clothes and books to pack, the Burlington house to rent, property to
dispose of, publishers to settle with, friends to bid good-bye, luggage to
send, passages to book. Neighbors, relatives, and total strangers begged
Marsh's help in getting jobs as postmasters, clerks, consuls.

He was also besieged by applicants for his own Ministry. "I dare
say you take your son as Secretary of Legation," inquired Lieber. "If
not, take mine." Lowell endorsed their mutual friend Charles Eliot Norton, long familiar with Italy; Marsh himself wanted Norton, but the
appointment was the president's, not his own.[64] If Marsh could not
pick his assistant, Caroline at least could chose hers. Her widower
brother Thomas's daughter, twelve-year-old Carrie Marsh Crane of
Terre Haute, joined them as companion and amanuensis. Carrie was the
first of a flock of Cranes the Marshes brought to share their life in Italy.

As Marsh got ready to leave, tension mounted between North and
South; all hope of compromise was extinguished at Fort Sumter on
April 12. Five days later Marsh said farewell to a thousand fellow
townsmen dazed by the advent of civil war—an "eloquent, earnest, patriotic, statesman-like speech," recalled a hearer; "nothing could have
been more timely."[65] But it grieved Marsh to contrast Italy with America. His prediction that Italians would burst their fetters was being
fulfilled, and no freedom was ever won with so little bloodshed. *To such
a people Marsh delighted to go; "from such a people as ours once was*, he
should also delight to go." American virtue defiled had now to be restored; the North must expiate the South's crime of slavery on the field
of war. To applause, Marsh urged Vermonters to pledge half a million
dollars and twenty-four regiments to the Union cause.[66]

Half an hour later he was on his way to New York. On April 27,
1861, he set sail with his family and with William Dayton, Anson Burlingame, and James S. Pike, ministers to France, Austria, and the Netherlands. Marsh thus left his homeland for the second and last time; at
sixty he began a new life of more than two productive decades in Italy.

11. Risorgimento
and Civil War

I cannot conceive of anything finer than to be sent to wish Italy joy on the fulfillment of her dream of many centuries. Truly, it is like being the first ambassador to the Sleeping Beauty after her awakening!

James Russell Lowell to Marsh, March 20, 1861

IT WAS SEVEN YEARS SINCE MARSH HAD LEFT EUROPE for America to battle creditors, Congress, and enemies who seemed bent on ruining him. They had been seven lean and gloomy years. Now Marsh looked to mend his fortunes abroad. The new Minister to Italy was by upbringing and temperament a New England optimist—someone sure that things are better than they are going to be. He was surprised to find himself happy and hale, if not wealthy. The reborn nation to which he came was racked by strife and penury, but the Old World was generous to Marsh, who reaped a rich personal harvest. He served in Italy for twenty-one years—a term unequaled by any other American envoy.[1] There he also fulfilled his destiny as a mighty prophet of environmental reform.

Marsh had left a state newly torn by secession and armed strife for one just uniting and, for the moment, at peace. The arms of Garibaldi and the diplomacy of Cavour, with the crucial might of France, had in 1859–60 added Lombardy and Tuscany to the Kingdom of Sardinia. Sicily and Naples (the ex-Kingdom of the Two Sicilies) and Emilia then broke from Bourbon and papal rule to throw in their lot with Piedmont; Vittorio Emanuele II of Sardinia and the House of Savoy became king of Italy. Prime Minister Cavour's astute and temperate guidance seemed to presage the early unification of the entire peninsula.

En route in Paris, however, Marsh learned that Cavour was seriously ill, and as he neared Turin the premier's death was announced. The capital was draped in mourning, the Piedmontese stunned by the loss of their matchless leader. But "the advocates of temporal and spiritual despotism throughout Europe," reported Marsh, greeted Cavour's demise with "ill-suppressed exultation." None could predict what would now become of Italy. One change soon became evident: an "age of poetry" gave way to "an age of prose."[2]

The Risorgimento's heroic epoch was over. Italian nationhood came to emerge in what seemed to many, Marsh among them, a halting, partial, even retrograde fashion—a *rivoluzione mancata*, in the canonical phrase, a limited political mutation that united a state but not a people. Italy was created, as reformers from Massimo D'Azeglio in the 1830s to Antonio Gramsci in the 1930s were fond of saying, but Italians had not yet been made.[3]

The earlier phase of the struggle had fired liberal hopes everywhere. Dormant since Dante and Petrarch, Italian national consciousness had been awakened by Napoleon, only to be suppressed after 1815; the Holy Alliance doomed Italy to remain, in Metternich's contemptuous dismissal, a mere geographical expression. The thwarted uprisings of 1848 left only the Kingdom of Sardinia free from Habsburg and Bourbon hegemony or papal suzerainty. The republican visionary Mazzini fled into exile; Risorgimento leadership centered thenceforth in monarchist Piedmont. The moderate and wily Cavour, premier under Vittorio Emanuele, gathered in exiles from other Italian provinces to help modernize and liberalize this state straddling the western Alps.[4]

Cavour's consummate flattery won Napoleon III of France as a self-interested ally. Piedmontese troops helped the French trounce Austria at Magenta and Solferino in 1859; Garibaldi's "Thousand" overthrew Bourbon rule in Sicily and Naples, which Garibaldi promptly relinquished to King Vittorio Emanuele; and the specter of an imminent Garibaldian march on Rome enabled Cavour to justify Piedmont's occupation of Umbria and the Marches as security for the papacy. Only Napoleon's need to placate French Catholics and his yen to keep a whip hand in Italy left Rome and Lazio in papal possession, Venetia in Austrian. Nonetheless, rulers of the new Italian kingdom of 22 million declared Rome its symbolic capital and began extending Piedmontese law throughout the peninsula.

Marsh was an ardent devotee of Italian unity and liberty, for him synonymous with democratic rule and the overthrow of papal and Austrian autocracy. His 1850s European stay had all but reversed his views of Teutonic and Latin merit. Italians now struck him as sympathetic, cosmopolitan, "inherently and collectively a civilized people, [whereas] stupidity, churlishness, and rudeness [are] as general among the German as they are rare among the Italian peasantry." He regarded Italy as "intellectually and esthetically . . . the most finely endowed nation of Europe, [Italians] the only race, whose character holds out any encouragement for . . . the political regeneration of the continent."[5] Political and religious freedom would recreate a splendid Italy. This "noble people who have groaned for a thousand years under the direst of curses—the tyranny of the soldier and the tyranny of the priest"— merited American support. Castigating congressional Catholics for "ill-disguised sympathy" with Austria and Rome, in 1860 Marsh urged Americans to send Italy "but the half of the million of muskets she asks for her unarmed militia."[6]

Squeezing his new view of Italy into his old determinist collar, Marsh still made a virtue of adversity. If fertile soil, gentle climate, and lovely landscapes tempted Italians to indolence, these bounties were offset by natural handicaps that needed extreme ingenuity to surmount. Few other lands, in his view, "task more highly the thinking and constructive faculties of the people in productive" enterprise. His estimate of Italian engineering works was hyperbolic: in bridging torrents, tunneling through the Alps, irrigating the Po plain, "resourceful" Italians had made "stupendous achievement[s] with a very moderate amount of mechanical aid."[7]

Politically above all Marsh saw Italians as pacesetters. Piedmont had initiated constitutional rule, free speech, religious liberty, secular schooling. The territorial accretions of 1859–60 offered "the sublime spectacle of radical revolutions . . . accompanied not with riotous license and wild disorder, but with an almost absolute cessation of private crime." How unlike Germany! Italian unifiers were pragmatic realists; German nationalism was "one of those unsubstantial, shadowy phantoms, half metaphysical conception, half poetic figment." Marsh feared that German union under Prussia would imperil European freedom. History showed the Teutonic race to be "grasping, unjust, aggressive, and disorganizing . . . arbitrary, unassimilative, unprogressive, and antidemocratic."[8] Quite an indictment, from one who fifteen years earlier had acclaimed the Goths as progenitors of New England democracy.

Marsh continued to personify national traits in judgmental terms. But new heroes and villains now peopled his stage. And he was by no means alone in reversing older stereotypes of Italians as shallow, immoral, decadent, unreliable. Many in America and England shared Marsh's new view of Italy. Sympathy for the downtrodden, hero worship of Garibaldi, renascent classicism, antipathy to papal reaction and, not least, political and commercial self-interest fostered an Italianism that conflated Roman antiquity with Risorgimento progress; Italy seemed "a land of second spring and fair promise." The British reformer Shaftesbury hailed Cavour's revolution as "the most wonderful . . . manifestation of courage, virtue and self-control the world has ever seen." Prime Minister Gladstone saw Italy as a canonical liberal creation. And Americans were flattered that Italians seemed to emulate their own model of self-fashioning.[9]

Of all the items on nascent Italy's agenda, the "Roman Question" seemed to many the most urgent. How long would Italy remain deprived of its historic capital? When would Napoleon III withdraw

Map 3. *Italy*, 1861

French troops from Rome? How could Pope Pius IX be persuaded to yield temporal power? These topics Marsh found uppermost among statesmen in Turin.

Other states were sometimes redeemed in Marsh's eyes, but the papacy remained a fixture of his demonology. He saw the evils of the Church incarnated in Pius IX. In 1847 Pio Nono had proclaimed political amnesty and agreed liberal reforms in tandem with secular Roman leaders. Encouraged by these signs, Congress opened diplomatic relations with the Papal States. Marsh was one of a minority who opposed this move. "Do not, for heaven's sake," he wrote Markoe at the State Department, "commit yourself to the belief in a *liberal Pope!* It is a contradiction in terms—an impossibility in the very nature of things. Whatever Pius IX may think now, he will find that he can't be both Pope and patriot."[10] Marsh's notion of liberalism, far more than that of the Italian leaders he admired, presumed a degree of personal liberty and democratic institutions that as yet existed nowhere in Italy, or for that matter anywhere in Europe.

Events in Rome soon bore out Marsh's 1847 warning. Having fled the newly proclaimed Roman Republic, the pope was restored to his realm by French and Austrian conquest in 1850. Henceforth obdurately reactionary, Pius IX ruled his remnant provinces in despotic style, here brutal, there benign, everywhere venal. To pleas for amelioration, Pius's Secretary of State Cardinal Antonelli rejoined that cleansing Rome was as ridiculous a notion as scrubbing the pyramids with a toothbrush. Marsh judged no previous pope "more obstinately wedded to all the traditional abuses of the Vatican [than] the treacherous and malignant" Pius IX.[11] The pope spurned liaison with Cavour's government, excommunicated king and cabinet *en bloc,* and never recognized the Italian kingdom, referring to it as long as he lived as "Piedmont."

Marsh was delighted to hear virtual unanimity in Turin about Rome. "A large majority of the leading minds of Italy," he advised Secretary of State Seward, "regard the temporal power of the Pope as a source of enormous and uncompensated political and social evils." Almost all the Minister's liberal confidantes, Catholic and secular alike, hoped and expected that Italy would soon acquire Rome. Like many, Marsh saw the pope's intransigence undermining "hereditary veneration" of the papacy; unless Pius relinquished temporal power, Marsh foresaw open schism. A series of seminary and clerical scandals, sexual and anti-Semitic, aggravated animus. Even the most devout Catholics, Marsh judged, "join in the contempt—[even] execration—in which the clergy is held by the middle classes." [12]

Marsh's opposition to papal temporal power did reflect the views of many liberal Italian spokesmen. But his forecast of total papal collapse was largely wishful thinking. Ingrained bias long hindered Marsh from understanding how Italians could at the same time deplore the papacy as an institution yet remain staunchly Catholic, even devoted to the pope himself. A New England Puritan perspective, binding morality, reason, and everyday affairs in a single consistent nexus, made him ill-equipped to comprehend Catholic, casuist, worldly Italians, for whom ambiguity was the norm. [13]

Just as Marsh held the papacy the embodiment of evil, so he at first found Italy's secular leaders flawless. On June 23, 1861, he was presented to King Vittorio Emanuele, a shrewd, roughhewn devotee of hunting, women, and cigars, famed among Piedmontese followers for bellicose patriotism and a commoner's passion for black bread and onions. Shelving Puritan prudery, Marsh praised the king as a patriot of commonsense courage who backed without demur whatever policies parliament (more precisely the cabinet) chose. The American envoy was little privy to the king's episodic efforts to overrule his own government with harebrained militarist schemes or to the royal squandering that crippled state budgets; cabinet officials concealed the king's follies for fear of republican upheaval. [14]

The talented ministry pleased Marsh even more than the king. "Few such cabinets," he wrote Sumner admiringly, "have surrounded an American president." Best to his mind was the new premier, Barone Bettino Ricasoli, the Florentine aristocrat who had ruled Tuscany after the flight of Grand Duke Leopold in 1859 and had then cajoled republican Tuscans into union with royalist Piedmont. Tall and angular, with piercing eyes, spade beard, and jutting chin, hands ever gloved, Ricasoli was a formidably charismatic speaker. Legendary for his austere life-style and puritanical inflexibility, Ricasoli held firm against radical and reactionary extremism alike. He and Marsh at once became very close, sharing political sympathies—Ricasoli was the firmest American Unionist in Italy—and a mutual absorption with agriculture and forestry. Ricasoli had pioneered silkworm and chianti production at his Brolio estate in Tuscany.[15]

Although a devout Catholic, Ricasoli was bent on separating Church and State and annexing Rome to the Italian kingdom. He was at first confident of a speedy settlement; to Marsh in September 1861 he said, "I invite you, Mr. Minister, to Rome. I invite you there *this year, this very year.*"[16] The most serious obstacle was Napoleon III, whose troops protected the Papal States. Rumors mounted that Napoleon wanted Ricasoli replaced by an Italian premier less eager to "settle" the Roman Question. Vittorio Emanuele certainly wanted one less censorious toward himself and more mindful of his royal prerogatives. Too haughty to play the courtier, Ricasoli would neither accept a salary nor don a court uniform; his ancestors had outranked the House of Savoy and had never worn anyone's livery. But August 1861 saw a new French ambassador in Turin, Napoleon's Corsica-born *intime* Vincente Benedetti. As Marsh surmised, this portended mounting French pressure to dissuade Italy from absorbing Rome. Unacceptable both to the French and to Garibaldians who wished to occupy Rome without delay, Ricasoli resigned in March 1862.[17]

The new prime minister, dexterous Piedmontese lawyer Urbano Rattazzi, struck Marsh as a man of "very pleasing address" but little more. If the uncompromising Ricasoli could not break the Roman impasse, Rattazzi's ingratiating maneuvers seemed to Marsh less likely to succeed. Decisions hung fire; nationalists pushed for taking Rome. In August Rattazzi suspended the radical journal *L'Opinione*; Marsh wrote in disgust that the government was "far behind the people."[18] Like many Italophiles, Marsh was partly taken in by the rhetoric of *rivoluzione nazionale*—the myth that unification was a mass movement, rather than the goal of a small elite.

Risorgimento sentiment united again behind Garibaldi, now enlisting riflemen in Piedmont and Lombardy. Marsh judged Garibaldi's influence "greater than ever, and no man doubts that he could overthrow the government, and make himself dictator, in an hour, if he chose to give the signal," but he remained loyal to the king. Several hours with the general reassured Marsh that for all "his apparently uncontrollable impetuosity," Garibaldi was "a man of consummate prudence." His dignity and purity of purpose impressed the American envoy. "Even at this giddy height of popular exaltation," he reported, "Garibaldi retains his simplicity of manner and habits." The general's unique charisma allowed Marsh in this rare instance to suspend—for a time—his congenital mistrust of popular heroes.[19]

Later Marsh became more aware how general, king, and cabinet used one another, trumpeting antithetical public postures while privately abetting similar if not identical ends. But in this instance Garibaldi was patently deceived. He was encouraged to foment uprisings meant to persuade Napoleon that revolutionary fever was uncontrollable; then royal forces would have to step in and take Rome in order to halt the uprising the king and Rattazzi had conspired to encourage.

Garibaldi landed at Palermo in June 1862 and crossed the straits to the mainland, ready to march on the papal capital with the cry "Roma o morte!" But this exceeded what Rattazzi's French mentors would tolerate. Disaster ensued: Garibaldi's troops were beaten by Italian government forces at Aspromonte in August, he himself being wounded and imprisoned. Fearful that the "feeble and slippery" Rattazzi would execute Garibaldi or let him die of his wound, Marsh offered him a haven in the United States. Garibaldi threatened to emigrate, the king granted amnesty, Rattazzi backed and filled; at last Garibaldi was freed and returned to his Caprera island retreat. As in the days of Cavour, Garibaldi had been a pawn to advance the aims of, and then be sacrificed by, the House of Savoy and Rattazzi's cabinet, themselves at odds.[20]

The episode gravely eroded confidence in king and cabinet. Rattazzi resigned at the end of 1862, his administration, in one diplomat's words, "a byword for corruption, jobbery, and intrigue unequalled in the parliamentary history of Piedmont."[21] Rumor was rife: there was talk of war with Austria, of a treaty with Austria against France, of Vittorio Emanuele's abdication, of a military dictatorship under his generals.

Over time Marsh came to feel that Italians, however skilled in oratory, were by nature averse to decision making and inept in constructive statecraft. His judgments on Italian politics became less roseate, more critical. The more ardent his hopes for the new Italy, the more he faulted premiers and cabinets for not realizing them. Even his admired Ricasoli "failed to achieve the success expected of him," Marsh judged on the Tuscan's death, because "his energies were always more conspicuous in resistance than in progressive action, and in preventing evil . . . by others" than in initiating polices of his own.[22]

Only later did Marsh appreciate that Italy's lack of progress stemmed as much from deep structural causes as from individual defects. Yet his assessments were not far off the mark. Most of Italy's leaders in the decade and a half after Cavour were high-principled patriots— Benedetto Croce's "spiritual aristocracy of upright and loyal gentlemen" —but none had Cavour's charm, fascination, sense of the practicable, and ingenuity in compromise. Preoccupied with personal rivalries, they were often narrow and unimaginative.[23]

Marsh's dispatches dwelt increasingly on general European affairs. As a multilingual Italophile in the hotbed of intrigue that was Turin, he was ideally placed to glean insights into Continental affairs. Secretaries of state read Marsh's detailed analyses "with special interest" as "luminous expositions" of the affairs of the day. His thousand-odd dispatches over two decades—pithy, colorful, often scathing, frequently prescient —comprise a European portrait of remarkable scope, penetration, and continuity. A nephew hardly overstated the case to Marsh in 1880: "you can in half a page compress more real matter and give a better statement . . . than others could do in ten pages."[24] His dispatches were esteemed not only for information but as inspiration: they expressed and championed archetypal American stances on morality, democracy, and institutional freedom.

Marsh found his fellow diplomats an agreeable lot. Most were friends of the new Italy; most were pro-Northern on the Civil War. Fully half had served in Turkey; "we are a very heathen Oriental set," noted Marsh. But turnover through death and dismissal was rapid; in 1863, when Sir James Hudson was supplanted as English ambassador, Marsh became dean of the diplomatic corps.

The Marshes preferred the frank and genial Hudson to any of his successors. Long resident in Italy, Sir James introduced Marsh to many Italian scholars. The Marshes were shocked when Hudson was replaced by Henry George Elliott, brother-in-law of foreign secretary Earl John Russell—a move they first attributed to nepotism, then to "the grossly immoral life led by Sir James and his suite." "Alas, alas, alas, for appearances," lamented Caroline Marsh; "whom can one believe and trust in this so-called high life!" Elliott, who as Minister at Naples had supported Garibaldi, Marsh found amiable though shallow. But most British envoys struck the Marshes as cold, reserved, arrogant, supercilious, condescending. "If you want an Englishman to be civil," Caroline concluded after encountering young Edward Herries, British Secretary of Legation, you must "treat him haughtily."[25]

Marsh got on well with Benedetti of France, a former Constantinople colleague. But many Italian leaders resented Napoleon III's support of the papacy; and his overbearing, choleric ambassador, though himself pro-Italian, was unpopular in Turin. Benedetti's successor, pro-Austrian Baron Sartiges, was soon followed by Baron Joseph Malaret, who helped engineer Garibaldi's Aspromonte fiasco and for six years wielded influence over Vittorio Emanuele. The Prussian Minister, Count Brassier de St. Simon, was a liberal lover of Italy and of women. He held bachelor dinners at Turin, maintained a wife at Nice, kept a mistress at nearby Piòbesi castle, where the Marshes later lived, and intrigued with Ricasoli to ally Italy with Prussia against Austria. Marsh was intimate with the scholarly and radical Swiss envoy, Abraham Tourte; the Turkish chargé Rustum Bey (in fact a Venetian, Conte Marini), talented linguist and gourmet; and the Russian Count Stackelberg, replaced in 1864 as too liberal. Baron Iver Holger Rosenkranz, the Danish envoy, was Rustum's linguistic antithesis. On reaching Turin, Rosenkranz announced that he would learn Italian at once, as "he did not wish to be as ignorant as an American Ambassador of every language but his own." The chagrined Rosenkranz became "a friend with whom," Marsh wrote Rafn in Copenhagen, "I can now and then exchange a few Danish words."[26]

*I*n this diplomatic circle Marsh was at once outstanding. But like many American envoys he bore a "legacy of disgrace" bequeathed by predecessors who drank to excess, brawled in public, failed to pay bills, or lived in squalor. He grumbled that "the character of the United States Foreign Service has been declining ever since the days of John Quincy Adams."[27]

13. *Therapia and the Bosporus, by W. H. Bartlett, ca. 1836*

THE CAMEL

HIS

ORGANIZATION HABITS AND USES

CONSIDERED WITH

REFERENCE TO HIS INTRODUCTION INTO
THE UNITED STATES

BY

GEORGE P MARSH

BOSTON
GOULD AND LINCOLN
59 WASHINGTON STREET
NEW YORK: SHELDON, BLAKEMAN & CO
CINCINNATI: GEORGE S BLANCHARD
1856

---

side, and are somewhat larger than the middle of the bone. The bit is nowhere used, nor am I aware that it has ever been even tried. It would seriously interfere with the animal's habit of feeding as he walks, but the mahari is seldom allowed to do this, and it might, perhaps, answer for that variety. There does not, indeed, appear to be any urgent necessity for its introduction, nor is there, on the other hand, any very obvious objection to its use. Perhaps a leather ring around the nose, like that which is often used at Naples and elsewhere in southern Italy instead of a bit, and armed with blunt points within, in something the same way, might be an improvement on the common camel-halter.

The pack-saddle, whether for riding or for burden, is made by stuffing a bag seven or eight feet long with straw or grass, doubling it and sewing the ends to-

ARABIAN CAMEL: PACK-SADDLE.

gether. This forms an oblong ring, which is

---

14. G. P. Marsh, The Camel (1856)

15. *Vermont State House, Montpelier, lithograph by J. H. Bufford, 1857*

16. *G. P. Marsh, photo by Matthew Brady, New York, 1861*

17. *Turin: baroque prototype at Venaria Reale, by Amadeo di Castellamonte, 1682*

18. *Turin: Piazza Castello, by Giovanni Battista Borra, 1749*

19. Turin: *Contrada di Dora Grossa (today Via Garibaldi), lithograph by Domenico Festa, 1835*

20. *Turin: Galleria Natta (today Galeria San Federico), lithograph by Giovanni Francesco Hummel, 1860*

21. *G. P. Marsh, Turin,
calling card, ca. 1863*

22. *Caroline Marsh, Florence, 1866*

23. *Alessandro Manzoni,*
calling card, ca. 1864

24. *Cesare Cantù,*
calling card, ca. 1864

25. *Giuseppe Garibaldi,*
calling card, ca. 1864

26. *Hiram Powers,*
calling card, ca. 1864

27. *Monte Rosa, by W. Brockedon, 1861*

28. *Val d'Ossola, by W. Brockedon, 1861*

29. *Piòbesi: The Castello, lithograph by Enrico Gonin, 1850*

30. G. P. Marsh in his library, Villa Forini, Florence, ca. 1870

31. G. P. Marsh with his adopted son, Carlo Rände, Rome, ca. 1880

32. *Caroline Marsh, Rome, ca. 1880*

33. *G. P. Marsh, by Jane E. Bartlett, ca. 1881,*
portrait at Phillips Andover Academy, Massachusetts

34. *Vallombrosa Abbey, by L. Giarrè, 1845*

Turin had fared better than most European capitals in its American agents. About the gravest issue faced by Edward Ambrose Baber, a doctor from Georgia who was chargé d'affaires from 1841 to 1844, was being offered a key to the Royal Theatre (but not a ticket to the show) in exchange for gifts to royal servitors; the State Department would defray the presents, Daniel Webster told him, but Baber must pay for his own tickets. His Kentuckian successor Robert Wickliffe, Jr. (1844–47), confronted few bones of contention. Nathaniel Niles, a Harvard-schooled Vermont physician on leave from his Vienna post to arrange a commercial treaty in 1838, advised Secretary of State John Forsyth that the kingdom of Sardinia had "by far the most enlightened, active and wealthy population" in Italy. When Niles returned to Turin as chargé (1848–50), he negotiated a lease of the port of La Spezia, south of Genoa, as a U.S. naval depot. His successor, New Jersey lawyer-journalist William B. Kinney, was an Italophile liberal who claimed to be often consulted by Cavour.[28]

But Kinney was followed as chargé in 1853 by John Moncure Daniel, editor of the Richmond *Examiner*, upgraded to Minister Resident the following year. Daniel's hauteur, wild parties, and disregard of local etiquette offended Piedmontese society. Daniel termed Piedmont "the most beautiful country I have ever seen" and its government "the best I know in Europe," but his abuse of its elite did not endear him when leaked to the press: "The people are nowhere as good as ours. The women are uglier; the men have fewer ideas." Ignorant of French on arrival, Daniel "jabbered bad grammar to countesses, and [was] sponged on for seats in my opera-box by counts, who stink of garlic as does the whole country." He survived a scandal in 1859, for which Cavour demanded his recall, only by having had his resignation already accepted in Washington, and stayed on in Turin until January 1861. Marsh was told of Cavour's pleasure that the "U.S. had now sent to Italy a man [i.e., himself] who would make amends for the discredit thrown on both countries by their last representative."[29]

Marsh was successful from the start in winning Italian leaders' confidence. But he did not so quickly allay critical doubts back home. While Catholics spread reports defaming him as antipapal, Greeley-ites circulated bizarre rumors that Marsh was pro-Austrian and anti-

Italian—rumors alarming to the American colony in Rome, according to the sculptor W. W. Story in September 1861. Two years later Story heard further tales "showing Mr. Marsh's want of tact"; Story concluded that "as a minister he is null." But he later found these rumors baseless and became a Marsh supporter and *intime*.[30]

Marsh's labors were gravely handicapped by the absence of a good secretary—or rather the presence of a very bad one. Daniel had left in charge a Catholic Southerner, Romaine Dillon, who was to remain Secretary of Legation until the arrival of the newly appointed William Fry. But where was Fry? The topic suffused Marsh's dispatches and Seward's muddled replies: Fry will sail soon; Fry is ill in Paris; Fry is on leave of absence; Fry is missing; Fry has returned to America. Meanwhile Dillon remained in Turin.

The keen-witted but brash young Dillon expressed views that put him beyond the pale. "No man who is a statesman can suppose Christianity would exist a single century if the Papacy were broken down," Dillon told Caroline Marsh one evening. "Some difference of opinion between him and his chef!" she commented. "How could I, an American-Puritan-Liberal-Union-Republican," Marsh wondered, "be expected to go [on] with an Irish-Papist-Bourbon-Secession-Democrat?" Marsh confided to Edmunds that he suspected Dillon was "kept here by [New York] Archbishop Hughes in order to counter-act any mischief I might do by my anti-Popish sentiments." He dared not ask to have Dillon recalled, "because I have already suffered too much from Catholic malice at Washington [over Dainese and his 1850s claims] to venture to provoke the hostility of [Pennsylvania Congressman Joseph R.] Chandler and Hughes and the rest of that gang, by new offenses."[31]

Meanwhile Dillon openly derogated Italy's claims to Rome and denounced Union goals in the Civil War. At length Marsh warned him that Italians were astounded that "a person so habitually unreserved in his condemnation of the course of his government . . . should be retained in its service."[32] Vexed, Dillon accused Marsh of seeking to speed his departure; Marsh's concern had leaked from the State Department to Greeley's *Tribune*.

This heedless disclosure left Marsh aghast. To avoid open conflict, he had borne with Dillon's indignities, only to have Seward "prejudice the usefulness of this legation, by still further embittering Mr. Dillon."

Marsh's "well-merited" complaint (Seward's own contrite word), backed up by the Italian correspondent of the *New York Times*, at last brought Dillon's recall. Fry resigned the post he had never filled, and in May 1862 Green Clay of Kentucky arrived as the new Secretary of Legation. A boy of twenty-one, Clay struck Marsh as having "no taste, and not a single qualification, for the duties of his position."[33] But Marsh came to like Clay, who matured in the job, and was sorry when he left six years later.

As in Turkey, Marsh suffered from having confidential dispatches carelessly, at times purposely, made public. "The notorious publicity of our government operations and our official correspondence" seriously impeded the "freedom of oral intercourse" essential in diplomacy. Shifts of party power in Washington put previous secrets at risk; "no official communication, from an American Minister abroad, however confidential in its nature or its form, is at all sure of being treated as such by a succeeding administration."

The Dillon episode bared another hazard: whenever an envoy left his legation, official despatches on file "would come into the hands of anyone temporarily in charge." Marsh heard much of import that was unsafe to put in the diplomatic record; might he send such matters to Seward, to be kept not in the State Department but in Seward's own hands?[34] Several private memos went directly from Marsh to Seward between 1862 and 1865, and to later secretaries of state via Marsh's nephew Senator Edmunds; but embarrassing disclosures, not always innocent, continued to plague him.

One such leak—noted further in Chapter 15—drew Marsh's retort that "it wasn't a blunder on the part of any body. It did itself; and it is a source of proud satisfaction to the parties *prima facie* responsible to find nobody is to blame." The State Department needed "an official scapegoat," whose salary would be more than compensated by what he saved. For example,

the inquiry about the betrayal of the British Treaty [in 1871] cost the Government $10,000 or more, all of which would have been saved, not to speak of the valuable time of various members of Congress and of Mr. Horace Greeley, if the Department had simply been authorized to say: "It was a blunder of Terence Thady O'Mulligan, official scape-goat of the Department." This . . . would have satisfied everyone, and the matter would have quietly dropped.[35]

Because he could not entrust Dillon even with copying dispatches, Marsh had no official secretarial help for a full year. At his own expense he hired Joseph Artoni, a naturalized American back in Italy in 1861 after a twenty-year exile. Trim and serious, "a dear good soul, as sincere as the light," in Caroline's description, Artoni remained Marsh's capable private secretary for two decades.[36]

The consular scene Marsh found as trying as the legation. He urged that Democratic incumbents be replaced at once by loyal Republicans. "I do not believe," he wrote Seward a month after reaching Turin, "there is an American consular agent in Europe, appointed by Presidents Pierce and Buchanan, against whom there is not grave cause of suspicion." In strategic Leghorn, Southerner Robert M. Walsh was "clamorously hostile to the policy of the Italian Government and a partisan of . . . papal misrule." Over Marsh's protest Seward next gave the consulship to a personal friend, Andrew J. Stevens, a gauche, self-important schoolmaster. In Genoa, Marsh got Southerner W. L. Patterson replaced by Reverend David Hilton Wheeler, "a plain, sensible, thoughtful and scholarly man." Later editor of the New York *Methodist* and president of Allegheny College, Pennsylvania, Wheeler thanked Marsh for much of the research, editing, and proofreading of his *Brigandage in South Italy*. At Palermo, ex-Harvard teacher Luigi Monti was so helpful to American travelers that Marsh had to warn him that "it is not the business of Consuls to suggest to American citizens methods of evading" customs duties.[37] At Florence, Consul-General Colonel T. Bigelow Lawrence, son of textile manufacturer Abbott Lawrence, devoted himself more to Italian society than to Italo-American commerce.

Marsh esteemed William James Stillman, America's artist-journalist consul at Rome. "We have never had a better agent, diplomatic or consular, who was more acceptable to the Americans who visit that city, or more creditable to the government," he wrote Seward, urging Stillman's promotion. But the anti-Catholic Stillman, in incessant conflict with American envoys at the Papal Court, was sidelined to Crete in 1865. In stark contrast, America's papal diplomatic corps were not only pro-Confederate but (owing to Archbishop Hughes, Marsh believed) "ultramontanists in politics and obscurantists in religion." In vain Marsh warned Seward in 1862 that sending another Minister to the Papal Court (future U.S. postmaster-general Alexander W. Randall having just resigned) would be "almost an act of discourtesy to the [pro-Unionist] Italian government."[38]

Seldom, however, did the State Department heed Marsh's strictures on personnel in or affecting his mission. Throughout his diplomatic career, legation and consular officers proved more of a hindrance than a help. Marsh was partly to blame. He inspired devotion and loyalty among some but did not conceal his contempt for second-rate subordinates. Only when he picked his own staff were his embassies run with efficient harmony.

To corral the support of European governments for the Union cause, or at least to deny it to the South, was at this time the most important task of America's Old World envoys. Antislavery was their trump card. It was Secretary of State Seward's "villainous" scheme, charged a slavery sympathizer, to spread abolitionist propaganda via his agents in Europe. Should William L. Dayton in France fail him, "the Secretary had instruments elsewhere, suited to his purpose. They were Mr. Cassius Clay in Russia, Mr. Fogg and Mr. Fay in Switzerland, Mr. Pike at the Hague, and, we regret to say, Mr. Marsh at Turin." Marsh seemed to have had most success.[39]

To boost crucial support for the North, Marsh prodded Seward to sway opinion through European journals. For a time it seemed that European intervention might be decisive. Winning the American Revolution had in the last analysis depended on European arms; Union victory in the Civil War might depend on keeping Europe out of it. Indeed, but for pledges of British and French sympathy and aid, Marsh thought that the South might have succumbed at once.[40]

Seward's propagandist envoys had no easy task. Europeans who deplored slavery had no other reason to favor the North; and many subordinated the slavery issue to contrary economic or imperial interests. To counter Confederate pressure, Marsh got Seward's approval "to furnish facts and arguments" to disabuse the English people "on the legal and moral aspects . . . of the right of the Southern States to secede from the Union." Marsh's unattributed tracts were widely circulated, but he was too censorious to pen anything persuasive for English readers. Privately Marsh labeled Britain the "most wicked and malignant of modern political societies," and in an official dispatch termed it generally reputed "the most selfish, unprincipled and perfidious Power of modern Christendom."[41]

Albion's perfidy had long dismayed Marsh. The Oxford Movement and other High Church trends twenty years before had shown "such a degree of intellectual decrepitude and of moral degradation [in church and state] that no iniquity or folly on their part has since seemed incredible to me." He lamented to Norton that his official post barred him from writing "freely" about the "decay" of England. Reproving Marsh's Anglophobe philippic as unfounded, his old friend Humphrey Sandwith was then forced to retract; it was "not to be denied, that the upper classes of England have taken the part of those damned slave drivers." Growing hostility to the North became evident in *The Times*, the recognized voice of Britain's elite.[42]

Amicable Italian leaders gave Marsh no such problems. "Our Italian friends are unanimous in favor of the North," he wrote Edmunds in the somber spring of 1863; no Europeans more warmly backed the Union cause. "In 1861 Italy regarded the United States as the paragon of liberty and the model of a union peacefully effected," concludes a chronicler of Marsh's mission; the South's rebellion was seen as "the very type of sectional estrangement" Italy was struggling to overcome. Abhorrence of slavery, Civil War reports from Vincenzo Botta and like-minded Italo-Americans, and Marsh's own prodigious influence kept Italy steadfastly on the Union side. This stand, Marsh stressed to Seward, was at some material cost; an embargo on Southern cotton forced several Italian factories to close, entailing severe hardship. Yet in Italy only the papacy and a few Bourbonites sided with the South. "All tyrannies sympathize," remarked Marsh; "the slave-driver and the priest are twin-brothers." He had "for many years hoped to live to see James Buchanan and Pius IX hanged," Marsh declared to Pike. "I shall never have full faith in the people of either country until it does justice on its great betrayer."[43]

Italian amity helped foil Confederate privateering in the Mediterranean. Rumors that Southern ships were loading arms at Genoa led Marsh to alert Ricasoli, in June 1861, to police the harbor. In October, Consul F. W. Behn mistook a three-masted schooner sighted off Messina for the Confederate *Sumter*; Marsh asked and Italy agreed to forbid rebel ships entry into any harbor except in stress of weather. In fact, the badly crippled *Sumter* reached Gibraltar only in January 1862, and was abandoned in April. But its plight was not generally known, and its presence in the Mediterranean roused "tremendous panic" among Northern merchantmen. No one would ship goods in the face of such a danger. Marsh repeatedly urged Seward to send out a ship and "clear the situation up." A frigate did come but was thought impotent against the *Sumter's* heavy fire. Not until June 1862—two months after the *Sumter* had been scuttled—did the arrival of the U.S. *Constellation* restore confidence.[44] Though other Confederate vessels were sighted from time to time, from then on the Mediterranean was felt safe for Union shipping.

But Union mishaps increasingly disheartened sympathizers in Europe, Marsh reported. The defeat at Bull Run (July 1861) "stamped us as cowards" and destroyed faith in early Union victory; delay in freeing the slaves "forfeited the confidence of European philanthropists, who had hoped we would show some signs of a national conscience." Instead, Confederate claims that the contest was simply one between free trade and protection gained credence. As long as the war seemed to be about freedom "we had the world with us, but at present," Marsh warned, "the current seems to be setting in favor of the South." Italians already thought the Confederacy evil; they needed proof that Lincoln and Seward were "honestly against slavery."[45] Not even in 1864 was it clear that emancipation truly mattered to Northern leaders.

How sincere *was* the North? Himself agonized by doubt, Marsh found it hard to convince others. Initially sanguine, he grew impatient, discouraged, despairing. At the start Marsh was heartened by Yankee men and money "offered without stint," and millionaires "serving in the ranks as privates." Within a few months idealism had evaporated. Congress failed to vote "the infamous Dred Scott supreme court" out of existence, passed John J. Crittenden's "vile resolution" asking for a "shameful" peace, and declined to punish Pierce and Buchanan. The North had "abandoned every principle" for which the Republicans had fought. Grieved, Marsh asked Republican leaders "what all this backing and filling of the party means," but he got no answer. "I know the Northern *people* are right enough," he fretted; "why are those who should lead them lagging in the rear?"[46]

As the months passed and McClellan's errors multiplied, Marsh's diatribes waxed incandescent. The general was a "pro-slavery politician and imbecile nincompoop"; the South would triumph "unless we are rescued by the strong arm of a military dictator." He would "rejoice to hear that a regularly organized mob had lynched Fillmore, Pierce & Buchanan, & given notice to Lincoln that he might expect the same fate if he did not dismiss Halleck, McClellan, Buell, & the other pro-slavery traitors in the Army. We have got beyond the reign of law," he exploded to Francis Lieber; "nothing but a reign of terror will serve us."[47] Marsh's faith in democratic forms was often sorely tested by his impatience with democratic consequences, notably what he reproved as an excess of charity.

The long-delayed Emancipation Proclamation scarcely raised the president in Marsh's estimate: he thought it "foolish in form" and "technically speaking, unconstitutional." He grudgingly conceded Lincoln's honesty and good nature but thought him feeble when right, obstinate when wrong—utterly unfit to be president. Later acknowledging Lincoln's renomination as "expedient," he censured the president's leniency toward Southern traitors.[48] To Marsh, remote from American realities, the proper path seemed clearly marked, the administration's dilemmas unreal and its equivocations inexcusable.

In this he differed little from many ardent Northern Republicans. And as a Unionist abroad, Marsh's uncompromising insistence on conscience and liberty and his tone of moral rectitude helped to shape and sustain Old World images of an exemplary America—an America patterned after libertarian ideals, rather than crass political profits—for decades to come.

While frustrating Confederate forays for Italian supplies, Marsh was also busy buying matériel for the North. Tens of thousands of muskets, pistols, and swords were shipped from Genoa and other Italian cities to Union forces—though Confederate competition for arms doubled prewar prices.

Men were more easily found than muskets. Thousands of applications flooded in from "men who have tried in vain to get knocked on the head in Italy, and now want a chance in America." But all wanted to be officers, and most were penniless; a would-be hospital surgeon demanded a life income and passage for his whole family. As the least encouragement would multiply petitions tenfold, Marsh was careful to offer no inducements. But Italians who could not imagine a "stingy" Uncle Sam kept thronging the legation. "They threaten to worry us out of our lives," complained the Minister. Finally he had notices published that no one's way to America would be defrayed. Migrants who paid their own way, though, were welcomed; Marsh looked to an infusion of Latin blood as an "antidote against the Celtic," his Irish animus reheated by Romaine Dillon.[49]

One Italian soldier whose way the North would gladly have paid was the great Garibaldi himself. The notion of enlisting the general in the Union forces was no mere pipe dream, but a serious proposal that nearly came to pass. Seward had heard from the American consul at Antwerp, J. W. Quiggle, that Garibaldi "might be induced to take part in the contest for preserving the Unity and Liberty of the American people, and the institution of Freedom and Self-Government." In July 1861, after the Bull Run debacle, America's envoy in Belgium, Henry S. Sanford, was ordered to join Marsh in Italy and—in secret—to offer Garibaldi a major-generalship in the U.S. Army.[50]

Marsh was deeply disturbed by "this worse than old-woman scheme, . . . calculated to prove our weakness and the imbecility of our leaders . . . and at the same time to excite against us the hostility of every power in Europe which does not sympathize with the Italian hero." But Marsh's opinion had not been asked. Late in August he and Sanford sent Artoni to Caprera to find out if Garibaldi was "at liberty to entertain propositions on the subject." But the flustered Artoni forgot both caution and instructions and told Garibaldi that Lincoln intended to make him commander-in-chief of the Union forces! Marsh and Sanford were staggered; the news of "this *undreamed* of offer," bolstered by a previous note from the bumptious Quiggle, might have dire effects. And Garibaldi indeed replied that unless his king needed him in Italy he was "immediately at your disposal, provided that the conditions . . . are those which your messenger has verbally indicated to me."[51]

The American offer also helped Garibaldi exert pressure on Italian affairs. He now wrote Vittorio Emanuele that unless His Majesty allowed him to march on Rome or Venice, he would go to America. The king faced unpalatable alternatives: to let Garibaldi attack Rome or Venice would call down great-power reprisals; to allow him to emigrate would be domestically unpopular, perhaps disastrous. At length, declining to countenance the Roman move, the king left the general free to follow his own conscience. Garibaldi now seemed certain to accept the supposed American offer.

Thus it was vital at once to correct the general's misapprehension about his promised rank and role. Marsh chartered a boat to take Sanford to Caprera on September 8. But Garibaldi brooked no compromise: he would serve only as commander-in-chief, with power to declare the abolition of slavery. As Marsh had surmised, the general's "constitutional independence of character and action, his long habit of exercising uncontrolled and irresponsible authority," and his pride made it impossible for him to accept any rank Lincoln could constitutionally offer him.[52] Marsh was deeply relieved the awkward affair had come to naught. But as he feared, word of Seward's initial offer to Garibaldi leaked out; Southern agents made the most of this seeming admission of Northern weakness.

A year later, after Garibaldi's capture at Aspromonte, he asked and Marsh in confidence offered him American asylum, as noted above. The need for a competent Union general seemed more urgent than ever, and Garibaldi, jailed and in trouble, was now less demanding. "It might possibly relieve the government of Italy . . . from embarrassment," Marsh suggested to Ricasoli, and "offer the prisoners an opportunity of usefulness to us without prejudice to the interests of Italy." Still haggling with the Italian cabinet, Garibaldi agreed in October 1862 to go to America and "appeal to all the democrats of Europe to join us in fighting this holy battle." But first the North would have to declare slavery abolished.[53]

A few days later Marsh sent Lincoln's preliminary emancipation proclamation to Garibaldi, with the hope that "we shall soon have the aid both of your strong arm and of your immense moral power in the maintenance of our most righteous cause." Seward confirmed this offer; Marsh should "inform the General that he and his friends will be welcomed with enthusiasm . . . and that a proper command will be assigned him." Garibaldi sent his aide-de-camp Colonel Gaspare Trecchi to discuss raising and arming 2,000 men.[54] But before Trecchi reached Marsh, Garibaldi made his peace with king and cabinet and once again declined the American offer. Five years later Marsh would one last time aid Garibaldi with an offer, gratefully received but declined, of American refuge.

12. Turin and the Alps

PIEDMONTESE LEADERS TOOK PRIDE IN TURIN. FOR three centuries the seat of the House of Savoy, this Alps-encircled city of 100,000 became a major diplomatic center with the Risorgimento. And when the Kingdom of Sardinia became the Kingdom of Italy, the chargés d'affaires of foreign states were elevated to plenipotentiaries and ambassadors. Turin itself, largely a creation of the seventeenth and eighteenth centuries, reflected the military and regal ambitions of successive dukes of Savoy, who had removed there in the late sixteenth century when their Savoy seat, Chambéry, proved too exposed and isolated. They transformed the small garrison on the river Po into a fortified showpiece of baroque grandeur. Along a grid of broad, straight streets stood stuccoed brick mansions of uniform height and decor, lending an overwhelming effect of rectilinear order. Eighteenth-century travelers judged Turin "the prettiest town" in Europe, admirable for "the beauty of its streets and squares," the "sociable temper of its inhabitants," and "all the conveniences of life." Subsequent visitors—save for Henry James, disgruntled by Turin's too "regularly rectangular . . . collection of shabbily-stuccoed houses"—echoed such praise.[1]

Marsh, too, rated Turin "a better suited, more convenient capital . . . than any political centre in Europe." He found its "broad streets and avenues bordered by magnificent porticoes . . . well-devised and commodious," and lauded the "unrivalled beauty of position" that gave the capital "a panoramic view of unsurpassed magnificence." The human scene struck him as no less impressive. Here lived most "of the patriots and statesmen whose wisdom and . . . devotion have wrought a great political miracle and realized the most splendid of historical dreams." The Marshes' praise of the Piedmontese capital was almost as perfervid as Nietzsche's accolade to this "dignified and serious city [of] aristocratic calm."[2]

From the balcony of the Hotel de l'Europe the newly arrived Marshes looked across the Piazza Castello to the imposing Palazzo Madama, a Renaissance castle enlarged by the eighteenth-century architect Filippo Juvarra, and now the royal family residence. Two months later the Marshes moved a few streets southeast to occupy two floors of the four-story Casa d'Angennes, a seventeenth-century palazzo on the via Teatro d'Angennes (today via Principe Amedeo), with their own passageway to the theater next door. The Marshes' home for most of their four years in Turin, the Casa d'Angennes was spoken of as "one of the few houses in Turin kept with sufficient cleanliness for an English or American family to occupy."[3] Evenings they drove along the broad riverside Corso Lungo Po, a miniature Bois de Boulogne, fashionable Turin's social rendezvous. "But oh the caprices of the monde!" sighed Caroline:

It does not do to drive beyond the limits of the Corso, unless one would at the same time put himself out of the pale of the best society. At the end of a short half mile every body turns round and goes back again and so to and fro till twilight . . . it must soon become the greatest of bores. After a few turns you recognize every carriage, every toilette and every face.[4]

Conservative, clannish Turinese society centered around the royal court, though the king himself stayed mainly at one of his hunting lodges or with his mistress Rosina Vercellana and their children at Millefiori, an hour's drive from town. In Turin, Vittorio Emanuele's Saxony-born sister-in-law, the Duchess of Genoa, oversaw etiquette and education for the widower king's daughters Clothilde and Maria Pia, until their marriages in 1859 and 1862, and for her own children Margherita and Tommaso. Buttressing the House of Savoy were the Piedmontese *codini*—Carignanos, Cisternas, Alfieris, Dorias, and other ancient interbred nobility, with palazzi in town and villas in the country. Among foreign diplomats and patrician statesmen from other Italian provinces, the Piedmontese aristocrats set the tone, both in Turin and on large and lucrative rural seats ornamented and tended like English landed estates.[5]

To name the statesmen and scholars Marsh came to know is to sound a roll call of Risorgimento elites. Among them were the ex-revolutionaries Conte Giovanni Arrivabene, Conte Gaetano de Castillia, Marchese Giorgio Pallavicino Trivulzio, and Marchese Giuseppe Arconati Visconti; formerly jailed and then exiled, all were now appointed senators. Frequent Marsh visitors were Barone Carlo Poerio of Naples, jurist brother of the martyred poet Alessandro; the aged Prince Emanuele Cisterna, diplomat-litterateur whose daughter was queen of Spain; folklorist Costantino Nigra, whose attachment to Napoleon III and Empress Eugénie served him well as Italy's perpetual envoy in France; the painter and belletrist Marchese Massimo D'Azeglio, who had preceded Cavour as premier. D'Azeglio's father-in-law, the legendary novelist Alessandro Manzoni, received the Marshes at home.[6] This cultivated set contrasted sharply with homespun Washington; democrat though he was, Marsh felt intellectually at home in the Italian capital.

Marsh enjoyed superb relations with successive ministers of foreign affairs. With Ricasoli (foreign minister in his own cabinet) he was on terms of frank intimacy. Rattazzi's first foreign minister, Giacomo Durando, founder of the influential liberal journal *L'Opinione*, was open and cordial. His successor, the gently persuasive, imperturbable Conte Giuseppe Pasolini of Ravenna, an early associate of Pius IX who had sought in vain to bridge sacred and secular in the 1848 Roman government, despite his clerical bent became one of Marsh's closest friends. With Pasolini, Marsh discussed agricultural and engineering projects both hoped would modernize Italy. Pasolini's secretary, young Chevalier Emilio Visconti Venosta, succeeded him in 1863. In charge of Italy's foreign office for much of Marsh's term as envoy, Visconti Venosta was ever prudent and obliging.[7]

Others he admired included Senate President Conte Federigo Sclopis, Italy's premier jurist and a renowned litterateur; two-time premier Marco Minghetti, Bolognese mathematician and economist; and —most congenial to Marsh—sagacious Quintino Sella, mineralogist and Alpinist, vital to Italy's solvency as frugal minister of finance. A regular visitor was the radical Angelo Brofferio, popular playwright and editor of the republican *Voce della Libertà*; Marsh thought Brofferio an unpractical visionary but the most eloquent speaker in Parliament.[8]

Formerly a place its own patriots had fled, Piedmont was now an exile refuge, and Turin a notable abode of outcast revolutionaries. There with his two sons was Kossuth, grown worn and sad in the decade since Marsh had aided him in Turkey. Cheerful, many-sided Ferencz Pulszky, foreign minister in Kossuth's short-lived Hungarian republic, became Marsh's confidant; when Pulszky was arrested with Garibaldi at Aspromonte in 1862, Marsh quickly secured his release.[9] His republican sympathies and candid warmth made him a recipient of many exiles' hopes and fears. But for all his revolutionary empathy Marsh remained a political realist, often dissuading visionaries from desperate impetuosity.

Kossuth was not the only old acquaintance the Marshes refound. Romualdo Tecco, former Minister to Constantinople, and the poet-historian Giuseppe Regaldi, a Marsh companion on the Nile, welcomed them to Turin. The belletrist Abbé Giuseppe Filippo Baruffi, friend of their New York well-wisher Vincenzo Botta, enlarged the Marshes' circle there. No week passed without visits from the amusing Baruffi, raconteur and gossip par excellence. An opponent of papal temporal power, he was, Caroline Marsh concluded, "a priest that even a Protestant can respect."

Through Baruffi they met the astronomer Barone Giovanni Plana, a liberal iconoclast famed for lunar studies. Though eighty and half deaf, Plana was full of fire and humor. A nominal Catholic, he mocked papal pretensions, but no faith escaped his barbs. "I came to see you a few days ago," he once told the Marshes, "but I forgot you were superstitious Protestants and came on Sunday—I did not see you of course."[10]

Another notable Baruffi brought to Marsh was the poet Cesare Cantù, author of the multivolume *Storia Universale* (1838–47). A striking, sardonic figure, with hawked nose jutting over great walrus mustache, Cantù had recanted youthful radicalism to become official Vatican historian. Marsh stigmatized Cantù as "bigoted, sectarian, obscurantist in religion and retrograde in politics," but he could not help enormously liking this witty raconteur.[11]

Among others Marsh came to know well were the physicist Carlo Matteucci, Genoese economist Gerolamo Boccardo, the Orientalist Michele Amari, Bolognese geologist Giovanni Capellini, the Sanskrit scholar Gaspare Gorresio (director of Turin's Biblioteca Nazionale), Italo-Swiss sculptor Vincenzo Vela, and the landscape painter Angelo Beccaria. Their range of expertise was as all-embracing as Marsh's own myriad interests.

The Piedmontese are really charming people, so simple & kindly. Only I wish they weren't all Counts.

Edward Lear to Chichester Fortesque, July 31, 1870[12]

Marsh regretted that protocol precluded his seeing more of such artists and scientists. The rigid Piedmontese social code barred bourgeois artists and scholars from titled and diplomatic circles— a selective rigor that endured well into the twentieth century. Even Cavour's right-hand man Nigra, now ambassador to France and much at home in the Paris court, when back in his native Turin could not set foot in the Marchesa Doria's salon, his father having been a mere army doctor. The officials with whom Marsh had to consort were "almost uniformly taken from the old nobility," a circle that excluded untitled literary men. He found it "almost impossible to get beyond a single exchange of visits," for the bourgeoisie would not risk meeting court figures at the Marshes'.[13]

Exclusivity was all but absolute. Cavour's battles with Piedmontese aristocratic prejudices were harder fought than any he had waged with the Austrians, Caroline was told. The fault, retorted the *codini*, was not all theirs:

Since the revolution of '48 and '59 the [noble families]—who for generations had treated the [haute bourgeoisie] with . . . studied insolence—made over-tures to these parvenus and sent them visiting cards. . . . These cards were in most cases entirely unnoticed, the bourgeoisie remembering old affronts. . . . Access to the Haute Bourgeoisie is far more difficult for a stranger now than admission to the oldest families. . . . Memory of ancient wrongs [thus helps] to keep up these absurd distinctions of caste.[14]

But the codini olive branch was a very meager twig, Caroline recognized. A distinguished visitor asked Conte Cesare Balbo to arrange for him to meet Angelo Brofferio, Lorenzo Valerio, and other bourgeois Turinese notables, and Balbo offered to invite them to join the visitor at his country house in about fifteen days.

The stranger replied that his time was limited. [Could he not see] these gentlemen at [Balbo's] house in town some evening during that week? . . . Balbo replied that . . . as these gentlemen did not socially belong to his own circle he could not ask them to his house in town, that they probably would not come if he did, and that at any rate it would give offense to his friends. But in the country he could receive whomever he pleased.

Caroline gathered that "this distinction was simply a question between title and not title." But not *any* title would do; those newly ennobled for Risorgimento services were ignored by the old aristocracy. Though forced to share political power, the *codini* had become not less but more socially remote.[15] Pride of birth and arrogant hauteur precluded any intimacy with the bourgeoisie. The Piedmontese aristocrat, in D'Azeglio's memoirs,

failed in nothing due to courtesy, uttered no word to which you could possibly object, . . . and yet at the same time gave forth from his whole person such a clear "Keep your distance," such an obvious "I'm what I am and you don't count," that, as there was no reason for getting angry and no possibility of putting up with it, one simply longed to get out of range and . . . never let oneself be caught again.[16]

Circumstances unique to Piedmont intensified the aloofness of its nobility. After 1861 they were the only aristocrats in Italy attached to a resident monarchy. Fealty to the House of Savoy, virtually feudal in its focus on military service and court ceremonial, enabled Piedmontese bluebloods to outnumber other Italians, titled or untitled, in ministerial, diplomatic, and army circles. Wealth from recently aggrandized agricultural holdings and from renting parts of their Turin palazzi made it unnecessary for these proud cavaliers to merge interests with the bourgeoisie. Unlike Tuscans and Lombards they continued to shun commerce and trade, banking and industry, and perpetuated anachronistic life-styles that stressed status, honor, and battlefield bravery. Unlike the effete Venetian nobles satirized in the plays of Carlo Goldoni, the Piedmontese had kept militantly energetic through perpetual warfare, "morally more salutary than long periods of peace," averred D'Azeglio, and fitting men "to act well and courageously" in other matters too. Yet D'Azeglio himself had been stifled by narrow *codini* conformity: "God had supplied one brain only for the nobility of Turin, which was kept at Court in a showroom, so that all could go and get any ideas they needed." [17]

Marsh felt "constantly surrounded by very agreeable people" and rated Turin society superior to any he had ever known "in culture, refinement, and manner." But the gentility was excessive, conversation often stilted, formal, trivial. Titled women set the tone. "I don't know how it is," sighed Caroline after seeing off a stream of crinolined callers, "all these Piedmontese ladies bewitch me with their indescribable grace and delicacy." There were Costanza Arconati Visconti and Emilia Toscanelli Peruzzi, famed courtesans and art patrons; Risorgimento heroines like the contessas Balbo, Confalonieri, Margherita Provana di Collegno, and Marchesa Pallavicino Trivulzio, the Florence Nightingale of Garibaldi's campaigns; Virginia di Castiglione, legendarily "the most beautiful woman of the century," and the spirited Marchesa Doria ("La Pomposa"), rival Turin doyennes; and the quizzical Genoese Marchesa Marina Spinola, who on meeting Marsh immediately asked, "Did you bring any parrots from America?" [18] Among these ladies and their consorts the language of discourse was, of course, French.

A frequent caller of another tongue was young Rosa Arbesser, Princess Margherita's governess, who at once made Caroline Marsh her confidante. Rosa chattered hours on end—court intrigues, royal gossip, tirades against the Duchess of Genoa's mentors, the marchesas and contessas Della Valle, Della Rocca, and Villamarina. Caroline hardly knew "at which most to wonder—at the things related, or at the imprudence of the narrator." But Marsh kept his distance from "our talkative German friend." He fancied Contessa Clara Novello Gigliucci, the renowned soprano; clearheaded, witty, frank to the point of bluntness, she was "a real live woman," a welcome antidote to Piedmontese ladies who, with all their sweet charm, were often inconsequential and empty-headed. Marsh shrank too from another alternative—evenings at cards with gentlemen in the elite Società del Whist. He abhorred exclusively male gatherings as much as ever.[19]

Gossip, slander, and power struggles were common in the social arena. Accounts of flagrant misconduct filled the air. Caroline could hardly credit "that the persons I meet here in society are guilty of the sins laid to their charge. But . . . if there are no evil-doers, there are a prodigious number of liars." Annals of adultery, poisoning, crimes of passion, and incest flowed from well-bred lips. Hearing of one lovely marchesa's lurid misdeeds, Marsh exclaimed to Caroline, "Well! well! I thank God we are not like these publicans!" Caroline was less censorious; Mediterranean skies had enlarged "the soft twilight [of] my charity." Contrasting New England with Italian schooling and matrimony—"the teachings of a Puritan preacher and a popish priest, the marriage of choice and the marriage of *convenance*"—she concluded that "*these publicans* were not worse than *we Pharisees* might have been under the same circumstances."[20]

Caroline herself throve in Turin. The mild climate left her stronger than for fifteen years past. She held receptions three times a week, received callers incessantly, even went to the theater (mostly French and "highly immoral"), and assiduously dictated her scintillating diary. Her practical good cheer, her readiness to listen and sympathize, endeared her to Italians of all classes.

But the endless round of polite chat, the frivolities of court life, the tedious ritual of cards and calls were often more than the Marshes could bear. "I grow very tired," Caroline wrote after a year in Turin, "of the everlasting sameness of this high society." Note-writing took half her time, "and all this work leaves nothing to show for it."[21] Marsh himself was by no means exempt from such vexations. He had to return calls; ministerial dinners and court balls took up most evenings; and at home the ceaseless flux of visitors put work and relaxation alike out of the question.

*T*urin social life occupied Sundays as well as weekdays. "How rational people live without one day in the week wherein to feed their own interior life" Caroline could not imagine. Once in a while the Marshes salvaged a quiet New England Sunday for reflection and reading aloud—F. W. Robertson's sermons, Adolphe Monod's *Les Adieux à ses amis et à l'église,* Wycliffe Stanley's *History of the Jewish Church,* Ausonio Franchi's *Religion of the Nineteenth Century,* Samuel Vincent's *Méditations religieuses.* Marsh also went—often reluctantly—to church, because he thought it good policy for a Protestant diplomat in Italy; too many Catholics equated Protestantism with atheism. But the sermons of both English and Vaudois clergymen were apt to be tedious.[22]

Pleasanter diversions were royal hunts specially held for the diplomatic corps. On his first outing, at Stupinigi forest south of Turin, Marsh joined several envoys, "all professed sportsmen . . . with their own guns." Weaponless for twenty-five years, Marsh "did not think of shooting . . . but they put a double-barrelled gun into my hands . . . & I marched forth with the rest." Peasants drove game toward the firing envoys who "nine times out of ten at the very least—missed." Not so Marsh. "When my attendant shouted, *la lepre,* a hare! and pointed out the animal, I 'winked and held out my cold iron,' not really meaning the poor beast any harm, but, to my astonishment and that of my man, he fell dead. . . . Notwithstanding the practice & self-complaisance of my fellow diplomates, I did as well as the average of them, but . . . any other Vermont boy . . . would have killed more than the whole nine of us." Marsh's proficiency with firearms made him persona grata with the king, who longed to hunt buffalo in the Wild West.[23] The American Minister regaled His Majesty with quite fanciful depictions of Western roundups.

American and English visitors leavened Turin society. Hiram Powers came from Florence for his daughter Lulie's wedding at the British Legation in May 1862; in June Caroline went to sit in Florence for a gravely classical bust by Powers. Caroline Estcourt came twice to Turin; so did Dr. Humphrey Sandwith, blithe and pungent as ever, back with his bride from a Mediterranean honeymoon. Marsh's younger brother Charles, a shy, kindly bachelor of forty-four, broken in health by Vermont toil and tuberculosis, stayed seven months in 1864.

Other American envoys in Europe often called. The Minister to Switzerland, erudite New Hampshire editor George Fogg, was much liked by Marsh; Henry Sanford at Brussels seemed able but overzealous and vain. Bradford Wood at Copenhagen was an old congressional crony; Marsh renewed enjoyment of Wood's witticisms. With James S. Pike, at The Hague, Marsh had become intimate on the voyage out; the two traded gripes on being overworked and underpaid. Many American travelers stayed with them: Robert J. Walker, Marsh's antitariff, expansionist foe of the 1840s, in Europe in 1864 to sell Union bonds; Alexander Dallas Bache of Smithsonian memory, now sadly aged; the English-born Blackwell sisters, journalist Anna and pioneering woman doctor Elizabeth, ever helpful to Caroline.

European visitors included the poet Richard Monckton Milnes; Matthew Arnold, much taken with Marsh and his "handsome" wife; Suez Canal builder Ferdinand de Lesseps, whose eloquence overwhelmed them; and the philologist Max Müller, who talked grammar with Marsh for hours. On an Alpine excursion at Simplon in 1862, Marsh encountered the Reverend Isaac Taylor; the aspects of nature were soon forgotten in the charms of philology. Taylor's revised *Words and Places* (1873) credited Marsh with help throughout and with virtual authorship of a chapter on Teutonic place-names in the Italian Alps, a topic of abiding interest to Marsh.

At the start of his Italian mission, Marsh was sixty but could have passed for forty-five, his face unlined and hair ungrizzled—including lately acquired beard and sideburns. He had also continued to add girth and now weighed 230 pounds. But his solid frame held up remarkably. He worked with ease fourteen hours daily in office and library and could climb all day in the Alps for weeks at a stretch.

Habitually reserved with strangers, Marsh impressed others with his vigor, candor, and unassuming erudition. He "looked like a Vermont farmer," wrote Genoa consul David H. Wheeler, "and talked with the trenchant force of a business man in New York." Matthew Arnold forgave this "tall, stout, homely-looking savant" his New World origins, terming Marsh "that *rara avis*, a really well-bred and trained American . . . redeemed from Yankeeism by his European residence & culture";

in Marsh, Arnold felt "the bond of race distinctly." His fellow-envoy Pike found Marsh *too* bookish; to Pike he lacked political acuity despite "great knowledge on a vast variety of subjects."[24]

Marsh had come to Italy expecting "little work and big pay," he half jested; "we shall have a fine, lazy time of it." He was wrong; between diplomacy, scholarship, and society, he was incessantly busy. "I have been entirely disappointed as to the rest and relaxation I looked for," he declared half a year on; he had "at few periods of my life, been obliged to work so hard."[25] On this point Marsh was a bad judge; every period of his life seemed the most arduous at the time. Official diplomatic duties were not too taxing—Italo-American relations, while friendly, were so embryonic that, Civil War matters aside, they required little high-level attention. But as noted, Marsh's diplomatic brief went well beyond routine Italo-American affairs. And what Marsh called "suits of fools" took up countless hours.

As their numbers mounted, American tourists grew more presumptuous. Marsh had no cause to amend the view he had formed in Constantinople that "the great mass of our countrymen, who visit the Old World, are ignorant, impudent, and vulgar." They used their envoys as bankers, hoteliers, and personal servants. Marsh was asked "to look up inheritances and investigate genealogical records; to search after the lost baggage of travellers; to act as mail agent [at his own expense] for reception and forwarding of letters; to aid in introducing patent medicines for man and beast; . . . to replenish the purses of travellers whose 'expected remittances' have failed to arrive"; to procure postage stamps and autographs and, to cap it all, a lock of His Majesty's hair. "I should as soon think of making a museum of the nail-parings or old shoe-heels of European sovereigns," he told Seward.[26] For such causes Marsh spent ten till three daily in the legation, besides the dinners and receptions where the real diplomatic work was done.

Engrossed by public affairs, Marsh transacted them with zest and thoroughness. Yet he still spent half his life in private study. He rose before five and did a day's work in his library before the legation opened. Back from the office, he dealt with a huge correspondence before dinner, then went calling, received guests, or read to Caroline till bed at ten if there was no evening function. Turin's mild climate was as good for him as for her. His eye ailment and neuritis cleared up. Rheumatism and sciatica he bore stoically as normal in a man his age and build; sporadic indigestion could hardly be charged to luxuriating in Italy rather than hibernating in Vermont.

He longed for his library. Not until August 1861 did six hundred books shipped from America reach Turin; then he regretted not sending "six times six thousand," he wrote Francis J. Child. "How endless are the wants of a scholar! Every time I cast a glance at my shelves, I say to myself, why did I bring *this* book, and why didn't I bring *that*?" Meanwhile he kept buying for himself and American friends: science for Baird, lyric poetry and dialect tales for Child, classics for C. C. Felton, Italian works of every kind for the Harvard library.

Marsh's work habits remained multivalent. Surrounded by books lying open on his desk and on the floor, he swiftly collated materials on a variety of subjects. He toiled without stopping; to relax he would turn to a text in another language or on a different topic. Completing *The Origin and History of the English Language* proved tedious. "Isn't it hard," Marsh moaned, "that I should be shut up in a closet 9 by 10, working over old lectures, instead of revelling in Italian literature and Italian nature?" Ignoring the protest of his nephew G. F. Edmunds that it was "just as sinful to commit suicide" by overwork as any other way, in spring 1862 Marsh sent the manuscript to the printer. But his royalties barely covered the postage costs.[27]

Although writing seemed slave labor, Marsh agreed to go on with it; his growing repute might in time pay him better. Even as he prepared his next book—*Man and Nature*—Charles Scribner urged new efforts. "You are so accustomed to hard work that you cannot be idle," wrote the publisher. Scribner wanted a text "in the department of English languages and literature of which you are the acknowledged head." Marsh projected a history of Mediterranean commerce and a text comparing American and European politics and religions.[28] But these envisaged volumes never materialized.

Weary of checking sources and reading proof, Marsh took more interest in the literary labors of Child and Norton than in his own. Child did not know "another man in America to whom I could go with a difficulty"; Marsh was helping him with Chaucer. "You are the editor [for Chaucer]," Child told Marsh, "if you had your eyes—and I should be satisfied to do small chores for you." Marsh championed Chaucer:

"I have formally proclaimed him to be (chronologically) the first, and all but the greatest dramatist of modern Europe." He advised Child on word meanings, style, and the touchy issue of Chaucer's vulgarity. Marsh lamented "works disfigured, stained, polluted by a grossness of thought and of language," but palliated Chaucer's lewdness as inescapable in the pervasive "moral and religious degradation of the fourteenth" century. "In publishing in America," he warned, "one must tread gingerly, you *can't* print his naughtinesses entire, and yet one hates sadly to mangle him." Marsh's counsel left Child "blazing like a rapt seraph with gratitude."[29] Norton, whose Dante translations Marsh took great pains to aid and promote, was also deeply indebted to the American Minister.

To Norton and Child, Marsh confided his "heretical" ideas about language. One was that Latin was not the parent tongue but a sibling of no greater antiquity than other Italian dialects, and that no "modern Italic speech is derived from *classical Latin*."[30] He marveled at the persistence of manifold French and Italian dialects, yet thought most dialects ultimately doomed by increasing mobility and intercourse among peoples everywhere.[31] In this Marsh was essentially correct, though he underestimated how long it would take to approach uniformity even of national language—above all in Italy.

Why did not Providence give us Alps and a good climate?
Marsh to Donald G. Mitchell, June 1865

When not studying the languages, Marsh was enjoying the landscapes of Italy. For total respite he turned to the Alps. So near did they seem to his balcony that "I often amuse myself," he told Baird, "knocking the icicles off the eaves of that respectable hillock, Mt. Rosa, by shying pebbles at 'em. Nay, when it is *very* clear, I can reach the walls of the mountain with my pipe-stem."[32] The view from Turin, ringed by snow-capped summits from Monte Viso in the southwest around to Mont Blanc, the Matterhorn, and Monte Rosa, made him "half mad with admiration," Marsh wrote Edmunds. "I did not know there was on earth any such beauty as that of . . . the Piedmontese Alps, and the air is as refreshing as the scenery is exquisite." Their vivid and delectable contrasts made the Italian Alps "far superior to the Swiss." They combined the "sublimity" of Switzerland with "the beauty and luxuriance of almost tropical vegetation."[33]

To admire the Alps from afar was one thing; to get to them quite another. Unless given explicit leave of absence, American envoys were expected to be constantly at their posts. Although court and cabinet deserted Turin all summer, Marsh could not quit the capital for more than ten days lest his salary be stopped. The jaunts of certain diplomats brought such odium that Seward sent out an "anti-vagabond" circular threatening to abrogate even the ten-day privilege. "I detest that rule," Marsh wrote Pike at The Hague. Perhaps Pierce's and Buchanan's envoys had been "too peripatetic," but Republicans should be trusted with some liberty (he did not advert to his own wanderings as a Whig envoy). For him the issue was clear: "Mrs. Marsh must go into the country, and I don't want to be divorced."[34]

No fussy scruples detained Pike. He went where and when he pleased and advised Marsh to do likewise. Marsh bent the rules as far as he dared. Legation business frequently recalled him to Turin—he grumbled at being summoned back from the Alps three times in one month to furnish passports—but conscience alone never confined him. Legation staff postdated his departures and predated his returns; letters written out of town were headed "Turin." Friends were sworn to secrecy; if the press got wind of his trips, "Greeley will have me sent to the penitentiary for absenteeism."[35] As noted, Marsh was not famed for compunction, having quit Constantinople for Egypt before getting Webster's *congé* in 1851.

Brief trips during the first Italian summer renewed Marsh's joy in mountain scenery. In July he visited Monte Rosa and Lake Maggiore; in August the whole family went north through Piedmont to the foot of the Matterhorn. Marsh scrambled up the moraine and across the glacier, wound around crevasses and crossed snow bridges to 11,000-foot Théodule Pass, then zigzagged down into Switzerland between Gorner and Zermatt. Carried by porters, Caroline fared so well on this trip that Marsh vowed her next venture would be a balloon flight from the top of Mont Blanc.

Thereafter Marsh went to the Alps every summer. Some trips were to nearby valleys, around Biella and Varallo to the northeast, and to Vaudois villages, refuges of the persecuted Waldensian sect, between Pinerolo and Paesana to the southwest. Others were farther afield; in October 1863, for example, they went to the Dauphiné Alps, down the tempestuous Durance to Embrun, thence west to the Rhone and Aigues Mortes, which "ought to be kept under glass," urged Marsh—a prescient preservationist—"for the benefit of those who love to pry into the life of the Middle Ages."[36]

But Marsh most adored the high Alps. He climbed the Becca di Nona (Pic di None) south of Aosta, among pastures so hard of access that goats were pulled up to them by ropes. He scaled the Schilthorn and the Faulhorn in the Bernese Alps. He crawled through the Col de la Traversette in the Alpine crest north of Monte Viso. Mesmerized by glaciers—Zermatt, Aletsch, Grindelwald—he risked life and limb studying ice flow structure. Marsh would have liked to perch his summer home "on the brink of a glacier," or perhaps even to "dig me a hole in a glacier and live there," he wrote Baird, "to escape the devilries which are going on around me."[37]

These expeditions were family affairs. Caroline claimed to conquer peaks and passes more easily than cities and dressing for dinner, though Marsh suspected she "feigned pleasure, to gratify her foolish husband who is ice-mad"; young Carrie Crane matched her uncle's hair-raising exploits. Marsh's ardor, sturdy constitution, and Green Mountain training made for feats of endurance. He had greater stamina at sixty than at forty. "Considering my age and inches (circumferentially), I am not a bad climber," he bragged, "going up and down many thousand feet in a day, without overmuch puffing and panting." He compared his Alpine exploits with the trifling efforts of most tourists. "Did you go to the Lake [Fiorenza]?" (the usual tourist goal at the source of the Po below Monte Viso), asked some travelers. "No," replied Marsh, "but we went a thousand yards above it."[38]

Whether such strenuous exploits were wise for an overweight sexagenarian, he returned from summer climbs ready to tackle "brain work" with renewed vigor. "As I am getting so strong at 63," he boasted, "I suppose I shall climb the Himalayas at 100." Even when rheumatism and old age precluded climbing, Marsh's love of the Alps never waned. Nothing in nature gave him more pleasure; he regretted that life was too short to visit all the mountains in the world. And everywhere else as well. "What do you think of Mrs. M. and me," exulted this dauntless traveler, "who haven't been out of our trunks since 1843, & are ready to start for Timbuctoo at an hour's notice as soon as any body will pay the cost!" [39]

Marsh's summer excursions gave insights into Italy's people as well as places. Poverty and malnutrition were the common lot, cholera and malaria rife. Inhabitants of Piedmont's Aosta valley seemed "filthy and miserable, & cursed with goitres & cretinism to a frightful extent." [40] Rural folk elsewhere were little better off. Agricultural modernization had raised productivity and landowners' profits but reduced landless laborers to penury. Worst was the plight of the Mezzogiorno (the south). But even "in North Italy," Marsh noted in a dispatch, "the houses of the poor have no floors but the bare earth, they are damp, without stoves, fireplaces or any means of drying or warming the laborer when he returns from the field in the evening [but] a small brazier of coals or a few sticks." Marsh judged "the wretched poverty of the agricultural laborers" a contributing cause of pellagra, the as yet undiagnosed dietary deficiency that wreaked such havoc where rice and maize were exclusive staple foods.

The "habitual misery" and "extreme destitution" of at least three-fifths of Italians seemed to Marsh the new nation's gravest menace. To this Piedmont's mean and stingy aristocrats seemed blind; "they expect the Church somehow to look out for their charities," noted Caroline, and "the poor are left to die as they may." [41] Italy was riddled with crime and fear, seething with unrest; insurgence was put down with brutal severity as "brigandage." Successive cabinets, obsessed by crusades to recover Rome and to suppress unrest, ignored the plight of ordinary folk. Epitomizing elite remoteness, a Piedmontese contessa told Caroline that an 1864 tax law would

ruin the landed aristocracy . . . they must sell, and the estate will soon fall piecemeal into the hands of the peasantry. [Caroline] asked whether this was likely to prove an injury to the prosperity of the country generally, or only likely to diminish individual wealth. "Oh, the poor will be better off, no doubt, but there will be no landed aristocracy to rally round the throne." [Contessa Maggiolini] had always supposed that where there was no aristocracy there must be anarchy.[42]

Urban laborers were seen as potential criminals. Marsh was dilating on "the hard fate of the poor here, forced to live in the garrets of these lofty palaces, without fire in winter and suffocated by the heat in summer, obliged to carry water and everything else up so many flights of stairs." A lady rejoined that it was "safer for the Government that the poor should live in this way in the garrets of the rich than that they should have houses in quarters by themselves," where the likes of Garibaldi would foment sedition and turbulence and "stir them up to mischief."[43] That common people might be upright citizens was unimaginable.

Marsh's initial infatuation with Italy had suffered some disenchantment: good intentions seemed crippled by its rulers' apathy, vacillation, insensitivity. But the defects of Italy's leaders, Marsh now discerned, were not the sole or even prime causes of Italy's plight. Face-to-face glimpses of rural life showed how limited and superficial were the reforms of the Risorgimento. Even in "progressive" Piedmont, barely half were literate, and fewer than 2 percent were entitled to vote—"a franchise which, narrow as it was, appeared to [their rulers] too wide," in Croce's mordant phrase, "considering the calibre of the constituents." Rural laborers were as hostile to rule from Turin as from Vienna or Paris; the freedom they yearned for was freedom from despotic landlords. The national unity dear alike to the exiled Mazzini, Garibaldi, parliament, and Cavour's successors was essentially an elite dream; it had little to offer the vast majority. For many, Risorgimento simply meant exchanging local for distant oppressors.

Nor had unification begun to overcome regional enmity and discord. Italy's dozen or so provinces had their own dialects, laws, culture, and genres de vie, a parochialism further fragmented by local allegiances. To most, unity felt less like Italianization than Piedmontization under a remote, not to say alien, House of Savoy (Savoy itself had been ceded to France). To the Turinese, on the other hand, the rest of Italy seemed a backward realm in need of ruthless improvement. Just as

Marsh reached Turin, Sir James Hudson was limning Piedmontese disdain. He told how Cavour, ignorant and contemptuous of the rest of Italy, went "patiently through a list of Lombards to find a Lombard [cabinet] minister. He named one after another, and looked wistfully in my face." The search was hopeless. "Next he turned to Neapolitans: *not one*, save Poerio (who won't accept). Next to Sicilians: here he brightened up a little; he has got two! . . . but the first perhaps won't accept, and the other is unknown to the public."[44]

Despairing views of southern Italy were not confined to Piedmontese such as D'Azeglio, who likened union with Naples to going to bed with a victim of smallpox. "In seven million inhabitants of Naples there are not a hundred who want a united Italy," wrote Luigi Carlo Farini, the Emilian who spearheaded central Italy's adhesion to Piedmont. As governor of the Neapolitan "hell-pit" in 1860, Farini had to contend with "12,000 tricksters . . . law-twisters, casuists and professional liars with the conscience of pimps. . . . What can you possibly build out of stuff like this?" A mere handful of leaders from "Italian Italy" could not "turn swine into heroes," he groaned to Marco Minghetti of Rome, like Farini a future prime minister. "And by God they will outnumber us in parliament. . . . If only our *accursed* civilization didn't forbid floggings, cutting people's tongues out, and drownings. Then . . . we would have a clean slate and create a new people." But instead of the greater Piedmont Cavour had envisioned, unified Italy came in many ways to seem an enlarged Naples. For all Marsh's earlier enjoyment of Sicily, he shared the anti-Mezzogiorno bias, terming Sicilians "deplorably ignorant, indolent, and vicious."[45]

Regional animus was not limited to the north/south hiatus. Lombards complained that Turin-imposed legislation was more regressive than the Austrian code it had replaced, that Piedmont was consigning Lombardy back to the Middle Ages. Tuscans, Cavour's "timid Etruscans," held their heritage superior to any other and were irked to play second fiddle to barbarous Piedmont. Others saw the Piedmontese as too little Italian, too provincial, too domineering. The elite embraced the House of Savoy not out of love but fear—fear that an aroused peasantry and an antimonarchist middle class might foment real revolution.[46]

Regional jealousies reflected discrepancies long hidden by mutual ignorance. Up to 1861, travel between provinces was not only physically onerous but illegal without a police permit. No one at the time, in Mack Smith's words, "truly knew the various Italian regions well enough to be able to compare" them objectively, let alone to predict how a uniform system of laws would affect their huge disparities. In the end, the extension of Piedmontese law throughout Italy, with many pernicious results, was driven less by hope than by anxiety lest a precarious unity succumb to papal reaction, to republican rebellion, or to foreign imperial dynasts. The American Civil War persuaded many Italian leaders that federalism was a risky precedent. Promises of regional autonomy reiterated by every prime minister were soon forgotten in the zeal to cement central authority, to collect taxes, and to stamp out crime and disaffection. And leader after leader clung to the delusion that national patriotism was indeed overcoming regional differences, thereby further aggravating regional rancor.[47]

One advance toward Italian national unity Marsh extolled was the educative role of the army. Illiterate recruits were taught to read, write, and compute, and still more, learned for the first time that "they have a country, a government, [with] reciprocal rights and duties." Military service was a pathway to patriotism: three years' service transformed "the raw, half-human, priest-fearing, and God-denying Calabrian conscript [in]to the trained Italian soldier." On the brink of the American Civil War, he had similarly enjoined federal allegiance on West Point cadets, who must rid themselves of all state and sectional partiality.[48] Marsh's vision of armed services as seedbeds of national patriotism, then common parlance, jibed with his notions of character building through hunting and fishing, discussed in Chapter 9 above.

As Farini's image of cutting out tongues suggests, lack of a national language was felt a major obstacle to Italian unity. Literary Tuscan—the tongue of Dante—was a lingua franca more in theory than in practice; local dialects were mutually unintelligible; the mother tongue of many prominent Italians was alien. Garibaldi's was Provençal, Cavour's French. When Cavour said, "Je suis Italien, avant tout," he said it in French. "In Italian he is embarrassed," said the Milanese Marchese Arconati of Cavour. "You see that he is translating; so is D'Azeglio; so are they all. [Piedmont's parliamentary] deputies all speak in Italian, but this is to them a dead language, in which they have never been accustomed even to converse."[49]

At least Italy was spared Piedmont's "bastard patois," closer to French and Catalan than to Italian, "the most awful jargon in or out of Christendom," in Sir James Hudson's view, echoed by the Marshes: "Oh that these unhappy Turinese had something like a language! If they try to speak French, they speak bad Italian—if Italian, bad French, and one is left to divine at least half of what they would say." As the Savoyard capital over two centuries, Turin had gained prestige for Piedmontese; some even dreamed it might become a national language. That dream was punctured by French invasion and annexation, followed by rebellion and Risorgimento, leaving French (though still essential in diplomacy) a symbol of oppression, but making Italian an icon of heroic resistance to foreign rule. Yet save on ceremonial occasions, court and other Turinese elites continued to speak Piedmontese.[50]

Despite dominating Italy after 1861, the Piedmontese communicated poorly with other Italians, for Turin bureaucrats detested standard Tuscan. "The aversion of a Piedmontese gentleman to Italian is intense," Caroline was told. The Piedmontese Barone Fava, crown prosecutor at Ancona, "perfectly understood Tuscan, yet he himself was *not* understood by the common people of Ancona," Fava confessed to Marsh. Fava's Piedmontese intonation was so thick "that they do not know what he is saying." Fava's experience was typical; Piedmontese officials were at linguistic loggerheads all over Italy. Antipathy to Piedmontese still lingers; many Italians' first reaction to Marshal Badoglio's 1943 radio announcement of the end of fascism was to resent his Piedmontese accent.[51]

Deploring Italy's Babel, the Lombard author Manzoni had rewritten his classic novel *I Promessi Sposi* (1825–26) in standard Florentine, "rinsing out his rags in the Arno," as he said—though Tuscan purists still found Lombardisms in it. Yet while eager to forge a national language, Manzoni condoned dialect too. Imagine speakers from Piedmont, Bologna, Venice, Naples, and Milan trying to commune in "our common Italian language," he said. Each would lack the "richness and sureness of vocabulary . . . and sure command of speech that he would have at home." Deprived of "our local idiom . . . we would find ourselves without a large number of lovely, subtle and appropriate expressions."[52]

Others felt literary Tuscan ill-suited to practical use. Cabinet minister Ubaldini Peruzzi (a Pisan) "often found himself tempted by the very beauty and grace of Tuscan expression to say more than he really meant to say," he told Marsh; he used "French wherever great precision was desirable." How pathetic, thought Marsh, that Peruzzi's "practical" tongue was not Italian but French. Marsh saw the sway of dialect crippling Italy. He longed to "crop and imprison any man who should write a line in any of these vile jargons. They are a great curse to the Italians, who will never be a nation till they begin to *think* as well as write in . . . a common cultivated language."[53]

A decade later, Marsh saw some advance toward linguistic unity. Moving the capital to Florence in 1865 had in his view accelerated the spread of the lingua franca: "accepted as the common tongue of the whole kingdom," Tuscan was now "exclusively" used in literature, education, parliament, law, journalism, commerce, and "general social circles." But he knew the battle was far from over. Champions of local and regional speech, Marsh wrote, held it "impossible, and even undesirable, for the whole Italian people to conform to the Tuscan"; grammar might be standardized, but each region must cherish its own "native vocabulary and idiomatic phraseology."[54]

With this view Marsh had some sympathy. Manzoni's "rich, expressive, and refined" dialects had to be conserved until an enriched Tuscan was familiar to all Italians. In the upper Val d'Arno Marsh had heard "Tuscan with a beauty of articulation—an elegance of phrase, and a picturesqueness of vocabulary" unrivaled in Europe. Here might be gleaned "the enrichment and variety of diction, the flexibility and versatility of construction, which the *lingua comune* needs before it can become truly an all-pervading vital medium of national thought and expression."[55]

In the main, though, Marsh stressed the evils of what he termed bilingualism: Italians read and wrote in one language while living in another. "An elegant writer [or] a fluent and correct speaker, habitually thinking, talking, . . . doing business, and even making love" in a dialect utterly remote from literary culture would be hard to imagine among users of English, yet most Italians routinely endured this schizoid existence.[56]

To this dualism Marsh ascribed an Italian inferiority complex marked by undue deference to French, "far too generally the habitual language of fashionable Italian society." Marsh held this doubly harmful. "A predilection for a foreign language and a foreign literature," he wrote, impedes "the cultivation and improvement of the national speech and national letters" and crushes "native originality and independence of thought." French was bad for Italians because it was *foreign*, "alien to the national heart, and disturbing to . . . the national intellect."

It may seem ironic that this eminent linguist, who had mastered so many languages, should urge Italians to cleave narrowly to their "national" tongue. But Marsh felt his own example unhelpful here. "With rare exceptions no man can freely use more than one language as a medium of intellectual or oral discourse," and habitual use of a foreign tongue reduced "mastership of our own." Rather than looking wistfully to French or Latin, Italians should cultivate their native vernacular. Yet a full century would elapse before national language usage spread much beyond the elites Marsh had enumerated in the 1860s. And even now a majority of rural Italians habitually use dialects opaque to outsiders.[57]

Marsh well understood why Italian linguistic unity was relatively retarded. Italy had lacked a popularizing Luther, a centralizing French monarchy, an Anglo-Norman vernacular potency; Italians had instead gone on revering fossilized literary forms. Above all, Italy remained socially fossilized: Risorgimento elitism blocked the public participation needed to turn Tuscan into "the universal popular speech" Marsh thought it promised to become in the 1860s.[58] To be sure, no Continental nation then boasted a standard language common to most of its people. But only Spain was as linguistically fragmented as Italy.

Most Italians in the 1860s and 1870s knew, with Marsh, how shallowly based was their political unity. But such doubts were seldom publicly voiced. It was not politic to note that some disparities of class and wealth had grown *more* intense under the Risorgimento. "The ignorance and degradation of the lowest stratum of society, and the obscurantism of the highest," wrote Caroline Marsh, left the chasm between "real" Italy and "legal" Italy as unbridged in 1880 as in 1850. In Marsh's final year, a famed socialist critic noted a grim irony: at the dawn of Italian unity, its leaders had forecast Italy would rank first among European states. Italian primacy had since been earned in but one respect alone: Italy had the Continent's highest rate of criminal violence.[59]

For all Marsh's bleak awareness of Italy's seemingly implacable miseries, he went on hoping that Italians might somehow surmount them. Stalwart folk in Alpine valleys braced his New World faith in rural rectitude and the virtues of adversity. Just as Marsh lauded self-reliant Italians by contrast with "dull and clownish" Austrians (the "Styrian or Carinthian peasant as solemn not to say stupid as the cattle he drives"), so he praised Italian plebeians in general as lively and imaginative.[60] Marsh singled out Italian coachmen for special praise. "In America, all stagemen & carriage-hirers, and the like are worse than bandits," whereas in Italy "we have never encountered a single unequivocal rogue." Not once in thirteen years' Italian residence had anything been stolen from them. Marsh extolled the overtaxed, underfed peasants whose "poverty is such," he wrote in his last long dispatch, "that but for the vague hope of some great change for the better it could scarcely be endured."[61]

Marsh avidly seized on and magnified every apparent sign of improvement. In 1865 he saw Italy's growing well-being "strikingly manifested in the decrease of mendicancy." Cultural progress was also evident: "education is rapidly spreading among the lower, and what is more surprising, even among the higher classes." At a Florence industrial exposition in October 1861 he noted a striking change from the 1850s; then he had seen mainly foreigners, now he heard scarcely a word but Italian. This "first common gathering of Italians for centuries," he wrote Seward, was giving "new impulse to the spirit of Nationality." Better still, that spirit "manifests itself most strongly & spontaneously among the peasantry, whose ignorance & hereditary prejudices, it had been feared, would prove an almost insuperable obstacle to the germination & growth of a large & generous patriotism."[62] Eager to be sanguine, Marsh's despair over Italy's plight became ever more bitterly voiced.

All in all, Marsh had never been happier than in these first Italian years. Absorbing tasks, delightful Turin, enthralling (if at times too numerous) companions, Alpine grandeur—he saw everything *couleur de rose*. Doubts about the wisdom of returning to Europe soon vanished. Remote from Civil War turmoil, Marsh at times felt he ought to be in its midst, yet Union disgrace and disrepute made him "feel as if I had no longer a country." He found ever more reason to stay where he was. Marsh could not earn a living at home; his family was scattered or gone; Vermont winters were lethal; the mere thought of an Atlantic voyage was unendurable; above all, "giving up the Alps seems to me like giving up all the material world." He had "such a passion for the *nature* of Italy," he declared after four years there, "that I do not see how I can ever live under another sky."[63]

The perennial flaw was shortage of funds. Marsh never found any salary adequate, and his $12,000 a year in Italy was no exception. Most American envoys grumbled over niggardly pay; as before, Marsh complained with woeful frequency. In staid Turin, which had mushroomed like a frontier town, the cost of living had more than doubled in five years. He was paying $2,000 a year for their half-furnished apartment in the Casa d'Angennes. The Turinese, he suspected, "know everybody's salary, and take the whole, generously letting us run into debt for any trifle we may want beyond our income."[64]

As in Constantinople, the disparity between the penurious American envoy and affluent European confreres was glaring and painful. The ten-day rule made matters worse, for it cost less to travel in the Alps than to summer in Turin. Marsh thought his pleas for leave of absence "would draw tears from the eye of a Cyclops," but the Secretary of State remained dry-eyed and flinty.

Marsh's troubles worsened at the end of 1861, when the Contessa Ghirardi, his Casa d'Angennes landlady, turned them out "because we would not submit to a small addition of $400 to the rent we already paid for an apartment hardly large enough to swing a cat in."[65] The Casa d'Angennes had been hard and costly to heat as well as cramped; they appreciated its virtues only after trying to find another home. Weeks of hunting turned up nothing. Wanting a mild climate for Caroline, Marsh sent his family to Lake Como, joining them on and off during late summer and fall. Finally they took rooms in a hotel at Pegli, a coastal village west of Genoa, four hours by train from Turin, where Marsh remained weekdays. On weekends in this then remote Riviera locale (now engulfed by the port of Genoa), in the winter of 1862–63, he turned in earnest to his most momentous work: *Man and Nature; or, Physical Geography as Modified by Human Action*.

13. Man and Nature: *The Making*

The face of the earth is no longer a sponge, but a dust heap, . . .
rendered no longer fit for the habitation of man.

Marsh, *Man and Nature*

MARSH INTENDED *MAN AND NATURE* TO BE "A LITTLE
volume showing that whereas [Carl] Ritter and [Arnold] Guyot think
that the earth made man, man in fact made the earth."[1] But in so doing,
Marsh warned, man might also destroy both himself and the earth. *Man
and Nature* bared the menace, explained its causes, and prescribed anti-
dotes. We must learn how we affect our environment, and restore and
husband it as long as we tenant the earth.

A large portion of the American envoy's time had gone into this
book's making—more than a diplomat might be supposed to spare. On
the very day he sent off his final proofs, Marsh told Secretary of State
W. H. Seward why he had written it: he wanted "to show the evils resulting from too much clearing and cultivation, and often so-called improvements in new countries like the United States." Several who had
seen the text, Marsh assured Seward, judged it "important to the inter-

ests of the American people."[2] The book's importance was, indeed, acknowledged almost from the start; revised and reprinted several times over nearly half a century, *Man and Nature* brought environmental awareness and reform not just to America but to the whole world.

More than Marsh had dreamed, *Man and Nature* ushered in a revolution in the way people conceived their relations with the earth. His insights made a growing public aware of how massively humans transform their milieus. Many before Marsh had pondered the extent of our impact on one or another facet of nature. But most took it for granted that such impacts were largely benign, that malign effects were trivial or ephemeral. None had seen how ubiquitous and intertwined were these effects, both wanted and unwanted. Marsh was the first to conjoin all human agency in a somber global picture. The sweep of his data, the clarity of his synthesis, and the force of his conclusion made *Man and Nature* an almost instant classic. Marsh had "triumphantly" investigated a subject "so abstruse, so vast, and so complex," attested his memorialists in 1882, "that it is fair to say he had no rival in the work."[3]

Marsh viewed the work in prospect in no grandiose terms. He made fun of himself as an idle fellow who set pen to paper only owing to his wife's incessant scolding:

Ever since we were married, my wife has been pestering me about making a book. . . . After fifteen years' punching, I made one about camels which took so surprisingly that the publishers actually sold not far from 300 copies in little more than three years. This I didn't find very encouraging, but Mrs. Marsh told me about Herschel, who tried 279 times before he was able to make a speculum [optical reflector]; Palissy, who labored 16 years & spent all his substance before he turned out a porcelain cup; Jean Paul [Friedrich Richter], who wrote 40 books before he sold one, &c, &c. So I made another book which did pretty well. . . . I hoped she would be satisfied. But no! she insisted I had not told all I knew, & said she would give me no peace till I had. I quarrelled with her as long as I could, compared her to the late Mrs. Albert Dürer of Nuremberg, of Xantippean memory, & the like, but the more I begged her to leave me alone, the more she wouldn't . . . and I have been enforced to begin another great volume, the object of which is to tell everything I know & have not told in the others. . . . I shall name it "Legion."[4]

He belittled *Man and Nature* as a "burly volume" on a "dreary subject" that would "ruin the printers." There was some basis for Marsh's self-deprecation. Like many, he was inclined to gather materials endlessly and put off synthesis and composition. Caroline urged him forward, fearing "that the accumulating process would be going on until the brightness of your intellect and memory would be dulled by age and the world would lose the results of so many years of study." The result was a volume not fully digested, nor easily digestible.[5]

Marsh had started *Man and Nature* in March 1860, in a glow of expansive creativity following his release from debt. "I began another book this morning," he wrote Spencer Baird from Burlington; "I won't tell you what the book is about, because you'll call me an ass (it's none of your *ologies*)." Marsh was anxious not to be charged with poaching on the domain of reputable scientists:

Now, don't roll up the whites of your eyes, and quote that foolish old saw about the cobbler and his last. I am not going into the scientific, but the historicals, in which I am as good as any of you. What I put in of scientific speculations, I shall steal, pretty much, but I do know some things myself. For instance, my father had a piece of thick woodland where the ground was always damp. Wild turnips grew there and ginseng, and wild pepper sometimes. Well, sir, he cleared up that lot, and drained and cultivated it, and it became a good deal drier, and he raised good corn and grass on it. Now I am going to state this as a *fact* and I defy all you speculators about cause and effect to deny it.[6]

Any settler might see that clearing drained his land and dried the soil. From such mundane facts Marsh would unfold his history of how human enterprise had both improved and impoverished the earth.

Other duties and deadlines intervened, and Marsh put *Man and Nature* aside. In 1862, on a wintry April afternoon in Turin, he sent off the finished text of *The Origin and History of the English Language*. An hour later Caroline went to his study to see how he was enjoying his newly recovered leisure. She found him

with a heap of manuscript, loose notes etc., about him: "What are you doing now?" I asked; "At work on my next book," he said in the quietest way in the world—and sure enough the projected "Physical Geography" was already in the forge. "Well," I said, "it cost me fifteen years of hard work to wind you up to the writing point, and now I believe you are likely to run on without stopping for the next fifteen." "Perhaps so," he answered, but did not look up.[7]

The following winter, in peaceful Riviera isolation, Marsh set to work in earnest. From his hotel window or pacing along the Pegli strand he saw an ever mutable Mediterranean coast; the splendors of that locale suffuse his pages. The sea was not always calm nor the face of the land genial. Penetrating northern winds and rain-laden southern siroccos brought insistent reminders of nature's violence. The rains bred torrents in narrow coastal ravines; the turbulent Torrente Varenna tore away soil and rock, wrenching trees from their footholds; slopes melted into landslides. The wind-tossed sea periodically churned up the shore, swamping hastily built bulwarks, splintering fishing smacks, threatening to engulf Pegli itself.

If nature diverted Marsh from his writing, man and his works did not. There was little to draw sightseers to the small fishing village or to the hotel fronting the gravel shore. Marsh spent every other week in Turin, leaving Pegli Monday night and returning Friday afternoon. At Pegli he wrote all morning, and in the afternoons strode along the beach, wandered in the Pallavicini gardens nearby, read aloud to the family, and listened to young Carrie's Italian lessons.[8]

As the book took shape, Marsh wrote the faster for fear some rival might beat him to it. "This would be rather hard," reflected Caroline, "when he has been studying the subject for so many years, and only being prevented by adverse circumstances from giving his thoughts upon it to the world long before. . . . An article in the London Times this morning on the very subject on which Mr Marsh is now preparing a book has made him a little nervous lest he might be anticipated." The Times article spoke of the exhaustion and erosion of "vegetable mould" (topsoil) as a growing global threat, triggered by the destruction of forest cover. An editorial rued that men had "burnt up the woods" and depleted resources; "we of this generation are recklessly wasting old mother Earth, . . . washing her away to bare rock and hungry sand." In time "good men will lament the folly and selfishness of their ancestors, who flung away their national patrimony and wasted what they did not use." This lament was indeed similar to Marsh's, but the human impacts held to blame—excessive use of fertilizer—were, at least as yet, little relevant.[9]

Marsh himself was not surprised by this coincidence of thought. He saw converging insights as a prevalent trait of the time. "In this day," he remarked, "if a man has any thing to say he must say it at once or he is certain to be anticipated."[10]

M*an and Nature* was half completed when Marsh quit his Riviera hotel in spring 1863 for a manor house in the Po plain, twelve miles southwest of Turin. Pleasant as Pegli had been, commuting to the capital was tiring and costly, and Marsh was glad to find a home much nearer. The locale was unenticing—flat, mosquito-ridden, "unhealthy," Marsh learned. Their house bordered the village of Piòbesi, containing "3500 peasants, and nothing else. It furnishes neither meat drink nor clothing, and we are as badly off as Sydney Smith when he lived '12 miles from a lemon.'"[11] Roads were rudimentary; the train to Turin required a two-mile walk to Candiolo station. Their old gray-walled castello, largely rebuilt about 1830, was dank and chill, the rooms dark. But this proved reparable. Fires banished the damp, sunlight dispelled the gloom, and the romance of inhabiting a "castle" with a tenth-century (not "Roman," as the Marshes at once made out) tower made up for other demerits.[12]

The compensations were many. Marsh's study opened onto a broad terrace facing the tower, the abode of countless starlings; another façade looked over cloisters festooned with climbing roses and wisteria. There was even a hint of treasure trove, for the previous tenant, Prussian envoy Count Brassier de St. Simon, had found 80,000 lire in gold buried in old Venetian glass vessels. And the terrace afforded an unobstructed view of the snow-covered Alpine wall stretching thirty miles west and north.[13]

Piòbesi villagers besieged them with offers of service and produce; Marsh had to hire a small army to tend their modest garden. The Piobesans fancied that all Americans were rich and happy except the slaves, who would be just as content when the war was won. They adopted the cause of the generous envoy who spoke with them in Italian and as equals. When Vicksburg fell to Union forces, villagers cheered the victory with red, white, and blue rockets on the meadow below the castello. Marsh provided wine and food; singing and dancing graced the garden until midnight. Marsh was moved by Piobesan courtesy and dignity. Dismayed by the daily hardships these villagers endured, he gave more money than he could well spare, provided meat at the *festa* of Santa Anna, and made a parting gift of beds and equipment to the local church hospital. Twenty years later the castello was still known by his name; visiting Piòbesi, art connoisseur James Jackson Jarves reported that Marsh was "worshipped even now like a god." [14]

"The quiet here," wrote Caroline after three months in Piòbesi, "is more complete than anything I ever hoped for. . . . I sometimes think that Mr. Marsh would never have got through this immense work on which he is now engaged if Providence had not kindly put us here for a few months." On some days Marsh was not needed in Turin at all, and the work went rapidly. As always, he rose early and wrote steadily all morning, putting down his pen now and then to watch the starlings sweep around the tower or alight at his low open window. The birds were as profligate as frontiersmen; every day Marsh found many small lizards lying near the tower which had "lost about two inches of the tail; this part the starlings gave to their nestlings, and threw away the remainder." The first nightingale sang the evening of April 27; two days later Marsh finished his first draft. These weeks were the least intruded on, the best period for work in Marsh's life. [15]

Fast as he progressed, new sources piled up. Wading through mounds of papers, Marsh grew so despondent that Caroline feared he might commit a "libricide" at some moment of exasperated fatigue. Interruptions dragged out the task. Marsh had to go more often to Turin on diplomatic business and also to negotiate a new lease of Casa d'Angennes with the Contessa Ghirardi, "as formidable a business as the taking of Vicksburg." [16] Winter had tired him; his health was poor; recurrent pain in his side made for sleepless nights. Even the weather now turned against him. From cool spring, Piòbesi passed to scalding summer; one could hardly walk on the sun-baked terrace flagstones. And flies were everywhere—on Marsh's eyelids, on his inkstand, on the very point of his pen. He could not do half so much in a day as before.

Despite these afflictions, Marsh finished his last revision in early

July 1863. On the same day, he signed the lease for Casa d'Angennes and went off to the Alps secure in the faith of a roof over their heads in the fall. But the Casa—promised for September 1—was not even ready when the Marshes, having lingered on in wet, wintry Dauphiné, moved in on November 9, fifteen months after they had left it.

It was a bleak Turin homecoming. During the intervening tenancy of the Countess de Solms—the courtesan Princess Marie Laetitia Studholmina Wyse-Bonaparte, who had become the mistress, then the wife, of sometime Prime Minister Urbano Rattazzi—furniture had been ruined, fittings wrecked or lost. Caroline accused the "witch de Solms" of turning the courtyard into a stable and leaving the marble staircase, kept spotless by the Marshes, looking "like the entrance to a gaming house." Casa d'Angennes was no longer the uniquely clean abode it had been. Damp walls and smoking chimneys made life miserable, and it was as dark and glum without as within. A year of quiet and peace was over; in busy Turin Marsh could expect little repose. Nor did city society compensate for the quiet calm of Pegli and Piòbesi. Fortunately, *Man and Nature* now needed little of his labor—though it called for a lot of his patience.[17]

M‍arsh had begun noting man's impacts on nature before he was six, when his father drove with him in Vermont's hills, taught him how to recognize various trees, and showed him what a watershed was. The need for forest cover was patent: trees intercepted rain and snow and slowed or stopped soils from being washed off hillsides; felling and burning the forest sped runoff, bringing floods in spring and drought in fall. Marsh confirmed what earlier observers—André Michaux, Samuel Williams, Timothy Dwight—had seen in montane eastern America. Many in his Woodstock years remarked on sorely changed landscapes. Near Burlington young Marsh saw what Yale professor Benjamin Silliman had noted in 1819: fields littered with stumps, trees "girdled, dry, and blasted, by the summer's heat, and winter's cold [and] scorched and blackened, by fire."[18]

As a Burlington mill-owner, Marsh himself had aggravated Vermont's forest depletion. Once a timber export center, Burlington by the 1830s had to import from northern Canada. As a sheep raiser, Marsh knew how flocks cropped scanty cover, baring slopes to sun and rain. As farmer, lumber dealer, and manufacturer, Marsh wrote Asa Gray, he "had occasion both to observe and to feel the evils resulting from an injudicious system of managing woodlands." Nowhere had forests been reduced and soils exhausted faster than on the flanks of the Green Mountains. Born a mere forty years after Vermont's settlement, Marsh saw its woodlands reduced from three-fourths to barely one-fourth of the state by the time he left for Italy in 1861.[19]

His assessment of these changes differed profoundly from his precursors'. They had not been blind to their impact on the landscape; but they judged it beneficent. Almost every change was to them an improvement: clearing the forest ameliorated winter cold and summer heat, draining swamps curbed disease, banishing wilderness rendered chaos into order. Vermonter Ira Allen praised the settler who "sees that man can embellish the most rude spot, the stagnant air vanishes with the woods, the rank vegetation feels the purifying influence of the sun; he drains the swamp, putrid exhalations flit off on lazy wing, and fevers and agues accompany them."[20]

Most inquirers before Marsh had trusted earth's plenitude, assumed resources inexhaustible, and never doubted that they could and should master nature; the conquest was God's command and national destiny. Only a few worried about dwindling timber supplies or heeded Franklin's caution that "whenever we [meddle with nature] we had need be very circumspect, lest we do more harm than good."[21] In the general view, the more nature was manipulated, the more fertile it became, buttressing the scriptural dictum that God had left the earth raw and incomplete for mankind to perfect. Subduing the wild became a hallmark of civilizing progress.

The classic statement was that of the eighteenth-century French naturalist Count Buffon: "The entire face of the earth today bears the stamp of the power of man," making his rude inheritance "perfect and magnificent." Land won from moors, fens, and forests, reclaimed from marshes and seas, ordered and embellished, showed both man's unique place in nature and unique power to improve it. "Wild nature is hideous and dying," felt Buffon along with most improvers; human effort made it "agreeable and living."[22]

American pioneers were prime enactors of Buffon's scenario. To them raw nature was repugnant, the forests "howling" and "dismal," the plains a "trackless waste" to be transformed into fruitful farms and flourishing cities. Seen through improvers' lenses trees became lumber, prairies grain fields. It was the mission of Americans "to cause the wilderness to bloom and fructify"; they invoked Genesis 1:28 to "subjugate" the "enemy"—savage nature. Even those who later venerated wilderness celebrated the civilizing impact. Walt Whitman's "Song of the Redwood-Tree" bade a joyous farewell to the arboreal giant who must abdicate his kingship so that "broad humanity, the true America," can "build a grander future."[23]

Adverse side-effects, such as excessive erosion and flooding, were surmised easily put right, alike in the Old World and the New. Environmental degradation was thought limited and reparable. Those who deplored it assumed that enlightened self-interest under government aegis would soon bring whatever reform was needed—a confidence that persisted even among conservationists well into the 1950s.

Fears of more serious damage first clashed with pioneer optimism in Marsh's young America. The devastations of just a few decades contrasted poignantly with previous plenitude. Vast forests were logged for fuel and timber, fencing and pasturage; wild game gave way to intensively grazing livestock; soils cultivated only lightly if ever before were exhausted by Old World crops for lack of regular manuring. All this depleted native flora and fauna, heightened extremes of flooding and drought, and hastened the erosion and exhaustion of lands less fertile, in any case, than at first supposed. Plenty had bred waste, nature's seeming abundance led to profligacy. American heirs found their legacy diminished by pioneering forebears' heedless husbandry.[24]

M arsh first publicly broached the central theme of *Man and Nature* in 1847. Within a single generation, he told fellow-Vermonters, "the signs of artificial improvement" had become "mingled with the tokens of improvident waste." Bald and barren hills, furrowed ravines, once fertile valleys had been converted "from smiling meadows into broad wastes of shingle and gravel and pebbles, deserts in summer, and seas in autumn and spring." What had caused this? A heedless "rage for improvement" that sped removal of forest cover, first for timber and fuel and then for sheep pasture. Steep slopes thus denuded ceased to conserve and equalize moisture. Rain and snow "no longer intercepted and absorbed by the leaves or the open soil of the woods . . . flow swiftly over the smooth ground, washing away the vegetable mould [to] fill every ravine with a torrent, and convert every river into an ocean."

Could Vermonters redress nature's balance? Yes, by felling trees only at stated intervals, as in Europe; it was "quite time that this practice should be introduced among us." The cost of timber and fuel taught that "trees are no longer what they were in our fathers' time, an incumbrance." Marsh did not yet urge public ownership or control of forests; he trusted "enlightened self-interest [to] introduce the reforms, check the abuses, and preserve us from an increase of [the] evils" he had sketched.[25] Time would dim his faith in enlightened self-interest.

Marsh soon elaborated a scheme of tree husbandry. Planning "an essay or a volume" on forest economy, in 1849 he sent the botanist Asa Gray a brief résumé of woodland benefits. Trees ameliorated microclimate, enriched undergrowth, replenished mold and humus. Above all, they checked "the rapid flow of rain-water and melting snows, thereby diminishing the frequency and violence of freshets, preventing the degradation of highlands and the consequent wash of gravel from above upon cultivated lands, and . . . acting as a reservoir, which gives out in summer the moisture accumulated in the wet seasons." Forests were also sites of recreation and natural beauty, animal and plant refuges, and havens for insect-eating birds. Marsh urged a national survey to ascertain proper ratios between wooded and arable land. He envisioned an experimental forestry program as yet untried even in Europe.[26] Before he ever saw the Old World, Marsh had stated his chief conservation premises.

Five years around the Mediterranean, when American envoy to Turkey, lent Marsh's insights geographical breadth and historical depth. Before, he had based comparisons on what he read of European practice; what he now saw vivified the contrasts. Marsh found the imprint of humanity in the Old World awesome in its antiquity: the very earth had a "hoary and ancient aspect that seems to belong rather to an effete and worn-out planet, the meadows levelled and the hills rounded, not as with us by the action of mere natural forces, but by the assiduous husbandry of hundreds of generations."[27] And the marks of degradation were just as clear. The same agents of decay, extirpating forests and fauna, overgrazing, exhaustive cultivation, betokened the fall of every civilization. Each desolated area—the sterile sands of the Sahara, the sinkholed Adriatic Karst, the malarial Roman Campagna, the rock-strewn ravines of Provence and Dauphiné—pointed a similar moral. Anciently fertile and populous, all were now barren lands, testimony to human improvidence.

Only gradually did Marsh shape his experiences into coherent synthesis. Ever on the move, he felt overwhelmed by new sights too multiform to encompass. "It is a shame," he wrote Baird at the Smithsonian, "that I have not the knowledge of *nature* that every traveller . . . ought to have. I see strange stones, plants, animals, geographical formations, and gaze vacantly at them, but what availeth it?" All he could do was "write a book" of curiosities designed to "show that one ignoramus is as good a traveller as any body." Such a book was *The Camel* (1856)—a potpourri of odd facts, learned lore, legends, and anecdotes. Yet it was also a solid study aimed at a practical goal of domestic adaptation. The dispersal of camels was a felicitous instance of human conquest of the "widely dissimilar, remote, and refractory products of creative nature."[28]

In his 1856 New Hampshire oration, Marsh reverted to Old World history as a cautionary lesson for the New. Centuries of loss had finally impelled Europeans to stem erosion and depletion. The environmental calamities of antiquity should open eyes in America, where similar damage was only recent and still limited. Yet Americans were "already beginning to suffer from the washing away of vegetable soil from our steeper fields, from the drying up of the abundant springs which once watered our hill pastures, and from the increased violence of our spring and autumnal freshets." The grim saga of destruction abroad should teach Americans to stem further loss by restricted logging, afforestation, and stream control, as Europeans were at last beginning to do. Tuscany's reclaimed Val di Chiana was for Marsh a prime instance of "remarkable triumphs of humanity over physical nature." There "a soil once used, abused, exhausted, and at last abandoned, had been reoccupied" and remade fruitful. (Vermont's hill pastures did later revert to woodland, though less because Vermonters took Marsh's advice than because potash and sheep had ceased to be profitable.)[29]

Marsh's 1857 report on Vermont's fisheries likewise stressed Old World modes of restocking. But the environmental cost of economic growth was his central concern. Fish stocks had plummeted owing to manifold causes. Some, like the rapacious gutting of fisheries, were deliberate; others, like changes in riparian milieus induced by tree felling, sawmilling, and dam building, were unintended, often unnoted until too late. The inadvertent consequences of human impact put the whole fabric of nature, organic and inorganic, at risk.

Alarm over such ill effects did not preclude controlling nature; on the contrary, it mandated more effective control. Presaging another theme of *Man and Nature*, Marsh's 1860 "Study of Nature" lauded science for furthering nature's conquest. He scorned those who inveighed against machines, held progress soul-destroying, hankered after some simple golden age. Nature was not sacred; man must rebel against its limits, subjugate it, impose human order. The view that all was for the best in a divinely ordered cosmos had long prevented man from seeing the damage he did. But human destructiveness was no more ordained than human felicity; technologies capable of wrecking nature could be used instead to mend it. By understanding the environment we might learn how to repair it.[30]

Man is everywhere a disturbing agent. Wherever he plants his foot, the harmonies of nature are turned to discords.

Marsh, *Man and Nature*

Marsh again found time and scope for environmental concerns when he reached Italy in 1861. He made less use of his surveyors' tools than in Egypt and Palestine, shipped back few dried plants, no pickled fish. But as in the 1850s he set up exchange programs with the Smithsonian and, later, the U.S. Department of Agriculture, to which he sent apple grafts, grape cuttings, and seeds of Algerian pine (*Cedrus atlanticus*). He helped Italians domesticate American conifers, agave, peas, lima beans, squash, sorghum, "Darling's early corn" (for Ricasoli, along with okra), wild turkeys (for the royal hunts), and flying squirrels (a wedding gift to Prince Umberto and his bride). Although he did not get the parrots coveted by Marchesa Spinola, Marsh did bring, via

Baird's taxidermist, a stuffed ruby-throated hummingbird to King Vittorio Emanuele.[31]

Most fruitful were Marsh's Alpine excursions. Studying glaciers, moraines, and avalanches honed his environmental insights. He took notes on farm tools, crop diseases, dietary deficiencies, and myriad aspects of rural life. He was an absorbed observer—now admiring, now alarmed—of huge engineering works: irrigating the Po plain, digging the Suez Canal, constructing Alpine tunnels, draining the Maremma swamp in Tuscany. But creation usually seemed less impressive than destruction—the erosive force of wind and water in the Alps and Apennines, avalanches at Eismeer and Grindelwald, landslides along the Riviera, torrents laden with soil stripped from Dauphiné valleys. Human agencies, above all tree felling and grazing, accelerated normal erosive processes into disaster: soils carried off faster than they could rebuild, denuded pastures gullied, stream channels deepened, dams and harbors silted up. Against such devastation, rural communities, often indigent and isolated, were helpless. Physical degradation eroded human will, portending the common doom of land and people—perhaps, indeed, the extinction of humanity.

Marsh's apocalyptic fears were, however, of little concern to most; they were the stuff of sacred prophecy, not secular enterprise. What worried people was some sudden shortfall, a failure of crops, a lack of fuel, dearth of this or that resource, notably of timber. Timber famine had long menaced European states and rulers. Once extensive forests had succumbed to the settler's ax, the herdsman's plow, voracious livestock, the demands of naval and other industry, the needs of growing numbers for fuel and shelter. But as modern monarchs replaced feudal magnates, control over land became centralized and absolute. State forests were strictly guarded, private woodlands kept inviolate as game preserves and sources of revenue. Taking wood for fuel, foraging for brush, even pasturing were severely punished in efforts to curtail forest attrition. Still the woods continued to dwindle. Imperial conquest and new manufactures multiplied demands for timber in naval supplies, smelting, and building. At long last, scarcity spurred silvicultural science: foresters began to harvest trees on a sustained basis, to replenish reserves, to select species suited to particular locales.

Forest protection and regeneration gained ground above all in nineteenth-century France. Although woodland management for the royal navy had been draconian policy since 1669, forests were more and more depleted by shipbuilding and other needs. After the Revolution, peasant incursions and extensive fires had further devastated forests maltreated, like chateaux, as emblems of aristocratic oppression and exclusion. Aiming to reimpose order over both land and people, the restored monarchy in the 1820s initiated a quasi-military forest code. French woodlands were managed and policed along lines set by the School of Forestry in Nancy, largely staffed by emigré German tree experts.[32]

In the New World, by contrast, forest improvidence was endemic. It stemmed from an ethos of free access, of limitless supply, and of wilderness subjugation. Yet even in North America some woodlands were managed—first by the Crown to ensure dependable timber supplies, later by states and communities to protect local fuel stocks. In the late 1820s a 60,000-acre live oak (*Quercus virginiana*) reservation in Florida was set aside for the U.S. Navy, "to lay up an article which may be perfectly preserved for many years [and to] keep timber out of foreign navies that might one day be our enemies." But the government was unable to protect these woods; "more than once," Marsh scoffed, it "paid contractors a high price for timber stolen from its own forests."[33] Forest legislation was scanty, seldom invoked, almost never enforced. There was no forestry research or education; Americans were too busy getting rid of trees to think of saving them, let alone the slow and lengthy process of cultivating them. "The *improvement* of forest trees is the work of centuries," wrote Marsh apropos Harvard's new Arnold Arboretum in 1879. "So much the more reason for beginning *now*."[34] "Now" came a full century after such beginnings in Europe.

Until the nineteenth century, almost all forest controls were designed simply to save wood, not to conserve soil or water or to protect adjacent arable lands and towns. Watershed management was absent even in Germany, where forestry was most advanced. To be sure, some French Alpine and Pyrenean communes, alive to the protective virtues of forest cover, had limited local tree felling as early as the 1500s. But the French Revolution swept away local restrictions along with the punitive national forest code. Only then were disasters ascribed to deforestation. In 1797 the engineer Jean Antoine Fabre described how rivers became torrents, tearing away fertile valley bottoms or burying them in silt. Fire, logging, and overgrazing in Alpine France ravaged millions of arable acres below. In the upper Durance and along the denuded tributaries of the Rhone, Marsh witnessed havoc next to which deforestation in Vermont seemed trifling. "In a single day of flood" in 1827, "the Ardèche, a river too insignificant to be known except in the local topography of France, contributes to the Rhone" more water than the Nile, a thousand time larger, to the Mediterranean. [35]

Not long before Marsh surveyed these devastated scenes, French experts readdressed Alpine erosion in the wake of new floods. Alexandre Surell's 1841 landmark study ascribed torrents in the Hautes-Alpes to the denudation of unstable sedimentary slopes. To regulate stream flow, conserve soil, and protect agriculture in the plains below, tree cover above was vital; mountain grazing should be prohibited. In the 1850s, Alexandre Moreau de Jonnès, Jean Baptiste Boussingault, Antoine Becquerel, and François de Vallès measured the effects of plant cover on runoff, stream flow, and soil temperatures. They showed that forests intercepted most precipitation and stored snow, slowing the pace of melting. And dead organic matter on forest floors absorbed up to ten times its weight in water, further reducing runoff and erosion.

These studies, reinforced by advice on tree planting and river regulation from Nancy foresters Bernard Lorentz and Jules Clavé, led France in 1860 to mandate watershed control by afforestation. Drainage management became a moral crusade; Napoleon III in 1857 pledged that "rivers, like revolution, will return to their beds and remain unable to rise during my reign." By the time *Man and Nature* appeared in 1864, France had begun to quell some torrential damage, though at the cost of beggaring peasants deprived of pasturage and communal rights.[36] Meanwhile, Karl Fraas in Germany and Antonio Salvagnoli, Adolfo Di Bérenger, and Giovanni Siemoni in Italy, confronting analogous timber shortages and torrent damage, urged state forest control.

Thus in Alpine Europe Marsh found concurrence with the concerns he had voiced in America. Scenes of desolation in Piedmont and Dauphiné reinforced his gloomy prognosis for Vermont; and Boussingault, Becquerel, Vallès, and Salvagnoli fortified his views about their causes. In modern as in ancient times man was the architect of his own misfortunes. What the Old World had suffered, Marsh warned, might become the fate of the New. At Pegli he penned his most telling omen:

There are parts of Asia Minor, of Northern Africa, of Greece, and even of Alpine Europe, where the operation of causes set in action by man has brought the face of the earth to a desolation almost as complete as that of the moon. . . . The earth is fast becoming an unfit home for its noblest inhabitant, and another era of equal human crime and human improvidence . . . would reduce it to such a condition of impoverished productiveness, of shattered surface, of climatic excess, as to threaten the depravation, barbarism, and perhaps even extinction of the species.

To repair past ravages and buttress future hopes, America must follow Europe's nascent lead and mandate state control of natural resources. "Man has too long forgotten that the earth was given to him for usufruct alone, not for consumption, still less for profligate waste." [37] The notion of usufruct—the view that each generation had rights only to current use, not perpetual title, which would forbid both hoarding and waste—had been advanced seventy years previously by Thomas Jefferson, fearful lest his Virginia fellow-planters squander finite resources. Marsh elaborated Jefferson's warning that individual entrepreneurs could not be trusted with long-range commitments. [38] Only by assuming public responsibility could Marsh's countrymen curtail private profligacy.

*H*eterogeneous as Marsh's materials were, in broad outline *Man and Nature* is straightforward enough. A brief preface explains the purpose of the book. A 211-item bibliography follows—more than half of recent date and of French or German origin. English-language works comprised under a third of the total, fewer than Latin, Dutch, and Scandinavian combined. In later editions Italian sources became more numerous.

The opening paragraphs of Chapter 1 limn the fruitfulness of the Roman Empire at its peak. What a contrast! Over half that realm was "now completely exhausted of fertility, or so diminished in productiveness" that it afforded no "sustenance to civilized man." Ever since Marsh,

Mediterranean deforestation has been cited as a classic instance of lamentable environmental impact.

What had reduced this once wooded, then bounteously fertile area to the arid deserts that now ringed the Mediterranean? Marsh assigns the decay to the crushing oppression of a destitute peasantry, forced by tyranny and misrule to abandon fields earlier reclaimed from forest. The soil was thus "exposed to all the destructive forces which [degrade] the surface of the earth when it is deprived of those protections by which nature originally guarded it, and for which, in well-ordered husbandry, human ingenuity has contrived more or less efficient substitutes." Both deforestation and subsequent abandonment were at fault. Once nature is tamed, we cannot relax our care of it. [39]

To arrest decay and restore nature called for both broader morality (social concern for future generations) and better science (comprehensive surveys of terrain and soils, climate and vegetation). But such surveys would require decades, and "we are, even now, breaking up the floor and wainscoting and doors and window frames of our dwelling, for fuel to warm our bodies and seethe our pottage, and the world cannot afford to wait till the slow and sure progress of exact science has taught it a better economy." Hence Marsh aims to derive from "the history of man's effort to replenish the earth and subdue it" some immediate practical lessons. [40] Successive chapters of *Man and Nature* treat impacts on plants and animals, woods, waters, and sands.

Chapter 2, "Transfer, Modification, and Extirpation of Vegetable and of Animal Species," is dauntingly manifold. The broadest scholar today would hardly have the temerity to review hemispheric transfers of flora, the economic role of plants, the origins of domestication, plant and animal interactions, the numbers of birds and insects, impacts of Old World creatures on New World lands, uses of earthworms, fish breeding, seed dispersal by birds and squirrels, and the role of diatoms in the economy of nature—all in seventy pages. But in Marsh's day most such topics were hardly considered scholarship, let alone science.

Marsh's insights were what are now called ecological, a term coined a few years later, by the German zoologist Ernst Haeckel, to denote relations among organisms and their milieus. [41] "All nature is linked together by invisible bonds," in Marsh's phrase; "every organic creature . . . is necessary to the well-being of some other." He instanced the

"house-that-Jack-built" structure of a familiar food chain: "the destruction of the mosquito, that feeds the trout that preys on the May fly that
destroys the eggs that hatch the salmon that pampers the epicure, may
occasion a scarcity of this latter fish."

Marsh's greatest contribution to nascent ecological awareness was
to include human impacts in the dynamics of nature. Men reshape their
environs by moving, killing, and breeding other creatures, also often
unwittingly affecting their predators, habitats, and sustenance. What
humans consume is seldom returned as waste to the soil it grew in. And
removing "a plant from its native habitat to a new soil [brings] a new
geographical force to act upon it. . . . The new and the old plants are
rarely the equivalents of each other, and the substitution of an exotic
for a native tree, shrub, or grass, increases or diminishes" how much,
and for how long, the ground is sheltered by plant cover. Humans and
their domesticates thus further tangle nature's network. "The existence
of an insect which fertilizes a useful vegetable may depend upon that
of another, which constitutes his food at some stage of his life, and this
other again may be as injurious to some plant as his destroyer is beneficial to another."

Because few recognized or heeded these interdependencies, nature often suffered from misjudged zeal for progress or purity. Marsh
instanced Boston's Cochituate aqueduct, which had become polluted by
"the too scrupulous care with which aquatic vegetation had been excluded from the reservoir, and the consequent death and decay of the
animalculae." It was crucial to maintain a balance among flora and fauna,
for "the excess of either is fatal to the other."

Species commonly seen as obnoxious were in fact often essential
to the ecological network. Like the orchids studied by Darwin, the
survival of most edible and otherwise useful plants depended on services rendered by "insects habitually regarded as unqualified pests." In
the wanton slaughter of robins and other insectivorous birds, Marsh
charged that "man is not only . . . depriving his groves and his fields of
their fairest ornament, but he is waging a treacherous war on his natural allies." In clearing litter from woodlots, farmers forgot that decayed
and fallen trees and undergrowth "furnished food and shelter to the
borer and the rodent, and often also to the animals that preyed upon
them." With stumps gone, insects and squirrels multiplied and "become

destructive to the forest because they are driven to the living tree for nutriment and cover."[42]

The cascading effects of such changes might depend on consumer trends half a world away. For example, beaver dams in pre-colonial North America had clogged and altered drainage patterns, created ponds and bogs, and accumulated layers of peat to an extent seldom remarked before Marsh. But so great was the demand for their fur, that within a half century hunting had all but extirpated beavers, drastically altering riparian landscapes over millions of acres.[43] But only for the time being. In Marsh's words:

So long as the fur of the beaver was extensively employed as a material for fine hats, it bore a very high price, and the chase of this quadruped was so keen that naturalists feared its speedy extinction. When a Parisian manufacturer invented the silk hat, which soon came into almost universal use, the demand for beavers' fur fell off, and this animal . . . reappeared in haunts which he had long abandoned.

And the multiplying beaver again began to modify water courses. "Thus the convenience or the caprice of Parisian fashion has unconsciously . . . affect[ed] the physical geography of a distant continent."

The ramified effects of human action impacted not just settled lands, but most of the globe. Many seemingly untouched locales were by no means devoid of human impress. The most casual intervention could alter nature profoundly. Thus the removal of dead matter from the forest floor accelerated drainage, and within a few years "the wood acquires, to some extent, the character of an artificial forest."[44]

Marsh stresses that most such impacts are unwitting. Species deliberately transplanted were but a small fraction of those accidentally introduced. Substituting new for existing plants and animals had unforeseeable results, mostly unintended and usually unwanted. "After the wheat, follow the tares that infest it." Introduced species often proliferate as uncontrollable pests in new locales because no longer held in check by their former natural predators.[45]

He next turns to forests—*Man and Nature*'s central focus. "As man and boy," Marsh confided late in life to Harvard forester Charles Sprague Sargent, "I knew more of trees than of anything else"—and preferred them to most people. He also valued trees as nature's prime stabilizing agents. Most of the inhabited globe had once been wooded, and "forests would soon cover many parts of the Arabian and African deserts" again, Marsh believed, "if man and domestic animals, especially the goat and the camel, were banished from them."[46] His goal was to assay how forest felling had affected climate, vegetation, and soils throughout the world.

Forest influence on climate was a much-disputed age-old topic of keen concern. Many contended that trees ameliorated climatic extremes. So-called desiccationists, harking back to Theophrastus in the third century B.C., argued that felling trees induced drought and that planting them brought rain. This theory long found favor among those eager for rain to augment fertility in arid lands.[47] For such effects Marsh sees no conclusive evidence. Forests might "promote the frequency of showers" and equalize rainfall over the year, but "we cannot positively affirm that the total annual quantity of rain is diminished or increased by the destruction of the woods." Claims that woods augmented rainfall in their vicinity Marsh dismisses as "vague and contradictory."[48]

Beyond all doubt, however, was the effect of forests—and their loss—on the rain and snow that *did* fall, and hence on streams, soils, and organic life. Equable stability was a prime virtue of undisturbed forest lands. In wooded areas, "tree, bird, beast, and fish, alike, find a constant uniformity of condition most favorable to the regular and harmonious coexistence of them all. [But] the disappearance of the forest" shatters this symbiosis:

The soil is alternately parched by the fervors of summer, and seared by the rigors of winter. Bleak winds sweep unresisted over its surface, drift away the snow that sheltered it from the frost, and dry up its scanty moisture. . . . Instead of filling a retentive bed of absorbent earth, and storing up a supply of moisture to feed perennial springs, the melting snows and vernal rains, no longer absorbed by a loose and bibulous vegetable mould, rush over the frozen surface, and pour down the valleys seaward. The soil is bared of its covering of leaves, broken and loosened by the plough, deprived of the fibrous rootlets which held it together, dried and pulverized by sun and wind. . . . Stripped of its vegetable glebe, [soil] grows less and less productive, and, consequently, less able to protect itself by weaving a new network of roots to bind its particles together, a new carpeting of turf to shield it from wind and sun and scouring rain. Gradually it becomes altogether barren.[49]

Marsh documents this broadbrush indictment. He traces the decline of European woodlands, assesses ensuing soil and other losses, and gauges the efficacy of efforts to restore them.

Transatlantic parallels next engage Marsh. Avaricious logging and careless fires had already wasted America's finest timber, reducing logged and burnt-over land to a scrubby second growth—as on Woodstock's Mount Tom. Yet wood for fuel and construction was more and more in demand. Wiser use of woodlands was imperative. Americans must learn both to husband and to improve their forests. Restricted cutting would conserve existing stocks; careful management through planting, culling, and pruning would enrich them. And on terrain where trees were crucial for soil moisture and plant cover, all cutting should cease. "We have now felled forest enough everywhere, in many districts far too much."

How could land laid bare be reclad? Farmers tended to shun tree planting because trees grew so slowly that "the longest life hardly embraces the seedtime and the harvest of a forest." Hence, notes Marsh, "the value of its timber will not return the capital expended and the interest accrued" for many generations. He faults such shortsightedness. Stewardship should be our mutual objective. "The planter of a wood must be actuated by higher motives than those of an investment. . . . The preservation of existing woods, and the far more costly extension of them where they have been unduly reduced, are among the most obvious of the duties which this age owes to those that are to come."

Aiding the future would benefit the present too; to insist on "an approximately fixed ratio" between woodland, pasture, and arable would reduce the "restlessness" and "instability" Marsh saw as major defects of American life.

It is rare that a middle-aged American dies in the house where he was born, or an old man even in that which he has built. . . . This life of incessant flitting is unfavorable for the execution of permanent improvements. . . . It requires a very generous spirit in a landholder to plant a wood on a farm he expects to sell, or which he knows will pass out of the hands of his descendants. . . . But the very fact of having begun a plantation would attach the proprietor more strongly to the soil for which he had made such a sacrifice.

What fraction of land should be forested Marsh does not specify; it would vary with climate, terrain, soils, and other factors. But woods ought to occupy no less than a quarter to a third of the surface. "The too general felling of the woods," he sums up, is "the most destructive among the many causes of the physical deterioration of the earth."[50]

Chapter 4, "The Waters," stresses more benign human impacts. After showing how England and Holland used drains and dikes to reclaim land from the sea, Marsh goes on to treat the long-term effects of aqueducts, reservoirs, canals, and irrigation. He lauds river and harbor works, the tapping of artesian waters, new techniques for reducing siltation. Along with afforestation, dams and embankments were essential to flood control. "The cost of one year's warfare, if judiciously expended" in reforestation and building conduits and reservoirs, Marsh ends on an exuberant note, "would secure, to almost every country that man has exhausted, an amelioration of climate, a renovated fertility of soil, and a general physical improvement, which might almost be characterized as a new creation."[51]

His penultimate chapter, "The Sands," surveys the genesis, structure, and geography of sand dunes, the damage done by their drifting, and their benefits as barriers against marine erosion. As with so much else, deforestation emerges as the prime culprit. Marsh had earlier noted the lack of an Old-Northern name for sand dunes, "for the reason, that it is only since the destruction of the forests of Jutland, in comparatively recent times, that the shifting coast-downs have excited any interest as a source of danger to the cultivated soil."[52] Dunes must be vegetated and protected by windbelts, for "when their drifts are not checked by natural processes, or by the industry of man, they become a cause of as certain, if not as sudden, destruction as the ocean itself." His laudatory and spirited account of dune stabilization in the Landes of Gascony, where sand had engulfed hundreds of square miles until the French planted them with maritime pine, remains a classic.[53]

A final chapter on future projected changes begins by assessing likely impacts of great canals then being built or projected—Suez, Panama, Cape Cod, Mediterranean–Dead Sea, and Caspian–Sea of Azov. Marsh then speculates on other means of improving nature: artesian watering of Arabian deserts, taming earthquakes by sinking deep wells, ameliorating volcanic eruptions by diverting lava flows. But as ever, the unwitting effects of human action engross Marsh more than intended results. Such effects may be far more consequential than most people dream. Following Charles Babbage's *Ninth Bridgewater Treatise* (1838), Marsh suggests that every human action, however seemingly trivial, may leave a permanent physical imprint. For example, erosion aggravated by clearing and cultivating displaces masses of matter, thickens the earth's crust beneath deltas and estuaries, and so may shift the center of gravity, perhaps the planetary motions, of the globe.[54]

The maxim that "the law concerneth not itself with trifles" Marsh thought a false guide for environmental understanding. Huge consequences often flowed from obscure and puny causes, as with the accretions of organic decay. "We habitually regard the whale and the elephant as essentially large and therefore important creatures," but rock formations owe little to their bones, whereas diatoms and other microscopic creatures make up strata thousands of feet thick over much of the earth. So might human impacts mount by innumerable tiny accretions to immense proportions. Such effects might be incommensurable, their outcomes obscure. But "we are never justified in assuming a force to be insignificant because its measure is unknown, or even because no physical effect can now be traced to it."[55]

Marsh's last (1884) edition of *Man and Nature* enlarged on this theme, still more patently salient today. "Cosmic forces of little comparative energy may, by long continued or often repeated action, produce sensible effects of great magnitude." For example, ordnance recoil and impingement might "accelerate or retard the rotation of the earth, or even . . . deviate the globe itself from her orbit." Research into such minutiae might prove crucial to the destiny of mankind; novel technologies if carefully harnessed could help to shape a well-husbanded world: incommensurable as such effects now were, "who shall say that the mathematics of the future may not . . . calculate even these smallest cosmical results of human action?" Finally, growing knowledge of how humans affect their surroundings might answer "the great question," posed almost at the same moment by T. H. Huxley as by Marsh, "whether man is of nature or above her."[56]

14. Man and Nature: *The Meaning*

AS MARSH'S IDEAS OF NATURAL AND HUMAN HISTORY matured, his environmental insights likewise grew and changed, often at odds with his previous views. Such ambivalence did not discomfit him. He saw inconsistency as an inevitable outgrowth of change. Reminded he had once held some contrary opinion, he would say: "A man who cares for the truth can't afford to care for consistency." Indeed, his ecological creed embodied an inherent contradiction. He wanted to treat man and nature as a unity, but his Calvinist, Enlightenment, utilitarian progressivism predisposed him to segregate and exalt humanity as above and at war with the cosmos. "The life of man," Marsh ever insisted, "is a perpetual struggle with external nature."[1] Yet from that same nature, it was abundantly clear, his own life drew unending inspiration.

The first command addressed to man by his Creator . . . predicted and prescribed the subjugation of the entire organic and inorganic world to human control and human use.

Marsh, *The Camel*

Mankind's unique potency was for Marsh a moral dictum as well as a fact of material observation. Free will, James Marsh had argued, came from "peculiar powers" of conscious intention and moral reasoning—traits possessed by no other species. In accord with his cousin, George Marsh aimed to demonstrate "that man is, in both kind and degree, a power of a higher order than any of the other forms of animated life." *Man and Nature* opens with an epigraph from the radical Congregational preacher Horace Bushnell: "Not all the winds, and storms, and earthquakes, and seas, and seasons of the world, have done so much to revolutionize the earth as MAN . . . has done since he came forth upon it, and received dominion over it."[2]

The title Marsh had first proposed was "Man the Disturber of Nature's Harmonies." His publisher, Charles Scribner, was dismayed. "*Is* it true? Does not man act in harmony with nature? with her laws? is he not a part of nature?" "No," retorted Marsh,

nothing is further from my belief, that man is a "part of nature" or that his action is controlled by what are called the laws of nature; in fact a leading object of the book is to enforce the opposite opinion, and to illustrate that man, so far from being, as Buckle supposes, a soul-less, will-less automaton, is a free moral agent working independently of nature.[3]

Though confined in part by their physical milieus, humans also transcended them, defying and manipulating nature for their own purposes.

Nature itself had no will, no purpose, no morality. Nor had any of nature's other habitants. Animals act only instinctively "with a view to single and direct purposes," whereas humans aim consciously at preconceived and often remote objectives. "The backwoodsman and the beaver alike fell woods; the man that he may convert the forest into an olive grove that will mature its fruit only for a succeeding generation, the beaver that he may feed upon their bark or use them in the construction of his habitation." Massive as were beavers' works, they self-consciously anticipated no beaver future.

On this crucial point, Marsh rejected Thoreau's anthropomorphism. Perhaps "the squirrel when it plants an acorn, and the jay when it lets one slip from under its foot," mused Thoreau, had "a transient thought for its posterity, which at least consoles it for its loss." Marsh would have none of the "pathetic fallacy." "When the bird drops the seed of a fruit it has swallowed, and when the sheep transports in its fleece the seed-vessel of a burdock from the plain to the mountain, its action is purely mechanical and unconscious, and does not differ from that of the wind in producing the same effect."[4]

Marsh's ecology embraced two hypotheses, one to do with the dynamics of nature, the other of culture. Nature, in his view, remained in balance until mankind interfered. Whenever otherwise injured, nature "sets herself at once to repair the superficial damage, and to restore . . . the former aspect." Nature's harmony was a state of overall stability. Geological and astronomical forces altered the earth only very slowly; other natural change was small in scale, often cyclical, tending toward balance, and broadly beneficial, maximizing the numbers and diversity of living forms. Organic life in turn tempered inorganic extremes, securing "if not the absolute permanence and equilibrium of both, a long continuance of the established conditions, . . . or at least a very slow and gradual succession of changes." The interactive fabric of climate, waters, rocks, soils, plants, and animals tended toward equilibrium.[5]

Marsh's vision of a self-regulating nature favoring diversity and equilibrium became, in its essential vision, the ecological paradigm of the early twentieth century. It survives to this day, along with its organicist premise, in popular environmentalism, as I discuss in Chapter 18. Modern ecologists, however, stress chaotic transitions no less than processual regularities. They see nature's evanescent equilibria punctuated by disruptive fluctuations over the whole of earth history, some as sudden or as catastrophic—meteorite impacts, seismic waves, flooding by dam-burst glacial lakes, sun-occluding ash from volcanic eruptions—as any man-made devastation. Marsh's stable verities are now circumscribed and precarious; equipoise is never reached, seldom even approximated. But where human impact is minimal (it is never wholly absent), change tends to be more cyclical, slower, and less sweeping in character, with extremes often buffered by diverse local circumstances.[6]

Marsh did not suppose nature's harmony perfect, the cosmos totally stable. Even unmanipulated nature never stayed *entirely* the same. For all his stress on stability, he remained alert to the flux of change within the ramified network of dead and living matter. For example, wood found in Danish peat bogs revealed shifting forest succession: "Every generation of trees leaves the soil in a different state from that in which it found it; every tree that springs up in a group of trees of another species than its own, grows under different influences of light and shade and atmosphere from its predecessors."[7] Plants continuously alter their milieus, often making them more conducive to other species than to their own. Most such sequences are not cyclical, but historically unrepeatable.

In short, organic life contains the seeds of destructive as well as constructive change. Just as "death is implied in the very idea of animated being," Marsh had earlier conjectured, so "every breath forever unfits for respiration a portion of the circumambient atmosphere." In a passage foreshadowing the concept of entropy, Marsh foresaw the ultimate demise of the earth's living fabric:

Every particle of matter, that has once entered into the constitution, or served the uses of a living being, becomes thereby less suitable for future organic constitution. The return to earth's bosom of a mouldering form of each of her children irrecoverably taints a portion of her soil with a poison destructive to similar organic life; . . . earth is continually growing unfit for the habitation of the living beings that animate her surface.

Quite apart from the human derangements detailed in *Man and Nature*, earth's decay was inherent in its very florescence. Marsh's insight that each living creature creates and destroys the ambience of its own and its successors' lives, "both producing and consuming the conditions of its existence," was extraordinarily prescient. Modern biology echoes Marsh: "Every bacterium uses up food material and excretes waste products that are toxic to it. Organisms ruin the world not only for their own lives but for their children as well."[8]

Though he saw all life destined ultimately to fade from the earth, for the sake of children in the nearer future Marsh was intent on husbanding the fabric of nature. Stewarding the globe was made doubly difficult by mankind's refusal to recognize the damage it uniquely wrought, and by reluctance to replace selfish greed with social foresight. Bent on this mission of reform, Marsh's rhetoric personified nature even while he denied it will and moral purpose. Human "ravages" subvert nature's "balance"; nature "avenges herself upon the intruder, by letting loose upon her defaced provinces destructive energies hitherto kept in check"; restoring disturbed harmonies was "repaying to our great mother the debt" imposed on us by prodigal and thriftless forebears. To harangue his reckless fellow men, Marsh reified for effect the nature they had ravaged.

"Forest-born," as Marsh reminisced, "the bubbling brook, the trees, the flowers, the wild animals were to me persons, not things." As a lonely boy he had "sympathized with those *beings*, as I have never done since with the *general* society of men, too many of whom would find it hard to make out as good a claim to *personality* as a respectable oak."[9] Fond of these beings as he was, Marsh never attributed intentions or morals to them, or supposed nature a person, much less a deity.

Man has too long forgotten that the earth was given to him for usufruct alone, not for consumption, still less for profligate waste. . . . The prodigality and the thriftlessness of former generations have imposed upon their successors . . . the command of religion and of practical wisdom, to use this world as not abusing it.

Marsh, *Man and Nature*

Marsh's most profound insights concerned the impress of culture on nature. Before he wrote, it was conventional Western wisdom that human influence was either benign or negligible. Nature had been designed by an omnipotent Creator. Mankind, as God's terrestrial steward, had dominion over other creatures and was empowered— indeed bidden—to subdue and cultivate the earth. This process had rendered the tamed globe ever more fruitful.

But as Marsh now showed, confidence in this immutably progressive saga accorded little with the facts of history and geography. God-given might had encouraged men to foul as well as to fructify the subjugated earth. *Man and Nature* limned more ruin than enrichment. Marsh's mentor Bushnell, citing human powers for evil ("Nature never made a pistol, or pulled a trigger") as well as for good ("Nature never built a house . . . or made a book, or framed a constitution"), saw nature "unstrung and mistuned, to a very great degree, by man's agency" and foresaw that the mounting forces of industry might "fatally vitiate the world's atmosphere." Marsh chronicled the course that devastation had taken to date.[10]

The prime cause, he stressed throughout, was the extent of human influence, accelerated by every advance in technology. Left undisturbed, terrain and soils, flora and fauna changed so slowly as to be virtually "constant and immutable." But "a self-conscious and intelligent will aiming as often at secondary and remote as at immediate objects" altered nature ever more drastically. Human influence was uniquely ferocious in scope and intensity, humans unlike other predators in the magnitude as in the purpose of their impact. "Wild animals have [n]ever destroyed the smallest forest, extirpated any organic species, or . . . occasioned any permanent change of terrestrial surface. . . . The destructive animal has hardly retired from the field of his ravages before nature has repaired the damages." By contrast, "man pursues his victims [un]limited by the cravings of appetite, [and] unsparingly persecutes, even to extirpation, thousands of organic forms which he cannot consume.

Human impacts were swift, extensively ruinous, and long irreparable; "the wounds he inflicts . . . are not healed until he withdraws the arm that gave the blow." And often not for years to come. Though little altered since its partial clearance early in the century, much of New England was still growing drier, springs disappearing, rivulets diminishing in summer, because surface drainage remained too rapid to enable aquifers to fill up; it might take centuries to restore equilibrium.[11]

Against such swift and massive derangements nature is largely impotent. Species wiped out are never reborn. Domesticated creatures do not revert to the wild. Woodlands denuded may never recover. "When the forest is gone, the great reservoir of moisture stored up in its vegetable mould is evaporated, and returns only in deluges of rain to wash away the parched dust into which that mould has been converted." Only human agency could repair the damage wrought by human agency. Here and there forests are replanted, inundations stemmed, swamps drained, sand dunes fixed with grass and trees, but such stewardship is rare. Instead, ignorance, carelessness, and greed lay waste the world. Arid lands of steep relief are most liable to be degraded, but no locales are immune. Enterprise already takes its toll not only in America but in newly settled Australia.

The more advanced the culture, the greater the destructive impact. Earlier, Marsh had thought primitive peoples earth's consummate wreckers, civilized men its saviors:

The arts of the savage are the arts of destruction; he desolates the region he inhabits, his life is a warfare of extermination, a series of hostilities against nature or his fellow man. Civilization, on the contrary, is at once the mother and the fruit of peace. . . . Savage man then is the universal foe . . . of all inferior organized existences; . . . civilization transforms him into a beneficent, a fructifying, and a protective influence.[12]

Two decades later, *Man and Nature* all but reversed this indictment. Farmers and herders were malignant, hunters and gatherers generally benign. "The wandering savage . . . fells no forest, and extirpates no useful plant, no noxious weed." The hunter's limited kills served to protect "the feebler quadrupeds and fish and fowls." Perhaps, Marsh suggested, "simpler peoples recognize a certain community of nature between man, brute animals, and even plants." This affinity might help explain why "purely untutored humanity . . . interferes comparatively little with the arrangements of nature," except creatively; almost all known crops and domestic animals had been tamed in primitive prehistory.[13]

Sedentary humanity was now Marsh's main miscreant. The "almost indiscriminate warfare [of] stationary life gradually eradicates or transforms every spontaneous product of the soil." Farming and permanent settlement subverted nature's balance—and doomed tribal indigenes who lived in greater harmony with nature. "In all quarrels between the civilized man and the savage," Marsh had come to think, "civilization is guilty of the first wrong." *Man and Nature* showed "the destructive agency of man . . . more and more energetic and unsparing as he advances."[14]

European occupation of the New World did not restore the balance of the Old; it revealed how fragile and easily shattered that balance was. Environmental damage in anciently desolated Old World lands was now in some measure restrained, but one American lifetime saw tracts larger than all Europe conquered, cleared, cultivated—and wrecked. New World colonizers seemed to Marsh to pose the gravest threats to nature, because ruthless in their conquest, transient in their local attachments, and unconscious of the damage they did.

The havoc wrought by natural agencies Marsh believed buffered by long-term equilibria. But "in augmenting the intensity of . . . mountain degradation," human action in some areas had "produced within two or three generations, effects as blasting as those generally ascribed to geological convulsions, and has laid waste the face of the earth more hopelessly than if it had been buried by a current of lava or a shower of volcanic sand." Left unlogged and ungrazed, Alpine slopes might in time revegetate, torrents dwindle, erosion diminish. But such natural regeneration required "not years, generations, but centuries; and man, who even now finds scarce breathing room on this vast globe, cannot retire from the Old World to some yet undiscovered continent, and wait for the slow action of such causes to replace by a new creation, the Eden he has wasted."[15]

Marsh's apocalyptic tone was novel. As noted in the previous chapter, few before him took alarm even when aware of the extent of alteration. Improving humanity's earthly home was divine intent, and the bounty yielded by clearing woods and draining swamps betokened God's approval. Successful at turning nature to account, Marsh's fellow-Americans felt themselves chosen partners in a sacred mission. The gains far outweighed the losses. Woodlands denuded seemed trivial next to timber empires beyond the horizon; soils eroded or exhausted were easily vacated for new lands to the West. Nature left behind would heal itself.

The great lesson of *Man and Nature* was that nature no longer healed itself. Once subdued and then abandoned, land did not revert to its previous plenitude but remained impoverished, unless taken into human care. Weakened by mankind's metamorphoses, nature needed perpetual stewardship. Although men were selfish and shortsighted, Marsh did not think them irredeemably evil or irrational. Greed was partly to blame for rapine, but much damage was unintended. Men seldom meant to destroy the balance of nature; they did not see they were doing so. Just because human impacts often caused grave damage did not mean the conquest of nature was wrong or foredoomed; failure was no more inevitable than success. Neither disdaining nor despairing, Marsh cautioned that civilization required enhancing, not abdicating, dominion over nature. If men could ruin nature, they might also mend it. To that end, intervention must become conscious and rational, instead of heedless and reckless.

Man and Nature is a diatribe, but not a jeremiad. While stressing the often calamitous side effects of human agency, Marsh details how the environment has often been improved. Forests were replanted, inundations checked, land won from the sea, swamps drained, fisheries restocked, Saharan sands fertilized by artesian waters. "These achievements are more glorious than the proudest triumphs of war," wrote Marsh, but they were not enough; "thus far, they give but faint hope that we shall yet make full atonement for our spendthrift waste of the bounties of nature." A concerted program of conservation and restoration was imperative. "In reclaiming and reoccupying lands laid waste by human improvidence or malice," man must "become a co-worker with nature in the reconstruction of the damaged fabric."[16]

Civilization did require reshaping the earth—clearing forests, regulating rivers, watering arid lands, quelling noxious creatures. But too much had been done too fast and without foresight. Farmland and pasture must now be restricted, forests replanted, torrents checked, wildlife protected, nature's aboriginal state restored—but only in part. The earth could never resume its primeval state; human derangement rendered it henceforth in need of ever-vigilant control. In place of nature's harmonies, new and greater human harmonies had to be forged.

Nature left untended, no less than nature crippled by heedless destruction, would subvert human destiny. "The life of man is a perpetual struggle with external Nature; it is by rebellion against her commands and the final subjugation of her forces alone that man can achieve the nobler ends of his creation." Man must create his own order by rebelling against nature, for "wherever he fails to make himself her master, he can but be her slave."

To be sure, conquering nature was for Marsh but a means to an end. "Our victories over the external world" were merely "a vantage-ground to the conquest of the yet more formidable and not less hostile world that lies within." But the second conquest required the first: only mastery of natural forces could release us "from the constraints which physical necessities now impose." Science had "already virtually doubled the span of human life by multiplying our powers and abridging" the time that had to be spent in gaining a livelihood. Even if "no agencies now known to man" could renew the denuded earth, new modes of harnessing the energy of waves and tides, hurricanes and volcanoes, might in time robe the Alps and Pyrenees and Taurus mountains "once more in a vegetation as rich as that of their pristine woods."[17]

Marsh had none of Emerson's faith in nature's beneficence or, as will be seen, of Darwin's in nature's boundless diversity. Yet he renounced the pessimism of those who later feared progress foredoomed by the exhaustion of new lands. "The multiplying population and the impoverished resources of the globe," urged Marsh, instead "demand new triumphs of mind over matter."[18] Although more at home in the woods than most Transcendentalists, Marsh saw no ethical virtue in wilderness and made no fetish of the primeval forest. In common with European foresters, he favored selectively planted woods with regularly pruned stands. His aim was not to restore pristine woodlands but to enrich and improve them by silviculture. Cultivation and civilization, not raw nature, enhanced diversity. "In a primeval forest, how closely the trees conform each to each," and wild animals of the same species could likewise hardly be told apart, in Marsh's view. "But in a well-tilled garden, what variety of fragrant flower and luscious fruit and nutritious bulb horticultural art derives from a few common and unsavory germs! In our domestic herds, what diversities of bulk and shape and hue."[19]

Yet for all Marsh's praise of human agency, *Man and Nature* is suffused with partiality for untamed nature. Marsh professed to "love better . . . the venerable oak tree than the brandy cask whose staves are cut out from its heart wood; a bed of anemones, hepaticas, or wood violets [better] than the leeks and onions which [men] may grow on the soil they have enriched and in the air they made fragrant." Though the then known medicinal uses of wild forest species did not furnish a telling argument for forest conservation, he judged that plants yet untested might prove valuable, so that their loss involved risk.

Although Marsh saw humans as superior beings ordained to rule all other creatures, he condemned the cruelty of those who ruthlessly tyrannized over subordinate species. He censured lack of "sympathy for those humble creatures which men too selfishly consider as at all times subject to their irresponsible dominion, as without individual rights and interests of their own." Even if devoid of conscious will and purpose, his wild animals and "respectable oaks" deserved tolerance as living beings in God's creation. "He whose sympathies with nature have taught him to feel that there is a fellowship between all God's creatures," wrote Marsh, "he who has enjoyed that special training of the heart and intellect which can be acquired only in the unviolated sanctuaries of nature . . . will not rashly assert his right to extirpate a tribe of harmless vegetables, barely because their products neither tickle his palate nor fill his pocket."[20]

So ingrained is *Man and Nature*'s ecological message today that one wonders how Marsh's intent could have perplexed anyone. Yet his English publisher, John Murray, turned it down, complaining of the book's want of definite purpose. "It is too general," he told the author; "it will be difficult to make the public understand what it is about." Marsh had switched to Murray only at Scribner's behest and now reverted to Sampson Low, Son & Marston. Further vexations came from Scribner. The New York publisher deleted some sharp comments about the Civil War as extraneous and offensively partisan; this dismemberment pained Marsh. Yet for every provocative remark Scribner excised, ten slipped through. Pungent asides on the evils of tobacco ("the most vulgar and pernicious habit . . . of modern civilization"), the iniquities of railroads, the follies of Catholicism, and a hundred other topics vivify the book.[21]

Man and Nature is a stylistic mélange, at once pedantic and lively, solemn and witty, turgid and incisive, objective and impassioned. A casual glance may discourage: one sees long sentences, interminable paragraphs, Latinate words, circumlocutory phrases, thickets of commas. But direct and evocative passages also abound. The striking metaphor, the scathing denunciation, the barbed precept, the polished rhetorical summation—in these devices Marsh excelled, and they infuse the work with life.

The immediate sources of *Man and Nature* are wildly heterogeneous. Interspersed with excerpts from French engineers on stream abrasion and German foresters on tree physiology are piquant anecdotes from Marsh's boyhood and travels. For all its history, the book has an up-to-

date vitality; fully half of the references are to articles and books published within the preceding five years. There are résumés from classical
authors, paragraphs from newspapers and personal letters, etymologies from all over, census statistics, snatches of poems and plays. So
higgledy-piggledy are these that *Man and Nature* seems less a finished
work than one in progress.

Yet this helter-skelter quality is just what makes it engrossing
and, above all, convincing. The lengthy quotes, the familiar asides, the
partisan diatribes, the confessions of ignorance, the pleas for new research are marks of an intensely personal book. They guide the captivated reader through a thorny terrain along the author's own paths of
discovery.

These attributes appear to best advantage in the footnotes. Their
bulk—as great as the text—makes the notes seem formidable. But to
neglect them is to miss the volume's unique scope and flavor. Here is an
epitaph for migratory birds in Vermont deranged by village lights; a
clue, deduced from ever thicker and less well-burnt bricks in buildings, to growing fuel scarcity and deforestation in imperial Rome; a
lament over the lack of forest care in Spain, whose people "systematically war upon the garden of God"; a harangue, provoked by grain storage hazards in Egypt, on "the slovenliness and want of foresight in
Oriental life." To the notes Marsh consigns most of his own recollections, enthusiasms, and crotchets. He has no compunction about di-
gressing. He admits that a long note on "cooking" railway surveys "is
not exactly relevant to my subject; but it is hard to 'get the floor' in the
world's great debating society, and when a speaker . . . once finds access
to the public ear, he must make the most of his opportunity."[22] Marsh
certainly did.

Publication delays vexed Marsh. Scribner promised November
1863, but the switch of English publishers and the loss of two ship-
ments of proof held things up. Marsh grumbled that his facts would be
out of date before they reached the public and fretted lest some rival
anticipate him. In March 1864 *Man and Nature* still hibernated in Scribner's den. Were it not too late, the disgusted author remarked to Norton, "I should be strongly tempted to suppress the book altogether, as
it will not now subserve some of the leading purposes for which it was
designed, and I have almost entirely lost my interest in it." Marsh won-

dered "did Calvin get his notions of total depravity from his dealings with his booksellers?"[23]

Man and Nature finally appeared in May 1864. No eager public stormed the shops. Sales were initially poor, Scribner thought, because times were bad. Doubting that the book would earn anything, Marsh donated the copyright to the United States Sanitary Commission, a Civil War charity. But his nephew Edmunds and his brother Charles, more prescient, quietly bought it back for $500 and returned it to the author.[24]

As it turned out, *Man and Nature* sold over a thousand copies in a few months; Scribner had to reprint. "The demand continues," he told Marsh; "I must confess I have been favorably disappointed." Reviewers found it "a mine filled with attractive treasure," inspiring hope, directing thought, instilling virtue. James Russell Lowell welcomed a book "which will lure the young to observe and take delight in Nature, and the mature to respect her rights as essential to their own well-being." Some thanked Marsh for championing resource stewardship, others for refuting materialism by showing the power of mind over matter. Critics praised his memory, powers of observation, sympathy for nature; he had "the eye of a traveler, the heart of a poet, the intelligence of a philosopher, the soul of a philanthropist." Marsh would achieve for geography, felt the writer-diplomat John Bigelow, "what Adam Smith did for Political Economy, what Buffon did for Natural History and what Wheaton and Grotius did for International Law": a synthesis of all available knowledge.[25]

If some wanted "fewer facts and more philosophy" or deplored Marsh's "pessimism," most agreed with Sir Henry Holland that "our American Evelyn" had written a uniquely important book. Charles Lyell confessed that Marsh had disproved his own view that man's geological impact was no greater than that of animals. Arnold Guyot, whose *Earth and Man* had incited Marsh to rebuttal, applauded Marsh's insights, as did Harvard geographers Nathaniel S. Shaler and William Morris Davis. Lauded by scholars and the general public, within a decade *Man and Nature* was an international classic. "One of the most useful . . . works ever published," declared a reviewer of the enlarged though not notably altered 1874 edition, it had "come with the force of a revelation."[26]

More than the 1864 book, it was this second edition, retitled *The Earth as Modified by Human Action*, that launched a radical reversal of American environmental attitudes. Initially "the matters of which Mr Marsh treats were only of curious interest," remarked Princeton President James McCosh. "Our woods: were they not exhaustless? Irrigation: what need had we to bring lands under cultivation . . . when the unsurveyed public domain amounted to fifteen million acres," and were they not believed to be "all of the same exuberant fertility with the prairies of Illinois and Iowa?" A decade later things looked far less rosy. "We have been brought very sharply to a realizing sense of our natural limitations." Railroad building had stripped the East of trees; the West had proved less and less fertile, with "huge barren plains, lava overflows, sterile and forbidding regions swept by tornadoes and devastated by winter torrents; huge tracts where nothing but sage-brush or chaparral grows." Marsh's timely book, noted McCosh, now taught Americans "to attribute unwelcome changes to our restless disturbance of the equilibrium of nature" and showed them how to protect their heritage from waste and abuse. In Wallace Stegner's phrase, *Man and Nature* was "the rudest kick in the face that American initiative, optimism and carelessness had yet received."[27]

In tandem with the tree-planting mania that swept the country in the Arbor Day movement, Marsh's warnings led Franklin B. Hough at the American Association for the Advancement of Science in 1873 to petition Congress for a national forestry commission. Hough, who became first chief of forestry in the federal agency set up in 1876, drew heavily on *Man and Nature* for the AAAS 1873 petition, held Marsh the pioneer crusader against excessive felling, and begged him to return to guide American forestry.[28]

Every leading forestry figure was inspired by the book and sought Marsh's advice. Frederick Starr of St. Louis and I. A. Lapham of Wisconsin in the 1860s printed excerpts that foretold disaster unless forests were protected. Big chunks from Marsh's "great masterpiece" filled a 41-page appendix in Shaler's 1875 Kentucky geological survey. Hough's successor, N. H. Egleston, credited *Man and Nature* with "awakening our attention here to our destructive treatment of the forests, and the necessity of adopting a different course." George B. Emerson held it "the best volume on its subject that has ever appeared," a view echoed by Egleston's German-trained successor Bernhard Fernow, by American forestry head John A. Warder, and by Arnold Arboretum's Charles Sprague Sargent, whose massive survey of American forests appeared in 1884. Terming Marsh his "ideal American scholar," the explorer Ferdinand V. Hayden carried "his splendid book . . . all over the Rocky Mountains." For Gifford Pinchot, America's self-styled "father of conservation," *Man and Nature* was "epoch-making."[29]

A federal forest reserve system emerged in 1891, then watershed protection, and eventually a conservation program for all natural resources. The U.S. Forest Service, the Sierra Club, and finally even the timber companies acted within an ecological mindset whose broad premises Marsh had set. Formerly an emblem of pioneering progress, the hewn tree stump came via Marsh to symbolize wanton destructiveness. And in Marsh's native Vermont, his Woodstock successor Frederick Billings turned Mount Tom into a model of reforestation, and a large Green Mountain tract, including Camel's Hump, was set aside in a state reserve as "a sample of the original forests."[30]

Europeans put Marsh's precepts to effective practical use sooner than Americans. The French geographer Élisée Reclus owed much to Marsh for his *La Terre* (1868), as their correspondence shows. Marsh commended Reclus's 1868 book as a complement to his own, treating "the conservative and restorative, rather than . . . the destructive, effects of human industry."[31] Italian foresters (Siemoni, Boccardo, Di Bérenger) found Marsh's work of huge value; a half century later Italian scholars still termed *Man and Nature* authoritative on forests and stream flow. Italy's 1877 and 1888 forest laws embody citations from Marsh, who helped Italy stress restoration over preservation.[32]

Man and Nature inspired Dietrich Brandis and others seeking to stem deforestation in India. Like Hayden, conservator Hugh Cleghorn told Marsh, "I have carried your book with me along the slope of the Northern Himalaya, and into Kashmir and Tibet." As early as 1868, New Zealand legislators cited the book to halt the "barbarous improvidence" of tree felling that threatened to turn their "land of milk and honey" into a "howling desolation." By the new century Marsh's insights had encouraged conservers in Australia, South Africa, and Japan.[33]

Even the modest Marsh recognized that *Man and Nature* had had some effect. "Though it has taught little," he wrote Baird in 1881, "it has accomplished its end, which was to draw the attention of better-prepared observers" to the subject. But not for another half century, if then, were any observers better prepared than Marsh himself, who went on updating his book to the day of his death. For decades it remained the only general work in the field. The third (1884) edition was last reprinted in 1907, on the eve of the White House Conference that led Theodore Roosevelt to create a national conservation commission under Gifford Pinchot. Yet some disciples scolded that *Man and Nature* was being ignored. Marsh's warnings had fallen "upon deaf ears," railed Charles Eliot Norton. C. S. Sargent, who owed his own interest in forests and forest preservation "almost entirely" to the book, feared "the younger generation apparently know nothing about it."[34]

M*an and Nature* became one of the nineteenth century's two seminal texts on the subject its title denoted. The other, more influential and more inflammatory, was Darwin's *On the Origin of Species*, published just five years earlier. Both illumined thitherto unrealized linkages between human and other forms of life. Together they put paid to traditional faith in a designed nature, and "knocked the props out from under the idea of a pre-established harmony between humankind and the natural world." Yet the two books have remarkably little in common. Their authors never met or corresponded. Nothing suggests that Darwin ever read *Man and Nature* or knew of Marsh's environmental, as distinct from his philological, scholarship—or even, as a Darwin informant on beaver building commended the American, simply as "a man remarkable for a large store of accurate and well arranged knowledge on many different subjects."[35] But Marsh's reaction to Darwin throws light on both, and on various uses of history in Victorian natural and social science.

Man and Nature is peppered with references to Darwin's publications, of which Marsh owned several well-thumbed volumes. Darwin's 1862 study of orchids helped underpin Marsh's broader conclusions, noted in the previous chapter, about mutual dependence among flora and fauna. Darwin's work on earthworms as geographical forces, showing how they aerated and raised soils, struck Marsh as absorbing and important, but he faulted Darwin for failing to note earthworms' crucial additional role in generating nutriment by their own decay. This omission Darwin repaired in his 1881 earthworm book, but he gives far less detail than Marsh on how earthworm excreta and casts enrich soils.[36]

But Marsh viewed the theory of evolution by natural selection with ambivalence. He hailed Darwin's concept as a major advance that brought scientific ideas of propagation, growth, and development into general consciousness. But whatever its validity in natural history, Marsh repudiated Darwinian evolution in cultural history. Whereas the forms of nature might tend toward increasing diversity, this did not seem to be the case with aspects of human culture.

For example, historical evidence showed languages converging rather than diverging over time. Supporters of evolution pointed to similar patterns of descent among biological species and families of speech, in line with Jacob Grimm's canonical theory of Indo-European linguistic dispersal. They viewed the early common ancestors of Dutch and Russian and Icelandic and Greek and Latin and Persian, in Dean Frederic W. Farrar's words, as all "living together as an undivided family in the same pastoral tents." Like his friend Max Müller, Marsh disagreed. "The proofs, or rather illustrations, adduced by Grimm amount to very little"; actual European language histories usually reversed Grimm's law. "The dialects diverge as we ascend." Around the North Sea, for example, "there is not a shadow of proof, there is no semblance of probability, that the inhabitants of these coasts spoke with more uniformity fourteen centuries ago than today, but every presumption is to the contrary."

The tree-branching metaphor might make sense for biological evolution, as demonstrated by Darwin, but it was only a small part of the picture of linguistic change. "*History* teaches us, the further back we go the wider was the diversity of speech among men," and present-day contact and amalgamation were making languages ever more similar. "Dialects have usually tended to uniformity and amalgamation as they descend the stream of time; and as we trace them backwards, they ramify like rivers and their tributaries, until the main current is lost in a dispersion as distracting as the confusion of Babel." The world was moving away from Babel, not back toward it. Progressive amalgamation and intermingling, not competitive differentiation, featured most realms of human culture. In short, the evolution of species was no guide to that of cultural traits; and Darwin's reliance on supposed linguistic parallels, elaborated by his *Descent of Man* (1872), weakened rather than fortified his theory.

Marsh also rejected Darwinian claims that common origin could be proven by structural similarities, despite gaps in the fossil and written records. Marsh mistrusted such speculative genealogies; as he had shown with Wedgwood's etymologies, supposed likenesses were often deceptive. For example, languages highly discrepant in vocabulary and syntax could be traced back through written evidence to the same parentage; contrariwise, "lexical or grammatical resemblances [among tongues] no more prove their original identity" than did apparent likenesses among all species of cats or of figs prove their consanguinity.[37] Only through explicit historical evidence could one learn origins and pedigrees, show what had converged and what had diverged.

Marsh paid tribute to Darwin's pioneering work in species modification by domestication (a field "virtually created" by Darwin), yet faulted him for giving too little due to human impacts on organic nature. To be sure, Darwin's *Descent of Man* termed man "the most dominant animal that has ever appeared on this earth," and *On the Origin of Species* began by detailing the cumulative effects of human selection on domestic species; but Darwin went on to find these effects infinitesimal compared with those produced by nature.

How fleeting are the wishes and efforts of man! how short his time! and consequently how poor his products will be, compared with those accumulated by nature, . . . infinitely better adapted to the most complex conditions of life, and . . . plainly bear[ing] the stamp of far higher workmanship.

Marsh demurred: no animals in a state of nature "have exerted upon any form of life an influence analogous to that of domestication upon plants, quadrupeds, and birds reared artificially by man," including both the unforeseen and the "purposely effected improvements" of *voluntary* selection.[38] To Marsh, the pace and scope of natural change paled before that of selective breeding under human aegis.

Humans had profoundly modified nature over a much longer time-span than Darwin took into account. *On the Origin of Species* noted massive changes in flora and fauna, including wholly new species that had colonized formerly barren Staffordshire heathland recently enclosed and planted with Scotch fir. "How potent has been the effect of the introduction of a single tree," Darwin underlined his point; except for the Scotch fir, the heath "had never been touched by the hand of man." "Perhaps not," Marsh rejoined, "*after* it became a heath." Marsh surmised that the seeds of these new plants probably "had been deposited when an ancient forest protected the growth of the plants which bore them, and . . . sprang up to new life when a return of favorable conditions awaked them from a sleep of centuries."[39] Darwin disregarded both the possible longevity of seeds in suspended animation and the ubiquity of human agency stemming from ancient times. That such heaths were not pristine untrodden realms, but products of neolithic deforestation, is now well understood.

If disappointed by Darwin's slighting of human aims and agencies, Marsh found his insights provocative. While Darwin focused primarily on plants and animals, Marsh on words and texts, they shared the omnivorous taste for highly eclectic and often anecdotal sources typical of Victorian natural history. But discordant philosophies underscore the fundamental disparity of their findings. Darwin's agnosticism made it congenial for him to stress human affinities with other living beings; Marsh's Christian faith spurred him to stress their differences. While no biblical fundamentalist, Marsh felt more at home viewing man as angel, however fallen, than as ape, however risen. Darwin's eagerness to locate humans within nature led him to posit "no fundamental difference between man and the higher animals in their mental faculties," and to assert that natural selection affected moral and mental like physical traits. Hence "any [social] animal whatever . . . would inevitably acquire a moral sense of conscience as soon as its intellectual faculties permitted." Marsh, as noted above, felt human moral consciousness divinely unique.[40] Finally Marsh's insistence on hard historical evidence led him to mistrust Darwinian conjectures in both natural and human history.

After a period of relative neglect, *Man and Nature* was resurrected by Americans made newly aware of the perils of floods and soil erosion by Dust Bowl and other disasters of the 1930s. As early as 1920 the Scottish visionary planner Patrick Geddes alerted the American critic Lewis Mumford to Marsh's work. Mumford and the historical geographer Carl Sauer led scholars from a score of sciences to reassess "Man's role in changing the face of the earth" at a 1955 "Marsh Festival" in Princeton. *The Earth as Modified by Human Action* was again updated (now as *Earth Transformed*, stressing the accelerated pace of impact and loss) at a Clark University symposium in 1987.[41] Marsh's boyhood home in Woodstock, Vermont, was designated a National Historic Landmark in 1967; the 1998 inauguration there of the Marsh-Billings National Historical Park marked an apogee of American conservation awareness, in thus celebrating its begetter.[42]

In forest conservation, watershed protection, stream regulation, and other specifics, Marsh had both forerunners and followers. But as he himself noted, no one before him had observed or traced the effects of human impact "as a whole." And not until the 1960s did anyone else effectively interrelate all these topics. Knowledge of and concern about environment have greatly advanced since Marsh's day; anxiety about our impact on nature extends to realms undreamed of by him. But his analysis of forest and watershed control remains largely valid, his broad cautions about the unintended consequences of human impacts still cogent.[43] Countless conservation texts, polemical and scholarly alike, salute *Man and Nature*, as former Secretary of the Interior Stewart Udall styled it, "the beginning of land wisdom in this country." And beyond its practical lessons, in Clarence Glacken's words, *Man and Nature* "gave the world a deeper insight into the nature of human history."[44]

For all Marsh's dire warnings, pragmatic optimism suffuses *Man and Nature*. Years before he wrote the book Marsh had made up his mind how he would answer "the great question" with which it concludes: "whether man is of nature or above her." Although he drew many remedies from Europe, his central themes—the urge to reform, faith that reform was possible—are typically American. "The work of geographical regeneration," Marsh urged, was likelier in America than in the Old

World, notably in ravaged and pauperized Mediterranean lands that "must await great political and moral revolutions in the[ir] governments and peoples."[45]

Marsh's reform hopes were interfused with another trait now less clearly American: commitment to the future. The whole force of the book lies in its conviction that the welfare of generations to come must outweigh immediate gains. Americans who disdained a better husbandry for themselves were morally obliged to practice it for their offspring. In this book's final chapter I assess the relevance of Marsh's environmental and historical insights for Americans and all humanity today.

When . . . we speak of the American scholar, we mean, not a recluse devoted to literary research, but one who lives and acts in the busy whirl of the great world, shares the anxieties and the hazards of commerce, the toils and the rivalries of the learned professions, or the fierce strife of contending political factions, or who is engaged perhaps in some industrial pursuit, and is oftener stunned with the clang of the forge and the hum of machinery, than refreshed by the voice of the Muses.

Marsh, *Human Knowledge,* 1847[46]

How did this virtually self-schooled nineteenth-century Vermonter, caught up in the myriad practical and political tumults of his time, come to seize upon and set forth a pioneering analysis of human environmental impact—an analysis that has revolutionized not just American but global awareness? Every aspect of Marsh's life is implicated: the early near blindness that deprived him of books and impelled him to observe nature, at the same time training his capacious memory; the rapid pace of change he saw transforming the local scene; the Vermont Transcendental stress on free will and human agency; the contrasts strikingly manifest in barren and deforested Old World lands; the language skills that brought him in touch with like-minded Europeans; his delight in toolmaking; his search for origins and obsession with domestic social history; his dismay, born of huge personal cost, at commercial greed and corporate selfishness; his zeal for national stewardship; not least, his love of nature.

It is a profound error to view Marsh as an aloof scholar who "had the luxury and liability of writing about nature from the easy chair."[47] Marsh lived so much in the world and was so thoroughly enmeshed in everyday affairs that he was, more than any scholarly contemporary, realistic, pragmatic, and, as he himself said in the last letter he ever penned, earthy. Extolling the north German poet Johann Heinrich Voss's "Der siebzigste Geburtstag" (seventieth birthday), a candid, unaffected depiction of everyday village life, Marsh added self-deprecatingly, "Behold, how material, how earthy, my taste is!"[48]

The academics of Cambridge and the scientists of Washington sowed no fields, spun no wool, logged no trees, sawed no marble. Emerson's essays are famed for down-to-earth phrases—"the meal in the firkin; the milk in the pan."[49] The homespun words that voiced Marsh's insights were not metaphoric; they were literal. Enforced intimacy with the world of enterprise patterned Marsh's language, tested his ideas against reality, and gave him a relish for hard facts, a zest for homely details. These are the qualities that forged *Man and Nature* and that long lent it authority.

Others said that the earth made man. Look around, said Marsh, and see what man does. He cuts down trees, clears land, tills the soil, dams rivers; all these things Marsh had done himself. Was the landscape the same afterward, did streams flow as before, were plants, fish, birds, animals as abundant or diverse? Assuredly not. Where these and other human endeavors had long gone on, the changes were profound and often ruinous.

In landscape as in language, rapid change was especially to be feared. Marsh accounted "love of innovation . . . more dangerous [than] ultra-conservative attachment" to the old, "because the future is more uncertain than the past, and because the irreverent and thoughtless wantonness of an hour, may destroy that which only the slow and painful labor of years or of centuries can rebuild." Marsh likened the stability of language to that of the environment. "Like the ultimately beneficial rains of heaven, social changes produce their best effect when neither very hastily precipitated, nor very frequently repeated."[50]

Anyone wielding an ax aims at known effects, but before Marsh no one had looked at the combined effects of all axes (let alone chain saws). Once he had done so, the conclusion was patent. Human livelihood depends upon soil, water, plants, and animals. But in using them we unwittingly alter and sometimes destroy nature's supporting fabric. To remedy this, we must learn how nature works and how we affect it. And we must take collective action to restore and maintain a humanized viable milieu.

Marsh preached no panacea. Nor did he profess despair, though gloomily persuaded that selfishness prompted most human action. Deeply engaged in life, for all his misanthropy Marsh was more concerned with mankind than with the cosmos. It was not for nature's sake that he wanted to protect it against human folly, but for humanity's. Nature was neutral; only man had conscious and moral force. Marsh always believed human powers superior, human purposes uniquely hallowed. These views suffuse his own career and mark his commitment to humanity.

15. Florence and Unfinished Italy

WITH *MAN AND NATURE* DONE, MARSH CAST ABOUT
for some utterly novel preoccupation. His secretary quipped that Marsh
was taking refuge from Man and Nature in Woman and Artifice;[1] he certainly needed a catharsis. Although he periodically updated the book
and responded to countless queries on related topics, it was ten years
before he embarked again on pivotal environmental work.

The social and political disruptions of the 1860s alarmed Marsh no
less than the physical disturbances he had chronicled in *Man and Nature*.
He felt ever more estranged from post–Civil War America and began
to look upon Italy as his permanent home. But home was no longer the
orderly Piedmontese capital to which he had grown accustomed. An
abruptly proclaimed shift of court and parliament from Turin to Florence presaged further Italian turmoil.

*Florence is a mighty fine museum and a mighty poor residence. Vile climate,
detestably corrupt society, infinite frivolity, servant's hall of Tophet.*

Marsh to Spencer Baird, August 2, 1865

313

The capital transfer was part of an initially secret Italian-French Convention of September 1864. Napoleon III agreed to withdraw his garrison from Rome in two years; in return, Italy pledged to secure the papal territory against invasion and internal revolt. France assured the pope that once Italians shifted their capital to Florence they would not seek to move it again, but Italy's leaders saw Florence as a face-saving step toward Rome. Napoleon did not discourage their hope. "You will eventually go to Rome," France's foreign minister promised the Italian envoy, "but a sufficient interval must elapse to save us from responsibility."[2]

But the Piedmontese were stunned that the king had consented to quit the House of Savoy's historic seat. Angry effusions filled the Turin press; mass meetings denounced the move and menaced king and parliament. Apprehending a riot, royal troops fired on a crowd in September; "the consequence of this weak and wicked mismanagement," Marsh reported, "was that between fifty and sixty unarmed and peaceable citizens . . . were shot dead in the streets." The next day Marco Minghetti's cabinet resigned; and General Alfonso La Marmora took over a shaky government. Customarily the mainstay of the monarchy, the betrayed Piedmontese scattered handbills proclaiming "Reggia da vendere— padrone da pendere!" (Royal palace for sale, proprietor for hanging). "When it leaves Piedmont," surmised Marsh, "the dynasty of Savoy slips the cable of its sheet anchor."[3]

It was not the fate of the monarchy, though, that worried him most. Even if the 1864 Convention presaged the acquisition of Rome, Marsh feared Italy would lose more by French-imposed concessions (shouldering papal debts, affirming papal temporal rights) than it would gain by territorial unity. "The fascination of the King and his leading advisors by the magic of the [French] Empire is so complete," Marsh wrote Seward, that they would "submit to any humiliation which Napoleon III may choose to inflict upon them," and the "government will become even more dependent than it now is upon the crooked policy and the mysterious caprices of the French emperor."[4]

Meanwhile, Florence was inundated with politicians, pickpockets, and riffraff. "The Florentines were quite wild with joy" when they heard they were to have the capital, Marsh reported, "but they are now pretty well sobered down, and I think the wisest of them wish they had been spared the honor." Revoking his earlier plaudits to Florence, he now held that Tuscany had been "a *caput mortuum*" (worthless) for three centuries: "No community ever produced so many men endowed with genius in the course of 300 years as the republic of Florence, [and] no Christian people—except the Greeks—so few in the 3 following centuries. The intellect of Florence died with her liberties, and the Tuscans are now the only Italian people who can be justly characterized as *stupid*."[5]

Marsh was only heartened that the "jealousy of Tuscans will unite and arouse other Italians as they would be on no other issue"; even the Turinese now enlisted in the quest for Rome. "Regional envies and jealousies . . . make every Italian city hated by all the rest. They all join to sacrifice Turin today, and would crush Florence tomorrow with equal satisfaction."[6]

For Marsh himself the move was burdensome. Washington did not defray its cost. From the "delightful" place he thought it in 1853, Florence had by 1865 become a "mighty poor residence."[7] Housing was scarce and dear, food prices had trebled. Marsh found no affordable lodging and for a year commuted awkwardly from Turin. Only in September 1865 did he move into the sixteenth-century Villa Forini, formerly the famed Soderini family seat and plant research center. The "Giardino dei Soderini" bordered a tributary of the Arno beyond the Porta alla Croce east of Florence, a mile and a half from his legation opposite the Medici-Riccardi palace on the via Cavour. The Villa Forini was thenceforth Marsh's true home, at first year-round, then summers, for his last seventeen years.[8]

M arsh had aimed to tide over the capital transfer with a few months in the United States. But the imminent end of Civil War left Washington in flux, and Seward thought it unwise to grant Marsh home leave—not least lest it jeopardize his post against a horde of job seekers. But Marsh had just had news so terrible that denial of his request was "a relief rather than a disappointment."[9]

His son George had done poorly in a Boston law office after recovering from typhoid in 1857. Visits to Scandinavia and the Midwest left him at loose ends; he drank more and more, and in spells of depression complained of parental neglect. "I would give my right hand," Caroline wrote her sister, "if I could but open his eyes to see the selfishness and sin of his present course"; by 1860 she was hoping George Ozias would not stay long in Burlington, since "we can do nothing for him and he is only a source of anxiety to his father." [10]

In May 1862 George enlisted in the Union army, but after two months with Benjamin F. Butler's regiment in Harpers Ferry, he was discharged in August for "chronic hepatitis and general debility." Far gone in alcoholism (and perhaps tuberculosis), George lived in a New York City boarding house on a trust fund doled out to him weekly by Marsh's lawyer. Befriended there by a Brakeley family, he told Mrs. Brakeley "all his misfortunes from his early childhood," she later wrote Marsh, "commencing at the death of his mother. He had longed for the affection that only a mother could feel but that great blessing had been denied him, as a child he was deformed and sickly and very unattractive and as a man he was no better." Following a hemorrhage in February 1865, George died on the 17th. "After severe invectives against yourself and all his friends," wrote Marsh's lawyer, "he leaves all his property to a Mrs. *Brakeley*." [11]

How far was Marsh responsible for his son's tragic life? "We all know the ardent love he cherished for his poor boy," wrote George's aunt Maria Buell Hickok, "and the affection with which he followed him under the most discouraging circumstances." Friends assured Marsh that George's "unnatural exhibitions of ill-feeling towards yourself and Mrs. Marsh were the misfortunes of a mind diseased." [12] But Marsh could not convince himself that he was wholly blameless, that there was no germ of truth in George's paranoia, that he might have helped at some crucial point.

George had cause to feel neglected. After his mother's death and his first years with grandparents in Woodstock, he had been raised by a busy and demanding father and a kindly but invalid stepmother. Marsh's chief solicitude was for Caroline's health; George came a poor second. And Marsh's lofty expectations made things no easier, as in the Siedhof debacle. "Devote yourself several hours a day to the law," George's father typically exhorted him in 1855, "& give your remaining daylight hours to general knowledge & your evenings to light reading & social intercourse."

In Washington as in Turkey and Egypt, George had been on affectionate terms with Caroline and her sister Lucy, even with his father.
But after the typhoid attack his outbursts grew hard to bear, and Caroline was relieved when he left home. "Till he has more reason and more
self-control," she felt, "we can do him no good, and he can be no comfort to us."[13] There were better times. George voiced concern for his
father's troubles and offered to help him in political or literary work;
Marsh thanked his son for New York errands. But he never took up
George's offers of help nor trusted him with any affairs. Father and son
had drifted apart even before Marsh left for Turin and George for the
army, his ensuing collapse, and his final illness.

George's vindictive testament was a crushing sequel. Marsh vented
his anguish against Mrs. Brakeley and insisted on contesting the will.
Hurt by Marsh's accusations, she spurned the bequest and detailed his
son's death at length. If George had been wrong "in believing that he
had been unjustly dealt with it was not the fault of the heart but the
head," and Marsh was "uncharitable and unchristian like in condemning my motives to your poor unfortunate son that seemed cast off by
yourself and the world."[14]

For many weeks Marsh could hardly eat or sleep; his eyes again
failed, and for a long time he did little work. The "heavy shade over my
whole personal life" cast by George's death at first impelled him to return "for the vindication of my son's memory and of my own character."
But he no longer wanted to go home. "The scenes I should be obliged
to revisit," he lamented to Charles Eliot Norton, "are too full of the
graves of disappointed hopes, for me to feel strength enough to return
among them."[15]

Many urged Marsh to come home, but he always had some excuse—ill-health, expense, lack of time, dread of the voyage. Caroline
went back alone for five months in 1869, but the ease of her trip did not
diminish Marsh's reluctance. "The accomplishing of a voyage to America and back, and then of a final return to the US, seems to me to require
the performance of 3 miracles." In fact he never wanted to leave Italy.
"In this climate, my chances of life are double what they would be at
home," he told Baird. "For an old man and a solitary, life is more en-
joyable and fruitful here than in America," glacial in winter, steamy in
summer. Marsh shunned the social as well as the physical climate of his

early years; "in Vermont I should never regain the elasticity that is as necessary to usefulness as to enjoyment."[16]

Not wanting to go home, Marsh had to ensure he would not be sent home. Many were clamoring for diplomatic posts, and Lincoln needed to be reconvinced that Marsh was still the best envoy for Italy. Energetic on his uncle's behalf even before the 1864 election, Edmunds was able to promise Marsh in March 1865 that his post was secure. But that assurance vanished with Lincoln's assassination in April. While in Congress with Andrew Johnson, Marsh had thought him sincere and honest even when they differed. But President Johnson now faced extreme pressures, and Marsh would not have been "thunderstruck" had he been supplanted. Few of Lincoln's appointees remained. Yet in July 1866, Edmunds, now in the Senate seat vacated by Solomon Foot's death, could again promise Marsh that his post was safe until the next election.[17]

Indeed, Marsh was so firmly ensconced in Italy that Grant's presidency in 1869 caused him no worry. Vermont's congressmen petitioned for Marsh's continuance "as much for the honor of the whole country" as for Vermont. Three days after his inaugural Grant assured Senate foreign affairs committee chairman Charles Sumner that "I do not propose to make any change in Italy, though the pressure for the place by half a dozen people is awful."[18] He replaced every other envoy in Europe save Bancroft in Berlin. In eight years Marsh had become a diplomatic fixture. As long as Edmunds was in the Senate and a Republican in the White House, it seemed Marsh might remain in Italy.

The Italy in which Marsh had elected to stay confronted hard tasks and grave vicissitudes. Save in the irrigated Po plain and Tuscany, agriculture was primitive, industry absent, disease rampant, isolation crippling. Most property was held by great landlords, lay and ecclesiastic. When the government confiscated monastic lands in 1867, Marsh predicted that "aristocratic privilege and priestly power will [soon] fall to the ground together." He was too sanguine. Deep in debt, the government exploited land sales to favor those with ready funds. The Church deterred peasants from buying seized lands on pain of excommunication; few anyway had the necessary cash. Speculators bought up much of the alienated acreage, which was then sold to middle-class entrepreneurs as a hedge against inflation. Marsh blamed the failure of land reform on "apathy in the Italian public, and want of moral courage and sense of duty in Italian statesmen." [19]

Italian statesmen beleaguered by financial need, French pressure, and Austrian and papal threats had scant means of hewing to a socially progressive agenda, even if they had wanted one. They sought instead to placate foreign powers and domestic rivals. Leaders formed broad coalitions that shunned extreme positions. Ministries changed "as often as the moon"; there were thirteen cabinets in the first ten years, all "so entirely occupied with keeping themselves and their seats well balanced," Marsh reported, "that they have no time to think of anything else." Each cabinet offered "the same music with a different conductor"—and a near-bankrupt orchestra. He shared—indeed, largely formed—*Nation* editor E. L. Godkin's judgment that "the great statesmen who brought about Italian independence have perished untimely, and their survivors have been displaying in these latter days enthusiasm rather than judgment, zeal rather than knowledge, the virtues of agitators and dreamers rather than those of soldiers and politicians." [20]

Government borrowed heavily to pay off provincial debts and to equip a modern navy and large army, befitting Italy's sense of its great-power status. By 1866 finances were desperate; in line with finance minister Quintino Sella's policy of "economy to the bone," harshly regressive taxes were levied. Worst was the 1868 *macinato*, a "highly impolitic, unjust and offensive" impost on flour and meal, "its object being," Marsh wrote mordantly, "to spare the rich by extorting . . . from those who have no property." Sella had wrongly supposed the tax would not cause hunger; the ensuing riots left several hundred dead and thousands in jail. [21]

Poverty bred crime; murder and theft were rampant. Italians were "the gentlest, kindest, most sympathetic nation in Europe," but, Marsh added, "they have a way of trifling with knives and sticking each other in tender places." Domestic police roused more wrath than had foreign tyrants. Tax collectors and constables were detested, confessed rogues pitied as scapegoats of state coercion; terrorism disposed witnesses to lie, juries to acquit, judges to pardon. Marsh felt crime endemic, fixed in the national character. Italians "regard the execution of the murderer with greater horror than they feel for his butcheries," he told his State

Department, "and sympathize with the criminal rather than with his victim." They "practically regard robbery and assassination as natural calamities, acts of God, which no more . . . call for preventive or punitive action . . . than the death of a citizen from a stroke of lightning."[22]

Yet a decade on Marsh still professed hopes for Italy. Commerce was growing, manufactures reviving, education slowly expanding. If only the government were in step with the people! He continued to view Italy's people, like its landscape, through a roseate lens. But too few played an active role—not until 1882 did suffrage rise from just 2 to barely 7 percent of the population, of whom scarcely half voted. Thus parliament seldom bothered to respond to public grievances.

The government was hamstrung and irresolute abroad as at home. The seeming renunciation of Rome in the 1864 Convention deflected Italian leaders toward acquiring Venice as a quid pro quo. Napoleon III hinted at persuading Austria to cede Venice, perhaps in exchange for German or Ottoman conquests. Austria, readying for imminent war against Prussia, was indebted to France for a huge loan on the Paris Bourse.

In Italy, Prime Minister La Marmora was pressed by assorted nationalists—Savoyards, Garibaldians, Polish and Hungarian exiles—to wrest Venice from Austria. Never, thought Marsh, had modern Europe seen "such universal, absorbing and generous devotion to a national cause as is presented by the Italian people today." Like many spokesmen for Italy, Marsh mistook the noise generated by a few zealots for the clamor of a widely popular cause. Bolstered by Napoleon's promise of friendly neutrality, Italy joined Prussia in declaring war against Austria in 1866. But Italian troops led by La Marmora himself were routed at Custoza, near Austria's Quadrilateral fortress redoubts in Lombardy.

Marsh was not alone in assigning the "imbecility [of this] insane attempt" to cross the Mincio River, as if on parade "with musicians in the lead," to La Marmora's "rash and thoughtless impetuosity and his intellectual incapacity." Vittorio Emanuele, whose insistence on being a heroic presence hopelessly muddled the chain of command, was no less to blame, but the cabinet hushed up the king's role in the imbroglio.[23]

As so often since Cavour, Italian enthusiasm had outrun prudent realism. What ensued was farce, elevating Custoza from a minor reverse (Italian casualties totaled only 725) into a major catastrophe, mourned

in Italy as "one of the most bloody battles of modern history"; the fes-
tering psychic wound inflamed hysteria for generations. Austrian troops
soon had to withdraw to defend Vienna against Prussia; eager to erase
the infamy of Custoza, Italian forces moved unopposed into Venetia
and Trentino. Austria now ceded Venetia to *France* for transfer to Italy,
but Italy claimed it by conquest. The morale-boosting fable that Italy
had won Venice by force of arms embittered Napoleon, who was jug-
gling incompatible roles as Italy's savior, as Italy's creditor, and (to as-
suage French Catholics who insisted he protect the papacy) as Italy's
policeman.[24]

Marsh viewed the affair in simplistic terms. He lauded the "wisdom, virtue and firmness" of Ricasoli (now premier again) for demanding the Tyrol to compensate Italy for "mortifications" caused by "the
subservience of Gallicized politicians to foreign dictation, & the imbecility of her general." He assailed Napoleon III for blocking the way
to Rome, for holding Italy in quasi-vassalage, and for shackling it with
an extravagant militarist heritage. Long after the emperor fell, Marsh
feared that "his old adherents" had not "rid themselves of the feeling
that Paris is the true seat of the government of Italy." So fixed a bête
noire was the French emperor that the Minister credited Neapolitans
with mental alertness for not being "drowned, smothered, crushed by
the baneful influence of French imperialism as are the minds of the
Northern Italians." Mistaking realpolitik for servility, Marsh thought
Italy strong enough to defy not only Austria but France. In actuality,
Italy was weak, divided, and dependent on help from dubious allies. No
European nation, an eminent Italian told Marsh in 1864, really wanted
to see Italy united and free.[25]

Venice gained, Italian nationalists pressed toward their main goal,
Rome. French troops left the city in December 1866, replaced by a volunteer force of papal Zouaves. Ricasoli offered the papacy moral and
fiscal inducements to cede secular power; Pius IX and Cardinal Antonelli remained intransigent. In November 1867 Garibaldi once more
marched on Rome. While Italy's army stood aside, papal troops backed
by French forces beat Garibaldi at Mentana. Arrested again by the Italian government, he sought American help; Marsh rushed back from
rheumatic therapy at a Swiss sanatorium to intercede with Menabrea
and Rattazzi. Since Garibaldi sat in Italy's parliament, Marsh could

hardly countenance his claim to American citizenship. But the Minister's intercession helped gain Garibaldi's release and return to Caprera. Marsh had also to scotch a rumor that Admiral Farragut was plotting to help Garibaldi overthrow the Italian government. "I shall remember all my life, with gratitude," Garibaldi wrote him, "your generous and courteous solicitude for me in troubled times." One historian judged Marsh's discreet aid "an immense service both to the Italian government and to Garibaldi himself." Again set up as sacrificial victim by Rattazzi and Vittorio Emanuele, Garibaldi at last lost trust in the king who had let him down four times within a decade.[26]

Incensed that Italy had scuttled the September Convention, Napoleon garrisoned Rome anew. But as with Venice, Italy reaped the fruits of great-power strife. The Franco-Prussian War forced the withdrawal of French troops from Rome in 1870; Napoleon's overthrow and France's defeat at Sedan finally resolved the Roman Question. Foreign Minister Visconti Venosta termed Italy's occupation of Rome requisite to the prestige of the monarchy and the spiritual well-being of the papacy; Europe's Catholic powers, piqued by the pope's assumption of Infallibility, held aloof. Although ratified by the usual spurious plebiscite, Italy's takeover left many Romans fearful of forfeiting the papal clientage they largely lived on.

Except for Trieste and Trentino, the Risorgimento was now territorially complete. "No nation ever saw its wishes and aspirations fulfilled so easily," commented *The Nation*. "Italy wanted Union and independence, and she found 50,000 Frenchmen ready to die for her. She wanted Venetia, and obtained it by two defeats. She wanted Rome, and her only obstacle, the French empire, vanishes into thin air." This "fairy tale" had come to fruition through the "masterly inactivity of Italians themselves."[27]

Marsh welcomed Italy's taking of Rome as at last making possible the wholesale reforms so often promised. There were no more French imperial watchdogs, no danger of foreign invasion, no fear of hostile reaction at home to the radical changes "so imperiously needed in the relations between Church and State." If Italy could only "free herself from the Old Man of the Sea, who had ridden her for so many centuries"—Marsh's Sinbadian phrase for the papacy—he saw "no reason why she may not be strong and prosperous." But the government's Law of Guarantees gave the papacy what Marsh termed "humiliating" concessions, preserving the principle of papal temporal rule to the detriment of Italian sovereignty. The main obstacles to progress in Italy now seemed to him those "the weakness of the Italian government has created." He wrongly supposed the people more anticlerical than the state.[28]

In the event, Pius rejected Italy's concessions, forbade Catholics to take part in Italian politics, termed himself a prisoner in the Vatican, and as in 1850 hoped to be restored to temporal rule by Europe's Catholic monarchs. To be sure, legal fictions allowed a modus vivendi between Church and State: religious orders continued to hold property, and even anticlerical Italians took pride in having the pope in Rome.

Rome was finally to be the Italian capital, but there was plenty of attendant confusion. "We Italians have got Rome at last," wrote Marsh, "but 'tis a large elephant, and a hungry. What shall we do with it?"[29]

*I*talo-American relations forged in the heat of Union and Unification causes now took a more prosaic tack. Problems of trade, tariffs, and tourism were uppermost, though one sorry sequel of the Civil War surfaced in 1866. John H. Surratt, an alleged accomplice in Lincoln's assassination, had fled America, enlisted in the papal army, been recognized, captured, and escaped to Naples. Marsh demanded his extradition to America, which humane Italy refused unless Marsh guaranteed that Surratt would not be executed. Meanwhile, Surratt fled again to Malta, then Alexandria. Extradited thence to America, he was tried and freed for want of evidence.[30]

Marsh had more success warding off international skulduggery. He deflected Italian support of French imperialism and, in 1866, of France's protégé Maximilian in Mexico. He stymied the nomination of "a most objectionable person" and advanced that of Italian Senate President Federigo Sclopis, as an arbiter of the sum due the United States for damage done by British-built Confederate ships—the *Alabama* claims case. The honey-tongued, plausible Sclopis became president of the five-man tribunal; it was a good choice for America, in view of the generous $15.5 million award.[31]

Marsh's ordinary diplomatic work was largely humdrum. A treaty

of commerce and navigation he proposed in 1864 required lengthy negotiation. Given Marsh's sagacity and "familiarity with the subject,"
Seward gladly left the mediation to him. Difficulties proved endless.
Italian duties were so low most American goods entered almost free,
whereas American duties varied suddenly and capriciously, high tariffs
virtually excluding many Italian goods. Italians wanted the treaty to
compensate for this; Marsh offered compromises. Not until 1871 did he
and finance minister Quintino Sella hammer out a treaty, ratified seven
years after it was first broached.[32]

Yet Italo-American trade remained trifling. In 1881 it amounted to
$11 million in Italian goods (oranges and lemons, sulfur, marble, and
straw hats) and $12 million from the United States (petroleum, cotton,
and tobacco). The United States took only 8 percent of Italy's exports,
Italy less than 2 percent of America's. In a late dispatch, Marsh explained that neither country needed the other's wares because "the
United States and Italy are, in most respects, geographical parallels to
each other." And dissimilar revenue systems—high internal taxes in
Italy, high tariffs in America—"oppose still more formidable obstacles
to any considerable extension of trade between them." Marsh judged
trade more apt to decrease than increase. His efforts to encourage investment in Italy were also frustrated. "Italy offers a wide field for American enterprise," he wrote a prospective investor; some hoped to see
Italian markets "invaded by Americans."[33] But Americans found more
profitable investment arenas elsewhere, and Italo-American commerce
continued to stagnate.

Immigration and naturalization were touchier issues. When America sought to regularize migration from Europe in the late 1860s, Italy
posed special problems. The Risorgimento had induced many 1848–49
refugees, now naturalized Americans, to return home; some then protested to the U.S. legation against being conscripted into the Italian
army. In the name of the president, Marsh was required to "make a
peremptory demand for the release of any American citizen unlawfully
detained abroad." But the Italian government would pay this no heed;
any Italian-born male, even if Americanized in infancy, was liable for
conscription. In 1871 Marsh warned one Italian-born American, who
had come over to dispose of family property, not to cross the border
but to transact his business through a lawyer. Marsh prudently avoided

any "discussion of the principle involved, because a strict adherence by us" to American "rights" would clash mightily with "diametrically opposed" Italian views. He preferred to curb such nuisances by securing "the *quiet withdrawal of the party into a foreign jurisdiction.*"[34] The issue was aggravated by a trade in passports and fraudulent papers that let thousands reap the benefits of citizenship in both countries, yet obey the laws and pay the taxes of neither.

Marsh's problem in this as in other negotiations, noted an analyst, was that American and Italian officials each "subscribed to a 'liberalism' which sometimes meant one thing, sometimes another." State Department officials, surprised that so "liberal" a government as Italy's should insist on applying Italian law to returned emigrants, ignored what was clear to Marsh: Italy treated international law as a matter of relations between "nations" (i.e., peoples) rather than states, a stance derived from Vico by Marsh's admired colleague, future foreign-affairs minister Pasquale Stanislao Mancini. In Italy, as in most of the Old World, formal citizenship meant less than did inherited identity, viewed as ineradicable.[35]

Meanwhile, needs for labor on transcontinental railroads spurred emigration among peasants left destitute by agricultural modernization in Lombardy and Piedmont. From a thousand a year before 1860, emigrant numbers reached 3,000 after the Civil War, 9,000 by 1873 (many more went to South America). But rumors of abuse—jobless Italians left destitute, pauper children coerced into beggary—soon spread. Beset by complaints from both sides of the Atlantic, Marsh broadcast warnings that emigration agents' promises "are not implicitly to be depended upon," that immigrants might not at once find work, and that many jobs in America were only seasonal: Italians were cautioned that "the severity of the winter climate causes the suspension of . . . outdoor labor for a considerable period." Italians then accounted for a mere 4 percent of all foreign-born in the United States. But after 1879, exodus from the Mezzogiorno became a torrent; by 1896, more Italians arrived than any other immigrants. Protestant American fears of Romanization, exemplified in Josiah Strong's best-selling *Our Country* (1886), echoed Marsh's anxieties about the 1850s Irish.[36]

Scores of minor negotiations preoccupied Marsh. He concluded consular, postal, and telegraphic conventions. Before 1870, when the

U.S. naval base at La Spezia reverted to Italy, he vainly sought a substitute anchorage. He wangled permission for the Smithsonian Institution to send publications to Italy duty free—no easy job, for "your Italian is ticklish in the purse." Italy had wanted the Smithsonian to pay freight both ways and "give everybody a small present" too. More distressing to Marsh was the Smithsonian's rejection of Cypriot antiquities offered by Luigi Palma di Cesnola, a naturalized American perturbed to find "very little taste, as yet, beyond that of money-making" in his adopted country. Ultimately, Cesnola sold his collection to New York's Metropolitan Museum of Art, whose first director he became in 1879. Another abortive exchange would have gained the Smithsonian antiquities from Pompeii, of which Italy had such an *embarras de richesse* that Baird feared "duplicates" faced destruction.[37]

Most bothersome were the grievances of American visitors. As the number in Italy increased (in 1873 Marsh estimated a hundred permanent residents and four thousand transients in Rome alone), vexing incidents multiplied. Complaints concerning religious freedom Marsh treated seriously, though ever careful to avoid offending Italy. But most American faultfinding seemed to him groundless or malicious. Indigent and eccentric Americans appeared in Florence and Rome with accusations of being molested, even tortured, by Italian police; Marsh had to ship them home, sometimes at his own expense.

Italian widows, wives, and mistresses of Americans haunted the legation, each believing "American officials to be bound to support her if her husband fails to do so." Spirited lasses who would "have the Romans know that a Yankee girl can do anything she pleases, walk alone, ride her horse alone, and laugh at their rules" caused much alarm; the "foolish simplicity" of American girls in Henry James's *Daisy Miller* recounts scrapes such as Marsh had to rescue them from. Young Preston Powers, Hiram's son, asked Marsh to protect his fiancée from her father's brutal beatings; he failed to see why the "somewhat rigid rules of official propriety" prevented Marsh from interfering in Italian family affairs. Another American demanded revenge after being arrested for publicly insulting the king and declaring allegiance to the pope; the State Department backed Marsh's refusal to intervene on her behalf.[38]

Marsh did not always take the Italian side. He censured Italy's notoriously corrupt and petty-minded customs service. When customs officers confiscated all tobacco on an American brig, including the plugs the sailors were chewing, Marsh got a "vague and grudging" back down. This was "one of the very few cases," he noted near the end of his tenure, "in which the Government of Italy is known to have disapproved of wrongs or abuses committed by its subordinates against foreigners." More often, though, Marsh had to defend Italy from importunate Americans, who in turn rebuked the Minister when they got home. Few American visitors shared his undeviating concern for Italy's "moral and material progress."[39]

Such solicitude so endeared Marsh to Italians that he survived without discredit the most embarrassing episode of his diplomatic career: the publication of a secret dispatch in *U.S. Foreign Relations of 1870.* Marsh had accused Italy's cabinet of being "so constantly in the habit of blindly following the dictation of the Emperor of France" that it had only been "forced by the fear of popular violence" to occupy Rome. "Its future course in this matter," he predicted, "will be characterized by vacillation, tergiversation and duplicity, as it has always been since 1864." The State Department held the leak "accidental," but this section of the dispatch had been in cipher. Marsh suspected that it was published "with the malicious object of making the Legation in Italy untenable any longer" by him.

He was dressing to dine with the Minister of Foreign Affairs when a servant gave him the newspaper with his dispatch in it. "He almost fainted," recalled an aide, "on reading this indiscreet divulgation and in the thought that in a few minutes he would have to face and receive the hospitality of the very men he had so severely handled." But far from snubbing Marsh, they took pains to put him at ease: Foreign Minister Visconti Venosta was "more gracious than ever"; Ricasoli crossed the room "to salute Mr. M. in a marked manner."[40] No one showed resentment. The high personal regard he was held in by the foreign office and the court, observers confirmed, precluded Marsh's recall.

His fellow diplomats, whom he served as dean and spokesman for fourteen years, also valued Marsh highly. "It is truly consoling," Garibaldi wrote him, "to hear an authoritative voice rising from the very midst of that Diplomacy which seems to busy itself with the affairs of nations only to involve them in a labyrinth of deception and despair." Comparing envoys of the Civil War era, the journalist W. J. Stillman judged that "as a representative of our country abroad, no one, not even Lowell, has stood for it so nobly and unselfishly; Charles Francis Adams alone rivalling him in the seriousness with which he gave himself to the Republic. Lowell was not less patriotic, but he loved society. . . . Marsh in those days of trial loved nothing but his country, and with an intensity that was as ill-requited as it was immeasurable."[41] At least he was kept in office.

Marsh's evident patriotism was crucial to his effectiveness as an envoy. So was the extension of his messianic Americanism to his hopes "for a new Italy, united, constitutional, free, secular, and prosperous." Along with the savoir-faire lent by Marsh's previous diplomatic experience, "the puritanical 'godliness' of his character, even its defects," forged a long relationship based on mutual friendship. "His very anti-Catholicism," concluded his Catholic memoirist, "encouraged officials who looked askance at the papacy, to place their confidence in him." Nor did Marsh's frequent aspersions against Italian statesmen dim his confidence in their essential caliber. There was "not a freer government under the sun than Victor Emmanuel's," he assured an American visitor in 1873, "& the gentlemen who constitute the Italian Parliament are infinitely more intellectual & full of general knowledge than the same number of senators and congressmen in America." The trust and friendship forged during his long incumbency survived a series of mediocre American successors in Rome. The like longevity of Barone Xavier Fava's Washington legation tenure (1881–1901) prolonged the tradition of "old diplomacy" in Italo-American affairs.[42]

*R*econstruction in America proved as perverse as unification in Italy, and Marsh was no less critical of his own government. "I have never feared the war," he wrote Charles Sumner, "but from the beginning, I have dreaded a peace which should involve the sacrifice of what it has cost in so many lives, so much treasure, to win." Before, he had blamed Union defeats on Lincoln's determination "to maintain slavery at all hazards"; now, he blamed the faults of Reconstruction on "the constant assertion," by Johnson and other moderates, of "the imperishability of State rights." Against states' rights he penned six short pieces in 1865 for E. L. Godkin's new weekly, *The Nation*. Not states but the "people" had created the Union, argued Marsh; lacking prior sovereignty, states that rebelled ceased to exist as states, and became federal territory.[43] Hence the Union had legal authority to deal with the South as it wished.

Marsh's animus against the South aligned him with ultra-Radical Republicans; Sumner was his *beau ideal*. Marsh wanted Confederate leaders hanged. "If such wretches as Davis and Toombs & Wise & Lee & Forrest & Mosby are to be pardoned and welcomed back with open arms," he wrote his confrere John Bigelow in Paris, "I am for opening all our penitentiary doors and pensioning & promoting the inmates." Sympathy with criminals seemed to him "one of the strongest proofs of the demoralization of the age." In America as in Italy Marsh demanded eye for eye and tooth for tooth.[44]

He also advocated Negro suffrage, since "we must come to that at last, and it is quite as well to come to it gracefully, and at once." Black Republicans would offset Irish Democrats: "I cannot see why a wild Irishman, who arrived at Chicago a year ago . . . should be allowed, by a state constitution, to vote for members of Congress, while a negro, born and bred in Connecticut and paying taxes there, should be excluded from the polls." Ex-slave voters of "intelligence and virtue" would help ensure that former slaveholders never held power again.[45] Indeed, ex-slave votes in the South enabled the Republicans to control the Senate, if not the presidency, as long as Marsh lived.

With passing years, though, Marsh took less and less interest in American politics. Disappointed by Johnson's softness toward the South, alarmed lest his Reconstruction program leave loopholes for states' rights, even re-enslavement, Marsh saw the expediency of Grant's nomination, but thought poorly of him. The scandals and corruption of Grant's tenure showed America to be mired in materialism. Ambassadorships went more and more to moneyed men, because they alone could afford salaries half of what they had been in real terms fifty years before. "The effect of this narrow policy," Marsh rightly foresaw, "will soon be to confine the diplomatic service to the rich, and thus to strengthen the aristocracy of wealth . . . which is so fast becoming a threatening evil in our American life."[46]

Marsh's state-sovereignty articles began a lasting bond with Godkin, for whose *Nation* Marsh wrote on a dozen topics, often under the pen name "Viator" (Traveler), to protect his post as envoy. He felt well suited to polemical journalism, "not to teach—I disclaim that—but to stir up people to teach themselves more than I know, and in that way, I do *some* good." Sharing Godkin's reform bent, he thought *The Nation* the most promising new aspect of the American scene. When Godkin pondered a history chair at Harvard, Marsh told him he could not be spared as editor: "History can afford to wait, the policy and political morality of our day cannot."[47]

Marsh's weightiest *Nation* essay comprised eighteen long "Notes on the New Edition of Webster's Dictionary." This did "not interest the multitude as much as it ought," Godkin was aware, but "the best of our readers follow it with great satisfaction." Marsh's fellow litterateur-diplomat James Russell Lowell demurred at his "Congregational" style, but thought "all he says is worth reading for its matter." A lone curmudgeonly nitpicker took exception to Marsh's use of "propose" for "purpose," "not less" for "fewer," and a possible alimentary confusion: in writing of "battered copper vessels, old brooms, cobwebs, appleparings, and the like, which the Flemish painters scatter so freely about their interiors," Marsh had failed to put "their interiors" in quotes.[48]

Most of Marsh's *Nation* pieces, which appeared from 1865 to 1881, were brief and pithy. He discussed "The Proposed Revision of the English Bible" ("very valuable . . . widely read," reported Godkin); promoted the Early English Text Society of London; reviewed Gerolamo "Boccardo's *Dictionary of Political Economy*"; wrote "Pruning Forest Trees," "The Education of Women," "Agriculture in Italy," "The Aqueducts of Ancient Rome," "The Catholic Church and Modern Civilization," "The Excommunication of Noxious Animals," and "Monumental Honors," an anti-Southern satire. "A Cheap and Easy Way to Fame" poked fun at biographical dictionaries that offered literary immortality at a stiff price. "If I can't go down to posterity on cheaper terms than these," Marsh decided, "I'll stay here!"[49]

Friends' projects engaged Marsh more than his own. "The hardest work in the world is *our* work," he concluded; "the easiest, other people's." He added etymologies to Wedgwood's revived *New English Dictionary*. He found materials for Columbus scholar Henry Harrisse. The Senate commissioned him to weigh "the substitution of the Phonetic for the Latin alphabet." He took up cudgels on behalf of the historian John Lothrop Motley, unjustly dismissed by Grant as envoy to Britain in 1870. For American teachers he collected data on deaf-mutes in Italy, harking back to his first public chore a half century before.[50]

For Henry C. Lea's *History of the Inquisition*—a topic dear to Marsh's heart—he secured official permissions and, helped by Pasquale Villari, biographer of Savonarola, trawled for assistants to copy archival materials. Ominously, the first two copyists engaged, Signori Bigazzi and Uccelli, each in turn died on the brink of starting work, "re infecta" (without accomplishing their object), as Marsh aptly put it. For some time he found no others, save librarians who could not lawfully be paid; "*pazienza* must be our motto," he counseled Lea in Philadelphia. Finally he arranged for a young Harvard scholar, Italian-born Francis P. Nash, to do Lea's archival work; Nash became a Marsh intimate. Lea recorded his gratitude for the "ready courteousness" of Marsh's aid.[51]

Through *Man and Nature* Marsh came to know the British geographer Henry Yule, who lived in Sicily from 1863 to 1875 while preparing his definitive edition of Marco Polo. Marsh and Yule exchanged scores of letters on word origins, Euro-Asian culture contacts, Indian forestry, international politics, and the ailments of their invalid wives. Yule was concerned lest this generous correspondence infringe on Marsh's official duties. "I am grown too lazy to work," the American Minister replied, "or to follow any consecutive study, & I would rather spend three hours in running down a word or looking up an idle question than in writing the best despatch ever penned by diplomate." His own current "scientific" interest was trying to find out why American Indians were almost always right-handed.[52] Settled and happy in Florence, Marsh could afford now and then to appear frivolous.

Marsh was well equipped for the demands of Florentine life in his roomy home on what was then the open plain of the Arno. Surrounded by hedges of laurel and bay, wide-spreading magnolias, and thickets of laburnum and acacia, the Villa Forini had massive portals, long, narrow passageways, slippery marble-coated floors, sculpted stone staircases, and bulky furniture placed all awry—"there is not a right-angled room, not even a room with *one* right angle, in the whole building." Marsh jested that this was "blind adherence to an old Etruscan superstition, to avoid the evil eye."[53] In the largest room he installed his books. Marsh had shipped out those stored in Vermont; most of his 15,000 volumes were at last together in one place.

At the Villa Forini the Marshes gathered a distinguished Anglo-American circle: Hiram Powers, Isaac E. Craig, Joel T. Hart, John Adams Jackson, and other artists; James Lorimer Graham (who became U.S. consul-general in Florence in 1869), crusading journalist Jessie White Mario, authors T. Adolphus Trollope and his wife Frances Ternan. In the Trollopes' magnificent Villa Ricorboli, close by across the Arno, Marsh reveled in the fine library, marveling most at "the 114 volumes written and published by Mr Trollope's mother [Frances Trollope] after the age of fifty."[54]

Permanent foreign residents were far outnumbered by throngs of visitors. Caroline held weekly receptions and gave innumerable dinners, but there was neither time nor energy enough for everyone. Many old friends came: Admiral Goldsborough, William Cullen Bryant, Caroline Paine in 1866; Justin Morrill and Bayard Taylor in 1867; George Bancroft, J. L. Motley, and Bayard Taylor in 1868, along with Willard Fiske and G. P. A. Healy; Asa Gray, Charles Eliot Norton, and Manning Force in 1869. Admiral Farragut made a fine impression, General McClellan a predictably poor one; Seward, alas, arrived in July, and found "no one" in town. Marsh made new British friends: through Matthew Arnold the parliamentarian and India expert Grant Duff, a devotee of *Man and Nature*; travel-guide writer Augustus Hare; the historian Lord Acton, a liberal Catholic whose antipapal views accorded with Marsh's. Introduced by Trollope to George Eliot and G. H. Lewes, the Marshes were unfazed to learn their relationship was not "normal."[55]

Court festivities were continual. The most "strenuous merrymaking" the Minister endured was Prince Umberto's wedding to his sixteen-year-old cousin Margherita at Turin in 1868, which Marsh decried as a "frightful extravagance" at a time of widespread economic misery. Back in Florence, he fed walnuts and pine-buds to the flying squirrels he had got from Baird as a gift for the royal couple. "What a foolish habit they have of sleeping all day and capering all night," remarked Marsh. "Well, that's the way princes do."[56]

The Marshes could not have borne these social demands but for Caroline's improved health. On the advice of Dr. Elizabeth Blackwell, Marsh took her to Paris in 1866 to the famed gynecologist J. Marion Sims, who removed a benign tumor of the womb. For the Marshes this was a miracle. "If she could or would hold still two months," he wrote, "she would be as good as cured."[57] Caroline never regained full mobility but was better than for twenty years past, active and often free of pain.

Marsh himself was less well. "The sorrows of the past year," he wrote after his son's death, "have made me suddenly an old man." Rheumatic attacks sent him every summer to spas—Lucca, Monsummano, Divonne-les-Bains, Bad Gastein, Aix-les-Bains, Wildbad. At the last, in 1869, Marsh joined two other American envoys, George Bancroft and Elihu Washburne, ministers to Prussia and France. "We astonished the natives," reported Marsh. "The Wildbaders are puzzled to know whether we are The Three Kings of Cologne [Cathedral] or the *drei kalten Heiligen*."[58] But neither warm salt baths, vapors, drugs, rest, nor exercise did much good; Marsh saw himself a hopeless case. "My medical advisors are trying to cheer me with fair words," he wrote Yule, "but I know, from their prescriptions, that they think me an ugly customer. Think of swallowing strychnine in heroic doses, and friction—*hard* friction, with bella-donna!" Worse yet, he was deprived of his beloved Alps, dying of boredom in dull resorts. Of a lameness that had moved from his left to his right hip, Marsh recalled his father's tale "of a boy who thought he could bear his boil better, if it were on the other leg." It didn't work with Marsh, "especially as the lameness is greater on the right side than it was on the wrong."[59]

Along with his wrong hip came a serious eye ailment in 1871 in Strasbourg. "The Dr. has prevented me from rushing, or rather tumbling into the street, and running amuck among the Strassburgers," scrawled Marsh rheumatically, "only by keeping me constantly stupefied by subcutaneous injections of morphine." Then a vein burst in his right eye, and his oculist tabooed "all reading and writing, now and hereafter." But Marsh committed "*felo de me*, or at least *de oculis*," reading and writing as much as ever. His disregard of "scarecrows [who] begin all their counsels with: Fee, faw, fum! To blindness you'll come! [and] wouldn't even let me revile the wicked, for fear of bursting a blood vessel inside the eye," was rewarded. A new oculist from Koblenz restored his vision in 1873. "Since I have seen Dr. Meurer," Marsh wrote Baird gleefully, "I do nothing but curse bankers . . . and booksellers all day long, and am neither sick nor sorry afterwards."[60]

Not only could Marsh for the first time in six years "read & write freely, pick up pins, tie my own shoelaces," but Caroline's eyesight too was restored after thirty-one years of privation. "The last book my poor wife read, in 1842, was Horace," Marsh wrote a friend; he now bought "Didot's beautiful edition of 1855 and it was the first she began after we returned [from Koblenz] to Rome."[61]

Although otherwise improved, Marsh rallied slowly from a typhoid attack in 1872. Deprived of exercise, he put on weight rapidly; dieting was of no avail, and he feared that he would "soon outweigh the colossal Bohemian girl now on exhibition with a singing fish in the Piazza Barberini [Rome] just below us." His hair and beard wholly white, in his early seventies Marsh began to seem an old man. His "only good" photo showed "the bristliest, frizzliest, grisliest and grizzliest" visage imaginable; "the photographer, hardened as his profession had made him to sights of woe, stood aghast at his own work," and Caroline consigned it to the flames.[62] But Marsh still had reserves of strength, and his wit and memory remained unimpaired.

I do not like Rome the better for becoming the capital of Italy. It is to me materially as detestable as it is morally, and I think a cataclysm, which should sweep it into the abyss, would be no evil!

Marsh to C. D. Drake, May 8, 1871

Marsh's objections to Rome as Italy's capital after 1871 were even more severe—and forgetful of past joys—than his strictures against Florence in 1865. The vexations of moving made him revile any new residence; he rarely realized how much he liked a place until he had to leave it. A hot and crowded "capital in the center of a desert 50 miles square," Rome offered grave grounds for complaint. It was "filthy, fetid, infectious"—defects remedied only slowly and in part. Like the transfer to Florence, the new move brought upheaval. Into Rome poured speculators, builders, demagogues, merchants, crooks, and beggars from all Italy, more than doubling the city's 200,000 within a few years. And all the government's difficulties were magnified in this new melting pot.[63]

For Marsh the move was financially crippling. "The change from Turin to Florence cost us a sacrifice equal to a quarter's salary," removal to Rome still more. Not until 1872 did he find a "tolerable apartment" in the Casa Lovatti, via San Basilio, at $1,000 a quarter ($50,000 a year in current money). A congressional grant of $6,000 "to defray the extraordinary expenses" of the two capital moves was lost in the bank failure of 1873 before Marsh could lay hands on it.[64]

The passage of time intensified his dread of penury. He was not miserly, but his letters show him obsessed with parsimony. Rome was his last stand. Old age at hand, he dreaded his inability to provide for retirement and for Caroline after his death.

16. Last Watersheds:
Rome, Cravairola, Vallombrosa

*The Europeans don't really understand how a government purely of law can
operate. A government which is never arbitrary or violent, is in their view
not really a* government *but only a quaker committee.*

Marsh to William Tecumseh Sherman, March 22, 1877

UNDER ROMAN SKIES THE YEARS PASSED TRANQUILLY
for Marsh. Changes were less abrupt, exciting, militant. In America, one
Republican president followed another; in Italy, conservative cabinets
gave way to liberals in 1876, but politics and policies changed little.
Meanwhile America's aging envoy stayed on, by now an almost legendary fixture.

With the material Rome, Marsh made peace; with the spiritual,
never. Here in the papal bastion of Catholicism he became once more
a Puritan iconoclast. In his last years he also turned again to environmental affairs, revising *Man and Nature*, guiding others' conservation
efforts, renewing his own affinity with mountain and forest. Nature solaced his growing gloom—gloom that obscured and corroded pleasure
in work well done, praise well merited, peace well won.

*L*ittle cheer could be found in Italian affairs. Marsh's dispatches reflected his pessimism. Gaining Venice and Rome made Italy at last a nation; yet poverty was endemic, injustice flagrant, crime rampant. Crop prices plummeted in the 1870s owing to American competition and a flood of cheap rice and silk from the East via the new Suez Canal. Physical calamities—frost, hail, drought, floods, volcanic eruptions, grape disease, silkworm blight—left rural Italians "more abjectly wretched," judged Marsh, than any elsewhere in Europe. Italy's crippling debt, large standing army, and other prodigal expenditures required oppressive taxes that evoked clamorous discontent.[1]

The deaths in 1878 of King Vittorio Emanuele and Pope Pius IX marked the end of an era but not of a tradition. Marsh expected little of the reputedly moderate Cardinal Pecci, now Pope Leo XIII; but he thought the new king, Umberto, had "all his father's good qualities and appears to be free from all his bad ones."[2] However, if Umberto aped his father's braggadocio he lacked Vittorio Emanuele's popular charisma. And his reign began with a general amnesty "which turned loose upon society a large number of the most depraved and dangerous inmates of the prisons," in Marsh's view wholly subverting "the moral sense of the nation."[3]

For much of Italy's plight Marsh blamed the successive right-wing cabinets, authoritarian yet irresolute, that had run the country since 1864. When the Left came to power in 1876, pledging lower taxes, broader suffrage, and state-aided schools, Marsh hailed the promise of reform. But he soon found the new cabinet both oppressive and feeble, the jobbery and expediency of the Left a sad decline from the high-minded rectitude of the old Right. Marsh soon wrote off premier Agostino Depretis as an opportunist of "habitual indecision, procrastination and inaction."[4]

Yet the new Left was much like the old Right, with personality-based coalition rule continuing as before. Endemic corruption and regional discord hobbled the economy and vitiated social reforms. Abolishing the grist tax failed to lower the price of bread; state monopolies and new high tariffs worsened the lot of the poor. As unrest and violence mounted, government clamped down on all opposition, jailing political agitators as common criminals. Charges of anarchist sedition seemed to Marsh, as to many, largely spurious, often trumped up by the police.[5]

Pervasive inequality was in his eyes the main problem. The huge gulf between rich and poor consigned most Italians to hapless, regressive servitude. Enterprise capital was scarce. Marsh saw "the superfluous means of the rich absorbed by a horde of undeserving retainers," instead of being invested in industry and commerce. In fact, lagging urban enterprise left redundant rural laborers without any prospect of work. The American Minister deplored widespread brigandage and sloth. The government restored *feste*, "days of idleness and vice." The Roman carnival of 1881, a "degrading spectacle" put on "to pick up a penny by amusing the gaping foreigners who flock to Rome to witness" it, struck Marsh as "more brutal than any for the last 30 years."[6]

He was dismayed but not surprised to find many Italians "weary of self-government." Two decades of high-flown talk had fashioned a Risorgimento only of the rich. Noting "the almost total indifference to the miseries of the poorer classes on the part of the higher ranks," Marsh in private caustically surmised that those "responsible for all this deserve, and will probably suffer, dire calamities for this betrayal of the plainest social & political obligation."[7]

Imperialist follies further beguiled Italian leaders, and acquiring Rome whetted the appetite for more. The costly army raised to counter French ambitions in North Africa seemed to Marsh preposterous. Irredentism in Trieste and Istria, not "Italian" since imperial Rome, and now largely Slavic, led to claims for the entire Adriatic coast. Having not yet secured life and property even in the Mezzogiorno, Italy had enough on its hands, he scolded, "without the importation of semi-barbaric thousands more." Italian cravings for Trentino, ventilated at the Congress of Berlin in 1878, were dismissed with the gibe that Italy must have lost yet another battle to be asking for yet another province.[8]

Only in the last year of his tenure did Marsh again sound a note of progress. Benedetto Cairoli's 1879–81 cabinet had abolished the grist tax, restored hard currency, and expanded the electorate. And the government was trying to widen access to schooling, still unaffordable for most Italians. But Marsh no longer saw education as an automatic good, let alone a panacea. "The most demoralized and dangerous of the lower classes," he observed, "are not to be found among the totally unin-structed so often as among those who have received that half-education which is all [they] can aspire to."9

Not Italy alone, he feared, but all Europe risked "relapsing into Medieval barbarism." War in the Balkans loomed in 1876. Unlike the Crimean War, here Marsh found little to choose between belligerents: "Satan's house is divided against itself." He agreed that "the Turks ought to be driven out of Europe, but when we ask: what better have we to put in their place? it is *not* easy to find a satisfactory answer." Concerned lest Turkish fury against America, as a suspected Russian ally, "endan-ger the lives and properties of all foreign residents in Turkey," he urged Secretary of State Hamilton Fish to provide "a display of at least moral force," if not an American naval fleet presence.10

Marsh judged world history at its nadir in 1878: "In the worst pe-riods of the French Revolution & Bonapartean Wars, men may have seen a sky as lurid as ours of today, but there were then *some* bright spots." But now "there are only degrees of darkness in the general pitchiness." The crises of the early 1880s seemed "to threaten more serious evils to the cause of modern civilization" than any previously. France's invasion of Tunis in 1881 led an alarmed Italy to ally with Germany and Austria. Consummated just as Marsh died, the Triple Alliance would have ap-palled him. In 1864 Marsh had rebuked "the rapacity of the Germans, whose aggressive character, & especially the national ambition to Teu-tonize the whole world" he saw as "one of the greatest sources of danger to the peace and social progress" of Europe. Now, in 1882, he suspected Bismarck and Kaiser Wilhelm I of plotting to dismember Italy, return Rome to the papacy, "and plunge Europe again into the miseries at-tending the Religious Wars" of the Reformation. As Marsh's Catholic chronicler concludes, "a merciful Providence prevented his seeing what might lie beyond."11

One of his last dispatches declared that he had "never seen the political horizon of Europe in so disturbed and so menacing a state." Marsh's political pessimism matched his environmental gloom. "The human race seems destined to become its own executioner," he had fore-cast in *Man and Nature*—"on the one hand, [by] exhausting the capacity of the earth to furnish sustenance to her taskmaster; on the other, compensating diminished production by inventing more efficient methods of exterminating the consumer."[12]

Given his morose outlook, it was fortunate that Marsh's diplomatic duties were now lighter. He still grumbled that no "ambassador at this court receives so little as four times my income, while their duties are not so onerous as mine on account of the large number of secretaries who really do all the work."[13] But aside from tidying up commercial and consular treaties, few matters called for much labor by him; Italo-American relations remained amicable and quiescent. Petty duties he delegated ever more to subordinates. Marsh quit Rome in the oppressive summers, using his Florence home as a retreat or as a base for mountain excursions. Even in winter he rarely spent more than two hours a day in the legation.

This degree of leisure was made possible mainly by George Washington Wurts, a Philadelphian who had arrived in Florence in 1865 highly recommended as "a young gentleman of fortune" and ambition. Taken on as unpaid attaché, Wurts promptly became indispensable. A dapper bachelor, elegant and snobbish, he sported two dozen pairs of gloves and a solid gold dinner service. But Wurts was cultivated, hardworking, discreet, intensely loyal—and he idolized the Marshes.

In 1868 President Johnson named as Secretary of Legation Reverend Henry P. Hay, "DD., LL.D., and M.D.," of Tennessee. The gauche, bumptious, monolingual Hay was *"in every subject* one of the most objectionable persons" Marsh had ever known; the Minister feared that Wurts and Artoni would be driven away, leaving him translator to his own secretary. "A man has no business accepting a position for which he is totally unfitted," Marsh grumbled to Caroline, "but, as I have done that same several times, I won't be noisy about it."[14] Luckily, disgruntled by his salary and his reception, Hay soon departed. In 1869 Wurts became Secretary of Legation, a post he held beyond the end of Marsh's term. Handling all the legation routine, Wurts was invaluable, leaving the old Minister to carry on diplomacy almost as well in 1881 as twenty years earlier.

With his consuls Marsh was less fortunate. Consul-General T. Bigelow Lawrence and his "charming and noble-minded" wife had guided the Marshes through the shoals of Florentine society, but Lawrence was no diplomat; confidences were not safe with him.[15] At Rome, Marsh got Consul-General Charles McMillan dismissed after he showed up at one of Caroline's receptions in a state of "palpable intoxication." McMillan was succeeded in 1879 by the brilliant but arrogant Eugene Schuyler. Known as a translator of Tolstoy and Turgenev, Schuyler had been Secretary of Legation at Constantinople, where his tirades against Turkish rule in the Balkans got him into trouble. Marsh's initial mistrust turned to aversion when his new consul-general demanded to be formally presented to the diplomatic corps. Marsh wrote in reply that Schuyler's post was not diplomatic but commercial, and owing to unhappy precedents the diplomats would take umbrage. The Marshes would personally do "anything we can to make Rome an agreeable residence to Mrs. Schuyler and yourself."[16] Schuyler was miffed; transferred to Bucharest in 1880, and later Minister to Greece, Serbia, and Romania, he never forgave the slight. As will be seen, he contrived to cast a pall over Marsh's final days in office.

One novel diplomatic task Marsh enjoyed more than any other during his final decade. He agreed to arbitrate a boundary conflict between Italy and Switzerland. Their quarrel over the Alp of Cravairola, northwest of Locarno, went back to the fifteenth century. It was the last remaining undemarcated stretch of the Italo-Swiss boundary.

The disputed area, some seven square miles of rugged rock with a patch of pasture and a stand of conifers, contained a few huts for summer herdsmen but no permanent dwellings. The dispute had festered so long because Cravairola lay on the east side of a ridgetop separating Italy's Valle Antigorio (drained by the Toce, a tributary of the Valle d'Ossola) on the west from the Swiss Val Maggia on the east. Easy of access from the Swiss village of Campo in Val Maggia, Cravairola was cut off from the Italian side by snow and ice for most of the year; even in summer a nine-hour climb over the pass was needed. Nevertheless, the Italian communes of Crodo and Pontemaglio had over centuries acquired title to Cravairola, renting parcels to villagers who pastured animals and occasionally cut timber there from June through September. But lands, huts, animals, and other goods were continually encroached on by the Swiss in Campo just below. Open conflicts had worsened relations in recent years.

Both countries claimed sovereignty. After many failures, an 1873 Italo-Swiss commission decided to arbitrate, both sides agreeing that private property rights should play no part in fixing the frontier. The Swiss wanted a boundary along the summit of the ridge. Italians believed the boundary should swerve eastward from the ridgetop to embrace Cravairola, then rejoin the summit chain three miles farther on.[17]

Marsh joined Swiss and Italian agents in Milan on September 7, 1874, departing for the alp the next day. On the 9th they left Crodo, climbing seven thousand feet by mule and on foot, to spend fourteen drenched hours crossing the debated land. From the Passo della Scatta at 2,500 meters a "grand spectacle" had greeted an earlier boundary inspector in 1868: "The Alp of Cravairola unfolds itself as an amphitheater at your feet, sweeping down a gentle and even slope towards the torrent of the Rovana, the outlet of the whole great basin of Cravairola."[18] On a small promontory at the bottom of the valley stood the chalet of Cimalmotto, an outlier of Campo.

Marsh's party six years later was sadly frustrated by the view. Immersed in heavy rain and fog, "we saw little of it except the entrance and the exit," and could not even halt at the knife-edge summit. Despite this strenuous wet hike on a mountain path "among the worst I have ever travelled," he was "neither stiff nor sorry the next day, which for an old cripple is pretty well."[19] (It was more than this biographer of Marsh, at the same age he had been, achieved 122 years later, even though the path to Cravairola is now well marked on a map in Crodo village center.)

By September 12 the party was back in Milan; Marsh left for Florence while the two nations' agents first sought to resolve the issue on their own. What ensued was paradoxical, as a later analyst put it. The Italian arbiter offered to accept the Swiss ridgetop boundary, if the Swiss would indemnify the Italian communes by buying their Cravairola properties. The Swiss arbiter rejected this proposal, noting that landownership had no bearing on the boundary settlement. This was a legal truth but a social fiction; the problems of Italian owners, already grave, would be magnified were the territory adjudged Swiss.[20] Hence Marsh had to be recalled to Milan on the 16th. The legal question, "certainly as thorny as the physical," required a week to study six centuries of documents annotated by both countries' agents. On September 23 Marsh issued his twenty-two-page decision.[21]

To Italy, Cravairola mattered most for safeguarding proprietary rights in Crodo and Pontemaglio; Swiss interests were different and far broader. Cravairola was part of the drainage area of the Val Maggia, vital for the protection of soils and crops in the lower Rovana and the Maggia. Policing, hydraulic regulation, and forest management had to be Swiss. In an 1877 addendum to *Man and Nature*, Marsh described the damage done to Campo, largely by flotation of logs but also by stripping away tree cover. This had so augmented the force of the Rovana that "in the course of four years it excavated below the village a new channel one hundred feet deeper than its ancient bed," causing 2,500 acres "including the village of Campo . . . to slide downhill in a body." The soil was now "so insecure," Marsh was told on the spot in 1874, "that meadow and pasture grounds, which, if safe, would be worth a hundred dollars per acre, cannot now be sold for ten."[22] Hence in his view "the ultimate best interests of both parties would be most effectually promoted by assigning the territory . . . to Switzerland."[23]

But Marsh's hands were tied by the terms of arbitration. He was not free to decide on the basis of "best interest," *ex aequo et bono* in legal parlance; the decision must be strictly *de jure*, based on historic title. Here documentary data strongly favored Italy. A mass of evidence— charters, land grants, treaties—showed that since 1367 Crodo and Pontemaglio communes had repeatedly affirmed not only ownership of Cravairola but administration over it, first under the duchy of Milan, later the House of Savoy. The people of Campo might originally have cleared and settled the Alp, but no proof of this or of any formal Swiss or cantonal claim to the territory was offered until 1641, when the Swiss sought to counter a House of Savoy claim. By historic right Cravairola was clearly Italian.

Ruling for Italy, Marsh regretted being denied a decision that would make geographical sense, improve land management, promote conservation, and benefit all parties.[24] Both countries thanked him for his unpaid labor, and the U.S. Congress grudgingly let him accept small gifts—from Italy, a marble-topped table inlaid with Florentine roses; from Switzerland, a plain gold watch. In receiving the watch from the Swiss envoy in Rome, Marsh again rued the best solution having been precluded; the land was physically Swiss and should be theirs politically.[25]

While both countries then voiced formal gratitude for Marsh's decision, dissenting views emerged much later. In 1954 the eminent Swiss jurist Paul Guggenheim assailed Marsh as wrong—wrong in ignoring many Swiss points; wrong in misconstruing the terms of reference that constrained him; wrong in lumping Italian property claims, which were uncontested, with Italian jurisdictional claims, which lacked proof; wrong finally in giving too little weight to general watershed principles. Citing the jurist Johann Kaspar Bluntschli's 1868 text, the Swiss held it widely accepted that, when a mountain chain separated countries in dispute, the watershed summit line would determine the boundary.[26] But in Marsh's judgment the watershed principle adduced by Bluntschli (for whom he had high regard) had been too recent, not well enough established in European practice, to guide international law.

If a Swiss jurist criticized Marsh for giving too little weight to watersheds, an Italian expert in 1975 was mystified that he had given it so much. Commending his acutely "European" (for an Anglo-Saxon) cultural *nous*, Ausonio Malintoppi nonetheless felt his devotion to watershed management bizarrely intense. Seemingly unaware that Marsh had devoted a lifetime's study to environmental matters, the Italian jurist ascribed to the American's childhood memory of being shown a *water-shed* by his father—"I never forgot that word"—to an obsession with water parting of manifestly "Freudian" character. Misled by his ignorance of Marsh's career, this legal guru contrived a psychoanalytic basis for the adjudicator's geographical rationale.[27]

That Marsh retained his vivid early recollection was indeed significant. He continued to dwell both on the meaning of the word watershed and on its consequences for land management. In an essay four years later, Marsh termed water*shed* preferable to, because more explicit than, water *divide*. He then discussed manifold exceptions to ridgetop watersheds: instances where underground drainage channeled rainfall to an opposite catchment area, or where seasonal change sent flows now one way and now another, or where streams from a single glacier debouched into different effluents.

Watersheds became most problematic where human intervention

reversed natural drainage. Marsh instanced Barton or Runaway Pond, Vermont, fearsomely drained in one hour into Lake Memphremagog by heedless channeling that reversed its outlet in 1810; the Val di Chiana; the Illinois Canal from Lake Michigan into the Mississippi River system; and Lake Biel, shunted by the Romans through a tunnel into the river Aar, a route latterly reversed by draining the Biel into Lake Neuchâtel. In doing so, just as Marsh was writing, engineers had uncovered pre-historic lake-dwellings on the lake's south shore.[28]

Marsh's Cravairola decision was most significant for setting forth the general benefits of watershed boundaries, commonly relied on in international law today for all the environmental reasons he adduced.[29] Yet modern jurists also cite Marsh for stressing historical proofs of sovereignty *over* topographic continuity or watershed principle.[30] And his decision pioneered in other respects too. This was not the first time territorial disputants relied on an external arbiter, but it was the first *successful* such occasion. The choice of a single neutral referee of Marsh's known probity and skills, rather than the usual joint tribunal or some ceremonial head of state, had clearly been crucial. Not least, Marsh set a useful precedent by adhering strictly to the terms of his arbitrage, against his own decided views of equity. By closely circumscribing his arbiter's powers, and avoiding any likely ambiguity, Marsh enhanced confidence in the entire process of arbitration.[31]

A Marsh was burned for heresy; that has confirmed our obstinacy. The whole race of us are desperate heretics to this day.

Marsh, *Address . . . New England Society of New-York*, 1844 [32]

*F*or the heretical Marsh to be stationed in the seat of the papacy made for an apt coda to the career of this New England Puritan. It also animated him to a last assault on his old enemy. His first years in Italy had roused Marsh's false hope that the Church's spiritual as well as temporal power was waning; his dispatches exude confidence in the triumph of "reason." He dismissed the 1864 Syllabus of Errors, a papal tirade condemning modern civilization and liberty, as "but the cooing of the dove" compared with previous "wolfish howl[s] against . . . scientific and social progress." Nor was Marsh alarmed by the prospect of papal infallibility becoming dogma. He hoped the 1870 Council would "stop nowhere, . . . elevate Joseph to the rank of Mary, in short declare Catholicism to be avowedly what it has always been practically, a polytheism, and finally gratify the pet wish of the drivelling old idiot who calls it together by proclaiming the pope personally infallible, God's alter ego."[33] The more arrogant the Church became, the sooner in Marsh's view it would collapse.

He was mistaken. It is true that papal fortunes in 1870 seemed at their lowest ebb in a millennium. Gone were the States of the Church that had long lent the papacy material leverage; many like Marsh expected Rome's moral authority to vanish along with them. Claims of infallibility had alienated even Catholics; anticlerical regimes in France, Germany, and Spain turned against the papacy; from Geneva, anarchists proclaimed the imminent demise of Christianity.

Yet far from weakening his Church, Pius IX set in train social and spiritual trends that gained the papacy greater spiritual influence than it had held for four centuries. Within a few years Marsh saw how wrong he had been. The rise to power of "organizations which surround and control . . . the chair of St. Peter," he wrote Secretary of State Evarts on the death of Pius IX, "render the personal will or character of the Pope a matter of little importance," and left "the moral influence of the Papacy . . . as formidable as ever."[34]

Italy's failure to subdue the papacy, resurgent proselytism, Protestant gullibility, and the flood of Catholic immigrants to America had already moved Marsh to discredit the Church by trumpeting its iniquities. Now in 1874 he compiled a book-length antipapal tract. He expected odium; "a torrent of abuse from the Catholic press" had hit Godkin's *Nation* after Marsh's essay on the papal syllabus.[35] And because his diplomatic status required anonymity, it proved hard to find a publisher. Not until 1876 did Harper's bring out *Mediaeval and Modern Saints and Miracles.* Few copies were sold, and it had little impact. "The public has looked in vain," the publisher told Marsh, "for the authority of a name to substantiate the startling statements made in the book." As a friendly reviewer warned, anonymity "will not only render him liable to severe criticism at the hands of the Church, but will also weaken the authority of his somewhat surprising statements with skeptical Protestants."[36]

Written in wrath, *Saints and Miracles* is a potpourri of diatribes—excerpts from the lives of saints and accounts of miracles interlarded with harangues on Mariolatry, the confessional, the martyrdom of Hus, the evils of Pius IX, Jesuits, Infallibility, bookburning, forgeries, Cardinal Newman, and the Inquisition. Anent an 1875 Vatican authority on Galileo who "denies the torture, while admitting . . . the menace of torture," Marsh found nothing to choose "between the robber who snatches my watch from my pocket, and him who holds a pistol to my head and threatens to blow my brains out if I do not give it to him." [37]

Marsh had sloughed off *some* prejudice. He now abjured his 1840s equation of race and place with religion and national character. Stressing that his quarrel was with Papists, not with Catholics, he had some good words for the latter. In Catholic Italy, he had met human beings as fine as in Congregational New England; "even the priesthood [displays] examples of piety, truth, honor, charity, benevolence." He felt these Catholics superior to Protestants in "the minor morals—the urbanities and amenities of mutual intercourse . . . that courteous regard for the sensibilities and the self-respect of others so characteristic of the Latin nations, and which contrasts so strongly with the bluff, oppressive address of the Englishman, the offensive self-sufficiency of the German, and the rude self-assertion of the American." Marsh termed Italian kindliness "genuine, as every stranger who has lived long in Italy will testify." [38]

But public vices offset private Catholic virtues. In largely Catholic lands like Italy moral standards were low, schooling minimal; the handful of enlightened citizens lacked the esprit de corps of, say, New Englanders—a spirit that proscribed "departures from the strict rules of morality" and outlawed "the indulgence of vicious propensities."

Reason alone, Marsh had come to see, would never free Italy from the incubus of the Church. Catholicism was for most a habitual way of life. Widespread hatred of priests did not weaken attachment to the faith. One "emancipated" scientist told Marsh that he could not "overcome the prejudices of my education, which prompt me to sustain a church in which I do not believe." Traditional social patterns, intermeshed Church-State elites ("the whole hierarchy of the Romish Church . . . is composed of the sons, brothers, nephews . . . of the very statesmen" who enacted government policy), and monopoly of schooling ensured the Church's continued supremacy in people's minds and hearts.[39]

That institution Marsh condemned wholly. Far from growing enlightened and humane, the Church seemed to him as tyrannical, cruel, and corrupt as in the darkest ages. The dogma of the Immaculate Conception, dividing the domain of grace "between Mary in heaven and the pope on earth," made Christ "an altogether superfluous personage"— save for his heart. Marsh instanced a picture in a Roman religious shop that showed God, the Virgin Mary, Saint Peter, the Holy Spirit, and— the central figure—Pius IX "with Europe, Asia, Africa, and America at his feet. *Christ does not appear in the picture at all.*"[40]

Marsh censured papal promotion of the Virgin for subverting morality and social order. Devotion to Mary, whose bizarre titles included "True and Sacred Lamprey of the Sea," gained those who witnessed her miraculous depiction in a Rimini church a plenary indulgence and remission of all sins, "transferable . . . to any of their friends in purgatory." An 1866 Papal Bull of Composition pardoned assassins and robbers in return for 3 percent of their plunder; since Mary thus forgave their atrocities, the Bull was displayed as a sacred talisman in the homes of Sicilian brigands.[41]

Devotion to the Sacred Heart of Jesus, lately magnified by Pius IX and the Jesuits, Marsh termed "the grossest case of purely material worship in modern times," its literature "surpassed in folly, vulgarity, and indecency by few chapters in the history of . . . religious aberrations." Bishop Languet's widely disseminated life of Marie Alacoque seemed to Marsh "fit for no readers but the inmates of an asylum for cretins."[42]

Yet the cult of the Sacred Heart had done most, he acknowledged, "to revive the flagging zeal of indifferent Catholics, and to secure perverts [sic] from Protestantism" in the United States. His countrymen's credulous ignorance and sluggish cynicism appalled Marsh. "A settled moral conviction of any sort is [considered] a weakness or provincialism; and any attempt at an exposure of the policy of Rome is triumphantly put down by classing it with the old vulgar mob-watchword of 'No Popery.'" Yet the papacy had "openly proclaimed itself the enemy of all human liberty" and was "assiduously laboring to sap" the foundations of freedom. Marsh declared "the danger of papal aggression on

American liberties is as real, as obvious, and we may almost say as imminent" as the threat of Civil War in 1860.[43]

Marsh's fear of and strictures against the Church did not abate in later years. He penned essays on the Inquisition, the Papal Index, and book-burning. He voiced alarm lest the Vatican foment revolution in the United States. "Our Popish friends are growing more and more bellicose," he wrote F. P. Nash; "nothing short of blood will satisfy them." It was Marsh's rule, whenever "Catholic Popality comes within my reach, to give him a rap on the back of the head with a sharp stone, or a poke in the eye with a stout stick; [I] always feel the better for it."[44] He reveled in combativeness; the Church was an enemy made to order.

More influential than Marsh's anticlerical fulminations were his renewed warnings on environmental risk. An enlarged edition of *Man and Nature*, retitled *The Earth as Modified by Human Action*, appeared in 1874. The same year brought his report, *Irrigation, Its Evils, the Remedies and the Compensations*, requested by U.S. Agricultural Commissioner Frederick Watts. Irrigation was then a topic of wildly extravagant hopes; Americans yearned for a miracle to transmute the arid West into a garden of Eden. Where rainfall was scanty and unreliable, farmers looked to irrigation for year-round water supplies that might multiply crop yields tenfold. It was said to be cheap and easy; the settler had only to dig a ditch and let the water flow.

Irrigation's potential benefits were indeed manifold. In *Man and Nature* Marsh had shown how water engineering enabled ancient civilizations to flourish in realms otherwise dry and sterile. It was still widespread, in both desert lands and semiarid Europe. "Few things in Continental husbandry," he wrote, "surprise English or American observers so much as the extent to which irrigation is employed in agriculture," even where it might be thought to injure plants more than benefit them. "In Piedmont and Lombardy irrigation is bestowed on almost every crop," multiplying yields on millions of acres.[45] The advantages were manifest. Were the Nile valley left unwatered, "a single year would transform the most fertile of soils to the most barren of deserts, and render uninhabitable a territory that irrigation makes capable of sustaining as dense a population as has ever existed."

Yet damage to soils, stream flow, plant and animal life, and crop quality was likewise enormous, potentially even catastrophic. Large-scale irrigation—and the tendency was always to overdo it—reduced fertility, made soils saline, diminished stream flow and navigability, and created breeding grounds for typhus, malaria, and other diseases, notably in rice and cotton plantations.[46] Irrigation brought economic, social, and political evils as well. It was hugely expensive, requiring great wealth or heavily taxed public works; it provoked incessant quarrels among proprietors and managers; it concentrated land into large holdings and squeezed out small proprietors.[47]

Mediterranean history and geography thus led Marsh to warn against hasty or excessive irrigation, "cautions which our head-long pioneers seem much to need." His handwritten addendum to the printed report counseled "farmers impatient to rush into costly canalling and grading for the introduction of irrigation" to heed the advice "often given with respect to new medicinal remedies: 'Let a friend try it first.'" He feared that Americans' "characteristic impetuosity and love of novelty" might wreak wholesale damage. Water drawn for irrigation deprived adjacent lands of their normal supply. Irrigation required reservoirs that were apt to burst their barriers and flood lands below. It exhausted soils, produced hardpan, concentrated lethal salts at the surface. It multiplied output at the expense of nutritive value and flavor. And in line with the ecological caveats of *Man and Nature*, irrigation works once undertaken had to be maintained at great cost, for human disengagement would leave desert conditions worse than before.[48]

For the dry American West, Marsh's main focus, climate and soils data were virtually nil. No one knew how much water was available, how much land was irrigable, or how much it would cost. No major work should begin, he cautioned, before hydrographic surveys had shown what each river basin got from rain, snow, and groundwater; what was needed for agriculture, industry, and domestic use; how far diverted water might augment supplies; and the costs and likely side effects of such transfers. After surveys, irrigation trials would be needed for each region "having marked peculiarities of climate, soil, and adaptability to special culture."[49]

The social effects worried Marsh as much as the physical ones. "Acquisition of the control of abundant sources of water by private individuals," he warned, led to "vested rights and monopolies liable to great abuse." Because irrigation demanded large capital outlay and skilled management, it tended to "accumulate large tracts of land in the hands of single proprietors, and [to] dispossess the smaller land-holders." In Europe the dispossessed became hired laborers lacking any "propri- etary interest in the land they till," with the ruinous results noted by Pliny for the Roman Empire. In modern Italy the rural middle class, "which ought to constitute the true moral as well as physical power of the land," was thus all but extinct.[50]

Hence irrigation demanded not just technical expertise but social control. "We must look to our rulers," Marsh insisted, "for such legisla- tion as shall prevent the [most] evil and secure the [most] good, from the introduction of a system so new to us." Most worrisome was the con- flict between private property and public welfare. "Like all attempts to appropriate to the use of individuals gifts of nature which have long been common to all, [irrigation] must clash with many rooted prejudices, many established customs, and many supposed indefeasible rights." One such right was ownership of water on one's own land. This com- mon-law riparian principle sufficed where rainfall was ample. But in the arid West a farm could perish if deprived of flow by upstream neigh- bors. A new water code was needed to protect the welfare of all.

Only public ownership, Marsh concluded, could resolve water is- sues in the West. "The first article of the water-code should be a decla- ration that all lakes, rivers, and natural water-courses are the inalienable property of the State, and that no diversion of water from its natural channels is lawful without the permission of the public authorities." Perpetual or even long-term "concessions of water-rights to individuals or to corporations" must be forbidden, because changing environments and needs might make an initially unobjectionable grant "highly in- jurious to the public interests ten years later." All irrigation should be "under Government supervision, from Government sources of supply," with water allocated by the state on the basis of local conditions and needs.

Marsh saw such controls as agents of social and economic equity. Not only must the state protect small landholders, it should promote "farms of relatively narrow extent" by giving them more water at lower rates than to large proprietors, and further aid "poorer occupants who build and inhabit houses upon these lands."[51]

This sweeping program, notably Marsh's advocacy of public ownership, "made considerable breeze in Congress." Though infuriating western land promoters, it had, as Spencer Baird predicted, "a decided effect in influencing national and state action." John Wesley Powell's *Report on the Lands of the Arid Region of the United States* (1878) echoed Marsh's warnings. As head of the U.S. Geological Survey a decade later, Powell began the irrigation assessment Marsh had proposed, meanwhile halting the sale or lease of irrigable public lands.[52] Although western landowners and speculators aborted Powell's program, under the Newlands Act of 1902 the Bureau of Reclamation regained some control over irrigable waters.

Thus with water, as with forests, Marsh's insights inspired and came to guide national conservation policy. Later pressures on western water resources attest his foresight. But even Marsh could not have anticipated that within a century three-fourths of California's agriculture would depend on irrigation.[53]

Marsh's passion for forests and forestry was renewed in his last years by friendship with Charles Sprague Sargent at Harvard's Arnold Arboretum. "I have long been a student of *Man and Nature*," Sargent wrote him in 1879, noting a few botanical errors, "and have derived great pleasure and profit from your pages." For three and a half years the old diplomat and the young botanist exchanged views on tree conservation, the acclimation of olives in Massachusetts, the uses of pine cones, the sex of cypresses, the grafting of figs and pomegranates, the meretricious essays of Ruskin, the vegetation of the Dolomites, the flora of colonial America, the sanitary effects of eucalyptus in California. As eager as a child, the eighty-year-old Minister strolled among trees, magnifying glass and measuring tape in hand. Sargent found Marsh's acuity of observation and knowledge of forestry "unexcelled even by that of the few specialists who devote themselves exclusively to the study of this subject."[54]

A sample suggests the flavor of Marsh's exchanges with Sargent.

At the hospice of San Martino di Castrozza in the Austrian Tyrol, in August 1879, Marsh learned from an "intelligent & charming" Irish priest (mirabile dictu!) that cattle allowed to roam in the woods "do not browse on the young vines, and therefore do little direct injury. At the same time, they secure the seed from destruction by treading the cones in the ground." Where cattle roamed, seedlings were more numerous and vigorous. "Yet in the cattle paths I notice that the roots are often bared & bruised"; how far, Marsh wondered, did the injuries offset the benefits of grazing?[55]

Marsh owed much to Sargent for the revisions of *Man and Nature* he made up to the day of his death. Changes from the 1864 to the 1874 and 1884 editions were many, but mostly minor. Marsh corrected errors, added fresh data, cited new instances of human impacts; but the book's central theme and spirit remained the same. Ravages of nature worsened: "the forests of the Adirondacks have continued to be the scene of ever more and more rapidly encroaching inroads from the woodman's axe; [only] a wiser legislation or a sounder public opinion" could avert their imminent "total destruction." But here and there science was vanquishing waste for wiser resource use; facilities for shipping refrigerated mutton from New Zealand in 1882 signaled the end of indiscriminate slaughter of sheep for pelts and suet alone.[56]

While Marsh found newly disturbing instances of heedless impact—the collapse of coal mines in Pennsylvania and Belgium; unwanted interchanges of marine life via the Suez Canal; untoward heat buildups in big cities—he also foresaw further benefits of controlling nature. Electricity opened "a prospect of vast addition to the powers formerly wielded by man" and presaged "new and more brilliant victories of mind over matter."[57] Deeper knowledge would enhance such power far beyond present limits. Although a political and social pessimist, Marsh saw no end to scientific advance.

Religion and nature were not all that engaged Marsh. In 1874 he joined Princeton geographer Arnold Guyot and Columbia president F. A. P. Barnard as an editor of *Johnson's New Universal Cyclopaedia*, for which—at $20 a double-columned page—he prepared thirty-nine literary, linguistic, and geographical essays, commissioning nineteen more (mainly on Italian cities) from his wife. Marsh wrote on Genoa, Girgenti, the Po, the Pontine Marshes, Sicily, the Tiber; sketched lives of six Iberian poets and chroniclers; discussed Catalan and Italian languages and literatures, Romansch, Index (book), Lexicon, Improvisation. In essays on the Index (Papal), Sicilian Vespers, and Legend he had to "tread gingerly, in regard to my position." He penned pieces on the Olive, Mulberry tree, Straw manufacture, Velvet, Watershed, Well, and several tunnels. Marsh most enjoyed these geographies. "As men please themselves more in their dilettantisms than in what really belongs to them," he told Nash, "I have taken more pleasure in writing on Irrigation, Inundations, Fireproof constructions, the Mt. Cenis and Mt. Gotthard Tunnels (the most important work of material improvement yet projected in Europe), the draining of Lake Fucino, etc. than in literary articles." A decade after *Man and Nature*, Marsh still fobbed himself off as a geographical amateur. And he made light of the whole enterprise as a waste of time. "What does any mortal want an Encyclopaedia . . . for, and where does any mortal find the money to pay for it? An alphabetical work, whereof A. shall be obsolete and forgotten before the litera longa, the thieves' letter, is arrived at!"[58]

Writing was irksome when he was in Rome with a deadline to meet and all his books were in Florence. Then Marsh dove "into the limbo of . . . 'unconscious cerebration' [to] fetch up enough brain dribble to fill some pages." As he told Yule, "needs must whom the devil drives, and Auld Sootie . . . is now after me with a sharp stick and howling: write! write!" Yet when the *Cyclopaedia* was done (it paid him $2,300 in all), Marsh sought further employment, which "the greatly increased expenses of living in Italy make a necessity for us."[59] A sorry admission for an eminent scholar and honored diplomat of seventy-five!

Old age and infirmity did prevent Marsh from publishing much thereafter, though to the end he went on revising his books—and reviling his critics. To an American scholar's charge that in *Lectures on the English Language* Marsh had understated Milton's vocabulary by more than half, the Minister retorted: "He must have counted *fool, fool's, fools*, as three words, and (if Milton had used it) *blunder, blunderest, blunders, blundered, blunderedest, blundering*, as six." (Marsh's repute as a word counter was vindicated by the lexicographer Francis March, professor of English literature at Lafayette College.)[60]

Marsh kept on with linguistics. He hunted for unique words lacking cognates in other tongues (the Italo-Swiss boundary foray yielded *inalpare* and *disalpare*: to drive cattle to and from Alpine pastures in spring and fall). He probed the origins of Italian dialects and promoted a lexicon of forestry, an Italian grammar, and a Bible translation. Intrigued by language mutations, Marsh enjoyed "watching the changing pronunciation . . . made *without the consciousness of the individual*" of friends met in Turin and again in Florence and in Rome. His penultimate essay, "The Biography of a Word," describes the birth, life history, and death of the word *hijo* (a donkey's bray) made up by a Swedish child in his own household.[61]

Art and architecture continued to claim Marsh's attention. Time did not fog his connoisseur's eye; to the contrary, old age was "gilded with the most exalted pleasures of the world of sense," as he held: "Perception of beauty . . . attains not its ripeness, save under the rays of an autumnal sun."[62] Marsh feasted on Roman art. He helped New York's Metropolitan Museum of Art acquire European paintings. He liaised American artists with Italian marble workers, found a sculptor for Rhode Island's state memorial, procured a cast of a head of Mercury for the University of Vermont, engaged W. W. Story to do a statue of Joseph Henry for the Smithsonian Institution, commissioned Preston Powers to make a marble bust of Jacob Collamer for the federal Capitol's Hall of Statuary, and advised New England's governors on the Bennington Battle Monument.

After fifteen years' urging from Marsh and John Pomeroy, a Carrara marble statue of Ethan Allen by Peter Stephenson was at last placed atop its granite column in Burlington's Green Mount Cemetery, sporting the same Ticonderoga pose as Larkin Mead's statue on the State House portico. A last-minute hiccup arose from a new-found "vicipitous photograph" (Marsh's epithet) of a purported Ethan Allen likeness. Marsh scorned the proposed "modification" as "a paltry vulgar expressionless head, and I would rather burn the statue in a lime kiln than see it on the column with such a mug." The stone-hewer's change was countermanded; the completed statue was locally lauded as "the most elegant monument in New England."[63]

Perhaps the most notable outcome of Marsh's aesthetic expertise was the completion of Charles Mills's Washington Monument, whose construction had been suspended in 1855. Held up by Civil War and lack of funds, the unfinished 300-foot obelisk stood forlornly truncated on the Mall. In 1879 Marsh learned with horror that it was to be left as it was, surmounted by a colossal statue. To him, who had "sketched every existing genuine . . . obelisk," this was heresy. To depart from forms and proportions "fixed by the usage of thousands of years [and pleasing] every cultivated eye" would be an "esthetical crime," he hectored his old crony Robert C. Winthrop, who headed the National Monument Society. A pyramid's height must be ten times its width, here 555 feet; a mere 300-foot apex would be "quite out of harmony with the *soaring* character of the structure." Above all, "the notion of spitting a statue on the sharp point of the pyramidon [was] supremely absurd." No ledge, molding, or other ornament should disfigure the façade.

Marsh's critique struck home. The Society adopted his strictures, abandoned the proposed statue, and agreed that the monument should "make no pretensions to illustrate the arts of 1880." U.S. Corps of Engineers' chief General Thomas Lincoln Casey stuck scrupulously to Marsh's diktat of purity and proportion. The world's tallest building when completed in 1884, this New World testimonial to antiquity was ironically overtopped just five years later by the Eiffel Tower, the Old World's testimonial to modernity.[64]

Many learned societies honored Marsh, and Harvard, Dartmouth, and Delaware awarded him LL.D. degrees. But no honor pleased him more than election in 1876 to the formerly papal, now Reale [Royal] Accademia dei Lincei, Italy's oldest scientific society, presided over by Marsh's old friend Quintino Sella. Joining former prime minister Luigi Menabrea and the historian Cesare Cantù, received together with the German historian Theodor Mommsen and Field Marshal Helmuth von Moltke, Marsh was only the second American (following Joseph Henry) to be so honored.[65]

*F*or the American Minister in the 1870s, Rome, though "happily restored to no small share of its ancient dignity and power," was a city of incongruous disharmonies. There was the moral Rome, which in Marsh's eyes the Vatican made the capital of evil. There were discrepant material Romes: one a fabulous aggregation of classical monuments and works of art; another a sprawling malarial slum, sultry in summer and bitter in winter. Gradually the slums were cleared away, sometimes at the expense of the monuments. Lauding the new apartments, drained swamps, works of sanitation, and restored aqueducts, Marsh allowed that Rome might be "a grand city to live in"—by 1900.[66]

In 1881 it was not an easy city to live in, as Marsh retailed his tribulations to Baird:

If I want to put up a new bell (not electric) in my house, I first buy a bell at a hardware store; second, I send for a mason to drill the necessary holes through the walls, 3d I send to the blacksmith to make the cranks, which are never kept ready made, 4th to the braziers for the wire, 5th to the upholsterers to put up bell and wire, 6th I go to the silk dealers to get the bell-pull. Each of these operations requires, at least, 3 days, or *18* in all—and as during these 18 days there are, at least, 3 *festas* on which none of the pious tradesmen will work, it takes fully three weeks to get my bell. This is a specimen of the difficulty of accomplishing what would be a very small thing at home.

Such travails confirmed the Minister's judgment that "acute as the Italians are about intellectual & social matters, they are in business more pig-headed & blockish than the wildest bog-trotters of Ireland."[67]

Social Rome loomed large. There assembled folk from divers realms—aristocrats, diplomats, politicians, priests, and, most visibly, tourists. Throughout the year Rome rang with English and, increasingly, American voices; in 1873 the city struck Henry James as "a monstrous mixture of the watering place and the curiosity shop." Sightseers Baedekered through palaces and churches, pestered Marsh for introductions to the Quirinal (once the papal, now the royal residence), and presented themselves to the pope "as they would to the Devil, if he were to give audiences above ground," grumbled the Minister. "I shall be truly thankful when we come to sackcloth and ashes, and get rid of some of the gaping idiots who come all the way from America to see such things." It seemed to him that "we are fast becoming the most tuft-hunting nation on earth."[68]

Quite apart from this motley throng, the Marshes' social duties increased enormously. Thrice a week they gave dinners for twenty, receptions for fifty to a hundred. Life for the Anglo-American colony in Rome in those days has been called a continuous picnic,[69] but it was the American envoy who often furnished the setting and footed the bills.

Everyone made the Grand Tour. Ralph Waldo Emerson stayed for a week and "was quite a lion," Caroline reported; but his now senile garrulity irked the taciturn Marsh. Bigelow, Bancroft, Lowell, R. H. Dana, Fanny Kemble, Willard Fiske, Bayard Taylor, and diplomatic colleagues often visited. Matthew Arnold returned in 1873, found Marsh ill and "sate a long time with him, because he liked it." To the German chronicler and diarist Ferdinand Gregorovius, Marsh appeared "a quiet man of great culture." The Minister enjoyed meeting British historian W. E. H. Lecky and French historian Jules Michelet, who amazed Marsh by declaring that he and Madame Michelet hardly ever went to the theater, "because theatre-going was incompatible with their literary occupations. What a sacrifice for Parisians!"[70]

As in Florence, many American artists lived in Rome—W. W. Story, Charlotte Cushman, Harriet Hosmer, Louisa Terry. Most were women, the "strange, white, marmoreal flock" immortalized by Henry James. Marsh found "the female persuasion [more than] slightly redundant" even among the transients. "Ladies come in trios and quartettes, with no male attendant. What can one do with such at a dinner table?" At one reception appeared "a *he* accompanied by eight *shes*. How despairing!"[71]

Worst were the "royal tramps," as Marsh termed touring nobility. "We are bound to believe that such entities have their uses," but they were "a grievous trial to poor innocent diplomates, and . . . ought never to be allowed to go abroad." High life bored Marsh; "gentlemen in swallowtails and ladies with trains are wonderfully alike all over the world." The longer he was abroad, the more assertive grew his republicanism. He hoped to "live to see the playing at foot-ball with coronets and mitres, crowns and tiaras."[72]

It was a "great boon" to Marsh when in 1876 several European countries promoted their Italian envoys to ambassador. Since the United States did not follow suit until 1893, Marsh as a mere Minister Plenipotentiary no longer had to officiate on behalf of the diplomatic corps, though he remained its dean.[73] But he still had to consort with crowns and tiaras, even mitres. Roman aristocracy was split into adherents of state or church, "Blue" nobles who had defected to the king,

"Blacks" who scorned the Savoyards. Why should a Borghese or a Barberini bow to parvenus whose ancestors a century ago had huddled in Alpine huts? Some noble Romans kept a foot in each camp, but diplomats had to take care not to invite Blue and Black to the same function. Marsh kept close watch on the shenanigans of the *nobili* and their papal patrons.[74]

To do proper honor to General Grant's 1878 state visit (which wore them out with problems of protocol), the Marshes rented the forty-room Palazzo Pandolfi, near Santa Maria Maggiore, "with more tinsel than gold in its fitting up." But the winter sun poured in through great bay windows, and there was a good furnace. These were major virtues, for Marsh suffered bouts of sciatica. The winter of 1880 was Rome's severest on record, "almost fatal to all old people, myself included"; his bronchitis lingered. "I have aged more in the last six months," he declared, "than in any previous six years of my life."[75]

In 1879 the Marshes had moved again, to the seventeenth-century Palazzo Rospigliosi, hard by the Quirinal. Their apartment was 126 steps up and needed three additional stoves, but Marsh was tempted by splendid rooms and furniture, a private gallery, a fine library, and "the most magnificent view of Rome and its environs I know" over the unbroken side of the Coliseum, looking, he fancied, much as it had from the same spot eighteen centuries before. What if the steps were many and steep! When he could not climb them, Marsh was carried in a sedan chair, but his 200-pound frame was not easy to hoist up and down the narrow staircase.[76]

Although active and cheerful, Marsh tired rapidly; his voice, always low-pitched, became less distinct. He disliked idle chatter, talking freely only with such intimates as Henry Yule, George Bancroft, Humphrey Sandwith, and Francis Nash. Then came a flow of drollery and piquant learning. In later years such flashes of wit were more rare, though memory, judgment, and invective power stayed with him to the last. But such traits needed nourishment seldom found in diplomatic milieus. During his whole tenure abroad Marsh felt sorely deprived of "intimate & constant association with men of culture," averred Caroline soon after his death. "We have had opportunities of knowing & of having at our own table *most* or at least *many* of the greatest men & women of our age, but these have been occasional banquets, not everyday fare, and we have had no one to come in & out freely, who was really interested in the subjects that touched him most nearly."[77]

Other proclivities ripened with the years. One was his taste for food and drink. He praised the bread made of white wheat, *grano di Santa Severa*, by Viennese bakers in Rome, but termed "that of Andalucia, the finest in Europe, remarkable for its light and spongy crumb & great thickness of a brown but still soft crust." Marsh became a bon vivant versed in *risotto* à la Milanese and Roman punch ("add rum or whiskey . . . to any common ice-cream"). Italian wines direct from the producer were fine, but he despaired of the Italian trade; "wines designed for sale are largely adulterated & the better [ones] extensively counterfeited," and Italian innkeepers "furnish only such native wines as it is impossible to drink, in order to compel their customers to order foreign wines on which they make a large profit."[78] He scorned the dry champagne then fashionable as "a nauseous and poisonous tipple adulterated with some nasty drug," and in print he castigated "the craze about dry wines":

Fruit which would have been neglected thirty years ago, as immature or diseased, is now subjected to the press without scruple, and the acid and corrosive juice obtained from it is doubly brandied and recommended to purchasers as dry wine. . . . People who habitually imbibe this caustic fluid will find in the long run that they have put an enemy into their mouths, if not to steal their brains, at least to ruin their digestion.

Marsh's digestion perhaps benefited from the Rhine wine he liked best. But he kept his nephew Edmunds in Washington supplied with claret "good enough to set before guests, and cheap enough to drink myself."[79]

The Marshes' burdens were eased by many young relations. Carrie Marsh Crane, Caroline's niece and namesake, "a very nice good girl," bright and eager to learn, was with them until 1865, by then more fluent in Italian than Marsh himself. Several Crane cousins followed as Caroline's amanuenses. Others came on short visits: her nephew Edward Crane, private physician to Empress Eugénie and later editor of the Paris *American Register*; Senator Edmunds and his family in summer 1873.

But Carrie was the favorite. "I have seen nothing on this journey [to Germany] without wishing that you were with us to share our pleasure," Marsh wrote her in 1869, "for nothing has ever increased my enjoyment of travel more than your intelligent curiosity in all that is worth the notice." In 1875, after Carrie suffered a nervous breakdown on the verge of entering college, Marsh arranged to bring her back to Italy. All these years the orphaned Carrie, whose life was otherwise one of dolorous adversity, had dreamed of returning. But on the voyage from America, Carrie was shipwrecked and drowned off the Scilly Isles. The calamity overwhelmed Marsh; "she was the comfort, the hope, the stay of my declining years, and the world has nothing which can supply her place," he mourned. "A childless old man needs some such support as this affectionate and strong and heroic girl was to me, some green and growing thing on which to fix his pride and earthly hopes."[80] It was a harder blow than the death of his son.

Other nieces and nephews now came: Marsh's "little Dutch niece" Carrie Wislizenus, "a perfect model of industry"; Silas Crane's daughter Mary; Alexander B. Crane, a "half worn-out New York lawyer" chased up and down the Alps by his septuagenarian uncle; the Edmundses again. Summer 1881 brought Alex Crane's six children, ranging in age from twelve to two, for a long stay in Europe. The delighted Marsh oversaw their schooling and chatted with them for hours. "Every night after tea Uncle asks us what we have been studying all day," wrote the eldest girl, "and then he tells us lots of funny and serious things."[81]

A singular mishap led Marsh in 1876 to adopt an orphaned Swedish two-year-old, Carlo Rände (later Simi). Carlo's Stockholm guardians had sold him to Mary Gilpin, a wealthy American termed by Marsh "a monomaniac, if not altogether insane." Word of her maltreatment of Carlo (they were now in Rome) reached the American legation, and after a police search Carlo was put in the custody of the Swedish Minister, Baron Essen. But since Essen was soon to leave, the child was taken to the Marshes', "where it could more conveniently be cared for," he wrote the Secretary of State, "and still remains with us."[82]

Young King Karl, as he was soon known, stayed for good. Carlo was slow to talk and read, but Marsh adored him, spoiled him inordinately, and left cross words to Caroline. "Though a sweet-tempered & affectionate child, his *ethical principles* are not of the soundest," Marsh wrote Sarah Wister. "'Why you not' said he to Mrs Marsh the other day, 'be good to me when I naughty, as *uncle* is? *He* pats my head and kisses me, but *you* cross when I naughty.'" The stocky, golden-haired boy was "a constant source of amusement." Fondness for Carlo and other youngsters was a great blessing to Marsh "because it tends to keep our moral nature from sinking into the torpor which so often accompanies old age."[83]

Old age was made more painful by the death of old friends. Of Marsh's immediate family none were left. Lyndon had died in 1872, Charles in 1873. The Marsh homestead had been bought in 1869 by millionaire railroad magnate Frederick A. Billings, who had returned to retire in his native Woodstock. Billings rebuilt the house in the gilt-and-gingerbread fashion of the time;[84] fortunately, Marsh never saw the result. In Burlington, too, the old order was gone. The Queen City had recovered from mid-century depression, but few of Marsh's cronies survived into its new prosperity. Only John Pomeroy lived on, spry and hearty in his late eighties. The years had taken their toll of other colleagues; Bache, Henry, Gilliss, Alexander, Choate, Sumner, Foot, and Powers were all dead. Among Italians all those Marsh had known in Turin were gone but Garibaldi. "We feel the loneliness of survivorship," he wrote in 1880, "and have comparatively few ties except those of duty."[85]

I have waited for leisure until it is too late. Ach, mein verfehltes Leben!
Marsh to F. P. Nash, June 5, 1879

The ties of duty sometimes galled Marsh sorely. He wondered on occasion if he could bear his burdens; "they are very heavy, and I feel I may at any time sink under them," he wrote during his 1880 bronchial winter. "To be a 'free man' for a little before I go hence, is my most earnest aspiration." He thought of retiring, but his nephew Alexander Crane deterred him: "Do not resign! . . . You cannot leave so good a reputation or do your country so much good as by holding on and bravely performing your duties at your post until the chord of life is snapped."

Caroline had all she could do to cheer him, and "at the same time to conceal . . . his condition from outsiders who were *gaping* for his place and eagle-eyed to see shortcomings." She would have liked to return to America but knew that he could not. Packing and shipping the library would take many months and leave him bookless still longer; and he would lose all sense of independence. Along with money, weather was the chief deterrent. "The *official* position I could well spare," Marsh wrote Yule in England in 1876, "but I do not look forward with pleasure to a return to our climate; of which it is enough to say that . . . it is a good deal worse than yours."[86]

But Marsh's imminent retirement was recurrently rumored, especially when he was too preoccupied or too feeble to meet tourist demands. An 1873 visitor charged that "scholarly pursuits" often excused him from "the ordinary courtesies or even duties of resident Ministers" and complained that the Rome legation was "inaccessible to the ordinary American traveller, as the Minister generally pleads ill-health for not receiving." This was promptly denied by a supporter, noting that whenever Marsh could not go to the legation he received visitors at home. It would disgrace the United States if "political intrigues or the petty spite of conceited travelers" caused his removal "from the responsible position which he has filled so long and so nobly." In fact Marsh did almost all his work at home. Half an hour in the poorly warmed legation brought on bronchitis and aggravated rheumatism.[87]

Hayes's election in 1877 again imperiled Marsh. As a faithful Republican Marsh had backed him, but Hayes promised civil service and other reforms. Marsh anticipated being replaced, "even if on no other principle than that a new broom sweeps clean." He knew he had enemies: "I have offended many, because I could not or would not aid them in the accomplishment of special objects." But friends assured the president that Marsh was as fit as ever, and to Marsh's delighted surprise, the Italian government formally requested his retention. Along with pressure from Edmunds, who had been influential on Hayes's behalf on the 1877 Electoral Commission, this appeal saved Marsh his post. But Edmunds warned his uncle, "if I should fail to see things exactly with Administration eyes hereafter, your merits might not seem to be quite so clear."[88]

Marsh hung on mainly owing to Edmunds's extraordinary prestige. In an era of flagrant corruption, the Senate majority leader was a man of legendary integrity. When Edmunds opposed Hayes on annexing the Dominican Republic in 1880, he again warned Marsh to "be prepared for the axe," but its fall was stayed by protests from Vermont and Secretary of State Evarts. Marsh had another narrow escape when Garfield was elected a year later; Secretary of State James G. Blaine "wants all the places for his friends," Edmunds wrote. But Garfield was assassinated and succeeded by Chester A. Arthur, with whom Edmunds had more influence. Working through Grant, whom Arthur admired, Edmunds convinced the new president (and Blaine's successor at State, F. T. Frelinghuysen) that it would injure the country to remove the still fully capable Marsh from Rome. Except for Marsh, Garfield and Arthur recalled all envoys save Lowell in London and John L. Stevens in Stockholm; "it may be," wrote Edmunds sourly, "that they would like to have one name remaining that represented something of the ancient character of the diplomatic service." Marsh seemed apt to go on forever, although, Edmunds jested in 1878, "when you will have held the post up to the average time that our ministers generally do, [your recall] may be thought of."[89]

In truth, Marsh was more hale in 1881 than for years past. He was "once more taking an interest in everything," Caroline wrote, "ready to perform all official duties . . . and writing now and then an article for a periodical." His nephew Edward Crane recalled Marsh finishing the "Dry Wines" piece in Paris, while waiting with hat and coat on for Caroline to come downstairs. On his eightieth birthday he scrawled with stiff rheumatic fingers that despite "many discouraging symptoms" he was "free from pain and in tolerable spirits tho most annoyed by all sorts of demands upon my time."[90]

Also formidable were the demands upon his purse. Marsh did not spend quite double his salary, but "I have *needed* it, and I have avoided many expenses which for the credit of my position and the country I should have incurred." But for a $7,300 legacy from his brother Charles, Marsh judged that after twenty years of service he would "be returning home a poorer man than when I left it."[91]

Lack of money and declining use prompted Marsh to try to sell his library. Harvard, Brown, and the New York and Boston public libraries nibbled at rare Catalan and Old Norse volumes but bought nothing. Marsh turned to the University of Vermont; he would "sacrifice" his books for $10,000, but President Matthew H. Buckham's trustees could not afford it. Their reluctance was augmented by Marsh's animadversions on fireproofing: "How the Trustees can consider an apartment *over* the Museum of Natural History *containing objects preserved in spirits* and connected by *framed* walls and *wooden* staircases with other apartments . . . as 'practically safe from fire,'" the outraged bibliophile could not fathom.[92] Only after his death did Frederick Billings buy Marsh's library for $15,000 as a gift for the University of Vermont, commissioning H. H. Richardson's handsome Romanesque building to house the volumes. Now somewhat dispersed, Marsh's collection still adorns the university.

I am not discouraged by the comparison of what all must admit to be a bad present with what I hold to have been a worse past.

Marsh to Henry Yule, July 10, 1876

The sorrows that have gone over our heads," wrote Caroline in 1873, "have not, I trust, either hardened or embittered our hearts."[93] But in later years Marsh's view of life grew more somber. This mirrored less the handicaps of his own old age, he felt, than the brutishness of the era. Selfish blindness seemed unending; men went on and on abusing the world and themselves. The marks of progress that had once meant much to Marsh now seemed trifling next to the decadence of human affairs.

Pessimism seemed to him apropos not just in Italy, where the fruition of union had bred only apathy and poverty, but also in America. Slavery was done with, Southern sedition vanquished; but to what good? Corruption and debauchery spread as never before. To Marsh as to other survivors from a more heroic and hopeful era, modern times lacked form, purpose, morality, conviction, courage; every apparent advance entailed new debacles. At each election it seemed vital to Marsh that the Democrats lose but no cause for rejoicing when Republicans won. In 1872 he was not pro-Grant but profoundly anti-Greeley, whose nomination by Liberal Republicans and Tammany Hall Democrats Marsh at first imagined a joke, a choice "as good as that of Tom Thumb, or the Siamese Twins, or Barnum, or Tweed." In sober truth, Marsh conceded Greeley's undoubted talents, but felt his "bitterness of prejudice & his wrongheadedness . . . rendered him a very pernicious blind leader of the blind," while his gullibility left him "an instrument in the hands of his political managers." In 1876 Marsh was not pro-Hayes, only anti-Tilden; he congratulated Edmunds on saving the country from this "embodiment of all that is worst and most dangerous in American politics."[94]

But was America worth saving? In the 1840s Marsh had judged that "we undoubtedly enjoy a more general and equal prosperity than this or any other country has ever known before." No more; America was now a morass of materialist greed. The reign of robber barons portended "the most pernicious of aristocracies," that of wealth, while the "scandalous failures of Jay Cooke" and others that brought on the panic of 1873 "inflicted an [abiding] stain on our commercial honor." Marsh doubted that Hayes's reforms, however radical, could overcome dollar worship and political patronage. From America came such an "uninterrupted succession of mortifying and discouraging moral & political news" that he wondered if "anybody believes in anybody's honesty these sad days." When in 1880 his niece Susan Edmunds feared that her husband might become the Republican nominee, Marsh was sure that "the chance of our Presidential chair ever being occupied by a man of so much ability and integrity is indeed small."[95]

For this grievous state Marsh blamed untrammeled corporate power. The shenanigans of Cornelius Vanderbilt and Jay Gould led him to inveigh once more against "great companies of transportation [that] defy all control, but that of the purse." Curbing corporations that blatantly defied the public interest was a task seriously addressed only after Marsh's death, in the Interstate Commerce Act of 1887 and the Sherman Antitrust Act of 1890. As chairman of the crucial Senate Judiciary Committee, Marsh's nephew Edmunds echoed Railroad Commissioner Marsh's 1850s tirades against the despotic power of corporate cartels.[96]

Engrossment of western lands by the few, like industrial monopoly, struck Marsh as socially divisive. "Small tenements have always been the salvation, large farms often the ruin, of European countries"; so too in America, as his irrigation report had warned. His ideal society, a nostalgized early Vermont, was a rural land of "small villages just large enough to support a blacksmith, a carpenter, a schoolmaster." Detesting urban concentration, Marsh deplored the "heterogeneous municipalities, which are doing so much to un-Americanize our people." Imperial expansion seemed to him likewise disruptive. "What have

Texas and Florida and New Mexico and even California profited us?" he inveighed in 1877. "Cuba would be still worse and it will be long before the $7,000,000 Seward threw away on Alaska come back into our wallet."[97]

Immigration was Marsh's paramount bane; "there *I* find the root of all evil." He saw America's social fabric fragmented by alien incomers. "When our fathers passed the naturalization laws, they administered to the commonwealth a fatal poison." Marsh no longer desired to expand the franchise; millions of immigrant voters made universal suffrage highly dangerous. He now feared a surfeit of democracy in America. Echoing his father's put-down of the Vermont Assembly, Marsh scored the popularly elected federal House of Representatives as "a disorderly mob."[98] And growing doubts about the efficacy of schooling, discussed in the next chapter, further narrowed Marsh's residual faith in progress.

*L*eaving the faithful Wurts at the legation in Rome, Marsh departed for Florence to join his young grand-nieces and nephews in early June 1882. A month later the whole family went to summer on the slopes of the Apennines at Vallombrosa, whose "Etrurian shades" had been famed, at least in fancy, since Milton's *Paradise Lost*. The mountains had remained Marsh's favorite recreation: in the 1870s he had two wonderful excursions in his beloved Dolomites, another in the hill towns of Tuscany, others in Switzerland, Bavaria, Auvergne, and in 1880 in the Austrian Tyrol—"stripped of its woods . . . near 2000 years ago, but now, by judicious planting, *wooded* more thoroughly than Vermont was by nature, 75 years ago."[99] Too ill to leave Florence, Marsh had spent the summer of 1881 in the Villa Forini. But during the past winter his health had much improved. He had been engrossed by European affairs, reminisced to Baird of boyhood fishing exploits in Vermont, arranged for Yule to excerpt *Man and Nature* for the benefit of Indian foresters, discussed Italian grammar and American politics with Norton, and revised *Lectures on the English Language* for a final edition.

Now he was glad to be back in the mountains. An ancient Benedictine monastery despoiled by Napoleon, suppressed in 1816, restored again and once more dissolved by Italy in 1866, Vallombrosa was in 1867 turned by Marsh's esteemed friend Adolfo Di Bérenger into the government's school of forestry, to which Marsh had donated both texts and seeds; he had long wished to visit the site. The ascent from the Arno was full of interest, and he marveled at "how very like Vermont" the hillsides were. The inn next to the monastery offered views of hills dark with fir and brilliant with chestnut, smooth beech woods on the mountains above, olive-covered slopes and vineyards below, the towers and domes of Florence in the distance. The massive thick-walled convent, gray amid murmurous wooded green, lent an air of solemnity.[100] As soon as the sun burned off morning mists, Marsh took walks through avenues of firs, talked with forestry students, and visited Di Bérenger, William Wetmore Story, and other *intimes* who summered nearby.

In mid-July, news came from Wurts that Congress had merged the post of consul-general in Rome with that of Secretary of Legation. Consul-General Lewis Richmond would now have both posts; Wurts was dismissed. Ostensibly a measure of economy, this move had been instigated by former consul-general Eugene Schuyler in revenge for Marsh's treatment of him in 1879. For the Minister it was decisive. "I hardly know how I could get on without him," he had said of Wurts some years before; now, at eighty-one, he knew he could not. He had too little strength to carry on the legation himself or to train a new secretary; "by putting it out of my power to perform the duties of my office," the government forced him to resign. Still chargé d'affaires, Wurts urged Marsh to reconsider, to await further word; and though the Minister's mind was made up, he did not yet send in his resignation.[101]

Instead he spent days in front of the inn watching Carlo and his nieces and nephews playing, reading and being read to, dictating letters, annotating *Man and Nature* for yet another edition. In pine groves Marsh measured the circumference of tree trunks, calculated rates of growth, and ruminated on the history of this long-managed forest. Nothing quelled his insatiate appetite for knowledge.

Vallombrosa left Marsh "more than ever impressed with the superiority of the timber of the artificial forest, both in quantity and quality, as compared with the natural and spontaneous growth." Man was a more purposeful and by and large successful creator than unaided nature. Vallombrosa's firs, "though so thickly planted as to leave almost no room for lateral growth," grew fabulously fast and big; Marsh thought "the famous ridge-pole of a barn in Waterbury, Vermont, which squared six inches, with a length of sixty-five feet, might readily be matched here." But the American Minister was "disappointed at not seeing a single *compatriot* among the forest growths" of Vallombrosa; he asked Sargent to send the forestry school seeds of "the *sugar-maple*, the American *ash* & *elm*, the *butternut*, the *black walnut* . . . and the *black birch*."[102]

Three days later, July 23, 1882, Marsh rose early, dictated letters, added notes to Wedgwood's *Dictionary* and to *Man and Nature*, and went outside with his family. The day was fine; the sun over the valley below bathed every detail in light, and scores of village spires stood out along the banks of the Sieve and the Arno. Thunderheads piled up to the west and south; Marsh told the children why the various cloud forms had been named and explained how they changed with the weather. Turning to the newspapers, he censured Britain's bombardment of Alexandria, but said of Egypt, as he had thirty years before, "I wish the English had complete possession of that wretchedly misgoverned country"[103]—a wish that came to pass less than two months later.

About six, Marsh suddenly tired and went in to lie down and sip a cup of beef broth. An hour later Caroline found him propped up, breathing with effort. She gave him sal volatile, then chloric ether, but without effect. Distressed, he asked to be raised up, to have the window opened, to be fanned. "Courage, dear husband," said Caroline, "you will be better soon." But by the time a doctor at the inn came, Marsh was dead. "*Ecco*," said the doctor, "*ecco la morte del giusto*."[104] A man of rectitude: a felicitous epitaph.

Before dawn two days later, Marsh's body was taken from the great hall of the old convent, wrapped in an American flag, put on a catafalque with wreaths of yellow immortelle, and carried down the mountain by forestry students. They thus honored the scholar whose work had awakened so many to the significance of their calling. Winding down through the dark woods, the cortege was met at sunrise by town officials at Pontassieve, and at the railway station in Rome by the Italian cabinet and the diplomatic corps. Marsh was buried in the Protestant cemetery in Rome, not far from the graves of Keats and Shelley. Vallombrosa was a fitting end to his long and distinguished life. "Could he himself have selected the manner, time and place of his departure," wrote Caroline, "he would have desired nothing different."[105]

17. Retrospect: Forming a Life

MARSH LED A FULL AND FULFILLING LIFE. NEVER wealthy, he never suffered actual want, though he always feared penury. Within a crowded political career he cultivated myriad enthusiasms in the arts, travel, and scholarship and enjoyed friendships with many of the most eminent men and women of his time. He often had to toil at mean or humdrum tasks, but seldom over prolonged periods. If diplomatic and other duties were at times taxing, they were often vivifying and thought-provoking.

While Marsh frequently had cause to complain, he tended to dwell obsessively on his grievances: his salary was too small; his masters were misguided, his minions remiss; fools and rogues robbed him of time and energy; no sooner was he settled in a place than he was ordered elsewhere, at huge trouble and expense. Despite his early claim to "never esteem any man the less for thinking evil of me," Marsh's enmity toward those who injured him was unremitting: his foes seem caricatures of perversity. Seldom loath to reveal his aversions, he later contended "a want of *prudence* to have poisoned my whole life." It was not only such outright scamps as Dainese and the Pecks whom Marsh flayed. "I do soundly hate a good many," he confessed, "who do not hate me."[1] His animadversions on many public figures—Napoleon III, Pius IX, Presidents Pierce and Buchanan, Horace Greeley, John Ruskin—are memorably scathing.

Marsh suffered numerous reverses and took them hard. But even during the hard times following his 1855 bankruptcy, his misery was seldom unrelieved. His laments were never simply self-pitying. "I don't know where I shall be," he wrote Baird from Burlington between bouts with creditors, "but if you blow a tin horn about once in half an hour all summer long I shall probably come within hearing of it, and will go to you."[2] Essentially solemn, Marsh was never without humor, sometimes frivolous and bantering, often ironic and self-deprecating. He turned the mishaps of his western lecture tour into self-mocking drollery. As an envoy he had to cry long and loud to get action from Washington. But when things went well Marsh was chary about saying so, lest he be seen as self-satisfied or overcompensated.

The best years were those abroad. Congressional life in Washington had its compensations, social and intellectual. But despite attachment to family and landscape, Vermont—which Marsh had considered leaving as early as 1835—appealed less and less with the passage of time. When he returned from Turkey in 1854, he had hardly lived at home for a decade, and Burlington seemed provincial, isolated, and unbearably cold in winter—that is, most of the year. "I find the period between November and May little better than an uneasy slumber," he told Francis Lieber, and "should like a climate where there was one good frost a year hard enough to kill the *ungeziefer* [vermin], eggs and all, and then perpetual summer." He missed the animated life of the mind he had enjoyed in Washington, New York, above all in Boston and Cambridge. "Nothing stimulates thought like an opportunity of uttering it where it is understood," his 1859 Boston sojourn moved him to reflect; "a man's wit and humor brighten under every responsive spark."[3]

It was on extended travels—in Egypt, the Alps, Sicily—that Marsh expressed fullest delight in life. Animated by new impressions of every kind, his accounts of Middle Eastern desert camps (unlike midwestern railroad depots) are vividly joyous, his depictions of glaciers, medieval churches, paintings and sculpture often ecstatic. Travel makes us "at once more patriotic and more philanthropic—cosmopolitan in the right sense," he wrote from Constantinople. Being in another country imposed a double obligation: "of drawing from it some contribution toward the material prosperity or the moral improvement of our own, & . . . making *to it* some return for the pleasure & advantage we have derived for ourselves or our country from it." To be sure, travel often lowered the envoy's estimate of his own land. Apart from his government's "meanspirited niggardliness" Marsh grew ever more disenchanted with American excess. In addition to pitying England's loss of manufacturing supremacy, "I am become so poor a patriot," he confessed, "that I don't quite like to see the poor Swiss driven out of the watch trade by cunning Yankees."[4]

It was not merely the stimulus of transatlantic travel that elated Marsh. Despite meager pay, he lived more comfortably in Europe and the Levant than in Burlington or Washington. And his livelihood was assured: diplomatic salaries may have been mean, but they were regular, lending a necessary security to his scholarly career. Marsh was ill-suited to private enterprise. He detested the sordid tension of competition and litigation. His desperate need for funds was ill-served by his abhorrence of money-making for its own sake. This revulsion sapped his business judgment and partly explains his many failures.

Marsh was a more adept public servant than entrepreneur. In Congress and its committees, in legations and with foreign diplomats, he was controlled, suave, efficient, quick to take in new circumstances—everything he was not in his own affairs. And once done with some public labor, he could relax, feeling free to turn to other pursuits. In business, by contrast, strain was never absent. Marsh spent days fretting over setbacks. As an envoy, he devoted little time to day-to-day matters; much more was done at social gatherings and in informal colloquy. But when necessary he worked with intense diligence, and official responses show how valued his dispatches were. They were valued as much, if not more, for hortatory prophetic vision as for practical utility. Appraising Marsh's bluntly antipapal Italian mission, even his Roman Catholic chronicler felt "forced to admit that with all his faults, Marsh was the right man in the right place at the right time."[5]

In none of his careers was Marsh personally ambitious. He sought enough money to ensure modest comfort, but wealth held no appeal for him. Still less did he crave position. He turned down many high political posts and viewed his diplomatic berths not as honors, but as means to a modest end—serviceable employment in congenial surroundings. Nor did Marsh seek fame from his scholarly work; indeed, he genuinely doubted that he deserved any. He wrote because he felt he had important and useful ideas to impart. Modest about his findings, diffident about his expertise, Marsh was pleased when people praised his books, but he never thought them masterworks nor gloried in the success they achieved. "In fact," he wrote Baird, "I never found out what I was good for, except miscellaneous volunteer work, odd jobs about nothing." He found his own *"Lectures [on the English Language]* worse & worse the more you read them." His conceits were confined to esoteric skills in glass-working and scholarly minutiae, such as this note on the flyleaf of a obscure medieval Latin volume: "I believe I am the only living man who can say that he has read this book, having perused it from title page to colophon."[6]

"If I can't go down to posterity on terms cheaper than these, I'll stay here!"
Marsh's reaction to being solicited by a biographical directory

Free of vainglory and lust for power, Marsh abhorred their effects on others. "With the majority of men not merely animal in life, the strongest passion is love of power; the strongest tie, attachment to party." Recalling a New York politico famed for saying "I would vote for the devil if he were *our* regular candidate," Marsh believed "he expressed a sentiment which, consciously or unconsciously, controls the action of most men in religion and politics." He shrank from the adulation showered on such heroes as Webster and Clay; "I hate man-worship in all shapes." Of one great man's religion Marsh remarked that "his prie-dieu is a mirror, and he serves the God he sees in it."[7]

Others might forgive the famous their feet of clay, never Marsh. On the death of Vermont Senator Phelps, Bishop J. H. Hopkins "admitted he (Phelps) had failings, & said the greater the man the greater his failings would be." Were that true, Marsh retorted, "the fewer great men the better." Elsewhere he expressed more caustic contempt:

S. S. Phelps died Sunday of "a complication of disorders" the newspaper says. I dare say he did. He was a great lawyer & judge. In regard to him as a *man*, 'tis well to let "expressive silence nurse his praise", seeing you can't land him any other way. . . . He has left a considerable estate. A prudent man he was, in that he never asked his best friends to drink with him, and generally imbibed at the cost of others. . . . I am a loser by him to the amount of something for lent money, & want his testimony in a lawsuit I have, and with this eulogy, I call him thrice, Samuel! Samuel! Samuel! as the ancients did, and the Moslems do, and bid him, Vale!"[8]

"He that is great wants to be rich; he that is rich wants to be great," Marsh cited the proverb. "Let our rich who want to be great address themselves—of course, with remittance post-paid" to promised enco-miasts, "and glory shall be dealt out to them in proportion to their tin."[9]

Along with vanity Marsh felt affectation, "the desire of seeming to be that which we are not," a besetting sin. "In all grades of society . . . from the wigwam to the saloon, the most natural thing in the world is to be unnatural." To learn how to be plain and simple "in speech, in ges-ture, in carriage," was for many a hard and painful task.[10]

Marsh's own talents he refused to acknowledge, let alone display. "He was always looked upon as a walking encyclopaedia," recalled his legation secretary George W. Wurts, "yet he always tried to hide his light under a bushel from an exaggeration of modesty." As a diplomat Marsh was chary of using the foreign tongues he knew well; his own presumed incapacity often let an interlocutor feel superior. "How pro-voked I was" at the royal wedding of 1871, added Wurts, when Marsh told Crown Prince Frederick of Prussia "that he had studied German, but not a word would he say in that language." While in Constantinople, Marsh made much progress in Turkish, Persian, and Arabic, "but his idea of what constituted *knowledge of a language* was so high," declared his wife, "that he would never allow these to be even spoken of as among his acquisitions."[11]

Marsh's linguistic humility was unfeigned. He stressed his mani-fold blunders in "awkward attempts" to speak foreign tongues. Told that after four months' study Charles Sumner could read any book in Italian without a dictionary, Marsh noted (not to Sumner) that although he had studied "Italian for more than sixty years, English wellnigh fourscore," it was "hardly an exaggeration to say that there is no book in either language I *can* read without a dictionary. So differently are people constituted." [12]

In fact, Marsh thought that those who boasted of dispensing with dictionaries had only a "very superficial comprehension" of any language. Not only could no scholar hope to retain a whole vocabulary, but "there are, in all languages, words that will not stick. I have looked out some of these a thousand times & don't know what they mean now." In Marsh's view, "to pass over words not thoroughly comprehended [was] a pernicious habit [that] leads to inexactness, slovenliness, and confusion in language and in thought." The greatest stylists in all tongues made habitual use of dictionaries and were given to careful, thoroughgoing, painstaking revision. "Habits of study," declared Marsh, should "oblige every scholar to unshelve and reshelve dictionaries twenty times in a day." [13]

Abhorring vanity, Marsh indulged in self-abasement. "Born of parents of marked characters," he claimed to have "inherited all the foibles of both, not one of the many and great virtues of either." Manifold sufferings had neither cleansed his soul nor improved his character. As it was "an impeachment of the Divine justice to suppose" forgiveness possible, Marsh never prayed to be forgiven. But he bemoaned misspent time. "I have given nine-tenths of my long years to sin, one tenth to remorse, nothing to repentance," he wrote F. P. Nash—"a dreary condition for a man of 76!" [14] He taxed himself with lack of charity, lack of patience, lack of faith.

Marsh was beset by religious doubts. Brought up a Congregational Calvinist, he never formally professed that creed but did not repudiate it either; he held it a "duty to adhere to the religion of one's cradle until one finds a less objectionable one, which I have not." In fact, Marsh mistrusted all organized religion. "I have never connected myself with any church," he explained, "partly because my recollection of the past & my consciousness of the present assure me that I am not fit for membership of the Church of Christ, & partly because I know no church which does not disgust me as inspired by anti-christian sectarianism. Miserable fragments are they all." The refusal of Burlington's Congregational minister to attend a funeral with the Unitarian minister, whom he stigmatized as non-Christian, appalled Marsh as socially and morally divisive.[15]

Marsh spurned Catholicism, both Roman and Episcopal, as authoritarian, obscurantist, and aloof. "I never liked the phrase: 'set apart to the service of God', and I believe we serve God most acceptably when we are at the same time serving our brother man." Like Emerson, Marsh felt that moral leadership required living one's precepts, and he deplored the detachment of clergy from secular pursuits. Nominally Trinitarian, Marsh drew inspiration from Unitarian tracts like James Martineau's *Endeavours after the Christian Life* (1858) and Edmund Sears's *Regeneration* (1853) and *Athanasia* (1858). The inconsistency did not distress him. "A living faith in a philosophically erroneous divinity," he reasoned, "is better than a dead conformity to confessions which, if even abstractly true, have lost their power over the hearts of men." But he himself could not achieve this "living faith." Although "all would be explained & compensated in a world to come," the God Marsh served was an unyielding judge. That this deity even existed could not be known with certainty, or "life would be only a passive waiting, not a discipline."[16]

Least could he fathom apparent divine injustice. While he and Hiram Powers had "been spared and prospered, our children who ought by the course of nature to have closed our eyes, have suffered and been removed to another world." How, Marsh asked Powers, could "pain and sickness and death [be] best for those who have not brought these things upon themselves by folly or by sin & who are called upon to leave the world before they have had time to enjoy its pleasures, to profit by its lessons, or to be led astray by its temptations"?[17]

Marsh's creed was a moral straitjacket. "A weak character may be generous," he wrote; "only a strong one can be just." He thought himself weak. Otherwise he must have denied free will; for he accorded the virtuous little choice of action. "Our means," he asserted, "moral as well as material, are so meted out to us that we never have more than enough to fulfil our obligations." Life was a treadmill driven by conscience. His articles of faith at length reduced to one, "the first clause of which I hope, the second I know, to be true: Christ Jesus came into the world to save sinners, of whom I am chief." Faced with this thin gruel of skepticism and self-abasement, Caroline Marsh termed her husband, in later years, "the last of the Puritans." [18]

If Marsh seemed a would-be Puritan to himself and a true Puritan to others, his liberalism and rationalism were peculiarly Victorian evangelical traits. "Compelled to live with the *real* Puritans," as his friend George Allen pointed out, he would soon have rebelled. Yet his often uncannily modern insights tempt one to move him forward as well as backward out of his own time. "Marsh had so many twentieth-century ideas," commented a reviewer of my earlier biography, "that we sometimes forget how very Victorian he was." [19]

Marsh often felt overwhelmed by personal tragedies that made a mockery of religious faith. He had much anguish to endure: the terrible deaths of his first wife and child, the premature loss of many siblings, Caroline's persisting maladies, the miseries and wretched end of his second son, the drowning of his most beloved niece Carrie. "Next to the loss of you," Marsh often told his wife, "there can be in store for me no greater affliction than the death of Carrie; this affliction has now come upon me!" His griefs were ineradicable. "Though it is now nine and twenty years since the death of my first-born," he wrote a similarly bereaved sister-in-law, "yet there is scarce a month in which this great affliction is not brought back to me in all its first keenness, by some accidental circumstance insignificant perhaps to all but me." As a reviewer of Marsh's letters put it, his suffering was "so acute and so frequent as to make one feel as though woe were always impending." Such domestic losses, much more frequent then than now, were not thereby more easily borne. Yet Marsh was far more devastated than his coeval Emerson, who likewise lost his first wife and adored son. Conscious of emotional coldness, Emerson was wrung by two years of grief; Marsh never ceased to grieve. [20]

Yet Marsh's personal life was far more felicitous than many of his own accounts suggest. He was specially fortunate in his second wife, Caroline. Intimacies are rarely revealed in the letters and diaries of these restrained New Englanders. But a clear picture emerges between the lines. For most of their married life Caroline was an invalid needing

Marsh's daily care. Such care seems always to have been gladly given, gratefully received. He derived satisfaction from looking after her, comfort from her fond faith in him. Because he was sixteen years her senior, he was often her mentor; because he was robust and she an invalid, he was literally her supporter. Marsh did well in and made light of both roles.

Caroline's ailments seldom stopped them from doing what they most relished; together they crossed deserts, climbed Alps, wrote books, kept open house for friends and relatives, and were assiduous diarists and correspondents. As dean and doyenne of the diplomatic corps in Italy from 1863 on, the multilingual Marshes were marvelous partners. At the same time, he often used her infirmity as an excuse for avoiding disagreeable jobs, places, and people.

Her physical ills masked strengths of character and responsive enthusiasms, for him founts of infinite support. She fostered his work while carrying on cognate projects of her own; she never hid her own strong and sometimes disparate views, though usually voicing them with tactful charity. She made friends more readily than he did; her easy rapport with folk in every walk of life enhanced Marsh's own experiences and affinities. "The most brilliant talker I have ever heard," reported an old schoolmate in Rome, Caroline "draws other people out which is more than half the art, [and] is so very young in her feelings for a woman who has never had children." It was as much for Marsh's sake as hers that he disliked spending time away from her. His niece Susan Edmunds never "knew a husband who depended more entirely upon a wife."[21]

Caroline had less success in leavening Marsh's ingrained pessimism. Seeing her husband "terribly given to grumbling at everything short of perfection and even that is not quite good enough," she urged him "take comfort thou sage bird; [remember] the world is less wise than thou." He did take some comfort. "Mrs. Marsh says I am a bloody old owl, a croaker, a humbug & the like and that everything is going on finely," he wrote Hiram Powers, in deep gloom over the Civil War. "Well, well, I hope she is right. She puts me in the wrong so often with her cheerful, hopeful temper, that I shan't mind being caught in error once more, if good news ever come."

In his last year Marsh penned a "loving and reverential" dedication

"To my dear wife, who devoted to me her youth and beauty and has given me her maturer age, having been, for more than forty years, my ever faithful and most affectionate companion, my wisest counsellor and my most efficient aid."[22]

All through life Marsh's family lent him succor to a degree unusual even in his day. The Marsh legacy in Vermont bonded him with his birthplace and with Puritan values. His paternal heritage launched and eased his start in life. His brothers Lyndon and Charles, Woodstock stay-at-homes, gave supportive affection and practical aid. In Burlington, cousin James Marsh was a radiant stimulus, cousin Leonard Marsh an indispensable adviser. His first marriage centered Marsh in Burlington's business community; Harriet Buell's family furthered his commercial and political fortunes, her sister Maria Hickok a stable support, her niece Maria Buell Hungerford an ever-helpful companion in Turkey and after.[23] Only Marsh's son George Ozias and his sister Sarah's husband Wyllys Lyman let him down. But Marsh relied immensely on the Lymans' son-in-law George F. Edmunds. Whatever Marsh's own diplomatic strengths, he could not have kept his Italian post so long without his nephew's powerful senatorial aid. Caroline's large family was also devoted to Marsh. Her sister Lucy Crane Wislizenus gave much needed help in Washington and Turkey and later in St. Louis; a dozen Crane nieces and nephews lightened burdens and enlivened life in Italy.

Love of these and other children—save for "stupid schoolboys"—was Marsh's "strongest passion," he averred, "next to selfishness." The feeling was often reciprocated. "Till I grew a beard, all babies would come to me, even from nurses' arms." Little Carlo, adopted in Rome, did so despite the beard. To the five-year-old son of H. A. Homes, in Constantinople in 1852, Marsh wrote from "Your old friend that you used to shut up in prison, and tie to his chair, and send out of the room, when he was naughty." The devotion of Marsh's niece Carrie did not lessen after she left Italy in 1865. "I love you more than I can ever tell you dear uncle," she wrote, "and think of you almost all the time. Sometimes I want you so much that it seems as if I could not wait another hour."[24]

Likewise profound and lasting were Marsh's friendships. For a man so much in public life, he had few casual acquaintances; he devoted himself to those he knew well. Playful affection animates his letters to Choate, Estcourt, Baird, Child, Dana, Nash, Norton, Pomeroy, Sumner,

Sherman, some from the 1830s, others met as late as the 1870s. Hearing of Estcourt's death in the Crimea, Marsh wrote the general's sister that "the three perhaps most valued friendships of my life" dated from Washington in the 1840s. Two—Colonel William Wallace Smith Bliss and a Mr. Seymour—had died while Marsh was in Constantinople; Estcourt had been "the last survivor, and I feel profoundly that at my time of life one can no more hope to form so close relations."[25]

Nonetheless he did. He was always finding new friends and enthusiasms. The Marshes "are so very kind and considerate and are so fatherly and motherly to us young people who are striving to do something," wrote the classical archaeologist Lucy Wright Mitchell; her *History of Ancient Sculpture* (1883) was "gratefully and reverently" dedicated to Marsh. Many echoed her, from Dana and Norton and Fanny Kemble's daughter Sarah, wife of Owen Wister, to Oliver Wendell Holmes, Jr., who felt favored to be remembered by such a great and busy man. What drew them was less Marsh's erudition than what Charles Eliot Norton called "his moral qualities. His modesty, his simplicity, his generosity, his patience . . . won and held my deepest admiration and respect." On his part, Marsh had "nothing in my expectations of life to which I look forward with so much pleasure" as visits from friends, old and new. "I had always heard of Mr. Marsh as a *cold* man," the wife of a Vermont political crony wrote home from Rome; yet she had "seldom met such sympathy and cordial sentiment."[26]

Marsh's letters to friends were "glorious and mirth-provoking, and soul-inspiring, and greatly refreshing," wrote Spencer Baird, the anointed son who got more of them than anyone else. Even when the Minister's "quaint original handwriting, once seen never to be forgotten," became rheumatically crabbed, the "Turkish loophole backhand" was well worth deciphering.[27] Baird at length made out that his Mary was not a "pious fishwoman" but a "precious penwoman," though "both perhaps applying well, the former as being the good wife of an ichthyologist, the latter as being handy with the quill." Unlike Marsh's often prolix published work, his private messages were pithy, colorful, explicit; letters written in awe at some new scene or in anger against some perfidy still seem warm with wonder or wrath. His formal works often obscured feeling and action, but his personal notes were vigorous and sparkling. Had Marsh "put his personality into literature," remarked his

Genoese consul D. H. Wheeler, "he must have been the most popular writer of his age."[28]

Other communities . . . may glory in the exploits of their fathers; but it has been reserved to us of New-England to know and to boast, that Providence has made the virtues of our mothers a yet more indispensable condition . . . both of our past prosperity and our future hope.

Marsh, *Address Delivered before the New England Society of the City of New-York,* December 24, 1844

Marsh was as hard on the world as on himself, seeing little prospect of improvement in either. The only notable gain he discerned in his time was the growing emancipation of women, "the most hopeful movement in the history of man since Christ prohibited arbitrary divorce at the husband's pleasure." He thought the best prospect for the future was women's release, through

mechanical improvements, . . . from a large share of those petty cares, that ceaseless round of household labors, which have hitherto so injuriously affected the health, the temper, and the intellectual life of the female sex . . . and made them the drudges, not the helpmates of the sterner sex.

Trusting women's judgment as less self-centered than most men's, Marsh was never sure "I am in the right, in any cause, which does not enlist the[ir] sympathies"; he was always "particularly solicitous of the approbation of intelligent women."[29]

Like Caroline, Marsh was an active feminist. Supporters of the Blackwell sisters and other women pioneers in medicine, education, journalism, and the arts, they devoted time, effort, and money to such causes as admitting women to practice before the U.S. Supreme Court and promoting women's careers in the arts. "When you write up the Munich Exposition," Marsh urged a young journalist, "don't forget the Duney portrait by *Madame* Jacquemart, & emphasize the fact that this remarkable portrait is by a woman."[30]

But while Marsh felt female suffrage "a necessary corollary of the true doctrine," he thought it a tactical error to press it before securing women's access to higher education; "if that reform is carried, the rest will follow of course." But even that reform was not easy, for "*men* show themselves more imbecile & ungenerous in their treatment of it than upon almost any other subject." Although American girls had long been schooled on a par with boys, equality ended at college. Schooling was essential for women to be good wives and mothers, but higher education was needed only by future ministers, professionals, and leaders—all men. Only after the Civil War did women's colleges become a significant presence. But other gender reforms, notably suffrage, did not, as Marsh had hoped, follow rapidly in the wake of higher education.[31]

Marsh wondered why men were so fearful of women as equals. He noted that women had "once voted in New Jersey," which gained neither wisdom nor credit by later barring them from the polls; "had women continued to vote in that state," he thought, "New Jersey would have been sooner redeemed" from slavery. Here Marsh parted company from his otherwise esteemed Horace Bushnell. Bushnell's *Women's Suffrage, the Reform against Nature*, lauded New Jersey's 1807 retraction of suffrage as essential to decorum. Not only had New Jersey women voted wrong, they had voted often, hoodwinking poll watchers with changes of dress; the exercise of political power led the frail sex astray. "A woman can not be as bad as a man in anything," opined Bushnell, "without being worse."[32] Despite growing equality of rights, the notion of separate spheres for the sexes remained sacrosanct for most men *and* women.

Separation was precisely what Marsh repudiated. Most of the supposedly immutable differences that led men to regard women as the weaker sex Marsh exposed as self-serving and self-generated myths. The platitude that "describes women as emotional and impulsive, men as rational and reflective," was empty and mean-spirited. Far from being the archetypal delicate Victorian wife she might seem, Caroline was tough-minded and venturesome. Marsh blamed men who shut their wives indoors for the "sickly sensibility . . . often developed in women."[33]

Both sexes were thereby deformed. Men who held aloof from matters of hearth and home hardened their hearts and deadened their own sensibilities, often "devolving exclusively upon women the burden of domestic duties, parental responsibilities, and personal charities." For this Marsh expressly blamed male-only schools and colleges that made no provision for social training, scarcely any for moral. "Youths, while passing through them, are deprived of the humanizing influences of domestic life, and the young man comes out of the hands of his professors a coarser, a more unsocial, too often a more vicious, being than the mother's boy he has grown out of."

Not that Marsh thought men and women the same; but "we know next to nothing at all about the relative powers and capacities of the two sexes. We have remained in blank but wilful ignorance of the whole matter." Nothing could be known until we "make woman legally and socially the peer of man, afford her equal if not identical means of education, give free scope" to her talents, and break down the "brazen network of arbitrary institutions which has everywhere enchained them." Here Marsh went well beyond many who, though eager to redress wrongs, were alarmed if women themselves took action. While lamenting their subordination, Emerson did not want "women to wish political functions, nor, if granted assume them," lest this curb their "legitimate influence"—that is, spiritual succor of home and family.[34] What was the good of such succor, Marsh would ask, to men devoid of moral quality?

Marsh also sought to protect women from coercion that ignored or denied clear biological differences. One appalling abuse was "the almost universal employment of women in field labor" in Europe, in his eyes harmful to their health, "tending to deprive them of the softness and grace of their sex, to assimilate them to the coarseness of the men with whom they work, to disqualify them for the duties appropriated to them by nature, and in short to debase and brutify their whole character." The greater "consideration, deference, and respect" given women in the United States owed much, Marsh believed, to "their exemption from field labor, and . . . the grosser and more repulsive cares and toils of husbandry."[35]

Italian residence strengthened Marsh's feminist views. Italy had early led in treating women as intellectual equals: several were professors in the fifteenth century, and by the early sixteenth, Marsh held, "the number of learned, genial, and liberal Italian women" exceeded those in all the rest of Europe. But the Counter-Reformation had reversed this advance: the convent and the confessional replaced the schoolroom, and few Italian women now were even literate. "The Catholic statesman who never enters a church and even encourages his sons to share in his skepticism," charged Marsh, "has continued to commit his daughters to clerical instructors and is not content unless his wife keeps on good terms with her confessor." [36] While one in five Italian men could read and write, fewer than a tenth of the women could; in contrast, 97 percent of New England women were literate by the early nineteenth century. "Women's higher rank as an object, not of passion, but of reverence, and the reciprocal moral influence [of] the two sexes" had earlier struck Marsh as a uniquely New England cause for pride. [37]

In Italy men ruled absolutely, barring women from public affairs. As the Marshes themselves were made aware, it was "disreputable for a woman to bring into active exercise the qualities which command respect & ensure success in a man." In Europe a prime impediment to schooling girls was the fear that they might want to learn more than strict obedience. Finding that some peasant girls aspired to become teachers, a European lady burst out to Marsh, "*Mais, Monsieur, elles aspirent à se déclasser!* They want to rise above their caste." Marsh saw the European tradition of female inferiority intrinsic to the general tyranny of class servitude. [38]

A major aim of the orphan asylum the Marshes helped found in Florence was to promote "the better education of women, which may teach Italian females to respect themselves by compelling men to respect them." He was confident that this asylum was "doing more *permanent* good than any other charity" they supported. [39]

Presumed lack of male respect forbade respectable females from going out on their own, as the American Minister often had to warn visitors. Well-born young women had to be properly chaperoned. On occasion the Marshes suffered the fetters of convention. When their young niece Carrie invited Manzoni's granddaughter to the Teatro Regio with Marsh in charge, "Madame Collegno said that she should like much to have her go if there was to be a matron in the box—otherwise it would not be *well received!*" Caroline was appalled. "That a man of over 60—of Mr Marsh's character and position—should not be a sufficient protection for his own niece and a little friend of hers, still a schoolgirl!" [40]

Marsh's feminism was not at odds with his own warmly observant and often gently flirtatious relations with women. "I have seen women in multitudes," he wrote Caroline on a lecture trip to Albany, "but of ladies, not above two or three. Verily Eve's daughters are waxed coarser and more uncomely in these parts." To a friend whose wife he admired, Marsh declared, "I have passed my 80th birthday, and am free to kiss any pretty woman I like." The company of the erudite feminist Emilia Frances Pattison, later Lady Dilke, "with whom my wife permitted me to fall in love, . . . did more to reanimate me than all my doctors and all their drugs." Marsh often found it easier to write to wives than to husbands; "one does not feel obliged to be quite so grave, judicious and didactic." [41]

Equal rights aside, he ascribed to women a "higher and more generous moral nature" than men; "women are stronger than we, partly no doubt because they are less selfish." Marsh viewed most men in his own image, as self-centered; and many women, like Caroline, as altruistic, forgiving, pliant. In his apothegm, "men are swayed by their prejudices, women by their partialities." Not all these partialities did Marsh find laudable. He had been disgusted by women's "want of self-respect" in idolizing Henry Clay. "How they followed him during the [1844] election campaign, begging for locks of his hair, etc. . . . How could a man be expected to treat women with true and dignified respect after such an experience!" [42]

To respect for women Marsh added respect for proscribed minorities. Those disadvantaged by race or creed (Greeks and Irish apart) found in him a stalwart champion. Several Marsh guests in Turin were members of the Jewish community, notably the lawyer-litterateur-parliamentarian Davide Lévi, who shared Marsh's antipapal animus. [43] Toward blacks in America Marsh felt compassion, shame, and future hope, though like many he thought that hope more apt to be realized in recolonized Africa than in racially polarized America. He helped his Washington servants, both hired-out slaves, to buy their own freedom; they declared Marsh's treatment of them color-blind. Not only were Negroes entitled to suffrage—as noted in Chapter 15 above, Marsh looked to freed slaves to buttress American liberties—but "the natural equality of rights of all men, . . . without distinction of caste or colour," should embrace "absolute liberty of thought, of speech, and of action."

Marsh was by no means exempt from the stereotypical views of his time: "I hate the Russians, & set little store by the Poles," Marsh wrote Francis Lieber, "but your Bohemian is almost as good as a white

man." Like many abolitionists, Marsh declared himself averse to inter-
racial marriage. Yet the mulatto journalist Edmund Quincy Putnam
declared that the Marshes had, over many years, "treated me and my
family with the politeness and courtesy due to their position & to ours,
& that quite spontaneously."[44] Nativist though he remained, Marsh
stanchly backed equal rights regardless of gender or creed or color.

*I*n many other ways Marsh was profoundly partial. But the biases
he voiced were often at variance with one another and with his conduct. Consider his oft-expressed contempt for utilitarianism. Despising
expediency as a motive for action, he professed to scorn the materially
useful. Yet Marsh was at the same time an utter devotee of utility; he
justified everything he undertook as demonstrably of use, whether to
his family, the community, the nation, or posterity.

Again, Marsh termed himself a man of the people. Contempt of
aristocracy and hatred of privilege suffuse all he wrote, and the longer
he was in Europe the more confirmed a democrat he became. When
justice failed or reform misfired, Marsh blamed rulers and elites and ac-
quitted the common man. He held equal access to goods and resources
not only morally just but socially and environmentally vital. In order to
feel "a strong interest in the protection of their domain against deterioration," rural laborers needed vested rights in land "as a perpetual possession for them and theirs." Marsh noted with scorn that what were
"'necessaries' when claimed by the noble and the rich were styled 'artificial wants' when demanded by the humble and the poor." He praised
workers' growing resolve not "to submit to a longer deprivation of a
share in the material enjoyments now monopolized by the privileged."
He similarly faulted Ruskin's "most pernicious doctrine . . . that beauty
is only for the rich, because costliness and rarity and inherent difficulty" were essential to it. Terming "everything vulgar that savors of the
common cares imposed by Providence on all save the members of the
favored class to which the critic himself belongs," in Marsh's résumé,
Ruskin held it "a profanation . . . even to plant flower-beds in the court
of a railroad depot." Marsh was as dismissive as William James of those
"trained to seek the choice, the rare, the exquisite, exclusively, and to
overlook the common."[45]

To the "ornamental elites" of England and their American imita-
tors Marsh ascribed "much sweetness and little light," rating them far
inferior to the middle class "in practical power of thought and judgment." In a rare accolade to the practice of law, he praised American
judicial processes as stimuli to everyday common sense. "Frequent attendance . . . in courts of law, where questions of fact, depending on
the comparative weight of a vast variety of modes of proof, are constantly subjected to searching examination," made ordinary Americans
"sound judges upon questions of fact" as well as keen logicians.[46]

Yet for all his populist advocacy, Marsh was no social egalitarian.
Even in the Vermont of his youth gentry were routinely differentiated
from common folk. Elites took it for granted that servants and laborers
were unlike themselves in origins, upbringing, and behavior. A moral-
izing aside on tobacco in *Man and Nature* conveys Marsh's ingrained ac-
ceptance of such distinctions:

There are some usages of polite society which are inherently low in themselves, and debasing in their influence and tendency, and which no custom or
fashion can make respectable or fit to be followed by self-respecting persons.
It is essentially vulgar to smoke or chew tobacco, and especially to take snuff;
it is unbecoming a gentleman to perform the duties of his coachman; it is in-
delicate in a lady to wear in the street skirts so long that she cannot walk without grossly soiling them. Not that all these things are not practised by persons
justly regarded as gentlemen and ladies; but the same individuals would be, and
feel themselves to be, much more emphatically gentlemen and ladies, if they
abstained from them.[47]

Of the many assumptions about decorum this passage reveals, most
important is Marsh's premise, presumably shared by his readers, that
gentlemen are unlike other men—a gentleman does not dress or act
like his "coachman."

In truth, Marsh scarcely knew the commoners he exalted. He may,
as he asserted, have found "the mudpiles better worth study than the
superstructure of the social edifice," but this did not necessarily make
them better company. Beyond clients and servants and coachmen, his
hoi polloi contacts were limited. He was truly at ease only with the cultivated. Reserved by temperament and aloof by upbringing, Marsh
scorned the stupid and shunned the vulgar. "'Tis a wicked world. The

sheep are in small proportion to the goats," he wrote Baird. "If it wasn't for me and you and our wives and two or three more, they'd all be burnt like Sodom and Gomorrah, bad luck to them." Putdowns of bumptious pomposity pepper his letters, as with an ignorant New Yorker in the Levant who horned in on a discussion of Dickens and Eugène Sue:

"I see little difference," quoth Jonathan; "they are both by the same author." Being reminded that Dickens was a living man and had actually visited New York, where our countryman might have seen him, "Well," he exclaimed, "at any rate, Dickens wrote Eugene Sue."[48]

Marsh deplored the impudence of the semieducated. Despite the spread of schooling and strides in literacy, he saw in his old age "no intellectual *general* gain" since his youth. "Some better scholars are now trained indeed, but the hoi polloi are below the standard of 60 years since." Educators confronted not a clean slate but a mass of ingrained superstition; ignorance was no "mere vacuous defect of knowledge, [but] a positive quantity . . . laboriously acquired, a darkness that can be felt."[49]

He had come to the sad conclusion that reason did *not* rule human affairs: "You cannot convert a savage who worships a serpent, by proving to him that his fetish is venomous." Few even of the literate could "form a legitimate judgment on any abstract moral proposition" or resolve any logical problem. Many "cultivated" people were totally unable to "weigh evidence, whether direct or circumstantial." What Marsh termed "the unreasoning classes . . . unhappily form a large proportion of the highest as well as the lowest circle in modern society." More contemptible than the untaught were the arrogant and aggressive quasi-schooled. The felt failure of education shrank Marsh's residual faith in democratic progress. The middle classes, "everywhere the true depositories of the strength, the intelligence, and the virtue of the modern world," were more and more beleaguered by religious reaction and by the newly enfranchised but easily manipulated masses.[50]

Hopeless almost by definition were folk whom Marsh saw as dedicated to evil. Against slaveholders, the papacy, the Irish, English aristocrats, corporations, and publishers he never let up. To these fixtures in his demonology he kept adding new targets of abuse. Even while absorbed in flaying the papacy, for instance, he took a sideswipe at French vainglory. "France, as is well known, always has been, and still is, 'the leader in the march of modern civilization and progress,'" Marsh quotes —and then he footnotes his source: "See any French book, *passim*."[51]

Of all the French, Marsh most detested the two Napoleons. Before he came to fear Bismarck as an even graver menace to Italy, the American Minister wrote his Berlin confrere George Bancroft that Bismarck "alone prevents what Manzoni said of Napoleon I being applicable to the third of that name." In Marsh's judgment,

Napoleon I was [not] a worse man, perhaps he was even a less bad man, than Napoleon III. [But] neither was ever surpassed in wickedness, and it is . . . proof of the improved tone & greater force of public opinion in our time, that neither of them dared to imitate the old Milanese tyrant in throwing their enemies into the kennel to be devoured by dogs, . . . from which not conscience, but fear of moral condemnation restrained them.[52]

So manifold were Marsh's bêtes noires that he had to assign them degrees of iniquity. He kept his choicest epithets on avarice for Americans, but "cannot admit we are more prone to the worship of the dollar than the English. Perhaps we labor harder to get the dollar than they," but the English never let go of it: "An American may forget the dollar, an Englishman never, with him it is omnipresent." At Caroline's urging Marsh classed Italians and French "also among the dollarators," concluding "that the American *pursues* the dollar with greater avidity, while the Englishman, Frenchman, & Italian *cling* to it with tenfold tenacity."[53]

Perfervid as his prejudices remained, Marsh soon abjured his early simplistic praise of Goths and censure of Romans. Travel showed the error of ascribing national character or religion to climate. "I do not assent to Montesquieu's views of the relation of Catholic & Protestant to national character, nor do I accept his facts," he wrote Nash. "Where are the most bigoted Papists in Europe? Among the Flemish." French Protestant strength lay in the south; the most ardent Catholics were in the north. Italians were essentially more democratic than Germans, who "cared nothing for . . . liberty among themselves" yet wanted to Teutonize the world. For any nation to impose its own ways on others violated the essence of human liberty. "The European Continent is to-day [1858] protesting against being Teutonized, as energetically as it did, at the beginning of this century, against a forced conformity to a

Gallic organization." Marsh hoped that Europeans could equally resist "the Panslavic invasion" he envisioned as the coming danger to Christian liberties.[54]

Marsh's group stereotypes flew in the face of his conviction that peoples did change—even if not usually for the better. The fallacy that traits of national character were permanent was disproved by what Marsh termed "history," though even here his generalizations now seem more moral than historical. For example, "the ancient Roman, though brutal was just," whereas modern Italians evince "the utter impossibility of even comprehending the notion of justice in private or public conduct." So too "the English were once a truth-speaking people, but [latterly] they have proved as unveracious as the Greeks—the only people I know among whom falsehood is the rule" (and the Greek Orthodox Church "the most *immoral* form of religion ever professed" by Christians). Marsh had refined his earlier verdict that "an innate and habitual disregard of the truth is eminently characteristic of the Oriental mind, Christian and pagan alike."[55]

Not all the changes Marsh noted were for the worse. And he began to realize that comparisons across epochs were not always apt or useful. He was delighted by Charles Eliot Norton's *Historical Church-Building in the Middle Ages* (1880), which set out not to condemn but to explain that epoch's moral tone. Marsh thanked Norton for helping him "to read history in a truer light, learning to perceive that what would be vices in our age were virtues in an earlier one." He had long felt it wrong to laud the past at the expense of the present, to compare the abiding good of former times with the supposed evils of his own. Nostalgists forgot that "antiquity, too, had its phantoms and its despairing delusions." Sixteenth- and seventeenth-century scholars had complained as loudly as moderns of the worthlessness of much new writing, while failing to honor such stalwarts as Shakespeare or Milton.[56] Harking back to "better" days of old was a habit Marsh himself was prone to, but in principle he knew it was an error.

To suggest that life in Europe had in some sense first regressed, then progressed, Marsh cited grammatical changes—notably in verb tenses. Why did medieval Gothic tongues have "a past tense, never a future? Why did the Romance dialects retain the Latin past forms, and reject the ancient inflected Latin futures" for compounds and auxiliaries of *will* and *shall?* A circumstance so widespread "must have some common psychological ground. . . . It is perhaps not an absurd suggestion," Marsh speculates, that medieval life was so unstable and insecure that most people could hardly think hopefully of the morrow. "The present was full of stern necessities, the past, of hard and painfully impressed realities. The future offered but dim uncertainties." Men lived not in "a dream-land . . . to be realized in the good time coming, but in a *now* which demanded their mightiest energies. [Because] the future was too doubtful [for] words implying prediction or even hope," they applied to it only "forms indicative of a present purpose . . . or duty, not of prophecy or expectation."

Though Marsh was as ever cautious in making inferences, this speculative flight of fancy was no means atypical. A verbal trait common among widely dispersed peoples in the past sparks a judgmental comparison of past with present.[57] While striving to view the past on its own terms, Marsh was ever determined to gauge it in present-day perspective. In chronicling the annals of languages, as of landforms, he deployed history to teach his contemporaries useful lessons.

Yet Marsh was not primarily a crusader. Although almost all he wrote aimed to further the welfare of the world or some cause of his country, the reader soon sees that if such motives justified Marsh's work, they seldom drove it; he was immersed in the topic at hand chiefly for its own sake. For instance, Marsh claimed he wrote *The Camel* to promote that beast's use in America. But this objective occupies only a few pages in the book and concerned Marsh very little while writing it. What really roused his interest was the exotic drama of dromedary life. Again, zest for detecting word origins stimulated—and saturated—his language studies more than any zeal to prove the superiority of Gothic to Romance tongues, of tradition over novelty, or of American English over English English.

Even where reform was clearly compelling, as in Marsh's environmental studies and antipapal tracts, didactic preachments occupy only a small fraction of the text. Fascination with subject matter swamps advocacy; he ever digresses to tell an anecdote or to dwell on some curiosity irrelevant to any hortatory aim. This discursive bent enhanced

his work: preconceived theories and goals seldom cramp Marsh's conclusions. While he was committed to his major premises, he gave many
disputed theses—such as whether forests generated rainfall—impartial scrutiny based on all available evidence.

Hence Marsh never fell captive to the programs he espoused.
However zealously he promoted this or that cause, they never monopolized his letters or his converse. And once a book or article was completed he put the subject aside, professing to have lost all interest in it.
This was evident in his environmental work. He never ceased to plead
for forest preservation, for resource husbandry, for flood control, for
other conservation measures; but after *Man and Nature* came out he
moved on to other matters. In his exchanges on forestry with Charles
Sprague Sargent and Henry Yule, it is research queries—origins of
plant names, ways of pruning trees, modes of sprouting from stumps—
that fill his long letters, far more than agitation for environmental reform. Advocacy is not absent: Marsh implores Sargent to send Portugal's
forestry commission seeds of American trees "of quicker growth well
adapted to . . . rewooding bare mountains." He himself had sent *Catalpa*
and *Persea* (laurel) seeds, though he was doubtful of adequate nutriment
on scorched rocks in Lusitania.[58]

It is thus inappropriate to term Marsh an environmentalist, any
more than an Americanist for espousing folk history, or a socialist for
urging public control of irrigation and railroads. Devoted to these
causes, he dedicated himself exclusively to none. Yet want of single-
mindedness did not make his work superficial, as he often chided himself. Had he had fewer interests and diversions, been less caught up in
earning a living or in savoring the world, reflection might have deepened his syntheses, but at the expense of vital breadth. Versatility was
his strength. If conservation or any other cause had ruled his life, his insights would have lacked their stunning comparative and eclectic scope.

The *process* of research, more than any particular topic or its practical uses, was Marsh's ruling passion. As a child, he found mastery of a
wide range of subject matter praiseworthy, but mastered it above all for
its own absorbing sake. As an adult, Marsh was famed for limitless curiosity in myriad realms of inquiry. His mainspring was a lust for learning. Marsh perfectly embodied Abby Maria Hemenway's depiction of
Vermonters as folk who "like naturally to know about a thing that in-

terests them, from beginning to end." Such knowledge should be harnessed, to be sure, to wisdom or progress or morality. Chagrined by his own magpie bent, Marsh held it "a mistake to suppose that all mental acquisition implies mental culture. Facts without end may be learned, familiarized, forgotten again, and leave the mind at last more inept than they found it." True knowledge was not detached descriptive details but residual affinities, as for the great lawyer recalled by Marsh of whom it was said that "his learning had passed out of his memory into his judgment."[59]

Knowledge was, for Marsh, emphatically public property to be shared. When he feared *Man and Nature* might be anticipated, he was less concerned for his own repute than intrigued by the speed of diffusion. "This coincidence of thought for which our age is so remarkable reminds one constantly of the theory that ideas are floating about and strike on the mental vision in a sort of random way—affecting now this brain now that. One advantage will come of it, at least," Caroline glossed his reflection: "our thinkers will be the more modest for it."[60] Some hope!

Marsh saw concurrent discoveries resulting from knowledge being popularized through printing and the dissemination of scholarship. Just as "almost every new mechanical contrivance originates with half a dozen different inventors at the same moment," so with literary creation: "if you conceive a striking thought, a beautiful image, an apposite illustration, which you know to be original with yourself, and delay for a twelvemonth" to put your claim "on record, you will find a dozen scattered authors simultaneously uttering the same thing."[61]

Marsh was only mildly distressed, seldom surprised, to find his own ideas—like his mechanical contrivances—anticipated. The last letter in his own hand reiterates Donatus's adage, *pereant qui ante nos nostra dixerunt* (plague take those who have said our smart sayings before us). Marsh's own dictum, that translators should use a period style consonant with that of their text, had been preempted, even expanded, in Maximilien Littré's *Dictionnaire de la langue française* (1863–69). "It is enough to discourage all invention," he jested in *Man and Nature*, "when one finds plagiarists in the past as well as the future!"[62]

Scrupulous in his own citations, he insisted on sources from others, scolding his friend Henry Yule for the naughty habit of quoting authors without chapter & verse. Pray reform, for the sake of your adversaries, if not of your friends. It is so hard to convict a writer of misquoting or misunderstanding his author, when he does not tell you where to find the passage! Even 'the wild Indian' leaves a scalp-lock in his crown for the convenience of his enemy. Shall not a Christian be as generous?[63]

When Marsh's acolyte Baird, "otherwise sinless, did great prejudice to his soul by following this evil example" of omitting references, "I chode him for it, as he well merited, . . . but he reformed not, and will howl in Purgatory accordingly." Marsh was ruefully aware he was too meticulous for many: in the common "dialect of criticism, an author who acknowledges his obligations is a *compiler;* one who conceals his thefts is an *original writer.*" In the second edition of *Man and Nature* Marsh takes sarcastic care to "acknowledge my obligations" to a Dutch writer "for assuming responsibility for many of the errors I may have committed in this chapter [on sand dunes], by translating a large part of it from [me] and publishing it as his own."[64]

Yet authors, like inventors, deserved recompense for their creations. Stealing texts seemed to Marsh as wicked as the U.S. Navy's efforts to cheat contrivers of rope machines in the 1840s. In his *Lectures on the English Language*, he made it "a point not to borrow" quotations from Richard Chevenix Trench "and thereby diminish the pleasure of readers" in Trench's own work. A friend of Marsh's in England found "his beeches were dying off, and discovered the cause to be that the squirrels girdled trees by tapping them for the sap," Marsh wrote Sargent. The friend "shot the squirrels and lost no more beeches. *Cessante causâ, cessat et effectus* [Remove the cause, and the effect also ceases]. This by the way is *private,*" Marsh added, "for Mr. Hamilton has promised to write an article about it for *Nature,* & I want him to have the first fruits of his observation, for *his* it was, as he caught the inventor of the process in the act of gnawing through the bark and quaffing the nectar."[65]

Marsh took infinite pains not only to assign credit where due but to rectify error. Working copies of his books are laden with corrections for new editions or just for the record. Printers ever played havoc with his work, their misprints often he thought deliberate. "Well, 'tis my fate," he wrote Baird when *The Camel* was thus mangled. The "grisly gnomes" of his Matthisson translation got transmogrified into *grizzly,* "I wrote *girth*—printer elevates it to *girdle.* . . . I said plain *Jackson.* Typo thinks this familiar, infra. dig. for an ex-diplomat, and scorning nick names prints *Johnson*." [66]

He was still worse served by the first Italian translator of *Man and Nature*, "a lady not acquainted with the subject, and with a very imperfect knowledge of both English and Italian." Marsh had the book destroyed owing to "not less than 6,000 . . . gross and often ludicrous errors." He was amused to find his translations from French authors faithfully retranslated into Italian *from his English*. "Did my translations correspond so closely to the French idiom," Marsh wondered, "that the French construction was visible in them?" [67]

That each language had its own untranslatable genius Marsh was certain. Not even the most fluent non-native could express her true thought in the alien tongue; as one Marsh guest put it, "I can never say in English *precisely what I mean*." Words seemingly identical often convey totally mistaken ideas or become vulgar in one language while remaining dignified in the other. Marsh instanced the French historian Jules Michelet's spurning of English literature as "satanic" and "anti-Christian," Shakespeare as "godless," to show how the most eminent scholar could ludicrously err in a foreign tongue.

Multilingual training availed little; indeed, it corrupted one's mother tongue. "With rare exceptions no man can freely use more than one language as a medium of intellectual or oral discourse," he felt, and "what we gain in power over a foreign tongue is compensated by a corresponding loss in the mastership of our own." Since "no man acquires a *mastery* of a foreign language without losing something of his command over his native speech, a partial knowledge of French, German, or Italian is purchased at a price which greatly exceeds its value." [68]

The best stylists and orators Marsh had met were monolingual. "The mightiest master of words the world ever knew was Demosthenes, who certainly was acquainted with no language but Greek." Shakespeare, Izaak Walton, and most peerless English writers knew little of other tongues. In sum, "the best method of acquiring a thorough knowledge of English is the study of English alone." [69]

English-speakers did need Latin, Marsh admitted. "We, whose native tongue has no grammar, can hardly get a clear notion of the grammatical relations of words, without . . . Latin *Formlehre* [grammar] and syntax." But trying to speak modern European languages was for most a waste of time and energy. He deplored "the fashionable American craze for sending young men and women abroad . . . to figure in society by assuming to speak two or three Continental languages." By then they had lost "the delicacy of ear and the flexibility of the vocal organs" of early childhood. "Though an adult may profitably study the literature of tongues not his own, in general the attempt to *speak* them only renders him ridiculous."[70] Moreover, accuracy was often at cross-purposes with fluency:

I once complimented an English lady on the ease with which she spoke German. She smiled and said: "I will tell you a secret; I could only stumble pitiably through a German sentence, till one day I resolved to perplex myself no longer with *der, die,* &c, or any of the cases, but to stick firmly to *das* . . . for all the articles, pronouns, etc., and nobody seems to have found me out."[71]

But if monolingualism suited the orator, it was anathema to the scholar. Familiarity with foreign tongues enhanced precision in one's own. "To express ourselves aptly in our native language," Marsh reasoned, we must "habituate ourselves to the utterance of thoughts and the portrayal of images . . . embodied in other tongues." Translating "authors whose thoughts run in channels not familiar" to us was a corrective to parochialism. "When we think another man's thoughts in our own words, we are forced out of the familiar beats of our personal diction, and compelled sometimes to employ vocables and verbal combinations, which we have not before appropriated and made our own by habitual use."[72]

Life-long study made Marsh acutely aware of the relative difficulty of languages. In common with most English and Americans, he found Italian particularly hard to assimilate. "German, the Scandinavian languages, possibly even French, we can learn after our beard is grown," but not Italian; "it is too disparate from our mother speech." Reading aloud in Italian was more tiring than in any Teutonic tongue, in modern Greek harder still. Marsh recalled the Jonas King case in Athens, when just ten or fifteen minutes of parsing Greek legal documents had exhausted him and his aide, because Greek accentuation was so utterly alien to them.[73]

Linguistic disparity bred cultural distance. Comparing Anglo-American and German with Italian library catalogues, Marsh was struck by how few books they held in common—at most a few hundred out of many thousands. The reason seemed clear: "The intellectual culture of the two races runs in different sets of grooves, and their literary artisans require quite different sets of tools."[74]

The idiosyncrasies of national scholarship fascinated Marsh. Germans began by refuting "all that has been advanced by other writers of all times and countries," then went on zealously to correct themselves. As revisions followed in rapid succession, "those who buy the first edition . . . soon find that the author has made that worthless, by refuting himself in the second." Marsh lauded German freedom of thought and "habits of persevering and continued research, which forbid scholars to [treat] even their own investigations as final." He credited much of their prowess to growing up with a homogeneous, rule-bound tongue: "the German boy comes out of the nursery . . . imbued with a philological culture which the Englishman and the Frenchman can only acquire by years of painful study."[75]

But German scholars had their defects. One was willful disregard of anachronism. Marsh reproved German "improvers" who, in restoring old texts, "so disguised authors [that] they would not be able to read, or even recognize, a line of their own works clothed in the orthography which modern scholars know they never used, but think they ought to have adopted." Excess theorizing was a more serious German flaw. Though the Teutonic scholar was keen "to amass all the known facts" on his topic, mere "want of sufficient data . . . rarely deters him from advancing a theory." As with Max Müller's comparative philology, "if he can obtain but one linguistic fact, he turns that one into a law . . . with scarcely less confidence than he sums up the results of a million." Marsh also poked fun at narrow German pedantry. "How various are the sources of happiness!" he noted of a coleopteron newly found in a cave of the Karst. "Think of a learned German professor, the bare enumeration of whose Rath-ships [honorific titles] and scientific Mitglied-ships [memberships] fills a page, made famous in the annals of science, immortal, happy, by the discovery of a beetle!"[76]

However much Marsh faulted Gallic chauvinism, he admired "the French gift of *method* in bookmaking, . . . the art of beginning at the beginning, going straight to the mark, & leaving off at the end; wherein all we Goths are grievously deficient. Most English and German books remind me of a criticism on a speaker in our provincial Vermont legislature, who talked much to little purpose: 'Mr. —— always begins his speeches in the middle, & comes out at both ends.'"[77]

Not that languages were fixed, let alone faithful, mirrors of national traits. Modes of expression might change while behavior stayed the same, or vice versa. Modern Icelanders spoke much like their medieval ancestors but renounced their habitual violence; on the other hand, Marsh judged Bedouin mores much the same as in biblical days, while their language had been totally transformed. In short, "history presents numerous instances of a complete revolution in national character, without any radical change in the language of the people, and, contrariwise, of persistence of character with a great change in tongue."

For Marsh, such incongruities pointed up a still greater contrast between physical processes and social history—a contrast crucial to understanding the role of humans in both realms. In his view—and here he parted company most markedly from Darwin—the creations of nature were far more uniform than those of culture. Hence humans, biologically fixed and "immutable, [had] passions, appetites, powers everywhere and at all times substantially the same." Conversely, all aspects of human culture—languages, artifacts, modes of life—were "infinite in variety of structure," frail, unstable, and transitory. Thus language decay and "dissipation into a multitude of independent dialects" were seemingly inescapable.[78]

Yet amalgamation, not fragmentation, was the most salient feature of linguistic change. Local, regional, and national tongues were in fact converging, differences among them dwindling. Renouncing his earlier reliance on the divine uniqueness of national character, Marsh came to view this convergence as a mark of progress beneficial to all. Some bewailed the "apparent uniformity" of modern civilization, alarmed lest it reduce "social man everywhere to one dead level of tame and unpicturesque mediocrity." Marsh repudiated such nostalgia: throwing off the confining shackles of nation and place fostered "the *general* improvement" of humanity, at the same time enabling *individuals* to become "more distinct, more consciously independent."

Diversified individuality was for Marsh the crowning glory of civilized life; "being more and more differently constituted, we diverge as we advance." But diversity of form went hand in hand with mutuality of purpose, with mastery of self and of nature for the common good. In lauding diversity Marsh comes close in spirit to Darwin, who ends by exalting the planet's progress from merely "cycling on according to the fixed law of gravity," to evolving "endless forms most beautiful and most wonderful."[79] The beautiful wonders of Marsh's vision of evolving cultural forms required insight into the conjoined histories of nature and human nature—propelled by disparate forces, yet intricately and indissolubly merged in their effects. This was the bedrock of his own work.

Marsh was no sedate scholar secluded from the world; his scholarly work fused with his public career and private life. Asked to take on some civic chore, he would write a report whose range always transcended the immediate issue. Added research then yielded further essays or a book. Meanwhile some new topic would engage him. Marsh used to complain that he was forced to switch from job to job because his writing did not pay. "My tread-mill . . . grinds out nothing," he complained late in life. "My work is of a sort that shows no results, & occupies my time & attention just enough to prevent me from doing any thing else." He felt he would have had a "broader life, if I could have chosen my vocation."[80] Marsh's biographer must demur. His genius lay precisely in *not* choosing his vocation, but following whatever path and taking up whatever task seemed suitable at each moment. Some proved onerous chores, but he enjoyed even those. He embarked on many jobs with a certain ambivalence, but once begun he gave each one his full commitment.

No intentional pattern shaped this sequence; every aspect of Marsh's career had some adventitious origin. His English studies arose from the supposedly lucrative post at Columbia, "an accident growing out of another accident," as he put it.[81] His camel book grew from diaries penned during his fortuitous mission in the Levant. A chance amalgam of Green Mountain insights with Mediterranean and Alpine observations led to *Man and Nature*. This mix of commissioned and other research on a variety of topics was more to his taste than sustained narrow study on any one; too-protracted labors on the English language and on *Man and Nature* led an exhausted Marsh to rebel against etymology and geography in turn.

Marsh justified scattering his energies and spreading his talents by the circumstances of his upbringing. Omnicompetence, an oft-remarked American trait, was much admired in his Vermont, which fostered talents as diverse as those of any frontier community. "An encyclopaedic training" was needed by all, explained Marsh, because "every man is a dabbler . . . in every knowledge. Every man is a divine, a statesman, a physician, and a lawyer to himself." Many-sided Americans were marvelously inventive: "the mechanical contrivances of the whole world, from the earliest ages to the American Revolution," in his view did not equal in "ingenuity or efficiency . . . those described in the records of the American Patent Office."[82]

The loss of such versatility was, alas, one consequence of social and scientific progress. Marsh well knew that "the days are gone forever when one human intellect could compass all recorded human knowledge." Already "the memory of the comprehensively learned man [was] less a repository of knowledge than an index to its archives." He was not happy about this. The aspiring scientist must "choose ignorance of some things well worthy to be understood," Marsh concluded as early as 1847, in order to "more perfectly know and appropriate those truths, for the investigation of which he hath a special vocation." Though "we must all turn specialists soon," Marsh could not help "envying scholars of the time when Crichtons were possible" (a reference to "the Admirable" sixteenth-century Scottish prodigy of learning).[83]

For Marsh the last true Crichton had been Alexander von Humboldt, whose learning "embraced the whole past history and present phase of every branch of physical research, and which was moreover graced with the elegances of all literature and dignified with the comprehensive wisdom of all philosophy." In noting how Humboldt's insights derived from "the mutual interdependence of apparently unrelated knowledges,"[84] Marsh might have been tracing his own trajectory.

Best placed to popularize this vital composite Marsh judged the "new geography" of Humboldt, Carl Ritter, and Élisée Reclus, which was not so much a science as a bond among the sciences:

No other [realm] has so many visible points of contact with the material inter-ests of human life, no other deals with subjects whose practical importance is so constantly forced upon our notice, no other so essentially consists in the in-vestigation of the relations of action and re-action between man and the me-dium he inhabits.

Not dependent on "accuracy of measurement or minute quantities," and free from forbidding nomenclature, geography "appeals to the widest circle" of thinkers. Because "its special phenomena are facts of hourly and universal observation," and it had an "intimate connection with the well-being and social progress" of all peoples, geography deserved a high place among "the *Moral Sciences*."[85]

Moreover, since geography "has not yet become a branch of for-mal instruction," as Marsh put it in *Man and Nature*, "those whom it may interest can, fortunately, have no pedagogue but themselves"; *fortu-nately*, since "self is the schoolmaster whose lessons are best worth his wages." For that reason Marsh stressed that he wrote not for special-ists—"professed physicists"—but for a general audience of "educated, observing, and thinking men." And his purpose was "rather to make practical suggestions than to indulge in theoretical speculations." As noted above, he condemned the "rage for causative speculation" he felt then common, especially among Germans.[86]

Man and Nature singled out for rebuke the false precision feigned by statistical scientists to impress or deceive—a "pretended exactness no better than an imposture," when accuracy was chimerical. Numbers then cloaked distortions of truth—as in a London study reporting "the mortality among manufacturers of children's toys [of a horrendous] fifty per cent. per annum. . . . There were *two* manufacturers of such toys, one of whom had died during the year, and this was the sole founda-tion for the conclusion that this manufacture was so destructive to hu-man life." Exactitude was very well in its place—and Marsh had shown himself no mean devotee of precise and accurate measurement—but he regretted its becoming de rigueur in all arenas. This tended to discour-age the study of fields like geography "as unworthy of cultivation be-cause incapable of precise results."[87]

As a linguist, Marsh knew that numbers in many languages seemed misleadingly to imply precision, when only some great but indefinite quantity was intended: 10,000 meant "myriad" to a Greek, 600 to a Roman, 40 or sometimes 15,000 to an Oriental. The last figure had been borrowed to apply in Rome: "Thomas de Quincey seriously informs us, on the authority of a lady who had been at much pains to ascertain the *exact* truth, that . . . the Vatican contains fifteen thousand rooms"—which would require a nine-acre building ten stories high.[88]

Marsh continued to think "it is better to taste a variety of scientific and literary viands than to confine ourselves to a single dish of stronger meat." Better both for individuals and for the body politic. Just as someone "who has mastered the ordinary use of a wide vocabulary [was often] a better speaker and even writer than the profoundest theoretical grammarian," so did diffusion of a wide range of knowledge make better citizens. Thanks to popularizers like Humboldt, "the aristocracy of science is divesting itself of its prerogatives and privileges," Marsh supposed in 1868; "the world of mind, like the world of politics, is becoming a democratic republic." Beset since then by more daunting obstacles to public understanding, science has yet to fulfill Marsh's uncharacteristically hopeful augury. Instead science has become "an aristocratic affair," in a poet's phrase, "not communicable to all men, nor to most men."[89]

Marsh also aligned scientific with spiritual advance. Unlike Schiller and other Romantics who blamed scholars for dimming nature's awesome mysteries, Marsh allied scientists with poets. Far from fearing that growing familiarity with the laws of nature made life more prosaic, Marsh like his mentor Humboldt saw new knowledge enhancing the beauty of the cosmos. "Wherever modern Science has exploded a superstitious fable or a picturesque error, she has replaced it with a grander and even more poetical truth." To Marsh, "the true story of the sage philosopher" (Franklin) chaining the thunderbolt with his kite was more stirring than the fiction of Jupiter's forked lightning, the new science of weather more thrilling than fables of Aeolus's fickle winds. Loftier than the skies of Aristotle and Plato was modern astronomy's cosmic remoteness, where "the rays reflected from the scenes of man's earliest labors have not yet reached," and whence an observer might "even now witness the rearing of the pyramids, the founding of the walls of Rome, . . . the triumphs of a Caesar."

Natural science had "vastly increased the wealth of imagery at the command of the poet," and popular "acquaintance and sympathy with nature [was] of infinite service to the cause of art" (notably landscape painting). At the same time, science was spread "by obliging its expounders to clothe it in more attractive forms." Fears that the cultivation of physical science fed materialism, skepticism, and sensualism seemed to Marsh chimerical.[90]

He considered himself a professional in no field, competent only in linguistics. By today's standards, Marsh was of course untrained and amateur. What was professional about his work was his life. He balanced avocation with occupation, work with pleasure, scholarship with experience, theory with practice, and activity with contemplation, to a degree remarkable even in his own age, so much more eclectic and less compartmentalized than ours. His virtue as a scholar was to see scientific and social issues from ever fresh viewpoints. All his work—linguistic, historical, environmental, political—synthesized wide reading and personal observation into provocative analysis.

Yet Marsh dismissed this immensely fruitful activity as secondary to life's main aim—self-understanding and self-improvement. Being, not action, he held in his 1857 Cambridge Phi Beta Kappa address, animated the highest human powers. Enlarging on this theme, in 1860 he adjured West Point's graduating cadets to remember that "the great end of human life is not to *do*, but to *be*." No individual's material or intellectual contributions to society were essential; "the ablest man can only do that which would probably be as well done without him." But the perfected human being was irreplaceable; "he who wisely and conscientiously employs the best methods of culture . . . may *be*, what the fewest only *become*—a fully developed, symmetrically expanded, and comprehensively instructed man."[91] All practical and pragmatic effort seemed to Marsh no more than a means toward this moral end, just as he saw the conquest of external nature merely a means toward a greater conquest—that of the far more intractable world within each of us.

18. Prospect: Reforming Nature

WHAT RELEVANCE HAVE MARSH'S ENVIRONMENTAL insights today? Why bother with this diplomat-linguist's 1864 *Man and Nature*, a book "full of facts that have since been shown to be erroneous [and] conclusions that went sour," as even an admirer said a century later, all in all, "a doctrinaire, maudlin, cant, overripe, moralistic cough-drop of a book."[1] Marsh's renowned environmental intuitions have been overtaken by time and superseded by other issues—issues that evoke alternative responses to still graver dilemmas. To some moderns, the Marsh depicted in my 1958 biography came across less as a polymath prophet than as a cranky Yankee.

Yet it is above all for his crucial role in environmental history that Marsh's life warranted retelling here. The lapse of half a century had done more than turn up much new data. His pioneering role had to be reexamined through the lens of new ideas, new attitudes, new insights on how humans impact their world and strive to mend what they seem to derange.

Moreover, a new account of Marsh had to address readers whose own assumptions about nature and human nature differ markedly from those of fifty years ago, when the geographer Carl Sauer and the historian Merle Curti first encouraged me to probe the making of *Man and Nature*. Marsh's very title underscores the change in decorum that, for many, makes even later classics hard to access. A student recently borrowed Clarence Glacken's seminal study of nature and culture in Western thought, *Traces on the Rhodian Shore*. She soon returned it. "I can't read this," she said; "the sexist language is too repulsive." If feminist propriety puts a 1960s Glacken beyond the pale, how can one impart an 1860s Marsh? Known in his day as a champion of women's rights, Marsh nonetheless couched his ecological inquiry in terms of that "great question, whether man is of nature, or above her," and concluded that "wherever man fails to master nature, he can but be her slave."[2] I must rely on readers to savor such phrasing in its historical context.

What notions of nature and culture were current when *Man and Nature* first appeared? How did they differ from subsequent views? The contrasts are crucial to our understanding. The ailments of the earth in the year 2000 seem as unlike those of the 1950s as the 1950s found Marsh's mid-nineteenth-century cosmos—unlike in the kinds and extent of perceived risk, unlike in outlooks for curtailing and repairing damage, unlike in modes and costs of reform, and above all unlike in assigning cause to natural or to human agency.

Marsh's own comparisons across time and space were fruitfully provocative. Vital for his alertness to human impacts, and the need to amend them, were the vivid parallels he drew: parities and disparities between ancient and modern environments, Old World and New World use and abuse of nature, Mediterranean and American reactions to degraded landscapes. Similar physical processes; differing cultural responses. The task Marsh set himself was first to account for the differences and then to bridge them, in order to engender awareness and spur reform on both sides of the Atlantic.

Every society's way of treating its landed legacy is spun in a web of custom; tradition resists changes that new circumstances ceaselessly mandate. Hence Americans schooled in a rhetoric of wilderness conquest found it hard to recognize the ill-effects of that conquest. In detailing the longer and more devastating Old World saga of unwitting destruction, Marsh sought to open Americans' eyes to their own impact on the fabric of nature. In this he succeeded far better than he had expected, though ensuing reforms were inadequate to stem the ever more rapacious gutting of resources. Since Marsh's time, Americans like Europeans have grown more aware, anxious, and pessimistic about slowing, let alone halting and reversing, processes of degradation. Tracing such concerns, I aim in this chapter to assess environmental views now current in the light of our precursors' views and deeds.

Marsh urged the New World to heed evils endured and reforms instituted in the Old. Similarly, we may learn from the triumphs and setbacks of those who earlier ravished and later sought to restore nature, pursuing both goals on the basis of perspectives quite unlike our own. We cannot regain the faith in technology and community that then made reform efforts seem radiantly worthwhile. But we may benefit from trying to understand what generated and sustained their faith.

Marsh in the 1860s castigated earth's greedy, reckless despoilers in invective as searing as Rachel Carson's in the 1960s. But *Man and Nature* spurred no such envenomed reactions as did Carson's *Silent Spring*.[3] No planter or hunter, entrepreneur or industrialist challenged Marsh's accusations as erroneous, defended the gutting of forests and the slaughter of wildlife as essential, or dismissed his ecological jeremiads as traitorous hysteria.

Why were there no such rebuttals? Because Marsh's world was in many ways unlike Carson's a century later. In the first place, Marsh framed his warnings within an accepted goal of environmental conquest; he disputed not the desirability of exploiting nature but the bungling way it was being done. Although Marsh's ecological cautions were radically new, their underlying philosophy inspired broad agreement. They stressed biblical and Enlightenment premises that mankind's mission was to subdue and domesticate nature.

Second, none took personal offense: for all Marsh's moralistic censures ("Joint-stock companies have no souls; their managers . . . no consciences"[4]), he inveighed against mankind in general, not against particular entrepreneurs. Third, no media broadcast his warnings worldwide or prompted rejoinders from accused malefactors, who are assailed in books like *Silent Spring* and by groups like Greenpeace. Fourth, Marsh's practical correctives—reforestation, controlled grazing, stabilizing sand dunes, monitoring environmental impacts—seemed to entail few economic burdens and to require no draconian remedies. Enlightened self-interest underwrote most of his prescriptions.

Fifth, no one in Marsh's day quarreled with his emphasis on stewardship, though few exploited properties in that spirit. So canonical was the credo of future good that the most avaricious get-rich-quick resource strippers deployed its rhetoric. Yet the industrial pillage and conspicuous waste of the late nineteenth century roused disquiet. Many feared the depletion of essential national reserves that might cripple productive enterprise—hence the zeal for forestry reform, water management, and stewarding public lands in the common interest.

The reforms then sought seemed much easier to achieve than their counterparts do now. Pioneer conservers were not preservers; they asked for prudent resource use, not for no use. Entrepreneurs often stymied or breached government controls, but they could not combat "sustained management" with the open fervor with which some developers today oppose "protection" as locking up resources. Moreover, many welcomed conservation as cheap and easy—cheap because public lands could be sold or leased to pay for it, and easy because technology and public agencies were trusted as competent and honest. Faith in applied science harnessed to public good was then at its peak.

In fact, conservers adopted only half of Marsh's analysis and a fraction of his reforms. Gleaning what they wanted from *Man and Nature*, they welcomed his positive messages—reforms that were clear-cut, widely beneficial, and allied with productive growth. They ignored or forgot his negative admonitions—watershed protection, inviolate woodlands, irrigation cautions, and warnings of irreparable damage from unintended impacts. Few took seriously Marsh's graphic preview of an earth as barren as the moon, threatening even the extinction of mankind.

Indeed, the effect of *Man and Nature* in some quarters was not to restrain but to rekindle optimism. For example, railroad promoters, land speculators, even scientists misread Marsh to support the fantasy that tree planting would drench the western plains, converting the Great American Desert into a well-wooded and hence well-watered land. "Marsh's remarkable synthesis of European literature, experience and observation was bent to give scientific respectability to the writings of rainmakers," notes Michael Williams; *Man and Nature* was "pillaged and distorted" for proof that afforestation increased rainfall.[5] Even now some encumber Marsh with the view that trees bring rain.[6]

To the contrary, Marsh saw no evidence that forests increased rainfall. *Man and Nature* stressed the uncertainty of meteorological knowledge. Rebutting a rain-follows-the-plough theorist in 1873, Marsh termed it "improbable" that forests exercised "any appreciable influence on the total amount of precipitation." He shared his friend Henry Yule's dismay that ignorant advocates "always begin by confounding two things, viz the effect of denudation on *rainfall*, which may be a myth, & its effect on the disposal of the water after it has fallen which is as clear as any fact in nature." In like fashion, Marsh's irrigation warnings were ignored by promoters who crossed the Atlantic and begged to be shown the wondrous benefits, but never the ominous side effects, of water management in Italy.[7]

*T*echnocratic optimism prevailed well into the new century. To be sure, reverence for wild nature gained ground, notably in America. Environmental concern now bifurcated into two camps: one primarily economic, the other inspirational. The former, dominant among policymakers, stressed resource management for present and future use. The latter focused on wilderness aims and areas that at the time impinged little on resource use—though when they did, as in the 1902–16 battle over damming Yosemite's Hetch Hetchy valley, antagonists did not spare their vitriol.[8]

But most who thronged to spectacular Yosemite admired wilderness as a source of occasional refreshment, not as a permanent way of life. The received view of untouched nature was that of T. H. Huxley and Herbert Spencer, who termed it ruthless, cruel, savage, wasteful. "Visible nature is all plasticity and indifference," agreed the philosopher William James; "to such a harlot we owe no allegiance." In taming the wild, humanity not only improved its surroundings but advanced morality.[9]

These social Darwinians echoed Marsh's notion of human primacy but rejected his vision of nature as an ally to succor rather than an enemy to subjugate, and ignored his fears of irreversible exploitative damage. Even conservationists saw environmental impacts as mainly benign; any injurious side effects could be cured by better management. Although nature was ever more drastically altered, most changes still seemed improvements, and few doubted that the triumphs of technology would soon rectify any damage. Earth scientists convinced that natural powers vastly exceeded human agencies buttressed this optimism. Mankind could not seriously harm the globe. However potent their tools, humans remained a minor geological force; the gravest man-made disasters were just tiny, temporary setbacks in progressive mastery of an infinitely resourceful earth. Secure in nature's overriding control, environmental determinists joined devotees of progress whose mission was to go on engrossing nature for human use.[10]

Typical of prevalent faith in technology was Sigmund Freud's 1929 accolade to the conquest of nature. "We recognize that a country has attained a high state of civilization when we find . . . everything in it that can be helpful in exploiting the earth for man's benefit and in protecting him against nature": the course of rivers is regulated, the soil "is industriously cultivated," minerals are "brought up assiduously from the depths," and "wild and dangerous animals have been exterminated."[11] How quickly and sharply our views change! Just one life span earlier, Marsh applauded a vision of mastering nature much like Freud's but cautioned vehemently against its manifold risks. Just one life span later, many now doubt Freud's progressivist premise and spurn his human-centered hubris as blindly arrogant.

Back in the 1930s, however, not even Dust Bowl calamities soured most conservationists against conquest. Faith that resources however maltreated were inexhaustible survived those and other disasters. The *past* was blamed; the present knew better. Successes like the soil conservation and shelter belts of the 1930s and 1940s seemed to show that heedless waste was a thing of the past. Although technology since Marsh had sped environmental change, most human impacts were still adjudged purposeful and beneficial, manipulation of nature ever more rational and farsighted.

Marsh's prescient ecological warnings were hailed at a 1955 Princeton symposium, "Man's Role in Changing the Face of the Earth." But few there heeded his warnings. In taking stock of "what man had done and was doing," most instead made light of the ill effects of human impacts. Optimism prevailed, along with complacency in the face

of what today seem massive threats. Humans could have no "significant" effect on global climate; nuclear-fission residues were wholly
benign; wise use would rebuild battered soils. Soil conservation was
already restoring land as no previous generation would have thought
possible. Why worry about such bogeys? Past fears had always come
to naught. In sum, environmental impact scared only scientific dim-
wits and crackpots. Doomsters like Fairfield Osborn (*Our Plundered
Planet*, 1948) and William Vogt (*Road to Survival*, 1948) were rebuked for
"lamentable lack of faith in man's ability to control his future with new
technology." [12]

"Man's Role" did not fairly reflect widespread 1950s qualms. Not
long since, Hiroshima had unleashed dread of global annihilation.
Many feared that "tinkering with nature" through antibiotics, destroying "uneconomic" species, introducing foreign predators, even prolong-
ing human life spans might wreak biological havoc. These gloomier
perspectives were consonant with the equilibrium model of ecology
developed by Frederic Clements in the early decades of the twentieth
century. Clements strongly echoed Marsh's view of nature—but not of
man. He and his followers saw nature most fruitful when least altered.
Ecosystems left undisturbed would gradually attain maximum diversity
and stability. Extractive spoliation thwarted the fruition or shortened
the duration of this beneficent climax; technology did not improve nature but corrupted it. [13]

Such evils, however, emanated only from so-called advanced cultures. Exempt from blame were primitives whose "nurturant tribal ways,
integrative communitarian values, and rich interplay with nature" respected environmental balance, in a later reformer's words. "*Harmony
with* rather than *exploitation of*" nature was not only "a guiding principle
for the Paleolithic mind [but] remains a cardinal commitment among
modern aborigines." Primitive peoples, like nature, existed in balanced
stability. Heeding their wisdom might help us regain "natural" environments fit to live in and to hand on to future generations. [14]

This organismic, biocentric mystique became an almost religious
tenet. Aldo Leopold's famous "Land Ethic" of 1949—"A thing is right
when it tends to preserve the integrity, stability, and beauty of the
biotic community. It is wrong when it tends otherwise"—became a token of right thinking even in government agencies. For many it is still

conservation gospel. The virtues of stability and passive noninterference mirrored reformers' views about human nature, too. The ecological utopia was—and is—a moral order. To "replace the chaos of a world torn by human greed and voraciousness with a well-ordered moral universe," we are adjured to limit our numbers, machines, and consumption habits.[15]

Ecological science had, to be sure, long since disowned much of the Clementsian paradigm. Although visions of enduring mature climaxes persisted in the writing of Eugene Odum even through the 1960s, by then it was mainly *non*-ecologists who extolled equilibrium, stability, and noninterference. Yet environmental writing still deploys these perspectives, and the succession-to-climax model still rules much biological thought. Nature is cast as normative good, technological man as evil destroyer.[16] This all but reverses the perspective of Marsh, who had exalted humanity's role as both conqueror and "co-worker with Nature in the restoration of disturbed harmonies."

In sum, the concept of nature as a finely balanced but unfinished fabric to be perfected by human ingenuity gave way, by the 1960s, to the view that technology debased nature and endangered its benefits. Interference was demonized, wilderness venerated. In the state of nature envisioned by the Enlightenment, by Marsh, and by his technocratic successors, rational managers would cultivate an ever more artificial environment. In the state of nature sought by twentieth-century "ecological" reformers, human impact would be curtailed until the environment regained stability.[17]

Many 1950s impact fears embodied animist or primitivist outlooks. As one celebrant of rural virtues put it, those who disregarded nature's warnings would suffer her revenge, for "nature always has the last word." Degeneration from previous harmony was commonly presumed: until the "immemorial unity . . . between man and nature is once again an everyday conviction, we shall continue to race toward material and spiritual disaster." Both human error and nature's purpose were invoked, lest "the taint of Man's World infect Nature beyond the limit of forbearance."[18]

Apprehensions about ecosystem loss, radioactive waste, and genetic mutations, though cushioned by lingering faith in science and technology, were already widespread fifty years ago. Their absence

from the "Man's Role" symposium highlights the disparity between expert and public opinion. Aiming to enlighten the future, "Man's Role"
envisaged no role for a concerned public, whose apprehensions this
"Marsh-fest" nowhere addressed. It reached out not to people in general but to "scholars in oncoming generations." [19] To Marsh himself, ever
eager to trumpet warnings among a broader public, this would have
seemed a signal defect.

Since the 1950s, public voices have gained influence in environmental discourse. Populist pressure has triggered research, monitoring,
legislation, global accords. The rise of environmental advocacy reflects
three postwar changes. Cogent warnings of the limits of spaceship earth
and the fragility of the biosphere have energized wide concern. Horrific mishaps have led to demands for environmental safety even at the
cost of economic productivity. And personal goals have persuaded
more people to take active reform roles. Many today feel environmental issues are too important to be left to experts—especially since experts have lost their aura of omniscience and integrity.

Awareness of human impact is no steadily progressing saga, however. Understanding and concern seem more episodic than cumulative.
Transient alarm triggered by environmental crises soon dissipates.
Chernobyl dominated global consciousness for some months; news of
most impact disasters endures only a few weeks. Between crises, levels
of general concern plummet.

Confusion and contradiction in attitudes toward nature shroud
long-term trends in doubt. The old view that "nature's bounty is infinite, humans are omnipotent" has been jettisoned, an American survey
shows, for the belief that "humans must live in harmony with nature in
order to survive." But most in the 1980s still believed that mankind was
created to rule over the rest of nature and that "technology got us into
environmental problems, but technology will get us out." To curb toxic
pollution, protect endangered species, and limit growth for the sake of
environmental health is conventional wisdom—but support for these
aims is weak and volatile. Public fears of impact are hard to convert into
electoral policy. "Unless God opens an ozone hole directly above the
Palace of Westminster, or melts the polar ice-cap just enough to bring

the Thames lapping round the chair-legs in the Members' Bar," concludes a critic, "'Green' issues will never rank with the central concerns of British politics."[20]

Yet the surge of environmentalism is patent. Doom-mongers and media scares have converted many. And impact issues are increasingly seen as global and interrelated, complex and unknowable, long-lasting and irreversible. None of these perspectives is wholly new—some echo *Man and Nature*. But only since the 1950s have they come to dominate public fears and to pervade environmental debate.

Environmentalists today think their concerns uniquely paramount. Yet Marsh and his disciples thought they faced no less dire problems. Consider his portent in *Man and Nature*:

There are parts of Asia Minor, of Northern Africa, of Greece, and even of Alpine Europe, where the operation of causes set in action by man has brought the face of the earth to a desolation almost as complete as that of the moon. . . . The earth is fast becoming an unfit home for its noblest inhabitant, and another era of equal human crime and human improvidence . . . would reduce it to such a condition of impoverished productiveness, of shattered surface, of climatic excess, as to threaten the depravation, barbarism, and perhaps even extinction of the species.[21]

That tirade was penned in the 1860s. What human impacts then menaced? Deforestation, overgrazing, erosion, flooding, and desiccation. These still haunt us—indeed, they are perhaps even more threatening[22]—but they are not now our prime concerns. New threats continually surface to overshadow old ones: in the 1960s pollution, chemical poisons, and the Bomb; in the 1990s acid rain, stratospheric ozone depletion, global warming, nuclear waste; in 2000 cloning and genetically altered crops.[23] Yet today's perils differ neither in imminence nor in apocalyptic immediacy from yesterday's. Past no less than present Jeremiahs enjoined instant reform against impending doom.

What is new are menaces that cannot be seen. The effects of soil erosion, even of DDT, soon were patent to any observant eye, but today's risks are clear only to arcane experts themselves often at odds on the magnitude of perceived risk.[24] Those afflicted by toxins that cannot be tasted, touched, smelled, or seen, like folk at Bhopal or Three Mile Island, feel petrified by poisons that "slink in without warning," in Kai

Erikson's phrase, "and then begin their deadly work from within—the very embodiment of stealth and treachery."[25] Bereft of confidence in a fruitful and manageable environment, victims lose faith in both the good sense and the good will of officials and scientists, themselves often baffled and impotent.

Longer time lapses between cause and effect make it ever more difficult to ensure precautions, assign liability, or provide compensation. "Years after the catastrophe," in Ulrich Beck's telling example, "the injured of Chernobyl are not even all *born* yet." At Yucca Mountain, Nevada, the American government plans to bury nuclear waste in containers guaranteed leak-proof for ten thousand years. But even assuming a civil stability without precedent, this span of time would be far too brief. For radioactive carbon-14 remains lethal in air or groundwater for up to a million years.[26]

Science is feared and resented both as remote and authoritarian and because its unintended consequences seem ever more ominous. Once luminous innovations now cast dark shadows. Nuclear power was but yesterday a glittering technological panacea. Today, health and safety fears all but throttle the nuclear industry in many lands. Science appears to be ever more costly, technology's benefits more dubious, environmental reform more difficult. These disillusionments induce despondency, impotence, and *après nous le déluge* escapism.[27]

The felt complexity of these problems hampers efforts to resolve them. Environmental changes are more and more seen to involve the whole interrelated biosphere. The most worrisome impacts—loss of biotic resources, carbon-dioxide buildup, ozone-layer depletion, toxic waste disposal—require global management that overrides local and national interests. The world is still a long way from becoming a single community, but world leaders know they must treat the globe as a single ecosystem.[28] That lichens in Lapland and Alaska contain radioactive iodine from Hiroshima and Nagasaki shows a world united by peril.

Nature's interconnectedness is now a central tenet not just of mystics but of mainstream science. A generation ago few biologists drew causal connections such as those routinely drawn today, between population pressure and rain-forest depletion, acid rain and heavy runoff, excess carbon dioxide and climate change. Echoing Marsh's concerns, environmental scientists now stress such linkages, pooling a wide spectrum of knowledge in interdisciplinary journals such as *Global Planetary Change* and *Global Environmental Change*, both launched in 1990.

But a wide spectrum of ignorance also persists. Marsh would have applauded the dawning awareness that we can never know enough, that developing technology always outruns our ability to monitor its impacts. Constitutionally "incapable of weighing their immediate, still more their ultimate consequences," as Marsh put it, we are bound to affect nature in ways that we cannot ascertain or assess. Many human impacts seem minuscule at the outset. But it is as unsafe now as in Marsh's time to assume "a force to be insignificant because its measure is unknown, or even because no physical effect can now be traced to it."[29]

Conservationists forgot Marsh's maxim during the ensuing century of managerial hubris. But radiation and toxicity issues of the 1950s and 1960s revived doubts about scientific omniscience; by the 1970s and 1980s biosphere derangements presaged incalculable damage from impacts too gradual to discern within a human life span. Because impact data typically come too little or too late for therapeutic reform, new cognizance often raises more doubts than it resolves.

The public meanwhile grows more fearful of dangers that mount with time yet can be detected only after precautions are futile. Scientists are berated for failing to predict adverse environmental effects with speed, precision, and certainty. Above all, people find intolerable even remote possibilities of long-term, low-level exposure that might cause cancer or birth defects.[30] Most fearsome is that some such impacts may already be irreversible, dooming earth's ecosystems. Current fears of extinguishing human life, perhaps all life, echo Marsh's 1864 warning cited above.

But that warning is all that many now know of Marsh's trailblazing role. Environmental texts pay almost obligatory homage to *Man and Nature*, then mention it no more. Marsh is now saddled with the managerial hubris of the 1955 symposium in his honor. As shown, the bent of "Man's Role" was optimistic, utilitarian, technocratic, manipulative toward nature, elitist in mode of reform. There's little to worry about; leave it to us experts. This mind-set is anathema to today's reformers, who find Marsh more an impediment than a role model for their cause.

That cause is idealistic, aesthetic, wilderness-bent. It consigns the resource-use philosophy of Marsh and his "gospel of efficiency" successors (Gifford Pinchot, Theodore Roosevelt) to complacent materialism. Marsh becomes "the founder of the interventionist, managerial school of conservation, which takes our disruptive presence in the natural world for granted," as opposed to the hands-off-nature ethos associated with Thoreau and Muir.[31] This ethos is de rigueur as popular ecology and has gained philosophical and religious support as well. Indeed, references to Thoreau and to Muir vastly outnumber those to Marsh even in the Science Citation Index.[32]

Thoreauvian litterateurs perfunctorily concede Marsh's "aesthetic dimension" only to vilify it as "pompous and sentimental." Defaming *Man and Nature* as "the materialistic obverse of *Leaves of Grass*," literary historian Cecelia Tichi notes that Whitman like Marsh thought "the landscape was made for man, not man for the landscape." This puts both men beyond the pale of today's biocentric poetics. Thoreau is preferred to Marsh both as a stylist and because "Thoreau's vision of . . . reciprocal interchange" is in tune with modern environmentalism—unlike "Marsh's more managerial . . . techno-fix."[33]

But this environmental divide is a latter-day construct. No nineteenth-century figure in either supposed camp would have condoned it. The same Thoreau who declared "in Wildness is the preservation of the World" shunned actual wilderness, finding nature devoid of man repellent and fearsome, "savage and awful" Mount Katahdin "vast and drear and inhuman."[34] Nor was he the reclusive devotee of primitive simplicity often painted; *Walden* aside, Thoreau admired the farmer who "displaces the Indian even because he redeems the meadow, and so makes himself stronger." He extolled the tools of man's "most important victories . . . the bushwhack, the turf-cutter, the spade, and the boghoe, rusted with the blood of many a meadow, and begrimed with the dust of many a hard-fought field." Like Marsh, Thoreau was given to rhetorical self-contradiction, now holding that "man's improvements . . . simply deform the landscape," later in the same essay lauding a quasi-vacated locale, "some retired meadow [where] the sun on our backs seemed like a gentle herdsman driving us home at evening." This Elysium is not pristine; it is pastoral, suffused with human as well as natural history.[35] Reading present credos into past tastes, today's Thoreauvians misjudge both.

Thoreau in fact had much in common with Marsh. Both shared their mutual hero Humboldt's passion for empirical natural history. Both obsessively measured natural phenomena. Both belonged to Spencer Baird's collecting network, Thoreau happily sending fish from Massachusetts, Marsh from the Middle East. Both were mesmerized by patterns of forest succession and modes of seed dispersal by birds and squirrels. Both felt environmental protection required curbs on private property and urged saving enclaves of "primitive" forest for conservation, instruction, and recreation. Thoreau's alarm at the ill effects of human agency paralleled Marsh's, as in "Thank God, men cannot yet fly, and lay waste the sky as well as the earth!"

Like Marsh, Thoreau prescribed a balance of tilled land, meadow, and forest. He hankered after no undefiled Eden. Though shaped by centuries of human violation, Thoreau's beloved Concord woods were for him "no less 'natural,' no less rich with the seeds of redemption" than any virgin wilderness. Indeed, the wildness Thoreau adored was no untouched terrain but a process of growth and decay, conquest and abandonment, in scenes made by both natural and human agency. Thoreau, no less than Marsh, routinely joined the two realms, writing of an artificial pond stocked with fish and lilies, of forests shaped by squirrels and axes.[36]

Thoreau died in 1862, too soon to read Marsh's book. But Muir keenly admired Marsh, drawing extensively on *Man and Nature* to secure Sierra soils and forests as watershed protection for Yosemite. Although Muir denied that the earth was made for man, it was for men's spiritual salvation that he sought to save wild nature. And Muir gratefully embraced Marsh's "economical" arguments to justify his "poetical" ones.[37]

To such links "poetical" environmentalists today seem strangely blind. Having "found no substantive references" in Marsh's papers, literary historian Lawrence Buell terms the utilitarian Vermonter no "more than idly interested in Thoreau, if that." Buell might be excused for failing to find the diary entry on Marsh reading aloud the "exquisitely poetic" Thoreau.[38] But how could he miss the manifold references to Thoreau in *Man and Nature?* Marsh revered Thoreau as "an *observer* of organic nature, in the old religious sense," remarking that "few men have personally noticed so many facts in natural history accessible to unscientific observation." Praising *Maine Woods,* Marsh was only surprised that Thoreau had never, until bivouacking there, seen "that common and very striking spectacle, the phosphorescence of decaying wood." Thoreau's essay "Autumnal Tints" gave Marsh a springboard to contrast fall colors in Alpine France with New England. In Dauphiné only small shrubs turned color; in New England "the foliage of large trees is dyed in full splendor, hence the American woodland had fewer broken lights and more of what painters call breadth of coloring."[39]

Marsh's delight in nature was no less ardent than Thoreau's and Muir's. *Man and Nature* notes how species are often transplanted by accident—"after the wheat, follow the tares that infest it." One recent inadvertent import for which Marsh was "not sorry" was a grainfield weed, the European scarlet wild poppy; "with our abundant harvests of wheat, we can well afford to pay now and then a loaf of bread for the cheerful radiance of this brilliant flower." Marsh upbraided Ruskin for the very insensitivity of which he himself is now accused. "Ruskin is . . . always in the wrong," he warned Charles Sprague Sargent. "Total want of sympathy with nature incapacitates him for the observation of natural objects & processes. . . . Sometimes indeed, the splendor of a landscape overwhelms him, so that he forgets his foolish theory [see Chapter 3 above], but in general he treads closely in the footsteps of the tailor who saw in Niagara only a fine place to sponge a coat."[40] To cast Marsh's own attachment to nature as one of mere utility is nonsense.

So is the delusion that Marsh took no interest in nature preservation.[41] In fact Marsh was an early and active advocate of setting aside part of the Adirondack wilderness as parkland. *Man and Nature* lists "poetical" as well as "economical" reasons.

Some large and easily accessible region of American soil should remain, as far as possible, in its primitive condition, at once a museum for the instruction of the student, a garden for the recreation of the lover of nature, and an asylum where indigenous tree, and humble plant that loves the shade, and fish and fowl and four-footed beast, may dwell and perpetuate their kind.

"Only in the unviolated sanctuaries of nature," wrote Marsh, out-Muiring Muir, could one gain "that special training of the heart and intellect" indispensable to the human spirit. Twenty years later Marsh profoundly regretted the impending "total destruction" of Adirondack forests.[42]

In severing an impetus they admire from concepts they deplore, today's environmentalists impose their own apartheid on the past. This disserves both history and their cause. Earlier views were far more complex and less polarized than many now suppose. The multivalent romantic values variously expressed by Marsh, Thoreau, Muir, and Frederick Law Olmsted, as Lewis Mumford noted, together gave rise to the politics of state and federal park reserves.[43]

Men of science in his day admired Marsh as a practical reformer. Marsh knew himself to be no scientist, rather a popularizer of science. But for all his reform ardor, he remained committed to empirical evidence and scholarly rigor—with footnotes.

Marsh thus repels some for being an economist rather than a poet, some for holding humans superior to other creatures, some for endorsing the biblical injunction to subdue the earth. Still others are dismayed by his belief that stewardship not merely restores but *improves* nature—that the artificial forest outshines the natural, the introduced crop outranks the native. Many mistrust Marsh's reliance on human agency, in their view delinquent by moral fiat.

Marsh's role is scanted by some environmental historians, too. It has become fashionable to dismiss Marsh in favor not only of free-spirited romantics like Thoreau and Muir but of unsung hoi polloi on the mainstream's margins. Marsh admirers from the 1920s through 1950s lamented he was forgotten;[44] some detractors today contend that he gets far too much credit. Just as conservation began as an elite enterprise run by and for the wealthy and the well educated, so, such critics claim, Establishment scholars wrongly attribute its pioneering insights to men like Marsh. In reality, these populist revisionists contend, we owe environmental reform to notions about nature and community long held by common farmers and herders and, indeed, by voiceless underlings all over the globe. In this view, Marsh simply copied conservation insights and observations already widely disseminated.

Here Marsh emerges as a mouthpiece of WASP America who saw eye to eye with the scholarly and political elites of his time, notably the Washington power structure, Boston Brahmins, and their patrician European counterparts.[45] *Man and Nature* served to justify managerial elites' draconian restrictions on resource use, further disempowering ordinary farmers and fishermen, peasants and pastoralists. Yet it was just these folk, runs this saga's populist version, who were most innately conservation-minded. An instinctive sense of community and of oneness with nature had fostered, among myriad common people, habits of environmental care that Marsh merely collated, often for coercive and autocratic ends.

Ordinary rural folk of northern New England, in Richard Judd's estimate, anticipated and nourished Marsh's insights. Persisting faith in a balance between nature and culture, traditions of resource sharing, and pride of locality spearheaded local reform in forest and fishery use and village planning, in what Judd chronicles as communal quests for livable and equitable landscapes. Hence Marsh's admonitions at Rutland in 1847, and later in *Man and Nature*, were more derivative than original. In Judd's opinion the true roots of environmental reform were neither academic nor urban; they stemmed from rural residents, whose enduring visions of a democratic commons conjoined perceptions of nature as commodity and as home.

This analysis has some merit; Marsh himself credited many of his intuitions to observant neighbors. But the evidence is overwhelming that his principal insights were original. The early popular concerns cited by Judd are largely fears of timber and fuel shortage. Only the most scanty ecological awareness antedates Marsh's own writings, and most later local sources trace directly back to his writings.[46] Unpalatable as it may be to patrons of folk virtue, environmental wisdom began with, and was for long confined to, an educated elite. In Wallace Stegner's words, "Marsh's warnings went unheeded except by a concerned few, upper-class bird-watchers probably, the kind who are generally well ahead of the mass public in these matters."[47]

Another Marsh put-down locates the origins of conservation not in Europe or North America but on their colonial margins. Long before *Man and Nature*, in Richard Grove's provocative thesis, farsighted officials on isolated islands—St. Helena, Mauritius, St. Vincent—learned resource management by adding non-Western folk insights to what they themselves observed. In such islands, and later in India and South Africa, French, German, and Scottish foresters melded European science with traditional modes of husbandry. Their prescient concern for protecting soils, water supplies, and local flora and fauna from unwanted impacts inspired later European and American forestry.

"Modern environmentalism emerged as a direct response to the destructive social and ecological conditions of colonial rule," Grove concludes. "Far more influential" than the work of Marsh and other Western scholars was "the experience of perceiving and countering deforestation and land degradation at first hand." In this reading, foresters like Hugh Cleghorn, John Croumbie Brown, and Dietrich Brandis were more Marsh's teachers than his pupils; *Man and Nature* served only to enhance their confidence in what they already knew from local inquiry. Western mainstream scholars, too prone to deem conservation a reaction to industrialization, have neglected these true pioneers, Grove maintains, wrongly according primacy to Marsh, Thoreau, and their successors.[48]

This fascinating claim is, however, ill-founded. Aghast at severe resource depletion in small islands and in subtropical India and South Africa, Europeans saw much that was new to them, just as Marsh did in Mediterranean and Alpine lands. But beyond their fears for local flora and fauna, there are few traces of ecological insight, none at all of Hindu, Zoroastrian, or Oriental holism. Grove's island sources refer only to dessicationist rain-making theories, to protecting forests for timber and fuel, and to soil exhaustion—all topics long presaged in Europe. Early Indian foresters were, like Marsh, indebted to Alexander von Humboldt and to Jean Baptiste Boussingault. But the importance of tree cover in retaining moisture and preventing excessive runoff—the crux of Marsh's cognition—is mentioned nowhere else.[49] To any reader of Marsh's correspondence with Yule, Cleghorn, Brandis, and Brown, Grove's contention that Marsh's insights "were of surprisingly little import" in India and South Africa will seem bizarre.[50]

"The roots of environmental ideas," a British historian concludes from Grove, lie "much further back in history than has ever occurred to the American practitioners, blinkered as they are by the nationalist obsession with George Perkins Marsh, John Muir, and Henry David Thoreau."[51] Indeed, some of these ideas go back to Theophrastus in the third century B.C., as Marsh himself noted. But to claim that Americans failed to appreciate Old World environmental ideas is to ignore their devotion to European mentors, from Buffon and Humboldt to Guyot and Agassiz, not to mention American followers of German silviculture and Italian flood-control engineering.[52] And to attribute the origins of conservation wisdom to oceanic islands and Oriental mysticism is without plausible foundation.

Another revisionist suggests that nineteenth-century environmental reforms, such as those lauded by Marsh, may have done more harm than good. Tamara Whited blames the misguided policies of technocratic foresters for much that went wrong in the Alps and Pyrenees. The French Forestry Service's draconian edicts of the 1860s drove local peasants out of forest pastures, with the aim of securing tree cover that would safeguard lowland settlements against devastating upland erosion. Pasturing, taking wood, clearing land for farming then became crimes. Inherited post-Revolutionary fears served to justify foresters' interventions in alpine hydrology. The dogma, stemming especially from studies by Alexandre Surell, much cited by Marsh, was that peasant deforestation caused flooding and landslides; replanting was mandatory. Only forests could stabilize these volatile mountains, and foresters saw every mountain pasture as potential woodland. But the foresters were wrong, Whited shows, on the absorptive limits of forested soils; wrong in seeking to afforest high and steep slopes; and wrong in anathematizing grassland, for grass consolidated many alpine soils better than trees. And on highly erosive slopes, stream-channel engineering was often a better solution than reforestation.[53]

With much of this critique Marsh would, in fact, have agreed. His initial delight in French forestry reforms was soon damped by their failures. He kept revising his own estimates of the proper mix of forest planting, pasturing, engineering works, and simply leaving things alone. But Marsh had no romantic faith in peasant practices. He thought folk myopia and improvidence major stumbling blocks to environmental reform. Yet his intimacy with Alpine life and landscapes made him fully aware of the injustices often done in the name of such reform, under the aegis of the state or of large landowners (see above, p. 281). Landlessness and lack of resources were what impelled most mountain folk to misuse milieus in which they were denied a permanent stake.

Thus some judge Marsh's environmental insights largely mistaken, others unoriginal or inconsequential. Still others fault the reform programs he fueled as technocratic, elitist, socially regressive, imperialist, or anthropocentric. Such criticisms seem to me unfounded or irrelevant.

Marsh's work might with more reason be held passé, obsolete, peripheral to modern times, modern viewpoints, modern environmental threats. To be sure, the issues Marsh tackled—deforestation, soil erosion, desertification—are still with us. Marsh's insights into their causes and consequences are widely agreed, his remedies still largely germane. But these are not the issues now uppermost in many minds. Fears of untoward impacts today focus more on global warming, depletion of the ozone layer, nuclear contaminants, a host of man-made pollutants. No one in Marsh's day was aware of any of these problems. Indeed, most of them did not then exist.

Many today share Marsh's concern about environmental impacts and salute his pioneering efforts to comprehend and contain their malign effects. But how we define and tackle these issues is now utterly different. The problems we face, our confidence in resolving them, our views of progress, nature, ecology, human transcendence, culture, and history—all have changed strikingly since Marsh's day, even since I first studied Marsh fifty years ago. It is not only the threats that are new, but our notions of what and whom to blame, how to cure present damage and cope with future risks, and whether we are apt to succeed. I cite four major changes:

♦ Marsh dealt with impacts for the most part visible and manifest within the span of a lifetime—"tokens of improvident waste" evident to "any middle-aged man who revisits his birthplace."[54] The impacts now most feared are largely invisible, their effects at first inconspicuous, their worst consequences often long-delayed, as with the untold future victims of Chernobyl and with nuclear wastes lethal for a million years.

♦ Marsh proposed reforms whose economic cost was thought, if not trifling, certainly less than the benefits soon to accrue from social stability, managerial thrift, and sustained yields—the replanted woodlot whose timber would in thirty years repay crop harvests forgone. The costs of today's most urgently needed reforms seem almost too staggering to implement: some warn that to curtail atmospheric contaminants as climate experts now advise threatens the entire global economy. No modern scholar or statesman would have the optimism or the hardihood to suggest, as Marsh did, that "the cost of one year's warfare" spent on environmental renovation might bring "an amelioration of climate, a renovated fertility of soil, and a general physical improvement, which might almost be characterized as a new creation."[55]

♦ Marsh judged attainment of a fruitfully sustainable earth not assured but probable, given foresight, collective will, technological advance, and expert guidance. Faith in progress has ebbed so far that many today contend technology does more harm than good, and judge enterprise unwilling, government unable, to implement the reforms needed for a livable globe. If Marsh was one of the world's great worriers, in Alfred Crosby's words, he was not, like so many today, convinced that perdition was our all-but-inescapable destiny.[56]

♦ Although Marsh's main aim was to alert mankind to the mess it was making of nature, he viewed many if not most human impacts as desirable. He appraised a managed earth as superior to raw untamed nature, human intellect and technology as not merely a source of global folly but the sole hope for nature's repair. In contrast, many environmentalists today deem untouched nature sacrosanct, and most suppose the less human impact the better.

Given such hugely unlike perspectives, how can Marsh's insights be of use today? By studying them we may see better how we relate to our own world, and hence begin to bridge the gulf between the environment we have and the environment we need. Before discussing Marsh's relevance, I must add a caveat. The past is a foreign country, not our own. We cannot simply borrow another era's lessons, however salutary. Marsh's views stem from a world whose memories and mindset, habits and hopes are remote from our own. They made sense in terms of their own time, not ours.

But even if we cannot directly profit from the wisdom of another time, we can benefit by seeing how and why other perspectives then seemed sensible. In becoming aware of outlooks unlike our own, we open the door to a wider range of alternatives, to the manifold ways a diverse humanity has lived in, used, and enjoyed ever-changing yet usually recognizable habitats. It is worth pondering certain Marsh perspectives that run counter to conventional wisdom today.

The inevitability of impact. We now see, more than Marsh ever did, how malign, even catastrophic, our environmental impingement has been. In revulsion against the horrific wrecking of habitats, many idealize nature devoid of human impress and yearn for a unimpacted world. It is an idle aspiration. We inherit a world indelibly altered by our precursors; we can neither erase their marks nor avoid adding our own. We may alter our influence, but we can neither halt it nor curtail its intensity. As long as contriving brain and skillful hand endure, the effects of our impress will be ever greater—and potentially graver.

To Marsh, the thought of relinquishing dominion over nature was a nightmare. It meant regression to a heedless, amoral life ruled by hunger, fear, and superstition. But short of total collapse, that could not happen. Every human act affects nature; every technical advance augments human impact. Marsh showed how such impacts derange natural balances by accelerating erosive and other processes. Resultant damage might be repaired, not by ceasing to alter nature but by exercising greater care in doing so. Growing human might demands not relaxing, but intensifying, purposive global manipulation.[57]

The primacy of the unexpected. Not even the best intentions can ensure sound environmental management. For as Marsh reiterated time and again, most human impacts are unintended. "Vast as is the . . . magnitude and importance [of] intentional changes," they are "insignificant in comparison with the contingent and unsought results which have flowed from them."[58] As global impacts proliferate, their unsought, unwanted, perhaps lethal consequences can never be fully foreseen, let alone prevented.

This insight has greater resonance in our time than in Marsh's, as William Meyer suggests. Impacts back then were more immediate and evident than now, when "the secondary, distant, and surprising effects of which Marsh spoke have become commonplace." We are today more alert than were Marsh's contemporaries to environmental evils that are invisible and unexpected. Natural history has ceased to be a saga of predictable regular processes, because ecologists now recognize that equilibrium in nature is rare and fleeting. But we have not yet schooled ourselves to accept the humbling awareness of uncertainty that pervades the pages and informs the insights of *Man and Nature*. Unlike most modern environmentalists, Marsh had come to terms with nature's "baffling complexity, its inherent unpredictability, its daily turbulence." He had also acknowledged the limits to human knowledge. "The equation of animal and vegetable life is too complicated a problem for human intelligence to solve," he concluded in *Man and Nature*, "and we can never know how wide a circle of disturbance we produce in the harmonies of nature when we throw the smallest pebble into the ocean of organic life."[59]

The singularity of human intention. The gulf Marsh posited between human and "brute" creation is abhorrent to most environmentalists today. Viewing organic life as a morally seamless unity, many now enjoin respect for all creatures from pandas to paramecia, for nature in toto, for a hypothesized Gaia.

Such respect may have spiritual, aesthetic, ecological, perhaps even practical benefits. But whatever its virtues, the idea of according rights to non-human nature is a value held in some cultures, not a universal truth. Such a view was alien to Marsh's time. His generation saw humans alone endowed with conscious will, a sense of morality, deliberate purposes. These attributes varied with culture but did not exist outside culture.

No evidence of moral intention is empirically discernable elsewhere in nature; humans alone, in Yi-Fu Tuan's phrase, are "congenitally indisposed to accept reality as it is." Even Gaia devotees recognize that humans are earth's only morally conscious beings. Environmental reformers who find nature's inarticulate indifference unbearable impute their own aims to nature and then purport to speak on nature's behalf. A reified natural world is then worshipped for being "virtuous."[60]

Marsh had no truck with nature as morality. To seek shelter in hidden higher wisdom, arrogated to self-appointed spokesmen, would have seemed to him impious. Impious or not, it is futile and self-defeating. To shoulder responsibility for our own actions and ambitions is more honest, and in the end probably more environmentally and socially efficacious, than to reify and take refuge in voiceless nature.

The necessity of stewardship. To deny nature a voice is not, however, to say that we may do with nature just as we please. If we are not accountable to nature itself, as social beings we are responsible for the world we hope our descendants will inherit. Social life requires communities, organisms that transcend the life spans of individuals and attach us to the heritage of our forebears and to the legacy we leave our descendants. Communities rely on compacts between the living, the dead, and the yet unborn. Faith in community, reaching into a past and future beyond our individual selves, is a necessary religion, as Durkheim put it; without it life would be shorn of meaning.[61] Only awareness of what we owe to those who preceded and concern for those who will follow enables us to care enough to plan ahead.

Active stewardship was for Marsh crucial to environmental health. But it required curtailing private property rights more stringently than most then realized or even now tolerate. Unless "the sacred right of every man to do what he will with his own" were rescinded, Marsh saw disaster certain. "Man has too long forgotten that the earth was given to him for usufruct alone, not for consumption, still less for profligate waste."[62] The milieus we inherit from myriad forebears include all their transformations, unwitting and otherwise. As temporary denizens we make the best of that environment according to our own lights. As stewards we pass it on to our heirs, trusting that they will also wish to be stewards.

Yet for all these transcendent benefits, stewardship is not an innate impulse; it needs to be induced and cherished. In modern post-industrial society, stewardship contends against many countervailing pressures. Immediate urgent crises, restless mobility, faceless corporate irresponsibility, the fraying of community ties, the democratic process itself—all impose a tyranny of the present that threatens to throttle stewardship. Already aware of many of these pressures, Marsh saw that such care had to be inculcated by education, training, and steadfast dedication to public affairs. Not least, regard for the future required active concern for *present* environments beyond their market value.

To be properly cared for, the environment must feel truly our own, not as a disposable commodity but as a locale integral to everyday life. As did our forebears, we make it our own by adding to it our own stamp, now creative, now corrosive. The environment is never merely conserved or protected; to use Marsh's word it is *modified*—both enhanced and degraded—by each generation in turn. We should form the habit of lauding, not lamenting, our own creative contributions to milieus we inhabit. By striving to praise, we are more apt to make changes that we and our heirs feel worthy of praise.

The primacy of the amateur. A consistent Marsh precept was the need to transcend narrow specialization. His own life is the supreme model. Celebrating the achievements of "this eminent man, who studied languages while he practiced law, who divided his time between business and politics, who wrote books and delivered lectures on literary subjects, and who investigated geographical problems while he elevated diplomacy," Harvard's William Morris Davis a century ago glumly contrasted "the breadth of his interests and the variety of his activities and duties [with] the high degree of specialization and the necessary narrowing of interests and activities" of scientists of his day. Davis strongly doubted that "advice on the treatment of national scientific problems can be as well given by intensive specialists of the modern school as by men of a wider experience, of whom Marsh was so admirable an example."[63]

In our yet more specialized present the need for generalists is greater still. In no realm is the tyranny of the expert more socially obnoxious than in environmental management. Marsh sought to breach not just the walls between academic disciplines, but also the boundaries that segregated academe from active life. The popularizing of science, some 130 years ago, led Marsh to hope that "the world of mind, like the world of politics, is becoming a democratic republic."[64]

Marsh's democracy of science seems a pipe dream in these days of impacts scarcely perceptible even to experts. But we must for our own sake endeavor, not to become Crichtons, but to be more broadly familiar with the congeries of forces that make and shape us. Only so armed can we play an informed and responsible role in using and controlling these forces and their byproducts. In the end *only* amateurs make sane decisions; specialization blinkers and biases specialists. Once it is realized that experts are often *more* irrational, defensive, and culture-bound than generalists, we amateurs may gain confidence in our ability, as it is clearly our need, to assess the most arcane and abstruse enigmas confronting us.[65]

Marsh's *Man and Nature* marked the inception of a truly modern way of looking at the world, of thinking about how people live in and react on the fabric of landscape they inhabit. Until Marsh, culture was widely assumed to be one thing, nature another; it was the task of the former to master and exploit the latter. That this worked better at some times and places was obvious, but no one until Marsh envisaged such mastery as more than simple impacts on this or that locale, this or that resource. Marsh showed how human culture acted in and reacted on a ramified web of plants and animals, soils and waters. Most such human impacts were unintended and unpredictable, partly because nature was too complex to fully comprehend, partly because their myriad cascading effects were often obscure and long persisting: "When something happened in one place, other things happened elsewhere in response," in Nancy Langston's gloss on Marsh's message. "Nature was no longer outside history," for the changes set in motion or otherwise altered by humanity "were now an integral part of natural communities. . . . People were part of an interconnected community of living things."[66]

Marsh was more than the pioneer observer of ramified interactions among people and locales. He also fashioned a compelling depiction of the damage wrought and a reasoned yet impassioned plea for reforms to stem the destruction and help restore a previously bountiful natural fabric. His advice embraced physical controls to maximize ecological

429

benefits, and social controls to minimize shortsighted private and corporate avarice—avarice that milked immediate profits at the expense
of long-term social goals. His union of ecological insight with social reform gives his arguments a lasting force four generations later.

Marsh was the first to show that human actions had unintended
consequences of unforeseeable magnitude. Future technology might repair previous damage, but science could never fully encompass the implications of present and projected behavior. The veiled complexity
of long-sustained actions made their cumulative impact impossible to
gauge. This led Marsh to conclusions in one sense fabulously optimistic, in another profoundly depressing. The energies of winds and waves,
of electricity and of solar heat might in time be harnessed to reclad
the denuded mountains of the globe, and thereby equalize current extremes of drought and flood. Yet technology at the same time bred ever
more ruinous modes of exploitation, ever more efficient weapons of
annihilation. Human impulses to magnify extremes pressed against a
framework of natural limits narrower than most realized: a globe grown
too hot or too cold, too wet or too dry, would forfeit the equilibria that
had brought into being and sustained the wealth of animate life and
landscapes.[67]

Thomas Carlyle's dictum that "the History of the world is but the
Biography of great men" was turned on its head by Marsh, who found
the cumulative lives of the humble more revealing.[68] Yet Marsh's own
life may be deemed especially worthy of study for the insights he signally added to our world view. How did this nineteenth-century American, growing up in northern New England a single generation after
the frontier, forge a self-conscious synthesis of ongoing environmental
change? How did he marry environmental insights at once Calvinistic
and scientific, predestinarian and progressive? There were other scholarly polymaths; there were other Americans of hugely various enterprise. Few, if any, combined these two proclivities as Marsh did.

Marsh's perceptive powers seem to me to derive from the creative
coincidence of his own special skills and circumstances with a habit,
then common, of contrasting Old World with New World perspectives.
Many Americans were conscious of enjoying a New World, largely
made by their immediate forebears, seen in manifold ways as superior

to the Old—more enterprising, more free, more egalitarian, more progressive, more hopeful. This was common rhetoric of the time, and readers of Tocqueville saw that Europeans concurred with, even envied, much of it.

Marsh was peculiarly equipped to observe and to test these differences, both from wide reading in European history and literature and from his own observation of European lands and peoples. Uniquely, he explored not only the cultural peaks of arts and letters, but the prosaic, everyday aspects of life he himself engaged in as farmer, carpenter, mechanic, road builder, logger, quarryman—the wellsprings of the "earthy taste" he avowed, as noted in Chapter 14, at the very end of his life.[69]

This earthy taste led Marsh to focus on how humans impacted the face of the earth. Like his eclectic breadth, his pragmatism was essentially American—unlike the mystical, abstract, theoretical bent of many European scholars, such as those etymological savants Marsh took to task. It is above all the everyday quality of *Man and Nature* "from which we experience a jolt like that we get from Alexis de Tocqueville's *Democracy in America*—a shock born of realization that what was seen years ago remains true today."[70]

Abbreviations in Notes and Bibliography

AHR	*American Historical Review*
B(D)FP	Burlington *(Daily)* Free Press
B(N)S	Burlington *(Northern)* Sentinel
BSI	Spencer F. Baird Correspondence, Smithsonian Institution
CBT	Charlotte Bostwick Thrall
CCM	Caroline Crane Marsh
CFP	Crane Family Papers, New York Public Library
CG	*Congressional Globe*
Cj	Caroline C. Marsh, Journal, 1861–65, MS, UVM
CM	Charles Marsh (George P.'s father or youngest brother)
DC	Dartmouth College
DCA	Dartmouth College Archives
DD	Diplomatic Despatches, National Archives
DI	Diplomatic Instructions, National Archives
FC	Ford Collection, New York Public Library
FM	Francis Markoe
FW	Frederick Wislizenus
GFE	George Franklin Edmunds
GMM	Galloway–Maxcy–Markoe Corr., Library of Congress

ABBREVIATIONS

GO	George Ozias Marsh
GPM	George Perkins Marsh
HP	Hiram Powers
HVG	A. M. Hemenway, ed., *Vermont Historical Gazetteer*
IADR	M. P. Trauth, *Italo-American Diplomatic Relations*
JCP	Jacob Collamer Papers, UVM
LC	Library of Congress
LEL	Marsh, *Lectures on the English Language*
LL	Caroline C. Marsh, *Life and Letters of George Perkins Marsh*
LW	Lucy Wislizenus
MHS	Massachusetts Historical Society
MMSM	Marsh, *Mediaeval and Modern Saints and Miracles*
M&N	Marsh, *Man and Nature* ([1864]; page references in notes to 1965 edn.)
MV	Marsh MSS, Vermont Historical Society
NEQ	*New England Quarterly*
NES	Marsh, *Address, Delivered before the New England Society of the City of New-York . . . 1844*
NYHS	New York Historical Society
OHSI	W. J. Rhees, ed., *Smithsonian Institution . . . Origin and History*
PAA	Hiram Powers Papers, Archives of American Art
PR	Post Records, Dept. of State, National Archives
SFB	Spencer Fullerton Baird
SI	Smithsonian Institution
SIJ	W. J. Rhees, ed., *Smithsonian Institution: Journals of the Board of Regents*
SM	Susan Perkins Arnold Marsh
SP	William Tecumseh Sherman Papers, Library of Congress
SS	*Scandinavian Studies*
UPa	University of Pennsylvania Library Special Collections
UV	University of Vermont
UVM	Marsh Collection, Bailey/Howe Library, University of Vermont
VGC	Vermont. Supreme Executive Council, *Records of . . . Governor and Council*
VH	*Vermont History*
VHS	Vermont Historical Society
VSA	Vermont State Archives
VSP	Vermont State Papers
WP	Wislizenus Papers, UVM

NOTES

Preface

1. Timothy O'Riordan, "Special Report: The Earth as Transformed by Human Action," *Environment* 30:1 (1988): 25–28 at 25; Lewis Mumford, *The Brown Decades* (1931), 78. See my "George Perkins Marsh and the American Geographical Tradition," *Geographical Review* (1953), and Daniel W. Gade, "The Growing Recognition of George Perkins Marsh," *Geographical Review* 73 (1983): 341–44. In its tenth printing in 2000, *Man and Nature* has sold more than 10,000 copies since 1965.

2. Carl Sauer, "Foreword to Historical Geography" (1941), in his *Land and Life* (1963), 351–79 at 356 (also 371–72); idem, "On the Background of Geography in the United States" (1967), in his *Selected Essays, 1963–1975* (Berkeley, Calif.: Turtle Island, 1981), 241–59 at 250; David Lowenthal, *George Perkins Marsh: Versatile Vermonter* (1958).

3. Lewis Mumford to the author, 15 May 1958; David Lowenthal, *The Vermont Heritage of George Perkins Marsh* (1960), 1; T. D. S. Bassett, "Matthew Buckham and the University of Vermont," in Robert V. Daniels, ed., *The University of Vermont: The First Two Hundred Years* (1991), 107–20.

4. Marsh [hereafter GPM] to Col. J. B. Estcourt, 1850, as reported in Caroline Crane Marsh [hereafter CCM], *Life and Letters of George Perkins Marsh* (1888)

[hereafter *LL*]; Dorothy Canfield Fisher, "Vermonters," in *Vermont: A Guide to the Green Mountain State* (1937), 3–9 at 3.

5. I am obliged to Jill Paton Walsh's jeu d'esprit on this transition (*A Piece of Justice* [New York: St. Martin's Press, 1995], 31–33).

6. George Bancroft to GPM, 16 Jan. 1869, Marsh Coll., University of Vermont.

7. GPM quoted in *LL*, 7; GPM, "Thoughts and Aphorisms" (1875), 4; GPM to Carrie M. Crane, 6 Oct. 1872 (CCM copy).

8. University of Vermont President Matthew H. Buckham in *Burlington Free Press & Times*, 25 Aug. 1888.

9. GPM, *The Origin and History of the English Language* (1862), 453.

10. Samuel Gilman Brown had given a memorial address on Marsh (*A Discourse Commemorative of the Hon. George Perkins Marsh, LL.D.* [Burlington, 1883]).

11. *LL*, v–vi; Elizabeth Green Crane, *Caroline Crane Marsh* (n.d.), 78–83.

12. Mary Philip Trauth, *Italo-American Diplomatic Relations, 1861–1882: The Mission of George Perkins Marsh* (1958).

13. Michael Holroyd, "The Ethics of Biography," talk at British Library Centre for the Book, London, 20 Apr. 1995.

14. Precise financial comparisons cannot be drawn. Over Marsh's lifetime the Consumer Price Index fluctuated between $\frac{1}{20}$ (in the 1840s and 1850s) and $\frac{1}{10}$ (in the late 1860s) of 1990s figures; wages were not equivalent. My estimates are from James C. Cooper and Karl Borden, "The Interpretation of Wages and Prices in Public Historical Displays," *Public History* 19:2 (1997): 9–29. See also John J. McCusker, "How Much Is That in Real Money? A Historical Price Index for . . . the United States," American Antiquarian Society *Proc.* 101 (1991): 297–333.

1. Woodstock and the First Watershed

1. GPM, "The Past, the Present and the Future of New England and Her Offspring and the Influence They Are Probably to Have on the World's Future," speech at Forefathers' Day, Middlebury, Vt., 2 Dec. 1859 (MS).

2. GPM to Rufus W. Griswold, 22 Nov. 1847, Park Benjamin Coll.

3. GPM, "Address [delivered before the American Colonization Society]" (1856), 11. Marsh held the then common view that prior to white settlement Indians had impacted this impregnable wilderness but lightly, though he later detailed the parklike "oak openings" made by Indians to attract deer to fresh herbage in the wake of fire (Marsh, *Man and Nature* [1864; hereafter *M&N*],

120n9 [page references are to the 1965 reprint]). In fact, the intervale grasslands were the product of recurrent burning by tribal Abenaki, who had also profoundly altered the character of the forest so as to control insect pests, improve access to game, and encourage fire-tolerant nut trees and berries (William Cronon, *Changes in the Land: Indians, Colonists, and the Ecology of New England* [1983], 38–58; Tom Wessels, *Reading the Forested Landscape: A Natural History of New England* [1997], 34–38). In eastern Massachusetts, Thoreau inferred equivalent Indian impact on bare pine plains (H. D. Thoreau, *Faith in a Seed* [1993], 157).

4. GPM to Charles Sprague Sargent, 16 May 1879. See D. C. Fisher, *Vermont Tradition* (1953), 163–84.

5. GPM to SFB, 12 Oct. 1881, BSI.

6. GPM, *M&N*, 85n. The streams Marsh mentions were probably the Quechee and its tributary South Brook, which joins it at Woodstock.

7. *M&N*, 235n180. H. S. Dana, *History of Woodstock* (1889), 552, dates the Mount Tom fire (caused by "a hunter's carelessness") as "about the beginning of the . . . century"; writing in 1863, Marsh puts the fire at "between fifty and sixty years ago," which dates it between 1803 and 1813. Marsh's description is that of an eyewitness not to the blaze but to the denuded slope, and perhaps to the next year's erosion, suggesting the fire's probable date as 1803 or 1804.

8. A. F. Hawes, "What Some Vermonters Are Doing in Forestry," *Vermont Agricultural Bull.* no. 9 (Apr. 1911): 16–17; U.S. National Park Service, *Marsh-Billings National Historical Park: Land Use History* (1994), 59. The Norway spruce flourished; the ash and birch died out.

9. Nathan Perkins, *A Narrative of a Tour through the State of Vermont from April 27 to June 12, 1789* (Woodstock, 1920), 26 (excerpted in T. D. S. Bassett, comp., *Outsiders Inside Vermont: Travelers' Tales over 358 Years* [Brattleboro: Stephen Greene Press, 1967], 42–45).

10. John Orvis, "Letter from Vermont, June 14, 1847," quoted in David Ludlum, *Social Ferment in Vermont, 1791–1850* (1939), 16; Timothy Dwight, *Travels in New-England and New-York* (1821–22), 2:316.

11. Gary J. Aichele, "Making the Vermont Constitution: 1777–1824," in Michael Sherman, ed., *A More Perfect Union: Vermont Becomes a State, 1777–1816* (1991), 2–37 at 14; Ethan Allen to Meshech Weare, 4 Mar. 1779, in John J. Duffy, ed., *Ethan Allen and His Kin: Correspondence, 1772–1819*, 2 vols. (Hanover: University Press of New England, 1998), 1:90–91. J. L. Rice, "Dartmouth College and the State of New Connecticut" (1876), remains the fullest account.

12. Roswell Marsh to Eliakim P. Walton, 23 July 1873, in Vt. Supreme Exec. Council *Records*, 1:236–38. On Joseph Marsh, see William H. Tucker,

History of Hartford, Vermont (Burlington, 1889), 10, 34, 113–14, 338–40; Dwight Whitney Marsh, ed., *Marsh Genealogy, Giving Several Thousand Descendants of John Marsh of Hartford, Ct., 1636–1895* (Amherst, Mass., 1895), 116, 130; Conn. Hist. Soc. *Colls.* 9 (1903): 249; *The Public Records of the Colony of Connecticut, 1757–1762*, ed. Charles J. Hoadley (Hartford, Conn.) 9 (1880): 513, 10 (1881): 140; Kate Morris Kane, "The Revolutionary History of a Vermont Town," *Vermont Antiquarian* 1 (1902): 1–28 at 18; Frank Smallwood, *Thomas Chittenden: Vermont's First Statesman* (Shelburne, Vt.: New England Press, 1997), 77–79. On the "long finger of Federalism" from Connecticut through eastern Vermont, see William A. Robinson, *Jeffersonian Democracy in New England* (New Haven: Yale Historical Publications, 1916), 166.

13. GPM to Eliakim P. Walton, 7 Apr. 1880, VHS MSS 22 no. 67.

14. Dana, *Woodstock*, 193.

15. Jacob Collamer to Frances C. Collamer, 5 Apr. 1862, JCP.

16. *Memoir, Autobiography and Correspondence of Jeremiah Mason*, ed. G. J. Clark (Kansas City, Mo.: Lawyers' International Publishing Co., 1917), 17, 20–22. Mason was U.S. senator from New Hampshire, 1813–17.

17. *Grandmother Tyler's Book: The Recollections of Mary Palmer Tyler (Mrs. Royall Tyler), 1775–1866*, ed. Frederick Tupper and Helen Tyler Brown (New York: Putnam, 1925), 151; D. W. Marsh, *Marsh Genealogy*, 133; James Barrett, *Memorial Address on the Life and Character of the Hon. Charles Marsh, LL.D.* (VHS Proc. 1870; Montpelier, 1871), 24–29.

18. Charles Marsh, *An Essay on the Amendments Proposed to the Constitution of the State of Vermont, by the Council of Censors* (Hanover, N.H., 1814); D. B. Carroll, *The Unicameral Legislature of Vermont* (1932), 46–56; *Spooner's Vermont Journal*, Windsor, 29 Nov. 1813; U.S. *Annals of Cong.*, 14 Cong. 2 sess., 1816–1817, 609–10, 637–38, 714; Claude M. Fuess, *Daniel Webster* (Boston: Little, Brown, 1930), 1:184–88; GPM to Sarah Butler Wister (CCM copy), 5 Apr. 1880. See Edward Brynn, "Patterns of Dissent: Vermont's Opposition to the War of 1812," *VH* 40 (1972), 10–27; Paul S. Gillies, "Adjusting to Union: An Assessment of Statehood, 1791–1816," in Sherman, *A More Perfect Union*, 114–49 at 139–41.

19. Charles Marsh "but for his singular modesty could have been for years our Governor" (University of Vermont President John Wheeler to J. H. Green, 16 Feb. 1843, in J. I. Lindsay, *Tradition Looks Forward: The University of Vermont* [1954], 145).

20. CM (1834) quoted in R. A. Roth, *The Democratic Dilemma: Religion, Reform and the Social Order in the Connecticut River Valley of Vermont, 1791–1850* (1987), 177; CM to Francis Brown, quoted in C. C. Marsh, *Life and Letters of George Perkins*

Marsh (1888) [hereafter *LL*], 16. See Ernest L. Sherman, "Meet Charles Marsh," Kimball Union Academy (Meriden, N.Y.) *Alumni Bull.* (Mar. 1947): 6–10.

21. *LL*, 8.

22. Elisha Perkins to Josias L. and Susan Arnold, 20 Oct. 1795, Letter and account book, 1788–1816, Yale Medical School Library; Benjamin Douglas Perkins, *Influence of Metallic Tractors on the Human Body, in Removing Various Inflammatory Diseases, such as Rheumatism, Pleurisy, some Gouty Affections, &c., &c.* (London, 1798), viii, 69–70, 84–85; idem, *New Cases of Practice with Perkins's Patent Metallic Tractors, on the Human Body and on Animals; but Especially on Infants and Horses* (London, 1802), 165–67; Jacques M. Quen, "Elisha Perkins, Physician, Nostrum-Vendor, or Charlatan?" *Bulletin of the History of Medicine* 37 (1963): 159–66; J. M. Quen, "Case Studies in Nineteenth Century Scientific Rejection: Mesmerism, Perkinism, and Acupuncture," *Journal of the History of the Behavioral Sciences* 11 (1975): 149–66; Eric T. Carlson and Meribeth M. Simpson, "Perkinism vs. Mesmerism," *Journal of the History of the Behavioral Sciences* 6 (1970): 16–24. Elisha Perkins's grandson George Perkins Marsh thought some truth might be "wrapped up in the poor charlatanry of animal magnetism" (*Human Knowledge* [1847], 21), but he nowhere mentions this grandfather.

23. Christopher Caustic [pseud. Thomas Green Fessenden], *Terrible Tractoration!! A Poetical Petition against Galvanising Trumpery, and the Perkinistic Institute,* 2d ed. (London, 1803), 63–64; Porter Gale Perrin, *The Life and Works of Thomas Green Fessenden, 1771–1837* (Orono: University of Maine, 1925), 50–70. Perkins's Metallic Tractors long lingered in American memory. "This country won't be able to get along without 'em," attests a character in Kenneth Roberts's *Oliver Wiswell* (New York: Doubleday, Doran, 1940), 15–16. "There ain't a rebel in America that wouldn't beggar himself to buy a pair." Magnetic therapy has lately resurfaced, along with charges of quackery; "we cannot explain the significant pain relief reported by our patients," write Carlos Vallbona et al. ("Responses of Pain to Static Magnetic Fields in Postpolio Patients," *Archives of Physical Medicine & Rehabilitation* 78 [1997]: 1200–3); see Andrew Bassett, "Therapeutic Uses of Electric and Magnetic Fields in Orthopedics," in D. O. Carpenter and S. Ayrapelyan, eds., *Biological Effects of Electric and Magnetic Fields* [San Diego: Academic Press, 1994], 2:13–48; for skeptical responses, *International Herald Tribune,* 11 Dec. 1997. Perkinistic fervor reechoes in Gary Null, *Healing with Magnets* (London: Robinson, 1998).

24. Newspaper quoted in Pliny H. White, "Early Poets of Vermont," VHS *Proc.* (1917–18): 91–135 at 105–7. Josias Lyndon Arnold's latinate *Poems* were published posthumously (1797). See also Edward T. Fairbanks, *The Town of*

St. Johnsbury, Vt.;—A Review of One Hundred Twenty-five Years (St. Johnsbury, 1914), 72–75.

25. *LL*, 324.

26 Attributed to Collamer, who lived in Woodstock from 1836 to 1865, in Peter S. Jennison, *The History of Woodstock, Vermont, 1890–1983* (1983), 4; *M&N*, 172–73. See James Walter Goldthwait, "A Town That Has Gone Downhill," *Geographical Review* 17 (1927): 527–55.

27. Dwight, *Travels*, 2:315–17 (based on a visit of 1803); C. W. Eldridge, "Journal of a Tour through Vermont . . . in 1833" (1931), 61. See Peter M. Briggs, "Timothy Dwight 'Composes' a Landscape for New England," *American Quarterly* 40 (1988): 359–77 at 372–74. In *Man and Nature* Marsh frequently cites Dwight.

28. Dana, *Woodstock*, 143–52, 248–57. Pearlash, a form of potash, was derived from wood ashes for the manufacture of soap and glass.

29. "A Friend to Real Order," Woodstock *Northern Memento*, 4 July 1805. See Ludlum, *Social Ferment*, 20; L. D. Stilwell, *Migration from Vermont* (1948), 109–12.

30. On Elijah Paine, see Russell F. Taft, "The Supreme Court of Vermont," *The Green Bag: An Entertaining Magazine of the Law* (Boston) 6 (1894): 31–32; T. D. S. Bassett, *The Growing Edge: Vermont Villages, 1840–1880* (1992), 32–34. On Paine's weather records, see Zadock Thompson, *Natural History of Vermont* (1853), 9, 433, and Marsh's comment in *M&N*, 50n52; on the impact of sheep, see Wessels, *Reading the Forested Landscape*, 57–59.

31. CCM in *LL*, 11; GPM to CCM, 24 Sept. 1856; Thomas Storrow Brown to CCM, 188?, recounting boyhood memories of 1809–11.

32. J. L. D., *Woodstock, Vt.: A Few Notes, Historical and Other, Concerning the Town and Village* (Woodstock, 1910), 16–17; Dana, *Woodstock*, 23, 42–46, 82, 174–76, 286–301, 459–75.

33. GPM to Rufus W. Griswold, 22 Nov. 1846, Park Benjamin Coll.

34. *Vermont: A Guide to the Green Mountain State* (1937), 152.

35. John H. McDill, "The Billings Farm: A Brief Historical Sketch" (1948), MS, Norman Williams Library, Woodstock, Vt.; J. A. Kerr, G. B. Jones, and W. E. McLendon, "Soil Survey of Windsor County, Vermont," USDA Bur. of Soils, *Field Operations 1916* (Washington, D.C., 1921), 18:175–94; C. H. Pierce, *Surface Waters of Vermont*, U.S. Geol. Survey, Water-Supply Paper 424 (Washington, D.C., 1917), 111, 192; interviews with S. E. Wilson, USDA Soil Conservation Service, and John McDill, Woodstock, Vt., summer 1951. On mills and rivers, Dana, *Woodstock*, 181–82, 306–17; William J. Wilgus, *The Role*

of Transportation in the Development of Vermont (Montpelier: VHS, 1945), 42–53; Lyman S. Hayes, *The Connecticut River Valley in Southern Vermont and New Hampshire* (Rutland: Tuttle, 1929), 37–38, 160.

36. GPM to "Tom," 11 Nov. 1847.

37. Lyndon A. Marsh to GPM, 31 Jan. 1868.

38. GPM quoted in *LL*, 11.

39. GPM to Henry Yule, 15 May 1871; see *LL*, 12.

40. Jones Very, *Essays and Poems* (Boston, 1839), 161; GPM to C. E. Norton, 24 May 1871, Norton Papers; GPM to Joseph Neilson, June–July 1882, in Neilson, *Memories of Rufus Choate* (Boston, 1884), 376–77; GPM to Charles Sprague Sargent, 16 May 1879.

41. *LL*, 7.

42. GPM, "Decision of Arbitration . . . fixing of the Italian-Swiss frontier" (1874), 2030–33; GPM, "Watershed" (1878). On the confusion generated by opposite meanings of the term, see Marcus Cunliffe, "American Watersheds," *American Quarterly* 13 (1961): 480–94 at 481, 494; James J. Parsons, "On 'Bioregionalism' and 'Watershed Consciousness,'" *Professional Geographer* 37 : 1 (1985): 1–6. Marsh's experience echoed Goethe's, who claimed, in Bavaria, he could "quickly get a topographical idea of a region by looking at even the smallest stream and noting in which direction it flows and which drainage basin it belongs to" (Johann Wolfgang von Goethe, *Italian Journey, 1786–1788* [tr. W. H. Auden and Elizabeth Mayer, Harmondsworth: Penguin, 1970], 3 Sept. 1786, p. 23).

43. John Powers Richardson to HP, quoted in R. P. Wunder, *Hiram Powers: Vermont Sculptor, 1805–1873* (1991), 1 : 31.

44. Dana, *Woodstock*, 449–51.

45. GPM to CCM, 7 Sept. 1856.

46. CM to Francis Brown, Apr. 1812.

47. Josiah Quincy quoted in Claude M. Fuess, *An Old New England School: A History of Phillips Academy, Andover* (Boston: Houghton Mifflin, 1917), 157, 173. See Catalogue of the Trustees, Instructors and Students of Phillips Academy, Andover, Aug. 1816 (O. W. Holmes Library, Andover); *Biographical Catalogue of the Trustees, Teachers and Students of Phillips Academy Andover, 1778–1830* (Andover, 1903), 18–19, 63; Charles E. Cuningham, *Timothy Dwight, 1752–1817: A Biography* (New York: Macmillan, 1942), 230.

48. GPM, "England Old and New" (ca. 1859/60), MS, p. 5.

49. CM to Francis Brown, in L. B. Richardson, *History of Dartmouth College* (1932), 1 : 347–48; J. M. Shirley, *The Dartmouth College Causes and the Supreme Court*

(1877), 81–115. See Lewis Feuer, "James Marsh and the Conservative Transcendentalist Philosophy" (1958): 6–10. The "Socinianism" decried by Charles Marsh was the sixteenth-century heresy that denied divinity to Christ; in early nineteenth-century America, Socinianism essentially meant Unitarianism.

50. Daniel Webster quoted in S. G. Brown, *The Works of Rufus Choate, with a Memoir of His Life* (1862), 1:515 ff.; George Washington Nesmith to B. F. Prescott, 17 Feb. 1885, DCA. See Dartmouth College, *General Catalogue, 1769–1940* (Hanover, 1940).

51. *Autobiography of John Ball*, comp. Kate Ball Powers et al. (Grand Rapids: Dean-Hicks, 1925), 19–20, 28; Benjamin Silliman, *Remarks Made on a Short Tour between Hartford and Quebec, in the Autumn of 1819* (New Haven, 1824), 416–18; Dartmouth College Treasurer's accounts, Ledger F (1810–29), 80, DCA.

52. Richardson, *Dartmouth*, 1:249–50, 272–78. See Henry Wood, *Sketch of the Life of President Brown* (Boston, 1834), 12–16.

53. GPM quoted in *LL*, 17; Cuningham, *Timothy Dwight*, 22–26.

54. G. W. Nesmith to Samuel Gilman Brown, 16 Nov. 1882. See *Autobiography of John Ball*, 25; *Course of Instruction, &c. at Dartmouth College* (Hanover, 1822), DCA; John F. Fulton and Elizabeth H. Thomson, *Benjamin Silliman, 1779–1864: Pathfinder in American Science* (New York: Schuman, 1947), 158. The same texts were central at Thoreau's Harvard in the 1830s (Robert Sattelmeyer, *Thoreau's Reading: A Study in Intellectual History* [Princeton: Princeton University Press, 1988], 16–19).

55. Perry Miller, *The Life of the Mind in America from the Revolution to the Civil War* (New York: Harcourt Brace, 1965), 276; Roger Stein, *John Ruskin and Aesthetic Thought in America* (1967), 158–59. See also Chapter 3 below.

56. James Barrett to S. G. Brown, ca. 1883, in *LL*, 17; Rufus Choate to David Choate, 5 Nov. 1816, DCA; J. V. Matthews, *Rufus Choate: The Law and Civic Virtue* (1980), 10.

57. Rufus Choate to David Choate, 16 Dec. 1816, DCA; *Dartmouth Gazette*, 5 Mar. 1817, in John K. Lord, *A History of Dartmouth College, 1815–1909* (Concord: Rumford, 1913), 120–21. See Clement Long, *Serving God with the Mind: A Discourse Commemorative of the Rev. Roswell Shurtleff, D.D.* (Concord, 1861), 22–48.

58. Lord, *Dartmouth*, 104–8; Marshall quoted in Shirley, *Dartmouth College Causes*, 201, 302–3, 411, 423.

59. GPM to S. G. Brown, in Brown, *Choate*, 1:11; GPM, *Earth as Modified by Human Action* (1874 ed.), 54n.

60. Faculty Records, May 1820, DCA.

61. Woodstock *Observer*, 4 Sept. 1820.

62. William A. Ellis, *Norwich University 1819–1911; Her History, Her Graduates, Her Roll of Honor* (Montpelier: Capital City Press, 1911), 1:1–11; J. T. Walker, Jr., "'Old Pewter': A Biographical Sketch of Captain Alden Partridge of Norwich, Vermont" (1965). See also K. R. B. Flint, "Alden Partridge," in Walter H. Crockett, ed., *Vermonters; A Book of Biographies* (Brattleboro: Stephen Daye, 1932), 165–68.

63. GPM to Arnold Guyot, 16 Nov. 1857; "Journal of an Excursion to Manchester, Vermont, by a Party of Norwich Cadets, 1823," in *Essays in the Social and Economic History of Vermont* (Montpelier: VHS, 1943), 185–99 at 195.

64. "Cousin Sarah" recalled the summer of 1824 in undated letter to CCM after Marsh's death.

65. Ferdinand Berthier, *L'Abbé Sicard, célèbre instituteur des sourds-muets* (Paris, 1873), 181–93; Mimi WheiPing Luce, "The History of Language Use in the Education of the Deaf in the United States," in Michael Strong, ed., *Language Learning and Deafness* (New York: Cambridge University Press, 1988), 75–98 at 76–80; Harlan Lane, *When the Mind Hears: A History of the Deaf* (1984), 206 ff.

66. GPM, ["Report on the Education of the Deaf and Dumb"] (1824). See Vt., General Assembly Journal, 1823, 11; 1824, 16, 61, 67; 1825, 35–38; 1826, 54–55; Ludlum, *Social Ferment*, 220.

67. GPM, "Report . . . Deaf and Dumb," 25–26; GPM, *Lectures on the English Language* (1861) [hereafter *LEL*], 32–33. Marsh does not mention Sicard by name, but bought his *Cours d'instruction d'un sourd-muet de naissance* (2d ed., 1803) in 1823; Sicard and Marsh are close in both phrasing and content. Marsh watched Sioux Indian visitors converse in sign language with deaf students in New York (*LEL*, 1885 ed., 27n). Deaf-mutes whose "mother tongue" is sign language have an added advantage: not needing to separate vocal from body gestures, they learn to construct complex sentences from referential symbols nine to twelve months before hearing children do (William Stokoe, "Sign Languages, Linguistics and Related Arts," in John Kyle and Bernice Woll, eds., *Language in Sign: An International Perspective on Sign Language* [London: Croom Helm, 1983], 264–68).

68. GPM, "Report . . . Deaf and Dumb," 23–24. But deaf-mutes deprived of language in youth could not, however well educated later, "recall their mental status at earlier periods"; they seemed, Marsh thought, "previously devoid of those conceptions which we acquire, or at least retain and express, by means of general terms; [for them] that alone which is precisely formulated can be clearly remembered" (*LEL*, 2). This accords with present-day views of language deprivation (Susan Schaller, *A Man without Words*, 2d ed. [Berkeley: University

of California Press, 1995], 105–8, 132–33, 187; Steven Pinker, *The Language Instinct: The New Science of Language and Mind* [London: Penguin, 1994], 37–39, 67–69).

69. GPM to E. M. Gallaudet, 14 Oct. 1870. On sign language in early America, see Douglas C. Baynton, "'A Silent Exile on This Earth': The Metaphorical Construction of Deafness in the Nineteenth Century," *American Quarterly* 44 (1992): 216–43 at 221–27. Discussing how the eye controls muscles used in writing, Marsh recalled Hartford's deaf-and-dumb pupils (*M&N*, 16n13).

70. GPM, *LEL*, 34–35. Marsh's example of gestural expression was conveyed to Charles Darwin; see Chapter 14 below.

71. GPM, *LEL*, 1885 ed., 354–55n; 1861 ed., 35. Marsh here rephrased Oliver Goldsmith's "The true use of speech is not so much to express our wants as to conceal them" ("On the Use of Language" (1759), in *The Collected Works of Oliver Goldsmith*, ed. Arthur Friedman [Oxford: Clarendon Press, 1966], 1:394–401 at 394); Goldsmith was in turn indebted to Robert South ("A Sermon Preached at Westminster-Abbey, April 30, 1676, " in his *Twelve Sermons Preached upon Several Occasions* [London, 1692], 434).

72. Windsor County Court Records, MSS, 14:53; H. S. Dana in *Vt. Standard* (Woodstock), 27 July 1882.

2. *Burlington: Blunders and Bereavements*

1. *BNS*, 4 July 1829, 2 Feb. 1832.

2. *BFP*, 25 Aug., Oct. 1830, 27 May, 1 & 8 July, 2 Dec. 1831. See G. F. Houghton, "Benjamin Franklin Bailey," in *Vermont Historical Gazetteer*, ed. A. M. Hemenway [hereafter *HVG*], 1 (1862):645–46; Eleazar D. Durfee and D. Gregory Sanford, *A Guide to the Henry Stevens, Sr., Collection at the Vermont State Archives* (ca. 1990), 33–34; T. D. S. Bassett, "The Rise of Cornelius Peter Van Ness 1782–1826," VHS *Proc.* n.s. 10 (1942): 3–20; *Proceedings and Address of the Vermont Republican Convention Friendly to the Election of Andrew Jackson . . .* (Montpelier, 1828), 3; D. A. Smalley, "C. P. Van Ness," in *HVG*, 1:608–14.

3. *BFP*, 2 Dec. 1831.

4. David Ludlum, *Social Ferment in Vermont, 1791–1850* (1939), 86–133; H. Nicholas Muller, "Early Vermont State Government: Oligarchy or Democracy? 1778–1815," in Muller and S. B. Hand, eds., *In a State of Nature: Readings in Vermont History* (1982), 80–85; William P. Vaughan, *The Antimasonic Party in the United States, 1826–1843* (Lexington: University of Kentucky Press, 1983), 70–

88; Paul Goodman, *Towards a Christian Republic: Antimasonry and the Great Tradition in New England, 1826–1836* (1988), 120–46.

5. R. A. Roth, "The Other Masonic Outrage: The Death and Transfiguration of Joseph Burnham" (1994).

6. GPM, "Thoughts and Aphorisms" (1875), 1–2; CM cited in GPM to CCM, 16 Feb. 1856, CFP.

7. Chittenden County Court Records, MSS, vols. 12 ff.; Vt. Supreme Court *Reports*, vols. 2 ff.

8. GPM to CCM, 1839, in *LL*, 30–31.

9. GPM, *LEL*, 311; GPM to S. D. Horton (CCM copy), 15 Oct. 1869.

10. GPM to LW, 14 Feb. 1875.

11. James Barrett to Samuel Gilman Brown, ca. 1884; Mrs. Levi Underwood to Gen. Rush C. Hawkins, 26 Dec. 1896, Hawkins Papers; GPM to LW, 14 Feb. 1875.

12. GPM to FW & LW, 13 Oct. 1862, WP.

13. Ibid.; GPM in *LL*, 25. See Henry P. Hickok, "Ozias Buell," in *HVG*, 1:592–93.

14. Henry James, "From Lake George to Burlington," *Nation* 11 (1870): 135–36.

15. T. D. Seymour Bassett, "An Inland Port: Burlington, Vermont in 1840," *NEQ* 44 (1971): 635–49; idem, *The Growing Edge: Vermont Villages, 1840–1880* (1992), 15–22, 167.

16. GPM, "Marsh, James" (1859–61).

17. Rufus Choate to James Marsh, 21 Aug. 1821, in S. G. Brown, *The Works of Rufus Choate* (1862), 1:16; James Marsh to Henry J. Raymond, 1 Mar. 1841, in George B. Cheever, *Characteristics of the Christian Philosopher: A Discourse Commemorative of the Virtues and Attainments of Rev. James Marsh, D.D.* (New York, 1843), 68–69; GPM, "Marsh, James," 217. See *Coleridge's American Disciples: The Selected Correspondence of James Marsh* (1973); Peter C. Carafiol, *James Marsh and the Forms of Romantic Thought* (Tallahassee: University Presses of Florida, 1982); Douglas Macrae, *James Marsh and the Transcendental Temper* (Ann Arbor: University of Michigan Press, 1983); Joseph Torrey, "James Marsh," in *HVG*, 1:529; John Dewey, "James Marsh and American Philosophy," *Journal of the History of Ideas* 2 (1941): 131–50; Ronald V. Wells, *Three Christian Transcendentalists: James Marsh, Caleb Sprague Henry, Frederic Henry Hedge* (New York: Columbia University Press, 1943), 14–48; Lewis Feuer, "James Marsh and the Conservative Transcendentalist Philosophy" (1958): 20–24; John J. Duffy, "James Marsh," in R. V. Daniels, *The University of Vermont* (1991), 63–77.

18. GPM to Prof. Edwards Amasa Park, 16 Oct. 1844, Charles Folsom MSS.

19. John Wheeler to J. H. Green, Feb. 16, 1843, in J. I. Lindsay, *Tradition Looks Forward: The University of Vermont* (1954), 145; variant wording in Wheeler to Samuel Prentiss, 4 May 1841, U.S. Dept. of State, Appointment Papers, NA. A "doughface," originally a false face or mask, was a pro-slavery Northerner. For Marsh's friends, see R. F. Taft, "Burlington," in *HVG*, 1:487–519; Wilbur H. Siebert, *Vermont's Anti-Slavery and Underground Railroad Record* (Columbus, Ohio: Spahr & Glenn, 1937), 19–21, 62–63; Bassett, *Growing Edge*, 39–41 (on Hopkins); G. F. Houghton, "Benjamin Lincoln," in *HVG*, 1:648–49; Benjamin Lincoln, *An Exposition of Certain Abuses, Practised by Some of the Medical Schools in New England* . . . (Burlington, 1833); *LL*, 21–22; Matthew H. Buckham, "George Wyllys Benedict," in *Services in Remembrance of Rev. Joseph Torrey, D.D., and of George Wyllys Benedict, LL.D.* (Burlington, 1874), 44–52; Joseph Tracy, *A Discourse Commemorative of Rev. John Wheeler* . . . (Cambridge, Mass., 1865), 10–23; Ezra H. Byington, *Rev. John Wheeler, D.D., 1798–1862, President of the University of Vermont 1833–1848* . . . (Cambridge, Mass., 1894), 16–18; Augustus Young, *Preliminary Report on the Natural History of the State of Vermont* (Burlington, 1856), 37–43; *BFP*, 20 Nov. 1829.

20. Ozias Buell, will probate, 20 Aug. 1835, & other papers, 12 Oct. 1836, Probate Office, Chittenden Co.; Burlington Grand List, 1825–32, 1837, 1843; Burlington Land List and Real Estate Assessment, 1837.

21. GPM in *BS*, 7 Mar. 1834. In all his 1830s travels, Marsh later recalled, "I did not meet five persons who approved of the removal of the deposits from the U S Bank, & yet the administration was soon found to have a popular majority" (GPM to R. H. Dana, Jr., 8 Oct. 1866, Dana MSS).

22. GPM bank petition, 10 Oct. 1834, VSP 64:19; U.S. Circuit Court, 20 Vt. 666 (May 1848).

23. Tom Wessels, *Reading the Forested Landscape* (1997), 58.

24. Vt. Supreme Exec. Council *Records*, 15 Oct. 1835, 8:229; *LL*, 29; *BFP*, 21 Dec. 1838; *BS*, 22 Jan. 1836. For Marsh's woolen interests, see Burl. Land Records, 11:331–36, 361; E. S. Stowell, "Merino Sheep Industry," in *Third Biennial Rept. of the Vt. State Board of Agriculture* . . . (1875–76), 199–226 at 213–14; *LL*, 28; Joseph Auld, *Picturesque Burlington* . . . (Burlington, 1893), 166; *BFP*, 28 Dec. 1838; Ezra A. Carman, "Special Report on the History and Present Condition of the Sheep Industry of the United States," 52 Cong. 2 sess., H. Misc. Doc. 105 (USDA, Bur. Animal Husbandry, 1892), pt I, ch. 4, 217–348. For Winooski River, see C. H. Pierce, *Surface Waters of Vermont*, U.S. Geol. Survey,

Water-Supply Paper 424 (Washington, D.C., 1917), 55, 215; R. N. Hill, *The Winooski* (1949), 83; *BFP*, 30 July 1830; Zadock Thompson, *Natural History of Vermont* (1853), 20; Bassett, *Growing Edge*, 23.

25. The mill resumed operations under new management and remained Vermont's largest, employing close to 500 women and men (*BDFP*, 7 & 23 May, 4 June 1849; Bassett, *Growing Edge*, 49), but Burlington never became another Lowell.

26. T. D. S. Bassett, "Vermont's Second State House: A Temple of Republican Democracy Imagined through Its Inventories, 1836–1856," *VH* 64 (1996): 99–107 at 99–100. Among the highest Green Mountain peaks, Camel's Hump rises midway between Montpelier and Burlington.

27. Ludlum, *Social Ferment*, 131; J. J. Duffy and H. N. Muller, *An Anxious Democracy: Aspects of the 1830s* (1982), 124.

28. C. W. Eldridge, "Journal of a Tour through Vermont . . . in 1833" (1931): 65–66; G. F. Edmunds quoted in H. L. Bailey, "Vermont's State Houses; Being a Narration of the Battles over the Location of the Capitol and Its Construction" (1944): 143. See Vermont, Directory and Rules of the House of Representatives (1835), 5; *HVG*, 2:262, 493–94.

29. *BNS*, 30 Oct. 1829; GPM in *BFP*, 13 Sept. 1830; H. J. Conant, "Imprisonment for Debt in Vermont" (1951); P. J. Coleman, *Debtors and Creditors in America: Insolvency, Imprisonment for Debt, and Bankruptcy, 1607–1900* (1974), 65–73.

30. GPM in Vt. Supreme Exec. Council *Records*, 10 Nov. 1835, 8:261. Marsh later recalled that jailed debtors, "commonly fed without stint . . . consumed a considerably larger quantity of food than common out-door laborers" (*M&N*, 80n60).

31. Montpelier *Watchman & State Gazette*, 17 Nov. 1835. See D. B. Carroll, *The Unicameral Legislature of Vermont* (1932), 96, 113, 152–58; Duffy and Muller, *Anxious Democracy*, 109–10.

32. GPM, *Address Delivered before the Agricultural Society of Rutland County* (1847), 21.

33. Daniel Crane to Caroline Crane, 25 May 1835; *BS* notice, 15 Apr. 1836 ("English Education; Latin, Greek, Drawing, French & Spanish $4, music $12 a quarter extra"); E. G. Crane, *Caroline Crane Marsh: A Life Sketch* (n.d.), 3–6; *LL*, 38–39; Betty Bandel, "Female Seminaries" [in nineteenth-century Burlington], Chittenden County Hist. Soc. *Bull.* 19:4 (Fall 1984): 3; Louisa C. Ingersoll Reid to CCM, 20 July 1868. Marsh was a trustee of the Burlington Female Seminary, chartered the previous year; the seminary shared Silas Crane's music teacher, Theodore Molt (J. K. Converse in *HVG*, 1:534; *BS*, 8 Jan. 1836).

34. *LL*, 29.

35. E. F. Brown, *Raymond of The Times* (1951), 31–32; GPM to CCM in *LL*, 35.

36. Caroline E. Crane, talks with the author, 1950; *BFP*, 4 Oct. 1839.

37. CCM to GPM, 25 May 1856.

38. Charles Lanman, *Letters from a Landscape Painter* (1845), 113–18; the book is dedicated to Marsh (pp. iii–iv). Marsh sold his house to J. S. Pierce when he went to Washington as a congressman in 1843 but kept rooms in it for family use. In 1895 it was reported on the brink of demolition (*BFP*, 6 Mar. 1895).

39. *LL*, 36–47.

3. *Puritans, Vikings, Goths*

1. GPM to CCM, 1839, in *LL*, 30–31.

2. GPM to F. P. Nash, 6 Feb. 1875; to CCM, 1838, in *LL*, 33.

3. GPM to CCM, in *LL*, 38.

4. GPM to HP, 25 Nov. 1847, PAA; to CCM, in *LL*, 35.

5. GPM, "The Study of Nature" (1860), 52; GPM to HP, 25 Nov. 1847, PAA. See V. G. Fryd, "Hiram Powers's *Greek Slave*: Emblem of Freedom" (1982). Marsh further extolled Powers, and American artists generally, in *The Goths in New-England* (1843), 25.

6. GPM, "Study of Nature," 43–45.

7. Ibid., 47. Ruskin found delectable a pastoral spot in the Jura, until he imagined it "a scene in some aboriginal forest of the New Continent: [then] the hills became oppressively desolate; a heaviness in the boughs of the darkened forest showed how much of their former power had been dependent on . . . the deep colors of human endurance, valor, and virtue," and on the human monuments that graced their borders (John Ruskin, *The Seven Lamps of Architecture* [1848], ch. VI, sec. I [New York: Noonday Press, 1961], 167–69). On Ruskin's loss of zest for nature in itself, see Robert Hewison, *John Ruskin: The Argument of the Eye* (London: Thames and Hudson, 1976), 125.

8. GPM, "Study of Nature," 49. I discuss these scenic tastes in "The Place of the Past in the American Landscape," in David Lowenthal and Martyn J. Bowden, eds., *Geographies of the Mind* (New York: Oxford University Press, 1976), 89–117 at 101–4.

9. R. B. Stein, *John Ruskin and Aesthetic Thought in America, 1840–1900* (1967), 164; GPM, "Study of Nature," 52–53.

10. GPM to Charles Lanman, 21 Apr. 1847, in Lanman, *Haphazard Person-alities* (1886), 100–101. See also GPM to William Coleman, 13 Aug. 1840, FC; to S. F. B. Morse, 14 Feb. 1844, Morse MSS; and Chapter 9 below.

11. GPM to Jewett, 24 May 1849; C. C. Jewett, "Report of the Assistant Secretary in Charge of the Library of the Smithsonian Institution, for the Year 1850," in *Fifth Ann. Rept. . . . SI* (Washington, D.C., 1851), 28–41 at 29–30 (reprinted in *LL*, app. 3, pp. 472–73); obit. of GPM, *Nation* 35 (1882): 75–76. A 94-page "List of Engravings" Marsh wrote for G. P. A. Healy in May 1846 attests his expertise. This "list" (which came to light only in 1999) was an annotated part-translation of the Milanese engraver Giuseppe Longhi's *La calcografia* (Milan, 1830), held by Marsh to be the best text on the theory and history of engraving, and of Adam von Bartsch's *Die Kupperstichkunde* (Vienna, 1821). Dismissing W. Y. Ottley's *Early History of Engraving* (London, 1816) as "a work of much more pretension . . . than merit," Marsh rated no English study "of any value whatever" beyond mere description (pp. 1–3). A fire at the Smithsonian destroyed Marsh's art collection in 1865.

12. GPM, "Study of Nature," 47n. For Marsh on Ruskin, see also Chapter 16 below.

13. Nineteenth-century Americans commended art and literature for their moral and civic utility (R. M. Elson, *Guardians of Tradition: American Schoolbooks of the Nineteenth Century* [1964], 231–35).

14. GPM to CCM, 28 Feb. 1838; *M&N*, 1884 ed., 12n. Marsh urged adults newly studying a foreign language "to listen long before they attempt to articulate, and to insist that the teacher, not the pupil, shall read the lessons" (GPM, "Old Northern Literature" [1845], 255). See Chapter 17 below.

15. GPM to Rafn, 21 Oct. 1833, in Benedikt Grøndal, *Breve fra og til Carl Christian Rafn, med en Biografi* (1869), 293–94; David Lowenthal, "G. P. Marsh and Scandinavian Studies" (1957).

16. GPM to CCM, in *LL*, 32; to Rafn, 17 Dec. 1834; Richard Beck, "George P. Marsh and Old Icelandic Studies" (1943), 202. Scandinavian volumes comprise about one-fifth of the 12,500 total (University of Vermont, *Catalogue of the Library of George Perkins Marsh* [1892]).

17. O. J. Falnes, "New England Interest in Scandinavian Culture and the Norsemen" (1937); Adolph B. Benson, "The Beginning of American Interest in Scandinavian Literature," *SS* 8 (1925): 133–41; Stefán Einarsson, *History of Icelandic Prose Writers, 1800–1940, Islandica* 32–33 (Ithaca: Cornell University Press, 1948), 16–21.

18. *M&N*, 56n4.

19. Rafn to GPM, 25 June, 29 Nov. 1834, 19 Nov. 1835. See GPM to Rafn, 17 Dec. 1834; George Bancroft to Rafn, 27 Dec. 1836, in Grøndal, *Rafn*, 181–82; ibid., 47–51.

20. C. C. Rafn, *Antiquitates Americanae* (1837); GPM to Rafn, 20 Aug. 1839; Rafn to GPM, 16 Jan. 1847; J. R. Lowell, *The Biglow Papers*, 2d ser. (1862), in *The Poems of James Russell Lowell* (London: Henry Frowde, 1912), 333–34. See Geraldine Barnes, "Reinventing Paradise: Vinland 1000–1992," in Barnes et al., eds., *Old Norse Studies in the New World* (Sydney: Sydney University, Dept. of English, 1994), 19–32 at 22–23. Helge Ingstad, *The Norse Discovery of America*, vol. 2 (Oslo: Norwegian University Press, 1985), and Erik Wahlgren, *Vikings and America* (New York: Thames & Hudson, 1986), summarize evidence on pre-Columbian Scandinavian settlement.

21. Robin Fleming, "Picturesque History and the Medieval in Nineteenth-Century America" (1995): 1080–84.

22. GPM, *A Compendious Grammar of the Old-Northern or Icelandic Language* (1838), iii–iv, ix–xi, 103, 140–42. See Rafn to GPM, 29 Nov. 1834, in Grøndal, *Rafn*, 295–99; GPM, "Origin, Progress, and Decline of Icelandic Historical Literature" (1841): 449–60; W. H. Auden, "Njal's Saga," *New Statesman and Nation* 52 (1956): 551–52. The chief basis for Marsh's translation was Rask's *Kortfattet Vejledning til det oldnordiske eller gamle islandske Sprog* (Copenhagen, 1832). Five years after Marsh's appeared George Webbe Dasent's *Grammar of the Icelandic or Old-Norse Tongue* (London, 1843); Marsh later extolled Dasent's edition of Guðbrandur Vigfússon's *Icelandic-English Dictionary* (Oxford, 1874) (GPM, "Lexicon, Dictionary, Thesaurus, Vocabulary, Glossary," *Johnson's New Universal Cyclopaedia* [1876], 2:1752).

23. GPM, "Summary of the Statistics of Sweden" (1851), and "The Origin and History of the Danish Sound and Belt-Tolls" (1844).

24. GPM to CCM, 1839, in *LL*, 31–32.

25. GPM, "Old Northern Literature" (1845), 256; GPM to CCM, 1839, in *LL*, 34.

26. GPM, "Swedish Literature: 1. Olof Rudbeck . . . and His Atlantica" (1841): 63, 68–74, 81; "2. The Life and Works of the Painter Hörberg" (1841); GPM, *The American Historical School* (1847), 19. See Gunnar T. Eriksson, *The Atlantic Vision: Olaus Rudbeck and Baroque Science* (Canton, Mass.: Scientific History Publications, 1994).

27. GPM, "Translations from the German" (1845).

28. GPM, "The River, from . . . Tegnér" (1841); CCM, "Axel," in her *Wolfe of the Knoll* (1860), 261–307; Andrew Hilen, *Longfellow and Scandinavia* (New

Haven: Yale University Press, 1947), 57–58; A. M. Sturtevant, "An American Appreciation of Esaias Tegnér," *SS* 16 (1941): 157–64 at 158; John B. Leighly, "Inaccuracies in Longfellow's Translation of Tegnér's 'Nattvardsbarnen,'" *SS* 21 (1949): 171–80 at 171–73; Leighly, letter to the author, 4 Aug. 1951. However, Marsh later praised the consonances in Longfellow's *Miles Standish* as marks of Old-Northern expertise (*LEL* [1861], 560–61).

29. N. H. Julis to Longfellow, 23 May 1838, in Samuel Longfellow, *Life of Henry Wadsworth Longfellow, with Extracts from His Journals and Correspondence* (Boston, 1893), 1:288; Ole Munch Raeder to Caroline [Raeder], 28 Dec. 1847, in *America in the Forties: The Letters of Ole Munch Raeder*, tr. Gunnar J. Malmin (Minneapolis: University of Minnesota Press, 1929), 163; GPM to Rafn, 1 Jan. 1864 [tr. Einar Haugen]; GPM to C. C. Andrews, 5 July 1875, Andrews Papers.

30. Beck, "Marsh and Old Icelandic Studies," 199. Fiske introduced himself to Marsh in the 1850s, had the run of Marsh's library, visited him in Italy, and after Marsh's death rented his villa in Florence. See Fiske to GPM, 13 Mar. 1851; GPM to Rafn, 5 Dec. 1851, Fiske to CCM, various dates, 1883; Horatio S. White, *Willard Fiske: Life and Correspondence* (New York: Oxford University Press, 1925), 12; Halldór Hermannsson, "Willard Fiske and Icelandic Bibliography," *Bibliographical Society of America Papers* 12 (1918): 97–106 at 97; *Nation* 35 (1882): 75–76; Bayard Taylor to GPM, 7 Jan. 1861.

31. GPM, *Goths in New-England*, 10, 14.

32. GPM, *Address, Delivered before the New England Society of the City of New-York* (1844) [hereafter *NES*]; Edward A. Freeman, *History of the Norman Conquest* (1867–70) and *History of the English Constitution* (1872); Thomas Jefferson, 4 July 1776, quoted in Merrill Peterson, *Thomas Jefferson and the New Nation* (New York: Oxford University Press, 1970), 98. See J. J. Ellis, *American Sphinx: The Character of Thomas Jefferson* (1997), 32–34.

33. D. G. Mitchell, *American Lands and Letters; Leather-stocking to Poe's "Raven"* (1907), 37; Samuel Kliger, "George Perkins Marsh and the Gothic Tradition in America" (1946); E. N. Saveth, *American Historians and European Immigrants, 1875–1925* (1948), chs. 1, 2, 8; A. A. Ekirch, *The Idea of Progress in America* (1944), 40–95; Merle Curti, *The Growth of American Thought* (1943), 247–53; G. P. Gooch, *History and Historians in the Nineteenth Century*, rev. ed. (London: Longman, 1952), ch. 4; H. Trevor Colbourn, *The Lamp of Experience: Whig History and the Intellectual Origins of the American Revolution* (Chapel Hill: University of North Carolina Press, 1965), 194–98; Dorothy Ross, "Historical Consciousness in Nineteenth-Century America," *AHR* 89 (1984): 909–28 at 917–20.

34. Isaiah Berlin, *Vico and Herder* (London: Hogarth Press, 1976), 194–

99; GPM, *NES*, 48–49; Alexis de Tocqueville, *Democracy in America* (1835), 1:
219; GPM, *LEL*, 5; *Goths in New-England*, 33, 7, 38–39. See also Merle Curti, *The
Roots of American Loyalty* (1946), 122–30. Marsh admired Tocqueville's insights
on France as well as America (UV, *Catalogue of the Library of George Perkins Marsh*,
674).

35. "The Persian Wars," in *The History of Herodotus*, tr. George Rawlinson (New York: Tandy-Thomas, 1909), 285–86; John White, *The Planters Plea*
(1630), in Peter Force, ed., *Tracts and Other Papers, Relating [to] the Colonies of North
America* (Washington, 1838), 18.

36. Charles de Secondat, Baron de Montesquieu, *The Spirit of Laws* (1748),
tr. Thomas Nugent (Chicago: Encyclopaedia Britannica, 1952), 126.

37. GPM, *NES*, 17–27. On environmentalism and Gothicism, see Samuel
Kliger, *The Goths in England: A Study in Seventeenth and Eighteenth Century Thought*
(Cambridge: Harvard University Press, 1952), 241–52; Thor J. Beck, *Northern
Antiquities in French Learning and Literature (1755–1855)* (New York: Columbia Uni-
versity Inst. of French Studies, 1934), 1:20–21.

38. GPM, *Goths in New-England*, 11–13, 19; "Old Northern Literature,"
250–51. For the seventeenth-century genesis of Anglo-Saxonism, see Christopher Hill, "The Norman Yoke" (1953), in his *Puritanism and Revolution* (London:
Panther, 1968), 55–125.

39. GPM, *NES*, 34; *Goths*, 19, 21. From the 1820s, the Pilgrims had begun
to replace the Puritans as exemplary Founding Fathers. Marsh admired both
without stressing the distinction. See Peter J. Gomes, "Pilgrims and Puritans:
'Heroes' and 'Villains' in the Creation of the American Past," *Massachusetts Historical Society Proc.* 95 (1983): 1–16.

40. GPM to John N. Pomeroy, 16 Dec. 1844.

41. GPM, *NES*, 48; *Goths*, 14.

42. *The Diary of Philip Hone, 1828–1851*, ed. Bayard Tuckerman (New York,
1889), 2:334 (22 Dec. 1847); GPM, *NES*, 48. See *HVG*, 2:334–38, 367.

43. GPM, *Mediaeval and Modern Saints and Miracles* [hereafter *MMSM*] (1876),
194–95. The tastes Marsh castigated are discussed in Fleming, "Picturesque
History and the Medieval in Nineteenth-Century America," 1062–72.

44. Lewis Stilwell, *Migration from Vermont* (1948), 184–85; Jeremiah K. Du-
rick, "The Catholic Church in the Diocese of Burlington, Vermont—a Centenary History," in *1853–1953: One Hundred Years of Achievement by the Catholic
Church in the Diocese of Burlington, Vermont* (Lowell, Mass., 1953), 20–36 at 25;
GPM to R. H. Dana, 8 Oct. 1866, Dana MSS. See Dale Knobel, *Paddy and the*

Republic: Ethnicity and Nationality in Antebellum America (Middletown, Conn.: Wesleyan University Press, 1986), 69–88; Oscar Handlin, *Boston's Immigrants, 1790–1865* (Cambridge: Harvard University Press, 1941), 191–99.

45. GPM, *Goths*, 37; Petition of John Shaw & others to the General Assembly of Vermont, 12 Oct. 1835, approved by Select Judiciary Committee (GPM, chairman), 31 Oct. 1835 (MSS VSP 654:119 [State Archivist D. Gregory Sanford to the author, 24 June 1998]); U.S. *CG*, 28 Cong. 1 sess., 674 (31 May 1844). The 1844 naturalization bill lost by 131 to 25.

46. *New Englander* 2 (1844): 490; "Episcopus," *Remarks on an Address Delivered before the New England Society of the City of New York* (Boston, 1845), 10–11.

47. George Allen to GPM, 19 June 1844; Rufus Choate to James Marsh, 14 Nov. 1829, quoted in S. G. Brown, *The Works of Rufus Choate, with a Memoir of His Life* (1862), 1:29. On George Allen, see *LL*, 66n. Marsh gloried in the fact that an English Nonconformist George Marsh (1515–1555) had been martyred (see p. 345).

48. GPM, *Goths*, 14; M. K. Cayton, *Emerson's Emergence* (1989), 73. See James Marsh, "Preliminary Essay," in his edition of Coleridge, *Aids to Reflection* (1829), xiii–xlvi; GPM, *LEL*, 189; Marjorie Nicolson, "James Marsh and the Vermont Transcendentalists" (1925): 35–37; Henry A. Pochmann, *German Culture in America: Philosophical and Literary Influences, 1600–1900* (Madison: University of Wisconsin Press, 1957), 88–95, 132–42; L. D. Walls, *Seeing New Worlds: Henry David Thoreau and Nineteenth-Century Natural Science* (1995), 55.

49. Walls, *Seeing New Worlds*, 29; GPM, *Goths*, 15–16.

50. GPM, *Goths*, 18, 6. See GPM, "Study of Nature" (1860), 57–58. Marsh used the word *Puritan* in its religious sense "as embracing all those sects, which hold, that the Bible is the *only* rule of Christian faith and practice, and reject the authority of tradition in rites, doctrine, and church government" (*NES*, 22n). On the popularity and felt menace of Paley's doctrine of moral expediency, see Wendell Glick, "Bishop Paley in America," *NEQ* 27 (1954): 347–54; Stein, *John Ruskin*, 158–59. On the reaction against Locke, see Merle Curti, "The Great Mr. Locke, America's Philosopher," in his *Probing Our Past* (1955), 85–93; Lewis Feuer, "James Marsh and the Conservative Transcendentalist Philosophy" (1958), 12–19.

51. GPM quoted in James Marsh to Nathan Lord, Dec. 1835, in James Marsh, *Coleridge's American Disciples: The Selected Correspondence* (1973), 188; CM quoted and Lyndon A. Marsh cited in Russell Streeter, *Mirror of Calvinistic, Fanatical Revivals, or Jedidiah Burchard & Co. during a protracted meeting of twenty-six days,*

in Woodstock, Vt. (Woodstock, 1835), 101, 41. (Burchard's reference to Lyndon Marsh is quoted from a variant edition of Streeter, in Ralph Hill, *Contrary Country: A Chronicle of Vermont* [New York: Rinehart, 1950], 112.)

52. C. G. Eastman, *Sermons, Addresses & Exhortations, by Rev. Jedidiah Burchard* (Burlington, 1836); Burchard, James Marsh to Nathan Lord, and GPM to CM, in J. J. Duffy and H. N. Muller, *An Anxious Democracy* (1982), 26–41. See Ludlum, *Social Ferment*, 56–57; Muller and Duffy, "Jedidiah Burchard and Vermont's 'New Measure' Revivals: Social Adjustment and the Quest for Unity" (1978), in Muller and Hand, *In a State of Nature* (1982), 117–26.

53. Vermonters' reactions quoted in Duffy and Muller, *Anxious Democracy*, 57–77.

54. GPM, "[Petition, Dec. 12, 1837] To His Excellency the Governor of Vermont" [with 22 others], and S. H. Jenison, "A Proclamation by the Governor" (13 Dec. 1837), *Vermont Chronicle*, 20 Dec. 1837.

55. GPM et al., "Petition"; James Marsh to David Read, and *BFP*, 15 Dec. 1837, quoted in Duffy and Muller, *Anxious Democracy*, 90, 100n5.

4. Congress and the Smithsonian

1. GPM in *BFP*, 10 Jan. 1840.

2. *BFP*, 2 Dec. 1831, 16 June 1843. See also Walter H. Crockett, *Vermont, the Green Mountain State* (New York: Century History Co., 1921), 3:307–12.

3. GPM in *BFP*, 16 June 1843; *Vermont Patriot* (Montpelier), 17 June 1843; *BFP*, 11 Aug., 8 Sept. 1843. Marsh may have been "exceedingly retiring and modest," but not to the point of being elected "almost without his knowledge," as a friend averred (Charles Lanman, *Letters from a Landscape Painter* [1845], 111).

4. GPM to Maria Buell Hickok, 5 Dec. 1843; slightly modified in *LL*, 59–60.

5. *Memoirs of John Quincy Adams* (1877), 12:57.

6. Foot to GPM, 4 Apr. 1860. See Mary Louise Kelly, *Woodstock's U.S. Senator: Jacob Collamer* (Woodstock: Woodstock Historical Society, 1944); Holman Hamilton, *Zachary Taylor* (1951), 165.

7. Alvah Sabin, Justin Morgan, and D. M. Camp to GPM, 16 Oct. 1844, VHS Misc. file 1080. See George F. Edmunds, *The Life, Character and Services of Solomon Foot* (Montpelier, 1866), 7–14.

8. GPM to CCM, 25 Aug. 1856 (UVM), and 1839, in *LL*, 31.

9. Caleb Atwater, *Mysteries of Washington City, during Several Months of the Session of the 28th Congress* (Washington, D.C., 1844), 19–20. See *LL*, 67; Wil-

helmus B. Bryan, *A History of the National Capital* . . . (New York: Macmillan, 1914–16), 2:294–95, 317; Samuel C. Busey, *Pictures of the City of Washington in the Past* (Washington, D.C., 1898), 143.

10. GPM to J. N. Pomeroy, 30 Mar. 1844.

11. GPM to Maria Buell Hickok, June 1844; Collamer to Frances Collamer, 4 Feb. 1844, JCP.

12. *LL*, 72.

13. GPM, "Speech on the Question of the Annexation of Texas" (1845), 319; GPM to Pomeroy, 15 Jan. 1844. On Clingman, see Hamilton, *Zachary Taylor*, 271. For Marsh's voting record, see Joel H. Silbey, *The Shrine of Party: Congressional Voting Behavior, 1840–1852* (Pittsburgh: University of Pittsburgh Press, 1967), app. II, 170, 182, 188.

14. GPM, *Speech on the Tariff Bill* (1844), 3, 9.

15. Ibid., 5, 13, 4. On New England's need for high tariffs, see Paul Varg, *New England and Foreign Relations, 1789–1850* (1983), 101–17, 204–5.

16. BFP, 7 June 1844; George Allen to GPM, 19 June 1844; GPM, 30 June 1846, in U.S. *CG*, 29 Cong. 1 sess., app., 1009–11.

17. GPM to Pomeroy, 30 Mar. 1844. U.S. senators were elected not by popular vote but by state legislatures until 1913. On Marsh's unsuccessful senatorial run against S. S. Phelps, see VSP 92:216–18.

18. GPM quoted in *LL*, 70–71.

19. GPM to Pomeroy, 16 Dec. 1844, 13 Jan. 1845.

20. GPM, "Speech on the Question of the Annexation of Texas," 316–19. See Varg, *New England*, 177; Michael Holt, *The Political Crisis of the 1850s* (1978), 42–44.

21. Rufus Choate to GPM, 30 Aug. 1845, in J. V. Matthews, *Rufus Choate* (1980), 70. Marsh had angrily rebutted the contention that Vermont, as the other "foreign power" taken into the Union, was a precedent for annexing Texas and seating its congressmen without delay. Not so, said Marsh: while Texas had broken away from Mexico, Vermont had always belonged to what became the United States, as delineated in the Treaty of 1783. Vermont's rebellion against New York had arisen owing to that state's failure to adopt the resolutions of the Provincial Congress (GPM, "Speech on the Question . . . of Texas," 315).

22. GPM to Pomeroy, 30 Mar. 1844; *LL*, 71.

23. GPM, Report on Petition of E. H. Holmes and W. Pedrick (1844), 1–2.

24. GPM, Report on Spirit Ration in Navy (1845), 1–2; Horace Greeley,

Recollections of a Busy Life (1868; reprint, Port Washington, N.Y.: Kennikat Press, 1971), 228. The Senate had killed a similar bill the previous year, George Mc-Duffie of South Carolina holding that they "might as well abolish the navy" as give up grog (U.S. *CG*, 28 Cong. 1 sess., 681–82). For later legislation, see *CG*, 29 Cong. 2 sess., 291; 12 Stat. at Large 565 (1862); for GPM on flogging, 15 June 1844, *H. Journal*, 28 Cong. 1 sess., 1135–36; GPM to CM, 4 Sept. 1852.

25. GPM to Edward C. Lester, 26 Aug. 1844, FC. See GPM, Report on Durazzo Library (1844).

26. G. B. Goode, "Genesis of the U.S. National Museum" (1897), esp. GPM to FM, 4 Apr. 1844, p. 133; GPM, Report on Memorial of the National Institute for the Promotion of Science (1844), 3. See Daniel Henderson, *The Hidden Coasts: A Biography of Admiral Charles Wilkes* (New York: Sloane, 1953), 201–18; Madge E. Pickard, "Government and Science in the United States: Historical Backgrounds," *Journal of the History of Medicine and Allied Sciences* 1 (1946): 254–89, 446–81; Benjamin Tappan, Senate Journal, MS, 23 Jan. 1844, p. 6.

27. GPM to CCM, 1845, in *LL*, 93.

28. W. J. Rhees, ed., *The Smithsonian Institution: Documents Relative to Its Origin and History* (1879), 2, 164, 218, 432–33, 447–48; Adams, *Memoirs*, 12:236; P. H. Oehser, *Sons of Science: The Story of the Smithsonian Institution and Its Leaders* (1949); A. H. Dupree, *Science in the Federal Government* (1957), 66–90; Pickard, "Government and Science," 446–81.

29. GPM to N. S. Moore, 22 Apr. 1848, FC.

30. Robert Dale Owen, 22 Apr. 1846, quoted in Rhees, *Smithsonian . . . Origin and History*, 277–78. See R. W. Leopold, *Robert Dale Owen* (1940), 220–26; Adams, *Memoirs*, 12:177, 235–36.

31. GPM, *Speech on the Bill for Establishing the Smithsonian* (1846), 4, 12–13.

32. Owen in U.S. *CG*, 29 Cong. 1 sess., 712.

33. GPM, *Speech on . . . Smithsonian*, 7–11; H. D. Thoreau, *Journal*, 16 Mar. 1852, vol. 4 (Princeton: Princeton University Press, 1992), 392. Difficulties in using European collections led Marsh later to discount the value of large public libraries.

34. Adams, *Memoirs*, 12:259; *New Englander* 4 (1846): 607; Isaac Edward Morse in H. Reps., *CG*, 29 Cong. 1 sess., 718–19. Morse was no unlettered backwoodsman, but a Harvard graduate.

35. Leopold, *Owen*, 227; J. Q. Adams and Andrew Johnson, 28 Apr. 1846, U.S. *CG*, 29 Cong. 1 sess., 738, 741.

36. W. J. Hough and GPM, U.S. *CG*, 29 Cong. 1 sess., 749; Rhees, *Smith-*

sonian . . . Origin and History, 469–72; R. V. Bruce, The Launching of Modern American Science, 1846–1876 (1987), 190–94; Matthews, Rufus Choate, 116–33.

37. W. H. Rhees, ed., The Smithsonian Institution: Journals of the Board of Regents (1879) [hereafter SIJ], 11–13. On Bache, see H. R. Slotten, Patronage, Practice, and the Culture of American Science: Alexander Dallas Bache and the United States Coast Survey (1994); on Jewett, M. H. Harris, The Age of Jewett (1975).

38. Bache to Henry, 4 Dec. 1846, in Merle M. Odgers, Alexander Dallas Bache: Scientist and Educator, 1806–1867 (Philadelphia: University of Pennsylvania Press, 1947), 165–66; Leopold, Owen, 244–45; SIJ, 23 Dec. 1847, pp. 27–47. See J. A. Boromé, Charles Coffin Jewett (1951), 23; Thomas Coulson, Joseph Henry (1950), 176–77.

39. GPM, SI Affairs, draft letter to C. W. Upham [then head of Board of Regents], Feb. 1855; GPM to John Russell Bartlett, confidential, 31 Mar. 1848, in Papers of Joseph Henry, 7 (1996): 298. See Coulson, Joseph Henry, 189, 211–12; Choate to GPM, 23 Dec. 1846; SIJ, 25–26.

40. SIJ, 448–51, 724–27; Cyrus Adler, "The Smithsonian Library," in G. B. Goode, The Smithsonian Institution, 1846–1896 (1897), 265–302.

41. GPM in the House, 11 Dec. 1848, U.S. CG, 30 Cong. 2 sess., 24.

42. Following the lead of veterans' organizations, Congress took umbrage at the Smithsonian's Air and Space Museum's projected in-depth exhibition on the ending of World War II in the Pacific, forcing its cancellation in 1994. See Martin Harwit, An Exhibit Denied: Lobbying the History of Enola Gay (New York: Springer-Verlag, 1996), 245–60, 424–25. As after reactions to the Smithsonian's 1991 "The West as America" show, the specter of congressional control over the Smithsonian was reinvoked. See Gary B. Nash, Charlotte Crabtree, and Ross E. Dunn, History on Trial: Culture Wars and the Teaching of the Past (New York: Knopf, 1997), 123–27.

43. J. G. Barnard, "Memoir of Joseph Gilbert Totten, 1788–1864," Natl. Acad. of Sciences Biographical Memoirs 1 (1877): 35–97.

44. Journal Exec. Comm., SIJ, 464–65.

45. GPM, 9 June 1847, in SI, First Ann. Rept. (1847), 22; GPM, "Advertisement," in E. G. Squier and E. H. Davis, Ancient Monuments of the Mississippi Valley (1848), x.

46. Squier and Davis, Ancient Monuments, xxxvi; GPM to Squier, 6 & 16 June 1848 (see also 23 Feb. & 6 Mar. 1847), Squier Coll.; Papers of Joseph Henry, 7:59n; Thomas G. Tax, "E. George Squier and the Mounds, 1845–1850," in Timothy H. H. Thoresen, ed., Towards a Science of Man: Essays on the History of An-

thropology (The Hague: Mouton, 1975), 99–124. The apparent similarity of trees on and off the earthworks later struck Marsh as evidence of the great antiquity of the mounds (*M&N*, 121–22n11).

47. GPM to C. C. Jewett, 8 Feb. 1851, in *LL*, 201–2; *M&N*, 40–41n37, 121–22n11, 199n8.

48. Squier's own speculative views on the mound builders' origin and disappearance were kept out of the book by Joseph Henry. See Robert Silverberg, *The Mound Builders* (New York: Ballantine Books, 1974), esp. 54–63.

49. GPM to Mary Baird, 10 Feb. 1847, BSI; Henry to SFB, 3 Mar. 1847, in W. H. Dall, *Spencer Fullerton Baird* (1915), 163–64; GPM to SFB, 19 & 14 June 1848, 25 Apr. 1849. See John Wesley Powell, "The Personal Characteristics of Professor Baird," *SI Ann. Rept. 1888*, 739–44; Oehser, *Sons of Science*, 60, 89, 92.

50. GPM to SFB, 6 & 10 Oct. 1848, BSI.

51. GPM to SFB, 25 Apr. 1849; Dall, *Baird*, 184–85; Spencer F. Baird, *Iconographic Encyclopaedia*, tr., ed., and updated from Johann Georg Heck, *Bilder-Atlas zum Konversations-Lexikon* (New York, 1851–52).

52. SFB to GPM, 9 July 1850; E. F. Rivinus and E. M. Youssef, *Spencer Baird of the Smithsonian* (1992), 133, 154–55. See G. B. Goode, "The Three Secretaries" (1897), in his *Smithsonian Institution*, 115–234 at 166–67; Bruce, *Launching*, 195–99.

5. American History from the Ground Up

1. CCM to Maria Buell Hickok, 29 Nov. 1845; *LL*, 72. Known as "Gedney's House," the Marshes' F Street home was owned by James M. Waterbury.

2. GPM, "The Late General Estcourt" (also in *LL*, app. 4, pp. 474–79); GPM, *Speech on . . . Smithsonian* (1846), 10; *LL*, 100, 103. Jean Paul was the pen name of the German novelist and litterateur Jean Paul Friedrich Richter (1763–1825), much admired by Marsh.

3. *LL*, 116; GPM to SFB, 4 & 8 Apr. 1848, BSI, 22 June 1848. See F. A. Wislizenus, Jr., "Sketch of the Life of Dr. Wislizenus," in Frederick A. Wislizenus, *A Journey to the Rocky Mountains in the Year 1839* (St. Louis, 1912), 11; Carl Wittke, *Refugees of Revolution: The German Forty-Eighters in America* (Philadelphia: University of Pennsylvania Press, 1952).

4. Jacob Collamer to Mary Collamer, 25 Dec. 1847, JCP; Mitchell to Mary Goddard, 2 & 28 Jan. 1847, in Waldo H. Dunn, *The Life of Donald G. Mitchell, Ik Marvel* (New York: Scribner, 1922), 167–70.

5. D. G. Mitchell, *American Lands and Letters: Leather-stocking to Poe's "Raven"* (1907), 38.

6. GFE to CCM, 13 May 1888; *LL*, 73.

7. GPM to Dr. Isaac Hayes, 2 Dec. 1845, 3 Jan. 1848, Hayes Papers; 28 May 1845, VHS; Charles Lanman, *Haphazard Personalities* (1886), 106.

8. GPM to Maria Buell Hickok, June 1844, in *LL*, 68–69; to GO, 30 Sept., 5 June 1848.

9. GO to GPM, dated only 1848; B. Sears to GPM, 29 Mar. 1848; J. V. C. Smith to GPM, 27 June 1848.

10. GPM to GO, 13 May 1848; GO to CCM, 4 June 1848; GPM to SFB, 9 June 1848.

11. GO to GPM, 12 June 1848; GPM to SFB, 19 June 1848; C. S. Richards to GPM, 4 July 1849.

12. GO to GPM, 1848; SM to Maria Buell Hickok, 29 Aug. 1848.

13. GPM to Henry Yule, 10 Jan. 1877; to Caroline Estcourt, 28 July 1847, in *LL*, 108; UV Trustees Minutes, 1843–49, 3:83–117.

14. *LL*, 105–6. See W. H. Dall, *Spencer Fullerton Baird* (1915), 169, for this occasion, and for Baird's benign view of snakes, 146–47.

15. Charles Sumner to GPM, 5 May 1847; GPM, *Human Knowledge* (1847); Edward Everett Journal, Everett Papers, 166:161; Ticknor to GPM, 6 Sept. 1847; Choate to GPM, 1847; Boston *Daily Advertiser*, 27 Aug. 1847; *New Englander* 6 (1848): 312. See Rutherford B. Hayes to Mrs. W. A. Platt, 30 Aug. 1847, in R. B. Hayes, *Diary and Letters*, ed. C. R. Williams (Columbus: Ohio State Archaeological & Historical Soc., 1922–26), 1:215–16.

16. GPM, *The American Historical School* (1847), and *Address . . . Agric. Soc. Rutland* (1847).

17. GPM to J. R. Bartlett, 9 June 1847, Knollenberg Coll.; GPM to Pomeroy, 27 Jan. 1845; Kenneth Olwig, "Historical Geography and the Society/Nature 'Problematic': The Perspective of J. F. Schouw, G. P. Marsh and E. Reclus" (1980): 32–38. Marsh did not find time to translate Schouw.

18. GPM, *Human Knowledge*, 12–14, 18–24.

19. GPM, *Goths in New-England* (1843), 33; *American Historical School*, 10. On Marsh's idea of history, see Charles A. Beard and Mary R. Beard, *The American Spirit; A Study of the Idea of Civilization in the United States* (New York: Macmillan, 1942), 264–73; Merle Curti, *The Growth of American Thought* (1943), 411; idem, *The Roots of American Loyalty* (1946), 50–51; David Lowenthal, *The Past Is a Foreign Country* (1985), 244.

20. GPM, *American Historical School*, 11; Job Durfee (1843) quoted in J. V. Matthews, *Rufus Choate* (1980), 93-94.

21. *American Historical School*, 26.

22. GPM, *M&N*, 203-4 (Adirondacks); Fisher, "Vermonters," in *Vermont: A Guide to the Green Mountain State* (1937), 3-9 at 3. For later efforts to market Vermont as nostalgia, see Clare Hinrichs, "Consuming Images: Making and Marketing Vermont as Distinctive Moral Place," in E. Melanie DuPuis and Peter Vandergeest, eds., *Creating the Countryside: The Politics of Rural and Environmental Discourse* (Philadelphia: Temple University Press, 1996), 259-78.

23. Edward Eggleston (1900), quoted in Herman Ausubel, *Historians and Their Craft: A Study of the Presidential Addresses of the American Historical Association, 1884-1945* (New York: Columbia University Studies in History, 1950), 314; Finley Peter Dunne, *Observations by Mr. Dooley* (New York: Harper, 1906), 271. See Edward Eggleston, "Formative Influences," *Forum* 10 (1890): 279-90 at 286-87; Michael Kraus, *The Writing of American History* (Norman: University of Oklahoma Press, 1953), 127-29; Merle Curti, "The Democratic Theme in American Historical Literature" (1952), in his *Probing Our Past* (1955), 5-11, 16-18; Charles A. Beard and Alfred Vagts, "Currents of Thought in Historiography," *AHR* 42 (1937): 460-83 at 482; James Harvey Robinson, *The New History* (New York: Macmillan, 1912), chs. 1, 5; Harvey Wish, *The American Historian: A Socio-Intellectual History of the Writing of the American Past* (New York: Oxford University Press, 1960), 134-37, 267-69; Eric Hobsbawm, "On History from Below," in his *On History* (London: Weidenfeld & Nicolson, 1997), 201-16.

24. GPM, *American Historical School*, 26.

25. GPM to E. G. Squier, 7 Aug. 1848; to C. C. Jewett, 8 Feb. 1851, in *LL*, 201-2.

26. *M&N*, 311n32; *Address . . . Agric. Soc. Rutland* (1847), 6, 8, 10-11, 17-19.

27. GPM to Albert Gallatin, 15 Dec. 1847, 3 Jan. 1848, Gallatin Papers. The pamphlet Marsh helped distribute was probably "Peace with Mexico" (1847) [*The Writings of Albert Gallatin*, ed. Henry Adams (New York, 1879), 3: 555-91]; the one Marsh helped write was "War Expenses" (1848) [see *Writings*, 2:666].

28. GPM, *Speech on the Mexican War* (1848), 7, 11, 4, 12.

29. Albert G. Brown (Mississippi), U.S. *CG*, 30 Cong. 1 sess., 333.

30. GPM, *Remarks on Slavery in the Territories* (1848), 9, 8.

31. J. G. von Herder quoted in William A. Wilson, "Herder, Folklore and Romantic Nationalism," *Journal of Popular Culture* 6 (1973): 819-35 at 822;

David Lowenthal, *The Heritage Crusade and the Spoils of History* (London: Viking, 1997), 234, 194. Despite the striking parallels, I find no evidence that Marsh had read Herder.

32. GPM, *Speech on the Mexican War*, 8–11; and *Remarks on Slavery in the Territories*, 12. See P. A. Varg, *New England and Foreign Relations, 1789–1850* (1983), 207–9; Curti, *Roots of American Loyalty*, 45. The threat of Southern secession later persuaded Marsh that "we are and must remain a united people, with a noble region of country stretching from the Atlantic to the Pacific shores" (GPM in *BDFP*, 9 Nov. 1860).

33. Sylvester Churchill to GPM, 16 Dec. 1846; GPM, *Speech on the Mexican War*, 11. See GPM to J. N. Pomeroy, 30 Dec. 1845; GPM to Gallatin, 6 Nov. 1847, and W. H. Emory to GPM, 25 & 30 Dec. 1847, in GPM to Gallatin, 31 Dec. 1847, Gallatin Papers. Emory, a Washington neighbor of Marsh's and subsequently a general, published *Notes of a Military Reconnaissance from Fort Leavenworth in Missouri to San Diego, in California* (1848), and *Report of Commissioners of the United States and Mexican Boundary Survey* (1857).

34. GPM to J. B. Estcourt, end May 1849, in *LL*, 138. From shipboard in Panama, Marsh's friend Gilliss limned these "gold-seekers [as] graceless, unwashed cubs from all the vile corners of our land; . . . preserve you and yours from the misfortune of being in close proximity to . . . this ocean of barbarians" (J. M. Gilliss to GPM, 27 Aug. 1849).

35. GPM to SFB, 15 June 1848, BSI.

36. GPM to Caroline Estcourt, 10 June 1848, in *LL*, 124; to SFB, 15 Sept. 1848; to Jacob Collamer, 11 & 14 Sept. 1848.

37. GPM to SFB, 6 Oct. 1848, BSI; GPM, *Remarks on Slavery in the Territories* (1848), 9; GPM, undated MS.

38. James Barrett to CCM, in *LL*, 129n; GPM, *Remarks on Slavery*, 11. For Marsh as a campaigner, see Merrill Ober, 28 Oct. 1848, in Wilson O. Clough, ed., "A Journal of Village Life in Vermont in 1848," *NEQ* 1 (1928): 32–40 at 39.

39. *BDFP*, 9 Nov. 1848; GPM to Caroline Estcourt, 3 Feb. 1849. Van Buren's Free-Soilers did best in Vermont, with 29 percent of the popular vote (Massachusetts 28, New York and Wisconsin 27).

40. GPM to Maria Buell Hickok, 27 Nov. 1848; to SFB, 7 Dec. 1848, BSI.

41. GPM to Caroline Estcourt, 3 Feb. 1849.

42. GPM to C. D. Drake, 7 Dec. 1848; to C. S. Daveis, 20 Mar. 1849, Daveis Corr.; C. C. Rafn to GPM, 21 July 1848.

43. *BS*, Baltimore *Patriot*, and Boston *Courier* quoted in *BDFP*, 17 Jan., 15 June, 13 Feb. 1849.

44. GPM to C. S. Daveis, 23 Mar. 1849, Daveis Corr. On the politics behind Hannegan's injudicious retention, see M. F. Holt, *The Rise and Fall of the American Whig Party* (1999), 422–25, 546.

45. C. C. Jewett to GPM, 3 July 1849. Marsh could hardly have served as Speaker; when invited to preside in the chair, he asked "to be excused, because," he said, "his eyesight was so imperfect, that he was unable to distinguish members in their places" (9 May 1846, U.S. *CG*, 29 Cong. 1 sess., 782). On the Patent Office prospect, see Joseph Henry to A. D. Bache, 13 Nov. 1848, in *Papers of Joseph Henry* (1996), 7:420.

46. GPM to Caroline Estcourt, 29 May 1849, in *LL*, 139; GPM in *BDFP*, 31 Aug. 1849; C. S. Daveis to GPM, 9 Apr. 1849, Daveis Papers; Solomon Foot to Zachary Taylor, 30 May 1849, U.S. Dept. State, Appointment Papers. See Holman Hamilton, *Zachary Taylor* (1951), 26, 221; B. P. Poore, *Perley's Reminiscences of Sixty Years in the National Metropolis* (1886), 1:359–60. Marsh complained that Vermonters had "not made an effort for me"; if he got a post, "I am not likely to owe it to the goodwill of the people of my own state" (to SFB, 21 March 1849, BSI). But State Department Appointment Papers show he was wrong.

47. Millard Fillmore to John M. Clayton, 10 May 1849, Clayton Papers; *BDFP*, 4 June 1849; GPM to J. B. Estcourt, undated [June 1849] in *LL*, 137, and 22 Oct. 1849, UVM.

48. GPM to SFB, 9 June 1849, BSI; GPM in *BDFP*, 29 Aug., 25 Sept. 1849. Meacham's majority for Marsh's House seat was over 1,900.

49. GPM, "Books and other things sent to Constantinople," 1850 notebook.

50. Henry felt generous in paying Marsh half the sum at once (to Bache, 23 Oct. 1848, in *Papers of Joseph Henry*, 7:608, 619).

51. SM to Maria Buell Hickok, 15 Oct. 1849.

6. *Constantinople and the Desert*

1. GPM to FM, 14 Jan. 1854, FC. See *LL*, 142.

2. GPM to the Estcourts, 22 Oct. 1849; to CM, 25 Oct. 1849, DCA.

3. GPM to Lyndon A. Marsh, 9 Nov. 1849. See GPM to Secretary of State J. M. Clayton, 9 Nov. 1849, U.S. Dept. of State, Turkey: Diplomatic Despatches [DD] 2.

4. GPM to CM, 3 Feb. 1850, DCA; to SFB, 23 Aug. 1850; GPM in *LL*, 148.

5. GPM to CM, 3 Feb. 1850, DCA; to HP, 4 June 1850, PAA.

6. GPM, "Notes on Vesuvius" (1852); *M&N*, 461n28; "Sicilies, The Two," *Johnson's New Universal Cyclopaedia* (1878), 4:266.

7. GPM to SFB, 23 Aug. 1850, BSI; to Charles Folsom, 17 Dec. 1850, Folsom MSS.

8. GPM, "Supplementary Statement of Complaint" (about Dubois's hotel), 4 Apr. 1850, MS; Caroline Paine, *Tent and Harem: Notes of an Oriental Trip* (1859), 6–8. Marsh inveighed against the local Claims Commission in his customary righteous tone: "While admitting that Dubois had completely failed of performing any one of the stipulations into which he had entered, & declaring that I had been misled by a *fraud* (deception), to which *Dubois must have been a party*, they have allowed him the full amount! When did a tribunal of justice ever before make the *fraud* of a plaintiff the ground of a claim against an innocent person?" On Constantinople and Pera, see Andrew Wheatcroft, *The Ottomans: Dissolving Images* (London: Penguin, 1995), 138–66.

9. GPM to HP, 4 June 1850, PAA.

10. GPM to Clayton, 14 Mar. 1850, DD 3; G. W. F. H. Carlisle, *Diary in Turkish and Greek Waters* (1855), 46.

11. GPM to CM, July 1850, DCA; to Edward Everett, 4 Feb. 1852, DD 26; GPM, "Future of Turkey" (1858). See Stanley Lane-Poole, *The Life of the Right Honourable Stratford Canning* (1888), 2:206–16. Ottoman tax-farming, ended by an 1839 decree, was soon resumed to avert imperial bankruptcy; religious equality and secularization were proclaimed but little implemented (Roderic H. Davison, *Reform in the Ottoman Empire, 1856–1876* [Princeton: Princeton University Press, 1963], 18–50; Carter V. Findley, *Bureaucratic Reform in the Ottoman Empire: The Sublime Porte, 1789–1922* [Princeton: Princeton University Press, 1980], 152–57).

12. GPM to Everett, 4 Feb. 1852, DD 26; to FM, 5 July 1852, FC. The 36 million in the Ottoman Empire in 1850 included 11 million Turks, 7 million Arabs, 6 million Slavs, 4 million Romanians, 2 million each of Greeks, Albanians, and Armenians.

13. GPM to SFB, 23 Sept. 1850, BSI; to SM, 17 Apr. 1850, DCA; to FM, 20 Dec. 1852, FC.

14. GPM to F. A. Wislizenus, 19 Dec. 1850, WP; to W. W. S. Bliss, Spring 1852, in *LL*, 261; to the Estcourts, Oct. 1851, in *LL*, 254. See D. M. Goldfrank, *The Origins of the Crimean War* (1994), 29.

15. GPM to C. D. Drake, 3 Aug. 1850; to Daniel Webster, 19 Aug. 1850, DD 13 (confid.). Marsh later withdrew some animadversions against Porter (GPM to J. W. Houston, 10 Nov. 1850; to George A. Porter, 23 Nov. 1850; to

Webster, 5 Nov. 1850, DD 19), but he perhaps recanted partly because his earlier criticism had leaked to Porter via President Fillmore, leaving Marsh open to charges of libel from a man with powerful friends in Washington (GPM to Robert C. Winthrop, 17 Dec. 1850, copy; FM to GPM, 26 May, 7 July 1851).

16. GPM to FM, 26 Apr. 1852, FC. See GPM to Webster, 24 Dec. 1851, DD 24.

17. GPM to Robert C. Winthrop, 4 June 1850, Winthrop Papers.

18. J. P. Brown to GPM, 26 May, 19, 22, & 26 Sept. 1850; Lewis Cass to GPM, 15 Jan. 1858; Daniel Webster, *Papers: Diplomatic*, 2 (1987): 36–37.

19. J. P. Brown to H. A. Homes, 10 Apr. 1851; Reshid Pasha to GPM, in GPM to Webster, 6 Jan. 1851, DD. See Leland J. Gordon, *American Relations with Turkey, 1830–1890; An Economic Interpretation* (Philadelphia: University of Pennsylvania Press, 1932), 12–47.

20. GPM to Clayton, 18 Apr. 1850, to Webster, 18 Dec. 1851, DD 5 & 22; GPM, "Persia Treaty Negotiations" (1852), MS.

21. GPM, "Constantinople and the Bosphorus" (ca. 1855), MS; to William Goodell et al., 24 Dec. 1853; Cyrus Hamlin, "Recollections of Turkey," undated MS, UVM. See Hamlin, *Among the Turks* (New York, 1878), 218–19; Bebek missionaries to GPM, 22 Sept. 1853; *LL*, 348–49n.

22. GPM to Clayton, 25 Mar. 1850, to Webster, 19 June 1852, DD 4 & 29.

23. GPM to Solomon Foot, 1 & 16 Jan. 1855, MV, and 3 Jan. 1855, in GPM, *Memorial . . . for Compensation of His Services* (1854). On Murad, whom Marsh regarded highly, see GPM to Marcy, 5 Sept. 1853, DD 53; Ruth Kark, *American Consuls in the Holy Land, 1834–1914* (1994), 102–8.

24. GPM to Sublime Porte, 13 May 1852, U.S. Dept. of State, Turkey, PR, Misc. Corr. Sent, 149; to H. E. Offley, 1 June 1852, PR, Misc. Corr. Sent, 156.

25. GPM to Webster, 25 Nov. 1850, DD 16; Kark, *American Consuls*, 55–65.

26. Kossuth to GPM, 1 Jan. 1851 (misdated 1850), U.S. Dept. of State, Turkey, PR, Misc. Corr. Rec'd, vol. 2, 1850–54.

27. GPM to Clayton, 18 Apr. 1850, to Webster, 19 Aug. 1850, DD 5 & 12.

28. Clayton to GPM, 12 Jan. 1850, DI (Turkey, vol. 1), 3, p. 338. See Merle Curti, "Austria and the United States, 1848–1852" (1926), 141–60.

29. GPM to Clayton, 14 & 25 Mar. 1850, DD 3 & 4; Kossuth to GPM, 1 Jan. 1851 (misdated 1850), PR, Misc. Corr. Rec'd.

30. GPM to Kossuth, 2 & 8 June 1850, PR, Misc. Corr. Sent, 28, 29; to Clayton, 15 May, 4 July 1850, DD 6 & 10; to Webster, 18 Nov. 1850, DD 15; Webster to GPM, 28 Feb. 1851, DI 15, pp. 346–49.

31. GPM to J. C. Long and to C. W. Morgan, 5 & 6 Sept. 1851, PR, Misc.

Corr. Sent, 101, 103; U.S., *Kossuth and Captain Long* (1852); Long to GPM, 11 Nov. 1851; GPM to H. J. Raymond, Apr. 1852 (copy). See John Bassett Moore, "Kossuth: A Sketch of a Revolutionist," *Political Science Quarterly* 10 (1895): 95–131, 257–91.

32. FM to GPM, 13 Jan. 1852, GPM; Webster, *Papers: Diplomatic*, 2:74–79. See B. P. Poore, *Perley's Reminiscences of Sixty Years in the National Metropolis* (1886), 1:404–6; Donald S. Spencer, *Louis Kossuth and Young America: A Study of Sectionalism and Foreign Policy, 1848–1852* (Columbia: University of Missouri Press, 1977), 2–4, 25–27, 34–37, 161–75. M. F. Holt terms Kossuth America's "most disruptive and politically embarrassing" visitor since Citizen Genêt (*Rise and Fall of the American Whig Party* [1999], 692–97). On Webster's anti-Austrian bellicosity vis-à-vis Amin Bey and Kossuth, see Kenneth E. Shewmaker, "Daniel Webster and the Politics of Foreign Policy," *Journal of American History* 63 (1976–77): 305–15 at 307–11; Robert Remini, *Daniel Webster: The Man and His Time* (New York: Norton, 1997), 698–705.

33. GPM to Solomon Foot, 26 Apr. 1852. Kossuth did eventually write Marsh but only to ask a favor for a friend (7 Aug. 1852, U.S., PR, Misc. Corr. Rec'd). Nonetheless, a decade later Marsh made Kossuth and his family welcome in Turin; see Chapter 12 below.

34. GPM to Charles D. Drake, 3 Aug. 1850; on Pera costs see GPM to Webster, 24 Sept., 17 Dec. 1851, DD 14 & 21.

35. GPM to SFB, 23 Aug. 1850, BSI; to C. D. Drake, 3 Aug. 1850.

36. FM quoted in LW to CCM, 7 Jan. 1851; Foot to GPM, 28 Mar. 1852.

37. GPM to Drake, 3 Aug. 1850; Glyndon G. Van Deusen, *Horace Greeley: Nineteenth Century Crusader* (Philadelphia: University of Pennsylvania Press, 1953), 126–28; U.S. CG, House Journal, 28 Cong. 2 sess., 446–47 (1845).

38. CCM, MS draft for *LL*, 163; GPM to CCM, 4 Mar. 1856, CFP.

39. N.Y. *Tribune*, 28 June 1850.

40. GPM to Drake, 3 Aug. 1850.

41. GPM to SFB, 23 Aug. 1850; to Webster, 24 Sept. 1850, DD 14; to FM, 25 Sept. 1850, FC, and 21 Jan. 1851, in Marsh letterbook, 1851–53; Webster to GPM, 25 Jan. 1851, DI 14, p. 344.

42. GPM to Caroline Estcourt, 21 Jan. 1851; to FM, 21 Jan. 1851; to H. A. Homes, 25 Jan. 1851. See Donald A. Cameron, *Egypt in the Nineteenth Century; or, Mehemet Ali and His Successors until the British Occupation in 1882* (London, 1898), 227–31; Ehud R. Toledano, *State and Society in Mid-Nineteenth Century Egypt* (Cambridge: Cambridge University Press, 1990), 108–48 (on Abbas), 181–95 (on rural conditions in the Nile valley).

43. GPM to J. G. Saxe, 7 Feb. 1851; also in *LL*, 200–201.

44. Ibid. *Ottava rima* are stanzas of eight eleven-syllabled rhymes in the form ababcc. Regaldi notes the Nile trip in his *Da Siene a File, memorie* (Turin, 1862), 14–16. Marsh elaborated on this "incredible performance" in *Lectures on the English Language [LEL]* (1861), 503n, and discussed Regaldi in "Improvisation," *Johnson's New Universal Cyclopaedia* (1876), 2:1129–30.

45. GPM to Homes, 25 Feb. 1851. Paine, *Tent and Harem*, 94–104, also describes their journey.

46. GPM to LW, 3 May 1851, WP.

47. Ibid.; GPM to J. B. Estcourt, 28 May 1851; to Dr. Taylor, 10 Oct. 1860, DCA; *LL*, 220–21.

48. GPM to SM, 16 June 1851.

49. GPM, *The Camel* (1856), 133–34. "Run to Quoddy" refers to the downwind journey by ship from Boston to Passamaquoddy Bay in Maine.

50. GPM, "The Desert" (1852), 49, 42. Paine, *Tent and Harem*, 237–41, was less enamoured of camels on this trip.

51. Paine, *Tent and Harem*, 248–52; GPM, *Camel* (1856), 73, 153–55.

52. GPM to the Estcourts, 4 July 1851.

53. GPM to SM, 23 Aug. 1851, DCA.

54. *LL*, 240–43. On Nachleys, see GPM to W. L. Marcy, 5 Sept. 1853, DD 53; Kark, *American Consuls*, 90.

55. James W. Kimball to Lyndon A. Marsh, 16 Sept. 1851, G. P. Marsh Coll., UVM.

56. GPM to SM, 23 Aug. 1851, DCA; GPM to Rev. Adolf Biewend, 1 Jan. 1852, in Marsh letterbook, 1851–53.

57. GPM, *Camel*, 159–61.

58. GPM, "Notes on Vesuvius, and Miscellaneous Observations on Egypt" (1852).

59. GPM to J. B. Estcourt, 23 Aug. 1851 (part in *LL*, 272–73); to SFB, various dates, 1849–1853, BSI. In the tradition of Alexander von Humboldt, early Victorian travelers routinely carted about huge quantities of measuring apparatus (Susan Faye Cannon, *Science and Culture: The Early Victorian Period* [New York: Dawson, 1978], 75–77, 95–100). On Thoreau's similar passion for scrupulous precision, see L. D. Walls, *Seeing New Worlds* (1995), 100, 109, 138.

60. E. F. Rivinus and E. M. Youssef, *Spencer Baird of the Smithsonian* (1992), 93–94; SFB to GPM & CCM, 2 May 1852, BSI (also in W. H. Dall, *Spencer. F. Baird* [1915], 275); GPM to SFB, 26 July 1849, BSI. See William A. Deiss,

"Spencer F. Baird and His Collectors," *Journal of the Society for the Bibliography of Natural History* 9 (1980): 635-45.

61. SFB to GPM, 9 Feb. 1851; GPM to SFB, 8 Feb., 3 May, 18 Nov. 1851; GPM to SFB and Mary Baird, 25 Dec. 1848, BSI. SI *Sixth Ann. Rept. 1851*, 61, acknowledges "Keg of fishes from Constantinople"; *Ninth Ann. Rept. 1854*, 45, Marsh's donation of "Keg of fishes and reptiles with shells, &c., from Palestine, Syria, &c." His zeal as a collector was widely known. Seeking an eastern turtle (*Trionyx euphratica*) to compare with the North American species, Agassiz testily asked Baird, "have you none from the Nile from George P. Marsh?" (13 Jan. 1856, in Elmer Charles Herber, ed., *Correspondence between Spencer Fullerton Baird and Louis Agassiz—Two Pioneer American Naturalists* [Washington, D.C.: SI, 1963], 107).

62. GPM to SFB, 4 Aug. 1852.

63. GPM to SFB, 14 Jan., 21 Mar. 1854, BSI.

64. Queen Amalie of Greece, June 1853, quoted in *LL*, 328-29. On exchanges, see John H. Hill to GPM, 11 Oct. 1852.

65. GPM to SM, 23 Aug. 1851, Marsh letterbook, 1851-53; to Caroline Estcourt, 28 Mar. 1851. Marsh later learned that "Lord Castlereagh . . . had the malice to anticipate me in my discovery of the tunnel," but was comforted to think "that he did not discover the *use* of it" (GPM to the Estcourts, Aug. 1851, in *LL*, 245). Marsh found remains of "a sluice by which the water was diverted to the tunnel" to prevent the damming up of the river Ain—also "not hitherto noticed by travellers" (*M&N*, 314n36). Castlereagh was Frederick William Robert Stewart, 4th Marquess of Londonderry, author of *A Journey to Damascus; through Egypt, Nubia, Arabia Petraea, Palestine and Syria*, 2 vols. (London, 1847).

66. GPM to the Estcourts, 18 June, 23 Aug. 1851.

7. Missionary Miseries, Mediterranean Jaunts

1. GPM to J. M. Clayton, 25 April 1850, unnumbered DD; J. P. Brown to GPM, 15 May 1850; GPM to Clayton, 20 May 1850, DD 7; CCM to LW, 1 Sept. 1850, WP.

2. Dainese to W. L. Marcy, 15 Mar. 1853, no. 115 in U.S., *Francis Dainese*, p. 72; GPM to Solomon Foot, 26 Apr. 1852 (draft). On H. A. Homes, see David H. Finnie, *Pioneers East: The Early American Experience in the Middle East* (Cambridge: Harvard University Press, 1967), 229.

3. Homes to GPM, 5 Aug. 1851; GPM to Daniel Webster, 5 Nov. 1851,

DD 19; acting Secretary of State William Hunter to GPM, 20 Mar. 1852, DI 20; GPM to FM, 26 Apr. 1852, FC; Aali Pasha to GPM, 18/30 Apr. 1852, in GPM to Marcy, 30 Apr. 1852, DD 27; acting Secretary of State C. M. Conrad to GPM, 26 Oct. 1852, DI 22.

4. GPM to W. L. Marcy, 15 Dec. 1853, DD 66.

5. GPM to FM, 26 Apr. 1852, FC; FM to GPM, 13 Jan. 1852, GMM. See GPM, "American Diplomacy" (1860).

6. GPM to Miner K. Kellogg, 5 May 1852, Kellogg Papers; to SFB, 24 May, 28 Oct. 1852, BSI. Marsh sent "The Desert" to Raymond's *New York Times;* the first half appeared in the *American Whig Review* (1852), which then folded; some of the rest was recycled in *The Camel.*

7. Webster to GPM, 29 Apr. 1852, DI, in U.S., *Jonas King, Communications . . . Relative to the Case* (1854).

8. *Hope* (Athens), 4/16 July 1852; *Jeune Hellas* (Athens), 12/24 July 1852; GPM to SFB, 28 Oct. 1852; to FM, 20 Dec. 1852, FC. For King, see GPM to Commodore Stringham, 25 Oct. 1852; John Henry Hill to GPM, 17 Aug. 1852; GPM to Hill, 31 May 1853; Hill et al. to Webster, 6 Apr. 1852, in Daniel Webster, *Papers: Diplomatic,* 2 (1987): 197–201.

9. GPM to Webster, 21 Aug. 1852, confid. unnumbered DD. Marsh attached a 60-page property case report with 29 appendices. See Webster, *Papers: Diplomatic,* 2:202–9.

10. GPM to Marcy, 20 Aug. 1853, confid. DD; to Paikos, 17 May 1853, in GPM to Marcy, DD, 2 June 1853; to FM, 24 Oct. 1852, FC. Appended to Marsh's 63-page report on the criminal trial were 152 pages of documents.

11. GPM to FM, 29 Oct. 1852, FC; to SFB, 28 Oct. 1852, BSI.

12. Everett to GPM, 5 Feb. 1853, DI 24, pp. 358–64. Paul A. Varg, *Edward Everett: The Intellectual in the Turmoil of Politics* (Selinsgrove, Pa.: Susquehanna University Press [1992], 143–51), gives the State Department view on the King cases.

13. GPM to Marcy, 27 May 1853, DD.

14. GPM to FM, 22 Jan. 1854, FC; to Marcy, 2 & 17 June 1853 (DD 23 & 24), in U.S., *Jonas King,* 162.

15. Jonas King to GPM, 5 May 1854.

16. GPM to FM, 14 & 22 Jan. [1854], FC; FM to GPM, 21 Feb. 1854. See GPM, 23 June, 6 July 1853, unnumbered DDs; GPM to Carroll Spence, 21 Jan. 1854.

17. GPM to HP, 10 Oct. 1852, PAA; to SFB, 23 Sept. 1853, BSI; to Caroline Estcourt, 19 Aug. 1850. Elgin had stripped off and shipped home the Parthenon frieze, still in the British Museum. Marsh's "hungry Greeks" were those

the satirist Juvenal depicted as infesting imperial Rome, who would stoop to anything to cadge a living.

18. GPM to Marcy, 16 June 1853, unnumbered DD; GPM, "The American Image Abroad" (1854); GPM, "Oriental Christianity and Islamism" (1858), 106. Marsh summed up King Otto as "a man of fair intentions, so far as he is capable of conscious aims, but . . . of very feeble intellectual powers . . . and incapable of apprehending a principle or grasping a whole, easily accessible to falsehood and flattery, and at the same time of unreasoning and inflexible obstinacy" (to Marcy, 20 Aug. 1853, unnumbered DD).

19. GPM to HP, 10 Oct. 1852, PAA; to CM, 8 Sept. 1852.

20. CCM and GPM to LW, 13 Sept., 10 Oct. 1852, WP.

21. GPM to CM, 28 Oct. 1852; to SFB, 28 Oct. 1852, BSI; to CCM, 2 Nov. 1852; to LW, 14 Nov. 1852; to FM, 20 Dec. 1852, FC. See LL, 299.

22. GPM to FM, 1 Jan. 1853, FC.

23. E. B. Browning to Mary R. Mitford, Feb., 15 Mar. 1853, in The Letters of Elizabeth Barrett Browning, ed. F. G. Kenyon (New York, 1898), 2 : 102–5; E. B. Browning to CCM, 19 July 1853; LL, 324–25; Cj, 30 July 1861, 1 : 25.

24. CCM to LW, 3 Apr. 1853, WP; Wise to GPM, 26 Mar. 1853; Goldsborough to GPM, 5 June 1853.

25. GPM, "Martin Koszta and the American Legation at Constantinople" (1853), MS, 4; J. P. Brown to Capt. Ingraham, 28 June, Offley to Brown, 4 July, Ingraham to Schwarz, 2 July 1853, in U.S., Martin Koszta (1854), 11, 20–22; CCM to LW, 13 July 1853. For the whole episode, see Andor Klay, Daring Diplomacy: The Case of the First American Ultimatum (1957), 27–100.

26. Ingraham to GPM, 4 July 1853; GPM to Karl Ludwig von Bruck, 30 July 1853; GPM to Marcy, 7 July, 4 & 17 Aug. 1853, DD 43, 48, & 50 (in U.S., Martin Koszta, 23, 31–38, 44–45); Bruck to GPM, 14 Sept. 1853; Marcy to GPM, 26 July 1853, DI 27, pp. 371–74.

27. GPM, "Martin Koszta and the American Legation," 12; GPM to Offley, 26 Sept., 5 Oct. 1853; to Marcy, 20 Oct. 1853, DD 60 (in U.S., Martin Koszta, 58–59); Koszta to Offley, 14 Oct. 1853; Klay, Daring Diplomacy, 146–55.

28. Alan Dowty, The Limits of American Isolation: The United States and the Crimean War (1971), 47–48; Klay, Daring Diplomacy, 186 ff.

29. Paul W. Schroeder, "Bruck versus Buol: The Dispute over Austria's Eastern Policy," Journal of Modern History 40 (1968): 193–217 at 195–96; Ann Pottinger Saab, The Origins of the Crimean Alliance (Charlottesville: University Press of Virginia, 1977), 55, 67–68.

30. GPM to Offley, 20 Oct. 1853, PR, Misc. Corr. Sent; GPM in U.S., Mar-

tin Koszta, 61–62; J. P. Brown to GPM, 19 Oct. 1853. See U.S. *CG*, 33 Cong. 1 sess., p. 313, app. pp. 50–51, 79–84.

31. J. M. Gilliss to GPM, 30 Mar. 1854.

32. GPM to SFB, 28 Oct. 1852, BSI; to FM, 1 Jan. 1853, FC; to LW, 10 Oct. 1852, WP; to SM, 1 Sept. 1852.

33. GPM to SFB, 13 Apr. 1853, BSI; to FM, 14 Jan. 1854, FC.

34. Healy to GPM, 5 Jan. 1853. See Marie De Mare, *G. P. A. Healy* (1954), 181, 208.

35. Homes to GPM, 12 July 1853; GPM, "Office-Holders Turned Out" (1854), MS.

36. GPM, "The Oriental Question" (1858), 393–97, 418. See Brison D. Gooch, "A Century of Historiography on the Origins of the Crimean War," *AHR* 62 (1956): 33–58; D. M. Goldfrank, *Origins of the Crimean War* (1994).

37. GPM to Marcy, 14 Aug. 1853, DD 49; to FM, 25 Sept. 1853; J. R. Herkless, "Stratford, the Cabinet and the Outbreak of the Crimean War," *Historical Journal* 18 (1975): 497–523 at 501–19.

38. Dowty, *Limits of American Isolation*, 76–104; GPM to F. P. Nash, 3 Aug. 1877. "By the time [the Crimean War] was over the United States was the only nation in the world that was neither ashamed nor afraid to acknowledge boldly her friendship for Russia" (Frank A. Golder, "Russian-American Relations during the Crimean War," *AHR* 31 [1926]: 462–76 at 474). See Stanley Lane-Poole, *The Life of the Right Honourable Stratford Canning* (1888), 2:228–328; Eufrosia Dvoichenko-Markov, "Americans in the Crimean War," *Russian Review* 13 (1954): 137–45.

39. GPM, "Oriental Question," 399–411; GPM speech on Turkish missions, extracted in *BDFP*, 1, 2, & 3 Oct. 1854.

40. GPM confid. dispatch to Hamilton Fish, in GPM to GFE, 27 Oct. 1876; to F. P. Nash, 3 Aug. 1877; to FM, 5 Aug. 1855, FC.

41. G. W. F. H. Carlisle, *Diary in Turkish and Greek Waters* (1855), 86. On Verney, see Cecil B. Woodham-Smith, *Florence Nightingale, 1820–1910* (1950; reprint, Glasgow: Collins, 1964), 304–5.

42. GPM to SFB, 25 Sept. 1853, BSI; to Marcy, 19 Dec. 1853, DD 67.

43. GPM to FM, 14 Jan., 21 Mar. 1854, FC; *M&N*, 22n; GPM to Professor Haynes, 21 Mar. 1881. Marsh's papyrus, not the Egyptian but the Syrian variety, had been reintroduced into Sicily only in modern times. Marsh had read of the "arrow-headiferous sands of Concord" in Thoreau's "May Days" (H. D. Thoreau, *Excursions* [1863] [Boston: Houghton Mifflin, 1899], 420); the latter

had found arrowheads sprinkled in almost every field (H. D. Thoreau, *Journals* [1906], 22 Oct. 1857, 28 Mar. 1859, 31 Oct. 1861, 10:118, 12:88–93, 14:206).

44. GPM to SFB, 21 Mar. 1854, BSI. See GPM, "Sicily, Island of," *Johnson's New Universal Cyclopaedia* (1878), 4:267.

45. GPM to FM, 22 Jan. 1854, FC; to LW, 26 Jan. 1854, WP.

46. GPM to FM, 21 Mar. 1854, FC. See *LL*, 353–54; Denis T. Lynch, *An Epoch and a Man, Martin Van Buren and His Times* (1929; reprint, Port Washington, N.Y.: Kennikat Press, 1971), 526–27.

47. CCM to LW, 18 June 1854, WP; GPM to SFB, 4 July 1854, BSI; GPM to Miner K. Kellogg, 5 May 1852, Kellogg Papers.

8. *Debts and Dromedaries*

1. *BDFP*, 24 & 26 Aug. 1854.

2. GPM to Samuel Adams, 23 Aug. 1855; to CCM, 26 Sept. 1856. For the Vermont Central, Burlington affairs, and Marsh's finances, see *BDFP*, 26 & 28 Dec. 1849, 17 & 18 Apr. 1852, 17, 19, & 10 Oct. 1854, 9 Dec. 1856; Vt. Central RR, *Proceedings of the Stockholders . . . at Northfield, Vt.* (Montpelier, 1852) and *Repts. of the Trustees* 1849 & 1851; E. C. Kirkland, *Men, Cities and Transportation: A Study in New England History, 1820–1900* (1948), 1:169, 176–80, 436–44; Raymond E. Bassett, "A Study of the Promotion, Building, and Financing of the Vermont Central Railroad," M.A. thesis, University of Vermont, 1934; T. D. S. Bassett, "500 Miles of Trouble and Excitement: Vermont Railroads, 1848–1861" (1981); idem, *The Growing Edge: Vermont Villages* (1992), 34–37; William Warner to GPM, 12 Apr. 1849; GPM to Vt. Central directors, 18 Aug. 1855; Samuel Adams to GPM, 22 Sept. 1855; Asa O. Aldis to GPM, 28 Dec. 1855; GPM to Wyllys Lyman, 14 Mar. 1850, 17 Dec. 1851; Henry P. Hickok to GPM, 27 Aug. 1850, 17 Feb. 1851; GPM to CCM, 19 Jan., 8 Sept. 1855, 18 & 22 Sept. 1856; *Life of Thomas Hawley Canfield* (Burlington, 1889), 12–13; Ira B. Peck, *A Genealogical History of the Descendants of Joseph Peck* (Boston, 1868), 301–2; G. P. Mayo to GPM, 2 Oct. 1856; Paris Fletcher to GPM, 1 Dec. 1856; *Michigan State Bank* v. *John Peck et al.*, 28 Vt. 200 (1855); *H. B. Stacey* v. *Vt. Central*, 29 Vt. 39; GPM powers-of-attorney to Wyllys Lyman, 24 Jan., 15 Sept. 1849, 27 Apr. 1852, & mortgages, 1 Nov. 1852, Burlington Real Estate Rec. 23:407–12 and Land Rec. 28:152; *Thaddeus Fletcher* v. *Lyman & Marsh & Vt. Central, John Peck* v. *Lyman & Marsh, Asahel Peck* v. *Lyman & Marsh & J. H. Peck*, Chittenden County Court Records,

24:131–33, 26:13–15; Wyllys Lyman Estate, Chittenden County Probate Office, 3 Mar. 1864, 48:423.

3. GPM to CCM, 26 Sept. 1856; GO to CCM, "Sunday"; GPM to LW, 17 Feb. 1856, WP; CCM to GPM, 26 Aug. 1854, 7 Dec. 1856.

4. GPM to Francis Lieber, 27 Apr. 1860; to Solomon Foot, 12 Apr. 1860; to CM, 25 Oct. 1862.

5. GPM, Memorial . . . for Compensation of His Services (1854), 1–2; GPM to Charles D. Drake, 2 Nov. 1857; Hughes to Dainese, 12 Oct., in Dainese to Marcy, 15 Oct. 1853; Chandler and Douglas in the Senate, 2 Mar. 1855, and House vote, 3 Mar. 1855, U.S. CG, 33 Cong. 2 sess. (1856), 1090–93, 1165.

6. R. A. Billington, The Protestant Crusade, 1800–1860 (1938), 380–436; Jenny Franchot, Roads to Rome: The Antebellum Protestant Encounter with Catholicism (1994); Michael Holt, The Political Crisis of the 1850s (1978), 120–21, 156–62; William E. Gienapp, Origins of the Republican Party, 1852–1856 (New York: Oxford University Press, 1987), 60–61; R. M. Elson, Guardians of Tradition: American Schoolbooks of the Nineteenth Century (1964), 51–53. The 14.7 percent proportion of foreign-born in 1910, in the wake of three decades of massive southern and eastern European immigration, was only fractionally higher than in the 1850s.

7. Mary P. Brown to CCM, 13 June 1855.

8. L. S. Finberg, "The Press and the Pulpit: Nativist Voices in Burlington and Middlebury, 1853–1860" (1993); GPM to Erastus Fairbanks, 19 Apr. 1855, in J. M. Lund, "Vermont Nativism" (1995), 18. Burlington's Catholic working force was vital to the local economy, but endemic Vermont nativism was again spurred by Senator William P. Dillingham, who spearheaded the restrictive quota system that drastically curtailed American immigration in the 1920s.

9. GPM to LW, 19 Dec. 1850, WP; to CCM, 4 Mar., CFP, 17 Dec., UVM, 16 Feb., CFP (all 1856); CCM to GPM, 24 Feb. 1856; GPM to CCM, 17, 9 Dec. 1856; CBT to GPM, 22 Mar., 9 Apr. 1861; CBT to CCM, 24 Sept. 1884.

10. GPM to CCM, 6 July, 22 May, 21 Jan., 4 Mar. 1856, CFP; to Solomon Foot (with quote from J. P. Brown), 4 Jan. 1855, VHS MSS no. 64. The "sick man" was already the stock appellation for the sultan, later for Turkey itself.

11. GPM to FM, 9 Oct. 1857, FC; Dainese in American Organ (Washington), 2 Aug. 1855; Foot and Brodhead in the Senate, 25 Apr. 1856, U.S. CG, 34 Cong. 1 sess., 818–21, 1019–21. Marsh again repudiated the Pacific territories in 1855 ("Principles and Tendencies of Modern Commerce," 161).

12. GPM, Reply to Mr. Brodhead's Remarks (1856); Foot in the Senate, 4 & 25

Apr. 1856, U.S. *CG*, 34 Cong. 1 sess., 818–21, 1020–22; Richard Brodhead in N.Y. *Herald*, 16 June 1856; Pennington in U.S., *Marsh Claim for Extra Compensation* (1860), 23 May 1856, pp. 4–5.

13. GPM to CCM, 5 Jan. 1856, CFP, 10 Aug. 1856, UVM; Dainese to Elisha Whittlesey (U.S. Treasury), 6 Aug. 1856 (copy, GPM Coll., UVM).

14. GPM to CCM, 19 & 23 Aug. 1856; GPM to SFB, 16 Mar. 1857; CCM to GPM, 27 June 1855, CFP.

15. Dainese to President Pierce, 30 Apr. 1853; to Whittlesey, 4 Sept. 1856; J. S. Morrill to GPM, 25 Jan., 26 Apr. 1857; GPM to McConnell (Treasury), in U.S., *Francis Dainese*, 255, 162, 102–4; Morrill to GPM, 18 Jan. 1858; FM to GPM, 2 Feb. 1858; GPM MSS refuting U.S., *Francis Dainese*; GPM to Foot, 4 Jan. 1858, MV.

16. Homer Elihu Royce, House Comm. on Foreign Affairs, 6 Apr. 1860, 36 Cong. 1 sess., H. Rep., 350; FM to GPM, 5 May, 6 June 1860; "An Act for the Relief of George P. Marsh," 13 June 1860, 124 Stat. at Large 857 (in U.S., *Marsh Claim*).

17. GPM to CCM, 7 Sept. 1856; Eugene Schuyler, *American Diplomacy and the Furtherance of Commerce* (1886), 181.

18. Samuel Marsh to GPM, 6 Jan. 1855; GPM to Caroline Estcourt, 31 Mar. 1857, in *LL*, 385; HP to GPM, 1 June 1855; GPM to HP, 13 Aug. 1855, PAA.

19. H. J. Raymond to GPM, 27 Aug. 1851, and ca. 1857, in E. F. Brown, *Raymond of the Times* (1951), 158; J. R. Lowell to E. L. Godkin, 16 Oct. 1866, in *Letters of James Russell Lowell* (1894), 1 : 372.

20. GPM to LW, 17 Feb. 1856, WP.

21. FM to GPM, 22 Mar. 1855; GPM to FM, 29 Mar. 1855, GMM; GPM to C. D. Drake, 27 Aug. 1856.

22. George Ticknor to GPM, 5 Feb. 1855; C. C. Felton to GPM, 9 Feb., 18 Mar. 1855. See James Walker to Ticknor, 16 Feb. 1855, College Letters, 4 (1853–60), Harvard Coll. Archives. The history chair had been occupied pro tem by Francis Bowen, editor of the *North American Review*, from 1851 to 1853.

23. GPM to Ticknor, 10 Feb. 1855. After Marsh declined the post, it was filled by the undistinguished Henry W. Torrey in 1856 (Samuel Eliot Morison, *Three Centuries of Harvard, 1636–1936* [Cambridge: Harvard University Press, 1936], 291–93, 347).

24. John Pennington to GPM, 15 Apr. 1859; SFB to GPM, 8 Apr. 1859; GPM to SFB, 25 Apr. 1859, BSI. On the University of Vermont, GPM, Report

to the UV on the condition of the library, 27 Nov. 1855, MS; Asa O. Aldis to GPM, 14 Aug., 27 Nov. 1855; UV Records, Trustees Minutes, 3:155–63, 172–73.

25. SFB to GPM, 6 May 1854; Gilliss to GPM, 12 June 1853; FM to GPM, 18 Sept. 1854.

26. Gilliss to GPM, 8 May 1854; N.Y. *Weekly Tribune*, 20 Jan. 1855; *SIJ*, 102–17; M. H. Harris, *The Age of Jewett* (1975), 53n11. See J. A. Boromé, *Charles Coffin Jewett* (1951), 81–94; Thomas Coulson, *Joseph Henry* (1950), 213.

27. GPM to J. R. Bartlett, confid., 31 Mar. 1848, in *Papers of Joseph Henry*, 7 (1996): 297–98; Gilliss to GPM, 23 Nov. 1854; GPM to SFB, 8 Sept. 1854, BSI; GPM on SI Affairs to C. W. Upham (draft), Feb. 1855; Stephen A. Douglas in the Senate, 18 Jan. 1855, U.S. *CG*, 33 Cong. 2 sess., 304–5; Henry to GPM, 3 Feb., 24 Mar. 1855. The first two of Marsh's Smithsonian lectures were on "Constantinople and the Bosphorus," the third on "The Camel," the fourth on Constantinople and Turkey (SI *Tenth Ann. Rept. 1855*, 34).

28. Morrill to Solomon Foot, 20 Sept. 1850, in Thomas D. S. Bassett, "Vermont Politics and the Press in the 1840s," *VH* 47 (1979): 196–212 at 206; Gilliss to GPM, 25 Jan. 1853.

29. FM to GPM, 22 Jan. 1853, 18 Sept. 1854; Gilliss to GPM, 25 Jan. 1853, 30 Mar. 1854. On the mountain rule, Sen. Ralph E. Flanders to the author, 29 Apr. 1955; Lyman J. Gould and Samuel B. Hand, "A View from the Mountain: Perspectives of Vermont's Political Geography," in H. N. Muller and S. B. Hand, *In a State of Nature: Readings in Vermont History* (1982), 186–90. To Marsh's pleasure, hard-working Solomon Foot had succeeded hard-drinking S. S. Phelps in the Senate in 1850 (GPM to R. C. Winthrop, 4 June 1850, Winthrop Papers); Phelps's appointment to Upham's former seat was disallowed because Phelps, like Foot, lived in western Vermont.

30. GPM to SFB, 13 Apr. 1853, BSI; to Drake, 27 Aug. 1856; to Winthrop, 6 Aug. 1856, Winthrop Papers. On the Council of Censors, *BDFP*, 29 & 31 Mar. 1855; VSA, 79:46–65; L. H. Meader, "The Council of Censors in Vermont," in *The Early History of Vermont* (Montpelier: VHS, 1943), 155–287; Bassett, *Growing Edge*, 122–26. On national politics, Holt, *Political Crisis of the 1850s*, 13, 40–44, 156–71; idem, *Rise and Fall of the American Whig Party* (1999), esp. 951–59; Gienapp, *Origins of the Republican Party*, 60–61, 92–102, 215–16. On Winthrop, Choate, and the Cotton Whigs, Kinley J. Brauer, *Cotton versus Conscience: Massachusetts Whig Politics and Southwestern Expansion* (Lexington: University of Kentucky Press, 1967); Tyler Anbinder, *Nativism and Slavery: The Northern Know Nothings and the Politics of the 1850s* (New York: Oxford Uni-

versity Press, 1992), 103–26; Thomas H. O'Connor, *Lords of the Loom: The Cotton Whigs and the Coming of the Civil War* (New York: Scribner, 1968), 122–23, 188–93.

31. GPM to CCM, 3, 16, 19, 23 Aug. 1856; to Foot, 18 Aug. 1856; Foot to GPM, 20 Aug. 1856; GFE to GPM, 8 Oct., 3 Nov. 1856. See also GPM to Charles Sumner, 1 Sept. 1856, Sumner Papers.

32. GPM to CCM, 23 July 1856; to SFB, 19 Feb. 1857.

33. GFE to GPM, 8 Mar. 1858; GPM to Foot, 23 Mar. 1858, VHS MSS no. 64; *BDFP*, 13 Feb. 1857, p. 2.

34. Gilliss to GPM, 17 Sept. 1857; GPM to F. J. Child, 26 July 1862.

35. GPM to HP, 13 Aug. 1855, PAA; to SFB, 2 July 1855, BSI; to FM, 23 June 1855, FC; to J. H. Alexander, 11 July 1855.

36. Alexander to GPM, 8 Aug., 21 Nov. 1855; GPM to CCM, 6 July 1855, CFP. On optical glass and vernier issues, see Susan Faye Cannon, *Science in Culture: The Early Victorian Period* (New York: Dawson, 1978), 100–101. J. H. Alexander's "Dictionary of the Language of the Leni-Lenape, or Delaware Indians" was never published.

37. *Laws of Vermont*, 1855, Act 81, pp. 112–13. On Read, *HVG*, 1:758–59; M. D. Gilman, *The Bibliography of Vermont* (1897), 229–30.

38. GPM to Maria Buell [Hungerford], 1 Oct. 1855; to CCM, 13 Sept. 1855.

39. GPM to CCM, 11 Sept. 1855, UVM, 2 Mar. 1856, CFP; J. F. Flagg to Gilliss, in Gilliss to GPM, 3 Dec. 1857. See Albert D. Hager, *The Marbles of Vermont* (Burlington, 1858), 8–9; GPM to Read, 21 Oct. 1861, in David Read, *Report and Statements Concerning the Winooski Marble, at Mallett's Bay, near Burlington, Vermont* (Boston, 1866), 8.

40. GPM to Read, 21 Oct. 1861; GFE to GPM, 29 June 1869.

41. GPM to CCM, 13 Nov. 1856, 7 Feb. 1857; *Boston Herald*, 15 Nov. 1854; *BDFP*, 4, 13 Nov. 1854; *LL*, 377.

42. GPM to CCM, 2 Dec. 1855, CFP, 30 Mar. 1855, UVM. Allan Nevins, *Ordeal of the Union* (New York: Scribner, 1947), 2:232–33, cites similar complaints by other travelers.

43. GPM to CCM, 17, 22, 20 Jan. 1855; to Maria Buell [Hungerford], 1 Oct. 1855.

44. GPM to Maria Buell Hickok, 10 Jan. 1855.

45. GPM to FM, 21 Mar. 1854, FC; GPM, "American Image Abroad" (1854); GPM, "American Representatives Abroad" (1860); GPM to C. S. Daveis, 23 Mar. 1849, Daveis Corr. In fact, none of these "lowbred, ignorant

clowns" stayed in Europe long, if at all; Hannegan left Berlin after a few months, Andrew Jackson Donelson never went to Frankfurt, Dabney Carr was replaced by Marsh himself. See Merle Curti, "The Reputation of America Overseas, 1776–1860," in his *Probing Our Past* (1955), 191–218.

46. GPM to SFB, 2 July 1855, BSI; GPM, "Oration" [New Hampshire] (1856).

47. Ibid., 42–43. I focus on Marsh's generalizations, but he took care to stress the diversity of European lands.

48. Ibid., 65–73.

49. GPM, Independence Day Oration, Woodstock, 4 July 1857, MS; GPM to Caroline Estcourt, 28 Mar. 1851.

50. GPM, "Oration" [New Hampshire], 75–77.

51. Ibid., 86–87.

52. GPM, *Speech on the Tariff Question* (1846), 1013; *Address . . . Agric. Soc. Rutland* (1847), 16; "The Camel" (1854), 116–18.

53. GPM, "Desert" (1852), 48; "Camel," 119–20; *The Camel* (1856), 177–96, esp. 188.

54. H. C. Wayne to GPM, 4 Oct., 3 Nov. 1856; H. D. Fowler, *Camels to California* (1950), 9–12, 18, 55–88. See U.S., 35 Cong. 1 sess., H. Ex. Doc. 124 (1858); A. A. Gray, "Camels in California," *California Historical Soc. Quarterly* 9 (1930): 299–317 at 301–2; Frank Bishop Lammons, "Operation Camel: An Experiment in Animal Transportation in Texas, 1857–1860," *Southwestern Historical Quarterly* 61 (1957): 20–50 at 24–25; Charles A. Carroll, "The Government's Importation of Camels: A Historical Sketch," USDA Bur. Animal Industry, Rept. 20, 1903, 391–409 at 392; Stephen Bonsal, *Edward Fitzgerald Beale: A Pioneer in the Path of Empire, 1822–1903* (New York: Putnam, 1912).

55. Richard W. Bulliet, *The Camel and the Wheel* (Cambridge: Harvard University Press, 1975), 254–55, 272–73; GPM, *Camel*, 26–27.

56. GPM to CCM, 9 May 1856. Like others of his time, Marsh erred in thinking camels stored water in their stomachs. Camels' water metabolism stems from their low rate of loss in urine and sweat, ability to withstand dehydration, and enormous and rapid drinking capacity. See Knut Schmidt-Nielson, *The Camel's Nose: Memoirs of a Curious Scientist* (Washington: Island Press, 1998), 124–25, 290–92; Hilde Gauthier-Pilters and Anne Innis Dagg, *The Camel: Its Evolution, Behavior, and Relationship to Man* (Chicago: University of Chicago Press, 1981), 50–77; Reuven Yagil, *The Desert Camel: Comparative Physiological Adaptation* (Basel: Karger, 1985), 19–36, 71.

57. *North American Review* 83 (1856): 561; J. H. Alexander to GPM, 7 July

1856; GPM to Charles Lanman, 21 Sept. 1856, in Lanman, *Haphazard Personalities* (1886), 104.

58. GPM, *Camel*, 40–41; "Desert" (1852), 46–47.

59. GPM to CCM, 21 July 1856; Gould & Lincoln (publishers) to GPM, 13–14 Oct. 1856, 17 Mar. 1860; GPM to CCM, 2 Sept. 1856, 13, 29, 31 Jan. 1856, CFP; CCM, *The Hallig* (1856). See William Morris Davis, "The Halligs, Vanishing Islands of the North Sea," *Geographical Review* 13 (1923): 99–106.

60. GPM to FM, 27 Aug. 1856, FC; to CCM, 12 Nov., 8 Dec. 1856; to FM, 8 Nov. 1856, FC.

61. GPM to CCM, 1, 4, & 5 Feb. 1857.

62. Marie De Mare, *G. P. A. Healy* (1954), 175–83; GPM to CCM, 8 & 9 Feb. 1857; to Maria Buell [Hungerford], 1 Oct. 1855. For similar Yankee sentiments about Chicago, see Stewart H. Holbrook, *The Yankee Exodus: An Account of Migration from New England* (New York: Macmillan, 1950), 71–72.

63. GPM to CCM, 11 Feb. 1857; LW to CCM, 8 Mar. 1857; GPM to CCM, 15, 22, & 18 Feb. 1857.

64. GPM to CCM, 28 & 13 Feb. 1857; LW to CCM, 8 Mar. 1857; GPM to SFB, 19 Feb. 1857, BSI.

9. Vermont Public Servant

1. GPM, *Address . . . Agricultural Soc. Rutland County* (1847), 18; GPM to SFB, 8 Apr. 1857, BSI; to LW, 20 Aug. 1857, WP.

2. *M&N*, 218–19.

3. GPM to Arnold Guyot, 27 Oct., 16 Nov. 1857; Guyot to GPM, 9 Nov. 1857; CCM to LW, 23 Feb. 1860, WP.

4. GPM to SFB, 6 Nov. 1857.

5. GPM to SFB, 12 Oct. 1881, BSI; R. W. Judd, *Common Lands, Common People: The Origins of Conservation in Northern New England* (1997), 49–52.

6. GPM, *Report, on the Artificial Propagation of Fish* (1857), 19–21. On Marsh and fisheries depletion, see Carolyn Merchant, *Ecological Revolutions: Nature, Gender, and Science in New England* (1989), 223–34, 239–40; G. G. Whitney, *From Coastal Wilderness to Fruited Plain: A History of Environmental Change in Temperate North America, 1500 to the Present* (1994), 308–9.

7. GPM, *Report . . . on Fish*, 12–16. See Zadock Thompson, *Natural History of Vermont* (1853), 19. Marsh would have relished hearing that "the first dam to be removed for purely environmental reasons," in 1996, was on the Clyde River at Newport, northeastern Vermont. "If I could catch a salmon," averred ten-

year-old Kate Grim 140 years after Marsh, "I'd turn my television off, my electric blanket and my stereo, [to] save electricity" (Pratap Chatterjee, "Dam Busting," *New Scientist*, 17 May 1997, 34–37).

8. Vt. House Journal, 13 & 22 Oct. 1857, 67, 123; S. F. Baird, *Report of the Commissioner of Fish and Fisheries for 1872 and 1873*, xlii, xlv; SFB to GPM, 12 Dec. 1874; D. C. Allard, *Spencer Fullerton Baird and the U.S. Fish Commission* (1978), 58–59, 84–85, 114–15. Marsh synopsized his *Report . . . on Fish* in M&N, 102–5.

9. GPM to SFB, 12 Aug. 1858, BSI; *M&N*, 105–6.

10. Roosevelt took the phrase from William James's essay of that title, championing military ideals of hardihood and discipline, or "toughness without callousness"; in James, *Memories and Studies* (London: Longmans, Green, 1910), 267–96 at 291.

11. GPM, *Report . . . on Fish*, 7–11. See Judd, *Common Lands, Common People*, 178. For related fears of loss of self-sufficiency, see Hal S. Barron, *Those Who Stayed Behind* (New York: Cambridge University Press, 1984), 123–28.

12. *M&N*, 71 & n39, 82n62, 257n217; David Lowenthal, *The Past Is a Foreign Country* (1985), 117–20.

13. GPM, "Oration" [New Hampshire] (1856), 50.

14. GPM, *Report . . . on Fish*, 10–11.

15. GPM to William G. Shaw, 8 & 9 Mar. 1857, VHS MSS 22 no. 64; to C. D. Drake, 23 May 1857. The sum allocated was later increased by the legislature and by private subscription in Montpelier. On the decision not to move the capital from Montpelier, see H. L. Bailey, "Vermont's State Houses" (1944); J. M. Lund, "Vermont Nativism" (1995), 18.

16. Norman Williams to GPM, 23 Dec. 1857; GPM et al., "Report of the Commissioners on the Plan of the State House" (1857). See Lawrence Wodehouse, "Ammi Young's Architecture in Northern New England," *VH* 36 (1968): 53–60; T. D. S. Bassett, "Vermont's Second State House: A Temple of Republican Democracy Imagined through Its Inventories, 1836–1856," *VH* 64 (1996): 99–107.

17. T. E. Powers to GPM, 20 July 1858; Winslow Ames, "The Vermont State House" (1964); Daniel Robbins, *The Vermont State House: A History & Guide* (1980), 32–45. See T. E. Powers to GPM, 3 Apr. 1857; T. W. Silloway to GPM, various dates, 1857–58; GPM to T. E. Powers, 22 Mar. 1858; GPM to Silloway, 9 June 1857 & 11 Feb. 1859, in T. W. Silloway, *A Statement of the Facts . . . Connected with the Rebuilding of the Capitol . . .* (Burlington, 1859); *Report of the Hon. Thomas H. Powers, Superintendent of Construction of the State House* (Montpelier, 1858); idem, *Vermont Capitol and the Star-Chamber. Testimony and Defense of the Superintendent of Construc-*

tion (Montpelier, 1858). With Thomas E. Powers, cousin of his friend Hiram Powers, Marsh remained on good political terms. As presiding officer at the Republican State Convention in 1857, Powers appointed Marsh to chair the resolutions committee (*BDFP*, 1 July 1857). See H. S. Dana, *History of Woodstock* (1889), 359–61; T. D. S. Bassett, *The Growing Edge: Vermont Villages* (1992), 125–27.

18. Silloway to GPM, 7 Aug., 6 Oct. 1858; GPM quoted in Robbins, *Vermont State House*, 46.

19. Stanford White quoted in R. N. Hill, *The Winooski* (1949), 206; Larkin Mead to GPM, Jan., 3 Feb., 12 May, 23 Aug., 11 Nov. 1858; Robbins, *Vermont State House*, 90, 124–26; GPM to HP, 20 May 1858, PAA. Mead's classical *Agriculture* was replaced by a folk-art "replica" in 1938, his *Ethan Allen* by a 1941 copy.

20. GPM to Charles Lanman, 29 Apr. 1847, in Lanman, *Haphazard Personalities* (1886), 100–101; GPM to CCM, 30 May 1856, CFP. Marsh may have been mistaken about Leutze's painting, for no Leutze canvas of Marsh's description is known from the 1840s.

21. GPM to Lanman, 21 Apr., 6 May 1847, in Lanman, *Haphazard Personalities*, 101–3. See V. G. Fryd, *Art and Empire: The Politics of Ethnicity in the United States Capitol, 1815–1860* (1992), 48–53; Russell F. Weigley, *Quartermaster General of the Union Army: A Biography of M. C. Meigs* (New York: Columbia University Press, 1959).

22. FM to GPM, 28 June 1855, GMM, paraphrased and amplified in GPM to HP, 13 Aug. 1855, PAA.

23. Fryd, *Art and Empire*, 205–6; idem, "Hiram Powers's *America*: Triumphant as Liberty and in Unity," *American Art Journal* 18:2 (1986): 55–75, and "Hiram Powers's *Greek Slave*: Emblem of Freedom" (1982).

24. GPM to HP, 16 Nov. 1857, PAA. Crawford's *Freedom* was cast in bronze, ironically by slave labor, and hoisted atop the Capitol in December 1863 in a ceremony aimed to inspirit Union troops; see J. M. Goode, *The Outdoor Sculpture of Washington, D.C.* (1974), 60; Fryd, *Art and Empire*, 177–200. On the Powers–Crawford rivalry, see Sylvia E. Crane, *White Silence: Greenough, Powers, and Crawford, American Sculptors in Nineteenth-Century Italy* (Coral Gables: University of Miami Press, 1972), 225–28, 239–42.

25. GPM to HP, 20 Apr. 1855, 15 Nov. 1857, PAA. See Goode, *Outdoor Sculpture*, 55–57, 522–23; Fryd, *Art and Empire*, 91–105, 125–26.

26. GPM to HP, 20 Apr. 1855, 20 May 1858, PAA. See Wayne Craven, *Sculpture in America*, rev. ed. (Newark: University of Delaware Press, 1984), 168–72; Goode, *Outdoor Sculpture*, 377–78.

27. *M&N*, 51–52n53.

28. Jacob Collamer et al., "Report of the Commissioners on Laws relating to Railroads," Vt. House Journal, 1855, app. 642–49.

29. "Report of Commission on Roads on the Foregoing Bill," Vt. House Journal, 1855, app. 650–52.

30. Charles Linsley, "First Annual Report of the Railroad Commissioner," Vt. House Journal, 1856, 583–92; *BDFP*, 5 & 6 Nov. 1857.

31. GPM to CCM, 8 Oct. 1858, in *LL*, 400.

32. E. C. Kirkland, *Men, Cities and Transportation: A Study in New England History, 1820–1900* (1948), 1:440 (see also 233, 241); GPM, *Railroad Commissioner... Third Ann. Rept.* 1858, 3–7 & app. A, and 7–9, 13–15; on lies about track gradients, *M&N*, 51n53. See GPM, "Oration" [New Hampshire], 71–72, and Chapter 8 above.

33. GPM, *RR Commissioner... Fourth Ann. Rept.* 1859, 5–6.

34. Edward John Phelps, "Lay of the Lost Traveller" (1865), in H. N. Muller and S. B. Hand, *In a State of Nature: Readings in Vermont History* (1982), 206–7. Phelps, a son of Senator S. S. Phelps, was U.S. Minister to Great Britain 1885–89. See Lucius E. Chittenden, *Personal Reminiscences, 1840–1890* (New York, 1893), 199–204.

35. GPM, *RR Commissioner... Third Ann. Rept.*, 9–15; *Fourth Ann. Rept.*, 5, 9; Vt. RR Commissioner, *Seventh Ann. Rept.* 1862, 10–11; Robert C. Jones, *The Central Vermont Railway: A Yankee Tradition*, 7 vols. (Silverton, Colo./Shelburne, Vt.: New England Press, 1981–95), 1:53–56. As noted by Jones (7:44), the spur remained perilous; in 1984 a culvert undermined by heavy rains south of Essex Junction gave way, killing five men.

36. Rodney V. Marsh to GPM, 27 Nov. 1858; *BDFP*, 25 Nov. 1858. See L. E. Chittenden to GPM, 4 Dec. 1858; GFE to GPM, 16 Oct. 1858; 1858 Vt. House Journal, 294, Senate Journal, 259.

37. GPM to SFB, 8 Jan. 1866, BSI; GPM to Alexander B. Crane, 21 Sept. 1876, CFP; GPM, *RR Commissioner ... Fourth Ann. Rept.*, 3, 10; Vt. RR Commissioner, *Fifth Ann. Rept.* 1860, 16. See Kirkland, *Men, Cities and Transportation*, 1:240–41, 254–64, 354–58, 432, 445–48; Bassett, *Growing Edge*, 49–54.

38. GPM, "Principles and Tendencies of Modern Commerce" (1855), 168; "Oration" [New Hampshire], 74.

39. GPM, "The War and the Peace" (1859), 266–67; *M&N*, 255n215.

40. *M&N*, 51–52n53; GPM to Giuseppe Pasolini, 29 May 1875 (CCM copy).

10. *The English Language*

1. Columbia College Archives, Minutes of the Trustees, 5 (1856–58): 375–78; Samuel B. Ruggles to GPM, 13 May, 21 June 1858.

2. GPM to CCM, 29 Sept. 1858.

3. GPM, *Lectures on the English Language* (1861) [hereafter *LEL*], v, 3, 643; Gerald Graff, *Professing Literature: An Institutional History* (Chicago: University of Chicago Press, 1987), 19–51; L. W. Levine, *The Opening of the American Mind* (1996), 78–82.

4. GPM to CCM, 12 Oct. 1858; to Charles Folsom, 22 June 1859, Folsom MSS; Francis Lieber to GPM, 12 Jan. 1859; Lieber in Frank Freidel, *Francis Lieber* (1947), 297; GPM to SFB, 1 Mar. 1859, BSI. An eminent student of Lieber's recalled that "as a teacher he was a definite failure" (John W. Burgess, *Reminiscences of an American Scholar, the Beginnings of Columbia University* [New York: Columbia University Press, 1934], 70).

5. GPM to SFB, 21 June 1858, BSI; to CCM, 1 Aug. 1859. Charlotte Bostwick's marriage failed, the Reverend Thrall turning out to be an opium addict (CBT to GPM, 13 Dec. 1870).

6. GPM to CCM, 26 July 1859; Lieber to GPM, 9 Mar., 22 Apr. 1860; GPM to Lieber, 12 Apr. 1860. See Freidel, *Lieber*, 286–316.

7. GPM to CCM, 25 July 1859; to SFB, 3 Jan., 26 Aug. 1859, BSI. Marsh's *Christian Examiner* essays were "Oriental Christianity and Islamism" (1858), "The Oriental Question" (1858), and "The War and the Peace" (1859); his poems for Hemenway (1860) were "Whole and Halfe" and "The River."

8. GPM to CCM, 20 July 1859. The spooky horrors of Ann Radcliffe's *Mysteries of Udolpho* (1794) and other romances were avidly consumed by American readers.

9. GPM to Lieber, 4 Jan. 1860; GPM quoted in CCM to LW, 7 Dec. 1859, WP.

10. Lieber to Ruggles, 17 Feb. 1860, Lieber Corr.; *Athenaeum* (London), 11 Aug. 1860.

11. N.Y. *Evening Post*, N.Y. *Tribune*, *Atlantic Monthly* (Apr. 1860), St. Louis *Universalist Quarterly* 17 (1860): 253–74 (all in Marsh scrapbook); *New Englander* 48 (1860): 532–33; *Critic*, 2 June 1860, 682–84; *Saturday Review*, 19 May 1860, 644–45. Ecstatic praise came from E. E. Hale, *Christian Examiner* (1860); scathing censure from E. B. Humphreys, *National Quarterly Rev.* 1:410–30.

12. Volume 6 of *Origins of American Linguistics, 1643–1914*, Ray Harris, ed.,

13 vols. (London: Routledge/Thoemmes Press, 1997), reproduces W. G. Smith's mutilated British variant (1862) of *LEL*.

13. GPM to C. C. Felton, 10 Jan. 1860, Felton MSS; to George Ticknor, 20 Feb. 1860.

14. George Ticknor, *Life of William Hickling Prescott* (Boston, 1864), 150n; George Bancroft to GPM, 4 July 1860; GPM to Bancroft, 12 July 1860, Bancroft MSS. For nineteenth-century linguistics, see E. F. K. Koerner, "Towards a History of Linguistics," in Herman Parret, ed., *History of Linguistic Thought and Comparative Linguistics* (Berlin: W. de Gruyter, 1976), 685–718; Pieter A. M. Seuren, *Western Linguistics: An Historical Introduction* (Oxford: Blackwell, 1998), 79–119; Anna Morpurgo Davies, *History of Linguistics, Vol. 4, Nineteenth-Century Linguistics* (London: Longman, 1998); S. G. Alter, *Darwinism and the Linguistic Image* (1999), 186–87.

15. GPM, "New Dictionary by the Philological Society of London" (1859); "Our English Dictionaries" (1860), 387–88; "Old English Literature" (1865), 778; GPM to F. J. Furnivall, 15 Oct. 1859; New English Dictionary to GPM, 13 Sept. 1879. See J. H. Alexander to GPM, 13 Dec. 1859; *New Englander* 48 (1860): 224; GPM to Norton, 2 June 1860, Norton Papers; Hans Aarsleff, *The Study of Language in England, 1780–1860* (Princeton: Princeton University Press, 1967), 259–60; Jonathon Green, *Chasing the Swan: Dictionary-Makers and the Dictionaries They Made* (London: Cape, 1996), 308–10. Not until after the First World War was American literature accorded a place of its own in American academe (Levine, *Opening of the American Mind*, 82–83).

16. W. C. Bryant quoted, Marsh cited, in H. L. Mencken, *The American Language* (1923), 241, 9. See also Joseph H. Friend, *The Development of American Etymology 1798–1864* (The Hague: Mouton, 1967), 85–88.

17. GPM to James R. Spalding, 26 Mar., 15 Oct. 1860; GPM, "The Two Dictionaries" (1860), expanded as "Our English Dictionaries" (1860), esp. 398–99. On Webster's etymology, Friend, *Development of American Etymology*, 75–79; Richard M. Rollins, "Words as Social Control: Noah Webster and the Creation of the *American Dictionary*," *American Quarterly* 28 (1976): 415–30 at 422–23. Webster had died in 1843, but the Merriams' new editor, Chauncey Goodrich, retained his etymologies.

18. The Merriams to GPM, 6 Apr., 25 May 1860; Noah Porter, in *Webster's Unabridged Dictionary* (1864), vi; GPM to Henry Yule, 8 June 1868.

19. Sheldon & Co. to GPM, 8 & 15 Mar. 1860; GPM, "Our English Dictionaries," 397, 404–5; GPM to Ticknor, 21 Jan. 1860; H. A. Homes to GPM, 7 Apr., 22 May, 28 Aug. 1860; GPM to Lowell, 18 Aug. 1860, Lowell Pa-

pers; to Lieber, 18 Apr. 1860; E. A. Sophocles to GPM, 13 Aug. 1860; GPM, "A Glossary of Later and Byzantine Greek, by E. A. Sophocles" (1860); GPM to H. S. Dana, 15 Feb., 5, 6, & 28 Mar. 1861; Dana to GPM, 2 & 12 Mar., 14 July 1861; GPM to Lieber, 6 Sept. 1860; to F. J. Child, 26 July 1862. Publication of Marsh's *Dictionary of English Etymology* (Wedgwood vol. 1) was delayed to 1862.

20. GPM to Hensleigh Wedgwood, May 1861; GPM, *Dictionary of English Etymology*, 246–47. Two decades later Marsh reiterated his view that early Dutch and Flemish offered an "almost unwrought mine of . . . invaluable information" about the early history of English (*LEL* 1885 ed., 88–89n).

21. GPM, *Dictionary of English Etymology*, 69, 112, 157, 158.

22. *Christian Rev.; North American Rev.* (1862), 285; *LL*, 424n; Marsh, *Dictionary*, 179; Hensleigh Wedgwood, *A Dictionary of English Etymology*, 2d. ed. (London, 1872), 150. Terms Marsh enlarged on in Wedgwood's volumes 2 and 3 include empeach, filch, flesh and flitch, fling, haberdasher, harangue, lord, luke-warm, pay, plough, provender, pry, queer, rogue, scabbard, scissors, shrewmouse, sir and sire, spick and span, spite, stain (Notes on . . . Wedgwood's Dictionary [1865], 2, 3, 8–10, 20 [as pages in the separate]). Wedgwood disputed Marsh on ballast and charcoal but accepted his emendations on abet, afford, average, baggage, balk, buoy, calibre, carboy, ceiling, cheese, to curry, fleam, fodder, harangue, pumpkin (a vulgarization of pompion), scabbard, sword, shrewmouse, sir/sire, spite (Wedgwood, "Corrections and Emendations," in Dictionary, vol. 3 (1865), 528–71; 1872 ed. passim). Neither Marsh's editorial role nor his later *Notes* are identified by Wedgwood, though the Philological Society, of which he was treasurer, had sponsored both.

23. GPM, "Notes on the New Edition of Webster's Dictionary, No. XVI," *Nation* 4 (27 June 1867): 516–17.

24. GPM, "Italian Cause" and "Italian Nationality" (both 1860); "Marsh, James" (1861); GPM to FM, 19 Mar., 20 June, 13 July 1860, GMM; to Francis Lieber, 27 Apr. 1860; to Sarah Butler Wister, 20 Jan. 1874 (CCM copy). The Markoe book was never published. See John Godfrey Saxe, "The Money-King and Other Poems" (Boston, 1860).

25. GPM to FM, 13 July 1860, GMM; *LL*, 424; Edward Everett, Journal, 22 Nov. 1860, Everett Papers 179:288.

26. *Origin and History* [hereafter *O&H*], 25; *LEL* 1885 ed., 5; F. M. Müller, *Lectures on the Science of Language* (1874), 45, 59, 70–72, 84, referring to *LEL*, 379, 667, 678; Child to GPM, 29 June 1862; *LEL* British ed. edited by William Smith, as *The Students' Manual of the English Language* (London, 1862); *O&H* re-

views in *North American Rev.* 96 (1863): 264; *Cornhill Mag.* 7 (1863): 138–39; J. H. Allen in *Christian Examiner* 74 (1863): 448–49. See Konrad Koerner, "Jacob Grimm's Place in the Foundation of Linguistics as a Science," in his *Practicing Linguistic Historiography: Selected Essays* (Philadelphia: John Benjamins, 1989), 303–17.

27. Alter, *Darwinism and the Linguistic Image*, 5, 21, 52–3, 167n100; Charles Darwin, *The Expression of the Emotions in Man and Animals*, ed. Francis Darwin (1890), *Works* (London: William Pickering, 1989), 23:110–271; Aarsleff, *Study of Language in England*, 229–30; Konrad Koerner, "The Natural Science Impact on Theory Formation in 19th and 20th Century Linguistics," in his *Professing Linguistic Historiography* (Philadelphia: John Benjamins, 1995), 47–76. The seminal comparative works, much-thumbed by Marsh, were Charles Lyell, *The Geological Evidences of the Antiquity of Man*, 2d. ed. (London, 1863) and August Schleicher, *Darwinism Tested by the Science of Language* (1860/63; London, 1869), of which Marsh had the 2d German ed. (Weimar, 1873).

28. *LEL*, 440–41, 644–45. Marsh had lamented the debasement of literature by hack writers and "enterprizing publishers" a decade earlier in *Human Knowledge* (1847), 14–16.

29. *O&H*, 564–65n; GPM to SFB, 8 Apr. 1857, BSI. "*Party*, for person, now an offensive vulgarism," had formerly been acceptable (*LEL*, 693n28).

30. *LEL*, 602.

31. *O&H*, 401; GPM to Charles Eliot Norton, 12 June 1862.

32. *LEL*, 647–48, 224–25. Marsh took many of these examples from Walter Savage Landor, "Dialogues of Sovereigns and Statesmen: V: Peter Leopold and President du Paty" (1828), in his *Imaginary Conversations* (London, 1891), 2:176–239 at 214–19. But Marsh repudiated Landor's indictment of Italians as "obsequious" (Cj, 22 June 1861, 1:9).

33. *LEL*, 225n; 1885 ed., 209n.

34. *LEL*, 119–31.

35. *O&H*, 26; GPM to George Ticknor, 21 Jan., 20 Feb. 1860; *O&H*, 40. Latham's *English Language* and Fowler's *English Grammar* were long standard texts. Marsh's sophistication as a word counter was noted by Müller (*Lectures on the Science of Language*, 84, 121).

36. *LEL*, 480, 599n; 1885 ed., 32n, 48n. The Müller text Marsh censured was *Origin and Growth of Religion* (Hibbert Lectures, 1878). Skeat's dictionary (Oxford: Clarendon) was far more comprehensive than Wedgwood's, but his word histories were abbreviated; Wedgwood and Marsh's *Lectures on the English Language* were among Skeat's sources (p. xxvi).

37. Alter, *Darwinism*, 135, 144–45; Hermann Osthoff and Karl Brugmann, *Morphologische Untersuchungen* (1878), quoted in R. H. Robins, *A Short History of Linguistics*, 2d ed. (London: Longmans, 1979), 184–85.

38. *LEL*, 267–68, 187n, 146, 412–13, 410; 1885 ed., 353–54.

39. *LEL*, 108, 449–50, 598. Mencken (*American Language*, Supplement II [1948]) thought Marsh misunderstood Latham; I think he exaggerated Latham for effect.

40. *LEL*, 273–74 (see *OED*, "Outsider"), 659; Mencken, *American Language*, 185. One Marsh admirer who opposed him on this issue was Richard Grant White (*Words and Their Uses, Past and Present: A Study of the English Language* [New York, 1871], 264–65; see also 46).

41. *LEL*, 202–3, 649.

42. *LEL*, 63, 176.

43. *LEL*, 646–49.

44. A. C. Baugh and Thomas Cable, *A History of the English Language* (1993), 288; R. W. Bailey, *Nineteenth-Century English* (1996), 223–61. See GPM, "Our English Dictionaries," 412–13; "Proposed Revision of the English Bible" (1870), 282.

45. *LEL*, 15–16.

46. Mencken, *American Language*, 8–9; *LEL*, 669, 265n; 1885 ed., 570.

47. *LEL* 1885 ed., 253n; 1861 ed., 454–56, 648–49, 672; GPM, "Old Northern Literature" (1845), 256. For similar views of superior American articulation, see Richard W. Bailey, *Images of English: A Cultural History of the Language* (Cambridge: Cambridge University Press, 1991), 151–58.

48. *LEL*, 671.

49. *LEL* 1885 ed., 568n.

50. *LEL*, 453–56, 666–71; "Our English Dictionaries," 412. Even in England, Marsh later noted, growing assimilation of the spoken to the written word had begun to proscribe such pronunciations as affectations (*LEL* 1885 ed., 572n).

51. G. P. Krapp, *The English Language in America* (1925), 2:19–22. Mencken cited many contrary cases (frontier, harass, mamma, papa, adult), where the English accent the second, Americans the first syllable (*American Language*, 215–16).

52. *LEL*, 675–76; 1885 ed., 571n; 1861 ed., 24–25, 18, 8–9. For other calls to strengthen Anglo-American cultural bonds, see Benjamin T. Spencer, *The Quest for American Nationality: An American Literary Campaign* (Syracuse, N.Y.: Syracuse University Press, 1957), 213. English has become more widespread

than even Marsh foresaw, spoken at the end of the twentieth century by 750 million people worldwide.

53. Einar Haugen, "The Ecology of Language" (1970), in *The Ecology of Language: Essays by Einar Haugen* (Stanford: Stanford University Press, 1972), 325–39 at 326; E. E. Hale, review of *LEL*, *Christian Examiner*, 2. For later views of Marsh, see Krapp, *English Language*, 1:46–47, 2:19–22; Robert C. Pooley, *Grammar and Usage in Textbooks on English* (Madison: University of Wisconsin, Bur. of Educational Research, 1933), 32; idem, *Teaching English Usage* (New York: Appleton-Century, 1946), 111; Charlton Laird, *Language in America* (New York: World, 1970), 442; Baugh and Cable, *History of the English Language*, 14, 288; Phyllis Franklin, "English Studies: The World of Scholarship in 1883," *PMLA* 99 (1984): 356–70.

54. GPM to GO, 22 Feb. 1861, 19 May 1860; GPM to Solomon Foot, 12 Apr. 1860, VHS MSS no. 64; GPM in *BDFP*, 9 Nov. 1860. See Eric Foner, *Free Soil, Free Labor, Free Men: The Ideology of the Republican Party before the Civil War* (New York: Oxford University Press, 1970).

55. Schurz to John Fox Potter, 24 Dec. 1860, in *Speeches, Correspondence, and Political Papers of Carl Schurz*, ed. Frederic Bancroft (New York: Putnam, 1913), 1:176.

56. M. P. Trauth, *Italo-American Diplomatic Relations, 1861–1882: The Mission of George Perkins Marsh* (1958) [hereafter *IADR*], 150; A. W. Salomone, "The Nineteenth-Century Discovery of Italy" (1968), 1384–89.

57. GPM to CCM, 14 Sept. 1856; to FM, 5 Aug. 1855, FC; to J. G. Saxe, 7 Feb. 1851; GPM, "American Image Abroad" (1854).

58. GPM to Norton, 3 Apr. 1860; to Lieber, 3 June 1859. See GPM, "American Diplomacy," "Future of Italy," "Italian Cause and Its Sympathizers," "Italian Nationality," and "What American Diplomacy Might Do" (all in N.Y. *World*, June–July, 1860).

59. GPM to GFE, 7 Mar. 1861, MV; Bryant to Lincoln, 7 Mar. 1861 & Collamer et al. to Lincoln, 8 Mar. 1861, U.S. Dept. of State Appointment Papers; Botta to GPM, 9 Feb. 1861.

60. GPM to GO, 22 Feb. 1861; to GFE, 7 Mar. 1861, VHS MSS 861207; Foot to GPM, 16 Mar. 1861.

61. GPM to Botta, 13 Feb. 1861, Norcross MSS; Botta to GPM, 24 Feb. 1861; Guglielmo Cajani to GPM, 19 Mar. 1861; Bertinatti to Seward, in Bertinatti to Cavour, 7 & 16 Mar. 1861, dispatches 70, 71, quoted in S. E. Humphreys, "Le relazioni diplomatiche fra gli Stati Uniti e l'Italia del Risorgimento (1847–1871)" (1945), ch. 4; Lincoln to Seward, 18 Mar. 1861, in *Complete Works*

of *Abraham Lincoln*, ed. J. G. Nicolay and John Hay (New York: Tandy, 1905), 6:218–19; Lincoln was embarrassed by having "too many" New Englanders as envoys. Marsh's zealous Vermont supporters included railroad moguls anxious to get him off their backs (*LL*, 427n). Schurz got the mission to Spain, which he termed "better than the Turin mission" (to Mrs. Schurz, 28 Mar. 1861, in *Intimate Letters of Carl Schurz, 1841–1889*, tr. Joseph Schafer, Wisconsin Hist. Soc. Coll. 30 [Madison, 1928], 253); Edward Joy Morris got Marsh's former post to Turkey; Anson Burlingame, appointed to but rejected by Austria after his arrival, became U.S. Minister to China.

62. Lieber to GPM, 19 Mar. 1861; N.Y. *Tribune* and Boston *Transcript* in *BDFP*, 20 & 22 Mar. 1861 (see *Vermont Standard* [Woodstock], 29 Mar. 1861); *New York Times*, 20 (quoting Greeley) & 21 Mar. 1861.

63. CCM to LW, 20 Mar. 1861, WP; GPM to LW, 3 Apr. 1861. See GPM to Drake and Ticknor, both 1 Apr. 1861.

64. Lieber to GPM, 19 Mar. 1861; Lowell to GPM, 20 Mar. 1861; GPM to Norton, 26 & 30 Mar. 1861, Norton Papers. Norton was as infatuated with Risorgimento as with Renaissance Italy (Salomone, "Nineteenth-Century Discovery of Italy," 1377–78, 1384–89). Marsh helped Lucius E. Chittenden get a Treasury job, secured the Burlington post office for George W. Benedict, recommended Charles Lanman as State Dept. Keeper of Rolls, put in a word with Seward for D. W. Fiske at Copenhagen, and aided Markoe, Maximilian Schele de Vere, and Bayard Taylor.

65. R. H. Howard in *BFP & Times*, 18 Aug. 1882.

66. GPM in *BDFP*, 18 Apr. 1861; "Departure of George P Marsh—The Meeting Last Night; Full Report of the Speeches of Prest. Pease and Geo. P. Marsh," *Burlington Daily Times*, 19 Apr. 1861. Five days later the Vermont Legislature voted half a million dollars for the Union cause—a sum larger than Vermont's total tax bill, and proportionately far above any other state's appropriation (D. C. Fisher, *Vermont Tradition* [1953], 235).

11. *Risorgimento and Civil War*

1. Marsh served from 20 March 1861 until his death, 24 July 1882 (21 years, 4 months, 4 days). Only Edwin Vernon Morgan, ambassador to Brazil 1912–33 (21 years, 3 months, 5 days), approaches Marsh's record.

2. GPM to W. H. Seward, 27 June 1861, DD (Italy, vol. 10) 3; C. C. Marsh Journal [hereafter Cj], 7 June 1861, 1:1; Croce attributed the "poetry to prose"

phrase to King Vittorio Emanuele (Walter L. Adamson, "Modernism and Fascism: The Politics of Culture in Italy, 1903–1922," *AHR* 95 [1990]: 359–90 at 367).

3. Henry Hearder, *Italy in the Age of the Risorgimento, 1790–1870* (1988), ch. 2; Richard Bellamy and Darrow Schechter, *Gramsci and the Italian State* (Manchester: Manchester University Press, 1993), 150.

4. Denis Mack Smith, *Modern Italy: A Political History* (1997), 21–22.

5. GPM, "The War and the Peace" (1859), 273–74; "Italian Independence," MS (abstract in *BDFP*, 26 Mar. 1860).

6. GPM, "Italian Independence," 24, 36–37; "England Old and New," 36. See GPM to Charles Eliot Norton, 3 Apr., 2 June 1860.

7. GPM, "The Future of Italy" (1860).

8. Ibid.; GPM, "The War and the Peace," 268–71. Marsh termed "the leaven of French democracy . . . a beneficent influence in the Italian peninsula" (*LEL*, 225n).

9. H. R. Marraro, *American Opinion on the Unification of Italy, 1846–1861* (1932), 305–13; A. W. Salomone, "The Nineteenth-Century Discovery of Italy" (1968); Trauth, *IADR*, 150; Shaftesbury to Cavour, 12 Sept. 1860, in Denis Mack Smith, *Victor Emanuel, Cavour, and the Risorgimento* (1971), 175. For pro-Italian British and American feeling, see Paul Ginsborg, "Il mito del Risorgimento nel mondo britannico: 'la vera poesia della politica,'" and Matteo Sanfilippo, "Il Risorgimento visto del Canada e degli Stati Uniti," *Risorgimento* 47 (1995): 384–99 and 490–510; Maura O'Connor, *The Romance of Italy and the English Imagination* (London: Macmillan, 1998), 80–92, 117–47.

10. GPM in the House, 3 Mar. 1848, U.S. *CG*, 30 Cong. 1 sess., 445; GPM to FM, 1847, in *LL*, 116. See Marraro, *American Opinion*, 5–15, 52–69, 123–54.

11. Antonelli cited in F. J. Coppa, *Cardinal Giacomo Antonelli and Papal Politics in European Affairs* (1990), 39–49, 68–71, 125–26; GPM, "The War and the Peace," 262–63, 280–82.

12. GPM to Seward, 27 June, 28 Oct. 1861, 2 Feb. 1864, DD 3, 28, 83.

13. Trauth, *IADR*, 66–72, 166–70.

14. Cj, 23 June 1861, 1:10; Mack Smith, *Modern Italy*, 27–29.

15. GPM to Sumner, 26 July 1861, Sumner Papers. On Ricasoli, see H. F. d'Ideville, *Journal d'un diplomate en Italie* (1872), 1:243–48; W. K. Hancock, *Ricasoli and the Risorgimento in Tuscany* (London: Faber & Gwyer, 1926); Enrica Viviani della Robbia, *Bettino Ricasoli* (Turin: Unione Tipografico Editrice, 1969); Giuliana Biagioli, *Agrarian Changes in Nineteenth-Century Italy: The Enterprise of a Tuscan*

Landlord, Bettino Ricasoli (Reading University, Inst. of Agricultural History, Res. Paper 1, 1970); Mack Smith, *Victor Emanuel*, 115–25, 264.

16. Ricasoli quoted in GPM to Seward, 4 Sept. 1861, unnumbered DD.

17. GPM to Seward, 5 Aug. 1862, DD 49; Cj, 18 Nov. 1861, 2:42; Mack Smith, *Victor Emanuel*, 280–85; idem, *Modern Italy*, 59. Ricasoli's resignation was also forced by his refusal to ban freedom of assembly in the wake of riots in Naples and Sicily (John Davis, *Conflict and Control: Law and Order in Nineteenth-Century Italy* [1988], 176–77). For Ricasoli's obsession with getting Rome, see Fiorella Bartocinni, *La "Roma di Romani"* (Rome: Istituto per la Storia del Risorgimento Italiano, 1971), 203–54.

18. GPM quoted in Cj, 5 Dec. 1861, 2:52; GPM to Seward, 5 Aug. 1862, DD 49.

19. GPM to Seward, 10 Mar., 4 Apr. 1862, DD 39 & 40. See also Derek Beales, "Garibaldi in England: The Politics of Italian Enthusiasm," in J. A. Davis and Paul Ginsborg, eds., *Society and Politics in the Age of the Risorgimento* (1991), 184–216; O'Connor, *Romance of Italy*, 149–85.

20. GPM to Seward, 6 Sept., 8 Oct. 1862, DD 51 & 52; Cj, 30 Aug. 1862, 6:20. See Cj, 4 Mar. 1862, 3:65–66; Garibaldi to GPM, 7 Oct. 1862; Seward to GPM, 5 Nov. 1862, DI 57, pp. 163–64; Mack Smith, *Victor Emanuel*, 289–93.

21. Sir James Lacaita quoted in Mack Smith, *Modern Italy*, 60.

22. GPM to W. M. Evarts, 25 Oct. 1880, DD 922.

23. Benedetto Croce, *A History of Italy, 1871–1915* (Oxford: Clarendon Press, 1929), 5; Mack Smith, *Modern Italy*, 49.

24. Seward to GPM, 7 Jan., 25 Feb. 1864, DI 88 & 90, pp. 187–88; Trauth, *IADR*, 161; A. B. Crane to GPM, 1 Nov. 1880, CFP.

25. GPM to GFE, 21 Apr. 1863; Cj, 26 Nov. 1863, 4 Nov. 1864, 12:23–26, 16:29.

26. Ideville, *Journal d'un diplomate*, 1:30–36, 270; Cj, 14 Aug. 1863, 11:7, 9; GPM to C. C. Rafn, 1 Jan. 1864 (tr. Einar Haugen). On Rosenkranz I am indebted to Kenneth Olwig (6 May 1997).

27. E. L. Godkin, "American Ministers Abroad" (1867); GPM to C. E. Norton, 12 June 1862.

28. Baber to Webster, 10 Mar. 1842, DD 3, and Webster to Baber, 18 June 1842, DI 2, in H. R. Marraro, *L'Unificazione Italiano vista dai diplomatici statunitensi* (1963), 1:113–14; idem, "William Burnet Kinney's Mission to the Kingdom of Sardinia," *New Jersey Historical Soc. Proc.* 64 (1946): 187–215; Niles to Forsyth, 18 Feb. 1838, in idem, "Nathaniel Niles' Missions at the Court of Turin (1838; 1848–50)," *Vermont Quarterly* 15 (1947): 14–32 at 14–18.

29. J. M. Daniel to W. L. Marcy, 4 Sept. 1854, and to Lewis Cass, 6 Aug. 1858, DD 13, p. 91; Cass to Daniel, 28 Aug. 1858, DI 23; Daniel to E. A. Peticolas, late 1853, in *Richmond Semi-Weekly Examiner*, 6 July 1854 (reported in New York and Boston papers); all in Marraro, *L'Unificazione Italiano* (1971), 4:32, 224–26, 485–86n238; Cavour quoted in CCM to Susan Edmunds, 9 June 1861. See Marraro, *American Opinion*, 192.

30. Story to Sumner, 2 Sept. 1861, 9 Aug. 1863, Sumner Papers; GPM to Story, 12 July 1882.

31. Cj, 27 June 1861, 1:13; GPM in Cj, 29 Mar. 1862, 4:10; GPM to GFE, 8 June 1861.

32. GPM to Dillon, 29 Apr. 1862.

33. GPM to Seward, 6 Feb. 1862, confid. DD; Seward to GPM, 8 Mar. 1862, DI 41, p. 151; GPM to Norton, 12 June 1862.

34. GPM to Seward, Aug. 1861, private & confid., Seward Papers.

35. GPM to John Bigelow, 2 Mar. 1872, in Bigelow, *Retrospections of an Active Life* (1913), 5:19–20. Marsh's use of an Irish moniker in this *jeu d'esprit* was maliciously intended.

36. GPM to Seward, 16 Feb., 21 Mar. 1863, DD 67 & 70; GPM to GO, 14 Jan. 1862; Cj, 4 Apr. 1863, 8:64.

37. GPM to Seward, 28 June, 22 July 1861, confid. unnumbered & DD 7; Cj, 2 Oct. 1861, 2:2; David Hilton [Wheeler], *Brigandage in South Italy*, 2 vols. (London, 1864); GPM to Monti, 2 May 1862, U.S. PR, Misc. Corr., 1:153. See GPM to Andrew Stevens, 12 May 1862, U.S. PR, Misc. Corr. Wheeler also "owe[d] much" to Marsh for his *By-Ways of Literature; or, Essays on . . . the English-Speaking Peoples* (New York, 1883), iii, 99–100, 122–29; see Cj, 7:28, 16:39. For Monti, see Howard R. Marraro, "Pioneer Italian Teachers of Italian in the United States," *Modern Language Journal* 28 (1944): 555–82.

38. GPM to Seward, 12 May 1862, DD 45. Richard M. Blatchford was succeeded at Rome by J. Clinton Hooker. See W. J. Stillman, *The Autobiography of a Journalist* (1901), 1:369–70, 2:388–91; Cj, 2:37–38, 10:67–69, 13:21–22. In 1867 the Senate voted to end the Roman mission, Justin Morrill noting that the Papal States were now reduced to Rome itself; the United States was the first country to terminate papal relations. See Marraro, *American Opinion*, 24–26, 64–65; L. F. Stock, *United States Ministers to the Papal States* (1933) and idem, *Consular Relations between the United States and the Papal States* (1945), xxxix; Howard R. Marraro, "The Closing of the American Diplomatic Mission to the Vatican and Efforts to Revive It, 1868–1870," *Catholic Historical Review* 33 (1948): 423–47 at 445–46.

39. [William Bradford Reed], *A Review of Mr. Seward's Diplomacy; By a Northern Man* (Philadelphia, 1862), 19–20. On Union propaganda efforts, see Margaret Clapp, *Forgotten First Citizen: John Bigelow* (Boston: Little, Brown, 1947), 149–61; R. F. Durden, *James Shepherd Pike* (1957), ch. 4; idem, "James S. Pike: President Lincoln's Minister to the Netherlands" (1956).

40. GPM to Seward, 14 Sept. 1861, DD 19; N. B. Ferris, *Desperate Diplomacy: William H. Seward's Foreign Policy, 1861* (1976), 177.

41. GPM to Seward, 29 May 1861, DD 2; to Norton, 12 June 1862; to Seward, 2 Feb. 1864, DD 83; to Norton, 17 Oct. 1863. Marsh essays arguing the illegality of secession, "Were the States Ever Sovereign?" and "State Sovereignty," appeared in *The Nation* (1865).

42. GPM to Norton, 14 Mar. 1864 (CCM copy); Sandwith to GPM, 25–28 Aug. 1862, 20 Dec. 1864; Martin Crawford, *The Anglo-American Crisis of the Mid-Nineteenth Century: The Times and America, 1850–1862* (1987), 106–38. See GPM to Harry Verney, 13 May 1861; GPM to R. H. Dana, Jr., 21 Dec. 1865, Dana MSS.

43. Trauth, *IADR*, 150; GPM to Seward, 30 Mar., 18 July 1863, DD 72 & 77; to GFE, 21 Apr. 1863; to J. S. Pike, 21 Nov. 1861.

44. GPM to Seward, 27 June, 24 Oct., 1 Nov. 1861, DD 3, 25, & 30; GPM to GO, 14 Jan. 1862. See also GPM, DD 39, 75, 77; GPM to Ricasoli, 26 Aug. 1861; to D. H. Wheeler, 5 Nov., 15 Dec. 1861; W. T. Rice to GPM, 12 Jan., 8 June 1862; Gideon Welles to Commander Thatcher, 28 Feb. 1862, & Raphael Semmes, Log, CSS *Sumter*, 4 Jan.-20 June 1862, in U.S., *Official Records of the Union and Confederate Navies in the War of the Rebellion*, ser. 1 (Washington, 1894), 1:332, 638–86; Ferris, *Desperate Diplomacy*, 187–88.

45. GPM to Norton, 16 Sept. 1861; to J. S. Pike, 14 Sept. 1861, Pike Papers; to Norton, 22 Jan. 1864 (CCM copy).

46. GPM to Harry Verney, 13 May 1861; to Pike, 21 Nov. 1861, Pike Papers; to Norton, 16 Sept. 1861.

47. GPM to F. J. Child, 26 July 1862; to Pike, 20 Aug. 1862, Pike Papers; to Lieber, 22 Aug. 1862, Lieber MSS, LI 2783.

48. GPM to CCM, 9 Oct. 1862; Cj, 11 & 18 July 1863, 7 Oct. 1864, 10:29, 37, 16:14; GPM to Norton, 14 Mar. 1864. Other American envoys echoed Marsh's views (Clapp, *Forgotten First Citizen: John Bigelow*, 179–80; J. G. Randall and Richard N. Current, *Lincoln the President: Last Full Measure* (New York: Dodd, Mead, 1955), 83–84; Durden, *James Shepherd Pike*, 109).

49. GPM to SFB, 8 Dec. 1861, BSI; to D. H. Wheeler, 9 Oct. 1862; Cj, 12 Nov. 1861, 2:36. See Seward to GPM, 21 Sept. 1861, DI 19, p. 133; GPM

to Seward, 19 Aug. 1861, DD 10; Howard R. Marraro, "Volunteers from Italy for Lincoln's Army," *South Atlantic Quarterly* 44 (1945): 384–96. During his own lifetime Marsh had little luck with his dilution. By 1880 there were only 44,000 Italians in the U.S., and almost 1,900,000 Irish.

50. Seward to Sanford, 27 July 1861; Sanford to Garibaldi, 20 Aug. 1861, in Sanford to Seward, 29 Aug. 1861; all in U.S. Dept. of State, Belgium PR, vol. 7. The Quiggle-Garibaldi letters are in Trauth, *IADR*, 10–11.

51. Cj, 17 Aug., 3–9 & 12 Sept. 1861, 1:31, 45–50, 55; Garibaldi to Sanford, 31 Aug. 1861, in H. Nelson Gay, "Lincoln's Offer of a Command to Garibaldi: Light on a Disputed Point of History," *Century Mag.* 75 (1907): 63–74 at 68, and in Trauth, *IADR*, 13–14n46.

52. Sanford to Seward, 18 Sept. 1861, Belgium PR; GPM to Seward, 14 Sept. 1861, DD 19. See Sanford to GPM, 13 & 17 Aug. 1861; James Mortimer to GPM, 22 & 27 Sept. 1861; W. L. Dayton to GPM, 21 Sept. 1861; Cj, 27 Oct. 1861, 2:20. For Washington reactions, see Jay Monaghan, *Diplomat in Carpet Slippers: Abraham Lincoln Deals with Foreign Affairs* (Indianapolis: Bobbs-Merrill, 1945), 136–37.

53. GPM to Ricasoli, 31 Aug. 1862, in GPM to Seward, 6 Oct. 1862, DD 51; Garibaldi to GPM, 7 Oct. 1862, in GPM to Seward, 8 Oct. 1862, DD 52.

54. GPM to Garibaldi, 11 Oct. (also 28 Nov.) 1862; Seward to GPM, 5 Nov. 1862, DI 57, pp. 163–64; Cj, 18 Oct. 1862, 7:6–7. Gay ("Lincoln's Offer," 73), says the second offer to Garibaldi embarrassed the U.S.; if so, Marsh was not aware of it. See S. E. Humphreys, "Le relazioni diplomatiche fra gli Stati Uniti e l'Italia del Risorgimento" (1945), ch. 5; idem, "Two Garibaldian Incidents in American History," *VH* 23 (1955): 135–43; Howard R. Marraro, "Lincoln's Offer of a Command to Garibaldi: Further Light on a Disputed Point of History," *Journal of the Illinois State Historical Soc.* 36 (1943): 237–70; Marraro, "American Opinion and Documents on Garibaldi's March on Rome, 1862," *Journal of Central European Affairs* 7 (1947): 143–61; Trauth, *IADR*, 21–29.

12. Turin and the Alps

1. Charles de Brosses (1739) and Thomas Nugent (1778), quoted in M. D. Pollak, *Turin, 1564–1680* (1991), 1–2; Henry James, "Italy Revisited" (1877), in his *Portraits of Places* (London: Macmillan, 1883), 39–74 at 41. James betrayed a medievalist bias in dismissing Turin for having "no architecture, no churches, no monuments, nor especially picturesque street scenery" ("A European Summer; VI: From Chambery to Milan," *Nation* 15 [1872]: 332–33). Matthew

Arnold to Mrs. Arnold, 22 June 1865, in *Letters of Matthew Arnold, 1848–1888* (1895), 1:329, exudes typical praise of Turin. See Geoffrey Symcox, "From Commune to Capital: The Transformation of Turin, Sixteenth to Eighteenth Centuries," in Robert Oresko, G. C. Gibbs, and H. M. Scott, eds., *Royal and Republican Sovereignty in Early Modern Europe* (Cambridge: Cambridge University Press, 1997), 242–69.

2. GPM, unpublished MSS, ca. 1867, slightly altered in CCM, "Turin," *Johnson's New Universal Cyclopaedia* (1878), 4:971–72; Friedrich Nietzsche to Peter Gast, 7 April 1888, in Lesley Chamberlain, *Nietzsche in Turin* (London: Quartet, 1997), 23. "No court in Europe now offers so much of ancient etiquette, so much of real stately aristocracy, as that of Turin," judged Caroline. Other "European courts have altogether a parvenu air when compared with this" (Cj, 13 Jan. 1862, 3:12).

3. Luigi Amedeo Melegari of the Italian Foreign Office, cited in Cj, 2 Aug. 1862, 6:3. See Augusto Cavallari Murat, ed., *Forma urbana ed architettura nella Torino barocca: Dalle premesse classiche alle conclusioni neoclassiche* (Turin: Unione tipografico, 1968), 2:306.

4. Cj, 14 June 1861, 1:5. For these drives along what is now the Corso Massimo D'Azeglio, see also GPM to Mrs. J. S. Pike, 19 June 1861, Pike Papers.

5. H. F. d'Ideville, *Journal d'un diplomate en Italie* (1872), 1:64, 133–34, 235–37; A. L. Cardoza, *Aristocrats in Bourgeois Italy: The Piedmontese Nobility, 1861–1930* (1997). The name *codini* came from the powdered queues, or pigtails, worn by Savoyard gentry after Napoleon; it took on the meaning of "die-hard reactionary" (Mme. Charles de Bunsen [Mary Isabella Waddington], *In Three Legations* [London: Unwin, 1909], 19).

6. Leone Carpi, *Il risorgimento italiano: Biografi storico-politiche d'illustre italiani contemporane* (1884–88); Carlo Moriondi, *Questi Piemontese* (1990).

7. P. D. Pasolini, *Giuseppe Pasolini* (1915), 2:92, 404; S. W. Halperin, *Diplomat under Stress: Visconti Venosta and the Crisis of July 1870* (Chicago: University of Chicago Press, 1963).

8. Cj, 6 June 1861, 1:4.

9. Cj, 28 Aug. 1862, 6:18–19.

10. Cj, 18 June 1861, 26 Feb. 1862, 1:7, 3:59.

11. "Viator" [GPM], "The 'Catholic Party' of Cesare Cantù and American Slavery" (1866); Cj, 7 & 13 Mar. 1864, 14:7 & 15.

12. *Edward Lear: Selected Letters*, ed. Vivien Noakes (Oxford: Oxford University Press, 1990), 223.

13. Anthony Cardoza, "The Long Goodbye: The Landed Aristocracy in

North-Western Italy, 1880–1930," *European History Quarterly* 23 (1993): 323–58; John A. Davis, "Remapping Italy's Past in the Twentieth Century," *Journal of Modern History* 66 (1994): 291–320; GPM to C. E. Norton, 14 Mar. 1864 (CCM copy); Cj, 8 Apr. 1864, 14:40. On Nigra, see Ideville, *Journal d'un diplomate*, 1: 239–40.

14. Carlo Poerio quoted in Cj, 25 Oct. 1861, 2:18.

15. Cj, 23 Aug. 1862, 6:15–16; Giovanni Montroni, "Aristocracy and Professions," in Maria Malatesta, ed., *Society and the Professions in Italy, 1860–1914* (Cambridge: Cambridge University Press, 1995), 255–75 at 256–59.

16. Massimo D'Azeglio, *Things I Remember* (1873), 6–7. D'Azeglio here describes the social scene of his Turin youth, ca. 1820, but goes on to affirm it little changed.

17. Cardoza, *Aristocrats in Bourgeois Italy*, 6–7, 13–14, 22–24, 66–75; Emilio Sereni, *History of the Italian Agricultural Landscape* (1961), 264–72; D'Azeglio, *Things I Remember*, 7, 136. Land ownership in Tuscany was no less concentrated among aristocrats, whose estates betrayed an Anglomania like that of the Piedmontese. But unlike the Turinese, Florentine nobles shared elite values with bankers, merchants, and entrepreneurs (Raffaele Romanelli, "Urban Patricians and 'Bourgeois' Society: A Study of Wealthy Elites in Florence, 1861–1904," *Journal of Modern Italian Studies* 1 [1995]: 3–21 at 12).

18. GPM to C. E. Norton, 14 Mar. 1864; to Henry P. Hickok, 14 Jan. 1862; Cj, 2 Jan. 1864, 13:3, 19 Apr. 1862, 4:26–27. See Moriondi, *Questi Piemontese*, 188–201; Renata Pescanti Botti, *Donne del risorgimento italiano* (1966).

19. Cj, 27 Feb. 1862, 3:62–64, 14 & 23 Jan 1864, 13:19–20, 33; GPM to LW, 21 Oct. 1876; CCM, "Voyage to Italy," MS; Anthony Cardoza, "Tra caste e classe: Clubs maschili dell'élite torinese, 1840–1914," *Quaderni storici* (n.s. 77) 26 (1991): 365–83. See Onorata Roux, *La prima regina d'Italia. Nelle vita privata—Nelle vita del paese—Nelle lettre e nelle arti* (Milan: Carlo Aliprandi, 1901), 21–34, and Moriondi, *Questi Piemontese*, 243–56 (on Rosa Arbesser and the royal family); Averil Mackenzie-Grieve, *Clara Novello, 1818–1908* (London: Geoffrey Bles, 1955), 274–75, 302.

20. Cj, 23 Jan. 1862, 3:26, 22 June 1863, 10:7–8.

21. Cj, 12 Apr., 17 Jan. 1862, 4:15, 3:18.

22. Cj, 10 Nov. 1861, 5 (quote) & 19 Jan. 1862, 20 Mar. 1864, 2:35, 3:5, 20, 14:26. "Ausonio Franchi" was the pen name of the ex-priest Cristoforo Bonavino. The Vaudois were an evangelical Protestant sect descended from the long-persecuted medieval Waldensian heretics of southern France, later (and still) centered in Alpine valleys southwest of Turin.

23. GPM to GFE, 26 Nov. 1861 (quote); Cj, 31 Dec. 1861, 1 Feb. 1862, 2:67, 3:35–36. In Shakespeare's *Henry V* (II, i, 7), Corporal Nym's iron is not "cold." Marsh ascribed skill in marksmanship to the cultivation of almost telescopic vision (*M&N*, 16n13).

24. D. H. Wheeler, "Recollections of George P. Marsh" (reprinted from *Christian at Work*), UVM; Matthew Arnold to Mrs. Arnold, 22 June 1865, in his *Letters*, 1:329; J. S. Pike, Notebook 18:60, MS, Pike Papers; R. F. Durden, "James S. Pike: President Lincoln's Minister to the Netherlands" (1956), 363.

25. GPM to Mrs. J. S. Pike, 19 June 1861, Pike Papers; to Henry and Maria Buell Hickok, 14 Jan. 1862.

26. GPM to J. G. Saxe, 7 Feb. 1851; to Seward, 12 May 1864, DD 93; to D. H. Wheeler, 4 July 1862 (CCM copy). See GPM to CCM, 9 Oct. 1862; Ralph Keeler to GPM, 18 Jan. 1869; Keeler, *Vagabond Adventures* (Boston, 1870), dedication to GPM; E. L. Godkin, "American Ministers Abroad" (1867), 132–33.

27. GPM to F. J. Child, 28 Sept. 1861; GFE to GPM, 28 Apr. 1862.

28. Charles Scribner to GPM, 25 Apr. 1861, 21 Oct. 1862, 1 May, 1 Dec. 1863.

29. Child to GPM, 14 Apr., 26 Nov. 1861; GPM to Child, 28 Sept. 1861; *LEL*, 22n; *O&H*, 437n; GPM to Child, 26 July 1862. See *O&H*, 381–401; *LEL* 1885 ed., 97n, 145n; F. J. Child, "Observations on the Language of Chaucer," Amer. Assn. for the Advancement of Science *Memoirs* 8, Part 2 (1863), 455 ff. Marsh found "too repulsive for quotation" much of another fourteenth-century poem, *The boke of curtasye* (Percy Society, London, 1841), which "discloses a coarseness of habits in the more elevated classes, strangely contrasting with the[ir] material luxury" (*O&H*, 291–92).

30. GPM to F. P. Nash, 31 Mar. 1874; see *O&H*, 38–39. Marsh here elaborates on an 1870 critique (of Frederic W. Farrar, *Families of Speech* [1870]) in the London *Athenaeum*, noting that other Italian dialects, not Latin alone, had given rise to the Romance languages (S. G. Alter, *Darwinism and the Linguistic Image*, 96, 174n65). See Žarko Muljačić, "The Relationship between the Dialects and the Italian Language," in Martin Maiden and Mair Parry, eds., *The Dialects of Italy* (1997), 389–93.

31. GPM to Child, 26 Sept. 1861; *O&H*, 51–54, 82–84; GPM, "Origin of the Italian Language" (1867). See F. M. Müller, *Lectures on the Science of Language* (1874), 58–71, referring to *LEL*, 379, 667, 678.

32. GPM to SFB, 21 Nov. 1861, BSI. This was purely rhetorical; "not that

I have a pipe or a nargileh, or even a cigar, or would smoke one if I had," Marsh added. He termed the use of tobacco "the most vulgar and pernicious habit engrafted by the semi-barbarism of modern civilization upon the less multifarious sensualism of ancient life" (*M&N*, 58).

33. GPM to GFE, 8 June, 26 Nov. 1861; to Child, 28 Sept. 1861; to Maria Buell Hungerford, 10 Aug. 1861.

34. R. S. Chilton to GPM, 4 Apr. 1861; Seward Circular no. 46, 6 Feb. 1864, DI; Durden, "James S. Pike," 359–60; GPM to Pike, 12 May 1864, 30 Apr. 1862, Pike Papers.

35. Pike to GPM, 9 June 1862; GPM to CM, 25 Oct. 1862, DCA.

36. Cj, 6–21 Oct. 1863, 11:64–107; GPM to Norton, 17 Oct. 1863. A century later Aigues Mortes became one of the first French "villes sauvegardées."

37. GPM to FW, 4 July 1864; to SFB, 2 Aug. 1865, BSI.

38. GPM to SFB, 21 Nov. 1864, BSI; to Susan Edmunds, 30 July 1863; to LW, 8 Sept. 1863, WP.

39. GPM to D. H. Wheeler, 19 Aug. 1864; to HP, 22 June 1862, PAA.

40. GPM to GFE, 4 Aug. 1862.

41. GPM to Evarts, 30 Oct. 1880, 2 Nov. 1877, DD 923 & 713; Cj, 6 Mar. 1864, 14:6. See Sereni, *History of the Italian Agricultural Landscape*, 273–74, 280–87, 304–13.

42. Cj, 30 Mar. 1864, 14:33. On elite aloofness, see Denis Mack Smith, "Francesco De Sanctis: The Politics of a Literary Critic" (1991), 253–55.

43. GPM and Mrs. Mayhew quoted in Cj, 3 June 1864, 15:23.

44. Benedetto Croce, *A History of Italy* (Oxford: Clarendon Press, 1923), 8; Sir James Hudson to Lord John Russell, 8 Feb. 1861, in Denis Mack Smith, ed., *The Making of Italy, 1796–1870* (1968), 339. See A. William Salomone, "The *Risorgimento* between Ideology and History: The Political Myth of *rivoluzione mancata*," *AHR* 68 (1962): 38–56; Lucy Riall, *The Italian Risorgimento: State, Society and National Unification* (London: Routledge, 1994), 76–82. That so-called Italian unification was in essence annexation to the kingdom of Piedmont is today commonly assumed (Alessandro Guidi, "Nationalism without a Nation: The Italian Case," in Margarita Díaz-Andreu and Timothy Champion, eds., *Nationalism and Archaeology in Europe* [London: University College London Press, 1996], 108–18 at 109).

45. Massimo D'Azeglio cited in Denis Mack Smith, *Victor Emanuel, Cavour, and the Risorgimento* (1971), 273; Luigi Farini to Marco Minghetti, 12 Dec. 1860, in Mack Smith, *Making of Italy*, 330; GPM, "Sicily, Island of," *Johnson's New Uni-*

versal Cyclopaedia (1878), 4:268. See Arrigo Solmi, *The Making of Modern Italy* (1925), 393; Stuart Woolf, *A History of Italy from 1790 to 1860: The Social Constraints of Political Change* (London: Methuen, 1979), 443–45, 455–80. On the South as viewed from Piedmont, see Nelson Moe, "'Altro che Italia!': Il Sud dei piemontesi (1860–61)," *Meridiana*, no. 15 (1992): 53–89; Lucy Riall, *Sicily and the Unification of Italy: Liberal Policy and Local Power, 1859–1866* (Oxford: Clarendon Press, 1998), 123–26 (and on subsequent regional discord, 138–55, 198–230).

46. Mack Smith, *Victor Emanuel*, 127–52, 258.

47. Mack Smith, *Making of Italy*, 339–40; idem, *Victor Emanuel*, 249–74; R. J. B. Bosworth, *Italy and the Wider World, 1860–1960* (1996), 18–21.

48. GPM, "Female Education in Italy" (1866); *Address Delivered . . . U.S. Military Academy, West Point* (1860), 5.

49. Arconati, 6 Nov. 1850, quoted in Nassau William Senior, *Journals Kept in France and Italy from 1848 to 1852* (London, 1871), 1:292.

50. Hudson to Clarendon, 4 Jan. 1856, quoted in Mack Smith, *Victor Emanuel*, 93n; Cj, 14 June 1861, 1:5; M. Mair Parry, "Ël piemontèis, lenga d'Europa," in W. V. Davies and R. A. M. Temple, eds., *The Changing Voice of Europe* (Cardiff: University of Wales Press, 1994), 171–92 at 174–75.

51. Mme. Giletta quoted in Cj, 3 Oct. 1864, 16:10; Fava to GPM, cited in Cj, 12 Sept. 1862, 11:36–37; on Badoglio's Piedmontese accent, Denis Mack Smith, *Modern Italy* (1997), 3. Though industry and immigration now make dialect in Turin more marginal than in most of Italy, it remains valued, on the one hand as a Piedmontese cultural front against all-engulfing Italian, on the other as a specifically local counter to the disunity of the Northern League (Tullio De Mauro, *Storia linguistica dell'Italia unita* [1963], 298–301; Parry, "Ël piemontèis," 176, 187).

52. Alessandro Manzoni, notes, ca. 1833–36, quoted in Mack Smith, *Making of Italy*, 71–72; see De Mauro, *Storia linguistica*, 293–95; Robert A. Hall, Jr., "19th-Century Italian: Manzonian or Deamicisian?" in Paolo Ramat et al., eds., *The History of Linguistics in Italy* (Amsterdam: John Benjamin, 1986), 227–35. The fascinated Marsh tracked verbal changes from Manzoni's "extremely rare" suppressed 1st ed. (Milan, 1825–26) to *Le correzioni ai Promessi sposi, e l'unita della lingua* (Milan, 1874) (University of Vermont, *Catalogue of the Library of George Perkins Marsh* [1892], 434–35).

53. Ubaldini Peruzzi quoted in Cj, 16 Aug. 1863, 11:10–11; GPM to D. H. Wheeler, 29 Sept., 10 Oct. 1864.

54. GPM, "Italian Language and Literature" (1875), 1330–31.

55. Ibid. Italy is still hailed as a treasure house of lexical diversity; "no other area of Europe [has so much] linguistic variety concentrated" into so small an area" (Maiden and Parry, *Dialects of Italy*, 1).

56. GPM, "Origin of the Italian Language" (1867), 29, citing Bernardino Biondelli, *Saggio sui dialetti gallo-italici* (1853). Marsh noted a similar speech/writing cleavage in England from the Norman Conquest to the mid-fourteenth century, and to some extent "in a large part of Europe to this day" (*O&H* [1862], 337n). In Marsh's day only urban middle-class Italians were bilingual; the duality he described is now termed *diglossia* (Alberto Sobrero, "Italianization of the Dialects," in Maiden and Parry, *Dialects of Italy*, 412–18 at 412–13). More complex than Marsh's stark dichotomy, Italy's language situation is termed "plurilingualism" by Tullio De Mauro ("Linguistic Variety and Linguistic Minorities," in David Forgacs and Robert Lumley, eds., *Italian Cultural Studies: An Introduction* [Oxford: Oxford University Press, 1996], 88–101).

57. GPM, "Italian Language and Literature," 1332. See Anna Laura Lepschy and Giulio Lepschy, *The Italian Language Today* (London: Hutchison, 1977), 36; Giulio Lepschy, "How Popular Is Italian?" in Zygmunt G. Baranski and Robert Lumley, eds., *Culture and Conflict in Postwar Italy* (London: Macmillan, 1990), 63–78; Sobrero, "Italianization of the Dialects," 413; Corrado Grassi, Alberto A. Sobrero, and Tullio Telmon, *Fondamenti di dialettologia italiana*, 2d ed. (Rome: Laterza, 1998), 243–69. The present spread of Italian stems from 1960s school reforms, war- and work-induced regional migration, and nationally pervasive sports and media. But today's critics continue to echo the debasements Marsh deplored in the 1870s: lazy articulation, enfeebled lexicon, flashy foreignisms.

58. GPM, "Origin of the Italian Language," 12.

59. CCM, "Naples," *Johnson's New Universal Cyclopaedia* (1877), 3:706–8; Filippo Turati, "Crime and the Social Question" (1882), cited in J. A. Davis, *Conflict and Control: Law and Order in Nineteenth-Century Italy* (1988), 262, 314. Open familiarity among Italians was based on secure consciousness of social distance, not on modern sentiments of equality (Adrian Lyttleton, "The Middle Classes in Liberal Italy," in J. A. Davis and Paul Ginsborg, eds., *Society and Politics in the Age of the Risorgimento* [1991], 217–50 at 228); English and American visitors mistook this as classless sociable ease (John Pemble, *The Mediterranean Passion: Victorians and Edwardians in the South* [Oxford: Clarendon Press, 1987], 134–40).

60. GPM, "Oration" [New Hampshire] (1856), 60; *O&H*, 300.

61. GPM to GFE, 4 Aug. 1862; GPM cited in Cornelia Underwood to Levi

Underwood, 5 Dec. 1873, in Tom Daniels, "In Italy with Mrs. and Mrs. George Perkins Marsh" (1979), 194; GPM to Frelinghuysen, 29 June 1882, DD 1030.

62. GPM to Charles Eliot Norton, 16 Jan. 1865 (CCM copy); to Seward, 28 Oct. 1861, DD 28. "The galleries of Florence, Bologna, Parma etc., which we formerly found frequented only by English, Americans, Germans & Russians, were now daily filled with Italians who at last are in a situation to enjoy what is their own" (Cj, 22 Oct. 1861, 2:13–14).

63. GPM to HP, 7 Aug. 1863, PAA; to SFB, 21 Nov. 1864, BSI; to Donald G. Mitchell, June 1865, in Mitchell, *American Lands and Letters; Leather-stocking to Poe's "Raven"* (1907), 43.

64. GPM to Mrs. J. S. Pike, 20 Aug. 1862, Pike Papers.

65. GPM to Seward, 12 May 1864, DD 93; to D. H. Wheeler, 24 May 1864; to GFE, 8 June 1861, 4 Aug. 1862.

13. Man and Nature: *The Making*

1. *M&N*, 187; GPM to SFB, 21 May 1860, BSI. Arnold Henry Guyot's *The Earth and Man* had appeared in 1849, Carl Ritter's multivolume *Die Erdkunde . . .* between 1822 and 1859; see *M&N*, 13. Marsh owned but did not refer to Friedrich Körner's *Der Mench und die Natur: Skizzen aus dem Kultur- und Naturleben* (Leipzig, 1853). For other like titles see Marcus Hall, "Restoring the Countryside: George Perkins Marsh and the Italian Land Ethic" (1998), 1.

2. GPM to Seward, 7 July 1863, DD 74.

3. B. H. and F. P. Nash, "Notice of George Perkins Marsh" (1882–83), 456.

4. GPM to Caroline Estcourt, 18 Feb. 1863.

5. GPM to J. S. Pike, 25 Apr. 1864, Pike Papers; CCM (ca. 1851) cited in Humphrey Sandwith to GPM, 9 Nov. 1864.

6. GPM to SFB, 6 Mar., 21 May 1860, BSI. See Guyot, *Earth and Man*, 29–34 and chs. 10–12.

7. Cj, 14 Apr. 1862, 4:16.

8. Cj, 7:28–53, 62–64, 80, 8:4–7; Bayard Taylor, *At Home and Abroad; A Sketch Book of the Life, Scenery, and Men* (New York, 1860), 1:314. See Fabio Calvi, *Villa Pallavicini, parco romantico di Pegli* (Genoa: Sagep, 1997).

9. Cj, 12 Jan. 1863, 7:82–83; A Correspondent, "Exhaustion of Vegetable Mould," and editorial, "Resources of the Earth," *The Times* [London], 6 & 8 Jan. 1863.

10. GPM in Cj, 12 Jan. 1863, 7:83; *LEL* (1861), 449. Darwin, all but beaten

to the draw by Wallace, differed from Marsh on this point. "It has sometimes been said that the success of the *Origin* proved 'that the subject was in the air' . . . I do not think this was strictly true," for Darwin had never "came across a single [naturalist] who seemed to doubt about the permanence of species." More likely, he felt, "innumerable well-observed facts were stored in the[ir] minds . . . ready to take their proper places as soon as any theory which would receive them was properly explained" (Charles Darwin and T. H. Huxley, *Autobiographies*, ed. Gavin de Beer [London: Oxford University Press, 1974], 73–74).

11. GPM to HP, 13 Apr. 1863, PAA; to GFE, 21 Apr. 1863. Forced to quit London for a remote rural parish, the Reverend Sydney Smith termed his Yorkshire abode "so far out of the way" as to be "12 miles from a lemon" (*Twelve Miles from a Lemon: Selected Writings of Sydney Smith*, comp. Norman Taylor and Alan Hankinson [Cambridge: Lutterworth, 1996],9).

12. Michele Tamagnone, *Piobesi nei dodici secoli della sua storia* (1985), 49–58; Cj, 2 Apr. 1863, 8:62–63.

13. CCM, MS (for *LL*, vol. 2), 59–60; Cj, 2–5 Apr. 1863, 8:63–65.

14. Cj, 22 May, 20 July, 20–21 Sept. 1863, 9:43–44, 10:39–40, 11: 47–48; Rinaldo Merlone, *Dalla Rivoluzione francese alle soglie del Duemila . . . in un comune della pianura torinese* (1996), 24–25; James Jackson Jarves letter to N. Y. *Tribune*, quoted in *BFP* & *Times*, 1 May 1883. In 1998 the Castello was acquired by the local municipality for possible use as a library, a kindergarten, a home for the elderly, and a public park (Comune di Piobesi Torinese, Verbale di deliberazione del Consiglio comunale no. 49, 24 Nov. 1998).

15. CCM to LW, 2 June 1863, WP; *M&N*; 98n89 (quote on lizards); Cj, 6 May 1863, 9:28.

16. Cj, 29 June, 1 July 1863, 10:16–18.

17. Cj, 29 Nov. 1862, 9 Nov. 1863, 7:55, 12:6. Laetitia Solms-Rattazzi was the illegitimate daughter of Princess Laetitia Bonaparte, wife of the Marshes' old friend Sir Thomas Wyse, British ambassador to Greece. The further shenanigans of Mme. Solms-Rattazzi, the Princess "Brouhaha," continued to scandalize Caroline Marsh (Cj, 25 Jan. 1864, 13:36–37). See Frédéric Loliée, *Women of the Second Empire* (London: Bodley Head, 1907), 64–82; James J. Auchmuty, *Sir Thomas Wyse, 1791–1862* (London: P. S. King, 1939); Magda Martoni, *Une reine du Second Empire: Marie Laetitia Bonaparte Wyse* (Geneva: Droz, 1957).

18. Chapter 1 above; Benjamin Silliman, *Remarks Made on a Short Tour between Hartford and Quebec, in the Autumn of 1819* (New Haven, 1824), 412.

19. GPM to Asa Gray, 9 May 1849, with "Economy of the Forest," MS;

Roland M. Harper, "Changes in the Forest Area of New England in Three Centuries," *Journal of Forestry* 16 (1918): 442–52 at 447; R. W. Judd, *Common Lands, Common People: The Origins of Conservation in Northern New England* (1997), 90–92. Tom Wessels, *Reading the Forested Landscape* (1997), 58, estimates three-fourths of Vermont was open by 1840. Recalling Green Mountain treks, Marsh wrote Arnold Guyot that "physical geography has long been my favorite pursuit" (27 Oct. 1857). The rapid transformation of Vermont from unbroken forest to cleared lands, the disappearance of trout and pickerel from the streams, and changes in flora and fauna had been noted by Zadock Thompson (*Natural History of Vermont* [1853], 15–19); also Samuel Williams, *The Natural and Civil History of Vermont*, 2d ed. (Burlington, 1809), 1:31–80; *BDFP*, 28 Feb., 3 Mar. 1857; L. D. Stilwell, *Migration from Vermont* (1948), 232–34. Carolyn Merchant, *Ecological Revolutions: Nature, Gender, and Science in New England* (1989), 1, 224–26, 240–41, discusses the degradations Marsh addressed. But observers tended to exaggerate the infertility of New England soils and the pace and degree of agricultural decline (Michael M. Bell, "Did New England Go Downhill?" *Geographical Review* 79 [1989]: 450–66).

20. Ira Allen quoted in Edwin T. Martin, *Thomas Jefferson: Scientist* (New York: Schuman, 1952), 204–5.

21. "In New England they once thought *blackbirds* useless, and mischievous to the corn. They made efforts to destroy them. The consequence was, the blackbirds were diminished; but a kind of worm, which devoured their grass, and which the blackbirds used to feed on, increased prodigiously; then, finding their loss in grass much greater than their saving in corn, they wished again for their blackbirds" (Franklin to Richard Jackson, 5 May 1753, in *The Life and Writings of Benjamin Franklin*, ed. A. H. Smyth [New York: Macmillan, 1905–7], 3:133).

22. Georges-Louis Leclerc de Buffon (1764) quoted in C. J. Glacken, *Traces on the Rhodian Shore* (1967), 668, 663.

23. Walt Whitman, "Song of the Redwood-Tree" (1874), in his *Complete Poetry and Selected Prose* (New York: Library of America, 1982), 351–55. See Gilbert Chinard, "The American Philosophical Society and the Early History of Forestry in America," and "Eighteenth Century Theories on America as a Human Habitat," *American Philosophical Society Proc.* 89 (1945): 444–88, & 91 (1947): 27–57; Peter N. Carroll, *Puritanism and the Wilderness: The Intellectual Significance of the New England Frontier, 1629–1700* (New York: Columbia University Press, 1969), 80–88; R. F. Nash, *Wilderness and the American Mind* (1967), 27–41; David Lowenthal, "The Pioneer Landscape: An American Dream," *Great*

Plains Quarterly 2 (1982): 5–19; idem, "Awareness of Human Impacts: Changing Attitudes and Emphases" (1990), 122; William Cronon, *Changes in the Land: Indians, Colonists, and the Ecology of New England* (1983), 3–6, 166–68. The idiom "to beat all nature" typifies the early American attitude (Mitford M. Mathews, ed., *A Dictionary of Americanisms on Historical Principles* [Chicago: University of Chicago Press, 1951], 2:1114).

24. Glacken, *Traces on the Rhodian Shore*, 690–98; Cronon, *Changes in the Land*, 116–48, 169–70.

25. *Address ... Agricultural Soc. Rutland County* (1847), 17–19.

26. GPM to Asa Gray, 9 May 1849; "Economy of the Forest," MS.

27. GPM to the Estcourts, 18 June, 23 Aug. 1851.

28. GPM to SFB, 5 Feb. 1853, BSI; GPM, *Camel* (1856), 16.

29. GPM, "Oration" [New Hampshire] (1856), 86–87, 75–77 (amplified in *M&N*, 353–59, 363–64); Hall, "Restoring the Countryside," 6; Judd, *Common Lands, Common People*, 92, 211; D. C. Fisher, *Vermont Tradition* (1953), 208–9.

30. GPM, *Report, on the Artificial Propagation of Fish* (1857); "Study of Nature" (1860), 34.

31. *M&N*, 36; U.S. Commissioner of Agriculture Horace Capron to GPM, 5 May, 13 July 1870; GPM to U.S. Commissioner of Agric. 27 Feb. 1880, U.S. Dept. of State Papers, Italy, PR, Misc. Corr., vol. 5; SFB to GPM, 23 Aug. 1867, BSI. Marsh letters to Ricasoli on exchanges of grape cuttings, hybrid corn, the American cranberry, the American white ash, white oak, walnut, and hickory (7 Apr. 1866, 1 Dec. 1868, 12 Dec. 1869, 10 Apr. 1870) are at the Villa Ricasoli, Brolio, Tuscany.

32. H. C. Darby, "Clearing of the Woodland in Europe" (1956), in W. L. Thomas, Jr., ed., *Man's Role in Changing the Face of the Earth*, 183–216 at 194, 199–204; T. L. Whited, *The Struggle for the Forest in the French Alps and Pyrenees, 1860–1940* (1994), 36–64.

33. Navy Secretary Samuel Lewis Southard, report to Congress, 29 Jan. 1827, quoted in Charles W. Snell, *A History of the Naval Live Oak Reservation Program, 1794–1880* (Denver: U.S. Dept. of Interior, Natl. Park Service, 1983), 36 (see also 26, 34–37, 41, 139–40, 149–50); GPM, *M&N*, 202n138.

34. GPM to C. S. Sargent, 12 June 1879. See *M&N*, 217ff., 278–328; Jenks Cameron, *The Development of Governmental Forest Control in the United States* (Baltimore: Johns Hopkins Press, 1928), 5–67; Michael Williams, *Americans and Their Forests* (1989).

35. Jean Antoine Fabre, *Essai sur la théorie des torrens et des rivières* (Paris, An VI

[1797]); Glacken, *Traces on the Rhodian Shore*, 698–702; GPM, *M&N*, 329; 1874 ed., 259.

36. Whited, *Struggle for the Forest*, 83–104; Andrée Corvol, *L'Homme aux bois: Histoire des relations de l'homme et de la forêt, XXVIIe–XXe siècles* (Paris: Fayard, 1987), 318–38, 373–410. Marsh assesses these French reforms in *M&N*, 195–97, 335–36.

37. *M&N*, 42–43, 36. The same year Marsh penned this omen, Henry Adams expressed parallel fears: "Man has mounted science, and is now run away with . . . Before many centuries more . . . the engines he will have invented will be beyond his strength to control. . . . The human race [may] commit suicide by blowing up the world" (to C. F. Adams, 4 Apr. 1862, in *A Cycle of Adams Letters*, ed. W. C. Ford [Boston: Houghton Mifflin, 1920], 1 : 134–35).

38. On Jefferson's notion of usufruct, see Herbert E. Sloan, *Principle and Interest: Thomas Jefferson and the Problem of Debt* (New York: Oxford University Press, 1995), 51–85; J. J. Ellis, *American Sphinx: The Character of Thomas Jefferson* (1997), 111–15.

39. *M&N*, 7–12; J. V. Thirgood, *Man and the Mediterranean Forest: A History of Resource Depletion* (1981), 1; J. D. Hughes, *Pan's Travail: Environmental Problems of the Ancient Greeks and Romans* (1994), 1. Following Marsh's footsteps in the ecologically depleted Mediterranean moved the forester-hydrologist Walter Lowdermilk in 1939 to proclaim an "11th Commandment" of conservation morality (*American Forests* 46 [Jan. 1940], 12–15; R. F. Nash, *The Rights of Nature: A History of Environmental Ethics* [1989], 97–98). Scientific opinion now minimizes the scale and consequences of classical deforestation relative to both neolithic and recent felling (Robert Sallares, *The Ecology of the Ancient Greek World* [1991], 34, 374).

40. *M&N*, 12, 36–37, 42–43, 52. When asked how a university in California might best spend an anticipated $30 million, Marsh advised that it "inaugurate as soon as possible a series of observations for the purpose of ascertaining the effects of human action on the physical conditions of the earth" (GPM to E. P. Evans, 5 Dec. 1870 [copy]; Evans to GPM, 24 Nov. 1870, and 20 Jan. 1871).

41. Haeckel's *Generelle Morphologie der Organismen* (Berlin, 1866) bases the need for a new science of ecology on Darwin's theory of natural selection (Robert C. Stauffer, "Haeckel, Darwin and Ecology," *Quarterly Review of Biology* 2 [1957], 138–44; Eugene Cittadino, *Nature as the Laboratory: Darwinian Plant Ecology in the German Empire, 1880–1900* [Cambridge: Cambridge University Press,

1990], 87). Marsh owned but nowhere referred to Haeckel's *Über Arbeitsheilung in Natur- und Menschenleben* (Berlin, 1869).

42. *M&N*, 92, 96, 91, 34, 34n34. The red-headed pileated woodpecker had "almost vanished" from New England because the dead stumps that housed the insects that fed these birds had been taken for firewood (*M&N*, 97).

43. *M&N*, 32–35. See Robert J. Naiman et al., "Alteration of North American Streams by Beaver," *BioScience* 38 (1988): 753–62; Nancy Langston, *Forest Dreams, Forest Nightmares* (1995), 57–59; Angela M. Gurnell, "The Hydrogeomorphological Effects of Beaver Dam-Building Activity," *Progress in Physical Geography* 22 (1998): 167–89 at 177–78, 186; Wessels, *Reading the Forested Landscape*, 99–111.

44. *M&N*, 54, 76–77; 1884 ed., 169. See Hugh Grant, "The Revenge of the Paris Hat," *Beaver* (Canada) 68:6 (1988–89): 37–44. The beaver was not restored to Vermont until 1921, however (Wessels, *Reading the Forested Landscape*, 110).

45. *M&N*, 61–62. After deforestation, the global spread of alien species is today adjudged the gravest threat to biodiversity (P. M. Vitousek et al., "Human Domination of Earth's Ecosystems" [1997]: 498).

46. *M&N*, 34, 36, 61–62, 94, 97; GPM to C. S. Sargent, 10 June 1879; *M&N*, 117.

47. R. H. Grove, *Green Imperialism* (1995), terms desiccationist views common among colonial administrators in Mauritius, St. Helena, St. Vincent, and elsewhere; see Chapter 18 below.

48. *M&N*, 170; GPM, MS, reply to Daniel Draper, "Has Our Climate Changed?" *Popular Science Monthly* 1 (1872): 665–74; GPM, *Earth as Modified by Human Action* (1874 ed.), 202.

49. *M&N*, 186–89. See W. M. Davis, "Biographical Memoir of George Perkins Marsh" (1909): 80.

50. *M&N*, 277–80, 189.

51. *M&N*, 381.

52. *M&N*, 412n42, 414–15; GPM, *A Dictionary of English Etymology . . . Wedgwood* (1862), 237.

53. *M&N*, 414, 419–20, 432–33. Marsh on the dunes of the Landes is cited as a classic by J. L. Reed, *Forests of France* (London: Faber & Faber, 1954), 253–66; S. J. Pyne, *Vestal Fire: An Environmental History* (1997), 559n49.

54. *M&N*, 463–64. Marsh's notion of such impacts "is still ahead of its time" (Wilbur R. Jacobs, "The Great Despoliation: Environmental Themes in

American Frontier History," *Pacific Historical Review* 47 [1978]: 1–26 at 15; see also W. B. Meyer, *Human Impact on the Earth* [1996], 5).

55. *M&N*, 111–12, 464–65. Emily W. B. Russell, "Discovery of the Subtle," in M. J. McDonnell and S. T. A. Pickett, eds., *Humans as Components of Ecosystems* (1993), 81–90 at 81, 85, classes Marsh's unknowns into (1) obvious activities with subtle effects (fossil fuels and pollution), (2) subtle activities with obvious effects (DDT and raptor numbers), (3) subtle activities with subtle effects (greenhouse gases and global climate change).

56. *Earth as Modified by Human Action* (1884 ed.), 616–17n. Huxley's phrasing is looser: "The question of questions for mankind—the problem which underlies all others— . . . is the ascertainment of the place which Man occupies in nature. . . . What are the limits of our power over nature, and of nature's power over us?" ("On the Relations of Man to the Lower Animals" [1860], in his *Man's Place in Nature* [1863], 71). Marsh's library contained two later Huxley texts, but not this essay.

14. Man and Nature: *The Meaning*

1. GPM quoted in *LL*, 64; GPM, "Study of Nature" (1860), 33.

2. GPM, *The Camel* (1856), 13; James Marsh, "Preliminary Essay" (1829), in his edition of Coleridge, *Aids to Reflection*, xl–xlii; *M&N*, Preface, 3; Horace Bushnell, "Sermon on the Power of an Endless Life" (in his *The New Life* [London, 1861], 287–307 at 293), slightly altered in GPM, *M&N*, iii. Bushnell's own supernaturalism owed much to James Marsh (H. Sheldon Smith, ed., *Horace Bushnell* [New York: Oxford University Press, 1965], 27).

3. Charles Scribner to GPM, 7 July 1863; GPM to Scribner, 10 Sept. 1863. Henry Thomas Buckle (*History of Civilization in England* [1857–61]) held human actions governed by fixed scientific laws and attributed disparate social and intellectual progress mainly to environmental differences.

4. *M&N*, 41; H. D. Thoreau, *Faith in a Seed* (1850–61), 130; *M&N*, 1874 ed. 42n. Marsh was intrigued by Thoreau's comments, much like his own, on the agency of squirrels and birds in planting and destroying nuts and seeds (*M&N*, 1867 printing, app. 552n5; Thoreau, "Succession of Forest Trees," in *Excursions*, 1863).

5. *M&N*, 29–30, 35–36.

6. Peter Van Voris et al., "Functional Complexity and Ecosystem Stability," *Ecology* 61 (1980): 1152–60; Daniel B. Botkin, *Discordant Harmonies: A New*

Ecology for the Twenty-First Century (New York: Oxford University Press, 1990), 9, 54, 108; B. L. Turner II and William B. Meyer, "Environmental Change: The Human Factor," in M. J. McDonnell and S. T. A. Pickett, *Humans as Components of Ecosystems* (1993), 40–50; Andy P. Dobson et al., "Hopes for the Future— Restoration Ecology and Conservation Biology," *Science* 277 (1997): 515–21 at 515; David Lowenthal, "Awareness of Human Impacts" (1990), 124–27.

7. *M&N*, 25n23. Marsh's interest in forest succession mirrored both Thoreau's and Darwin's (L. D. Walls, *Seeing New Worlds* (1995), 183–88, 197.

8. GPM, *NES* (1844), 10–13; R. C. Lewontin, *Biology as Ideology: The Doctrine of DNA* (Concord, Ontario: CBC/Anansi, 1991), 88–89.

9. *M&N*, 42, 13; GPM to C. E. Norton, 24 May 1871, Norton Papers.

10. M&N, 36, 13; Horace Bushnell, *Nature and the Supernatural as Together Constituting One System of God* (London, 1862), 23–25.

11. GPM, *M&N*, 41, 37, 1874 ed. 42, 1864 ed. 24n22.

12. *M&N*, 38–42; GPM, *Address . . . Agric. Soc. Rutland* (1847), 6.

13. *M&N*, 40, 39n36. By 1856 Marsh had already revised his earlier judgment of primitive man as wholly destructive (*Camel*, 15–16). He did not absolve them from all blame, however. The extinction of mammoths and mastodons seemed due to prehistoric hunters. Even "in his earliest known stages of existence, [man] was probably a destructive power upon the earth," though less emphatically than his successors (*M&N*, 70). On current assessments of Native American impact, see William M. Denevan, "The Pristine Myth: The Landscape of the Americas in 1492," *Annals of the Association of American Geographers* 82 (1992): 369–85; on current primitivist ecological pieties, see Chapter 18 below, and Gary Paul Nabhan, "Cultural Parallax in Viewing North American Habitats," in Michael E. Soulé and Gary Lease, eds., *Reinventing Nature?* (Washington, D.C.: Island Press, 1995), 87–101.

14. GPM to CCM, 15 May 1856, CFP; *M&N*, 39–40.

15. *M&N*, 226–28; spontaneous revegetation in the Canal della Piave, below Feltre, north of Venice, is instanced in the 1884 ed., 148n.

16. *M&N*, 44, 35. Marsh here pioneered what is now termed restoration ecology (Dobson et al., "Hopes for the Future," 518–21).

17. *M&N*, 35, 105; GPM, "Study of Nature" (1860), 33–34, 60, 58, 56, 61–62; *M&N*, 44–45.

18. GPM, "Study of Nature," 33–34. See *M&N*, v; Ralph Waldo Emerson, "The Man with a Hoe," *Complete Works* (Boston: Houghton Mifflin, 1903), 7: 135–54. This is one of Marsh's few references to population increase as a cause of resource depletion. T. R. Malthus is mentioned nowhere in his writings.

19. GPM, "England Old and New" (1859/60), 7. Well-managed woods can be as biologically diverse as old-growth forests (Ian R. Noble and Rodolfo Dirzo, "Forests as Human-Dominated Systems," *Science* 277 [1997]: 522–25).

20. *O&H* (1862), 425; *M&N*, 248–49. In censuring needless exploitation of animals and plants Marsh foreshadows such reformers as Albert Schweitzer, Aldo Leopold, and Arne Naess (R. F. Nash, *The Rights of Nature* [1989], 61, 71, 147).

21. Murray to GPM, 28 Sept. 1863; Scribner to GPM, 10 Oct. 1863; *M&N*, 58, 51–52n53. In one excised paragraph, Marsh railed against cotton as the "most virulent [of] vegetable poisons, [which] crazes the brain of those who deal in it, benumbs their moral faculties, fills their hearts with a lying spirit" (MS for *M&N*, 381–82). The publisher Murray was given to complaining about obscurity, as he did with Darwin's use of "natural selection" (Robert M. Young, *Darwin's Metaphor: Nature's Place in Victorian Culture* [Cambridge: Cambridge University Press, 1985], 95).

22. *M&N*, 85n (on migratory birds, quoted in Chapter 1 above); *Earth as Modified by Human Action* (1874 ed.), 320n (Marsh's Roman brickwork clue to deforestation is lauded in John Perlin, *Forest Journey: The Role of Wood in the Development of Civilization* [New York: Norton, 1989], 129); *M&N*, 240, 165n91, 51n53.

23. GPM to C. E. Norton, 14 Mar. 1864; to Henry Yule, 7 May 1871.

24. Scribner to GPM, 15 Dec. 1864; GPM to U.S. Sanitary Comm., Oct. 1864; John Sherwood to GPM, 25 Jan. 1865; CM to GFE, 18 Nov. 1864; GFE to CM, 27 Dec. 1864; GFE to GPM, 11 Jan. 1865.

25. Scribner to GPM, 17 Jan. 1865; J. H. Allen in *Christian Examiner* 77 (1865): 65–73; J. R. Lowell in *North American Review* 99 (1864): 318–20 at 320; *Round Table*, 7 May 1865(?), in Marsh scrapbook; Bigelow to GPM, 4 June 1863.

26. Henry Holland in *Edinburgh Review* 120 (1864): 469–82; Charles Lyell to GPM, 22 Sept. 1865 (but Lyell's 1868 edition of *Principles of Geology* fails to note Marsh's insight); Guyot to GPM, 5 May 1868; David Livingstone, *Nathaniel Southgate Shaler and the Culture of Science* (1987); W. M. Davis, "Biographical Memoir of George Perkins Marsh" (1909), 80; review of *Earth as Modified* (possibly by S. F. Baird), *Nation* 19 (1874): 223–24. The diarist John Evelyn's *Sylva* (1664) was a classic often cited by Marsh.

27. James McCosh, "Ideas in Nature Overlooked by Dr. Tyndall," review of *Earth as Modified by Human Action* (1874), *International Journal* 2 (Jan. 1875): 120–25 (I am grateful to Donald C. Dahmann of the U.S. Census Bureau for alerting me to this review); Wallace Stegner, "It All Began with Conservation," *Smithsonian* (Apr. 1990): 35–43 at 38.

28. H. A. Homes to GPM, 31 July 1872; F. B. Hough, "On the Duty of Governments in the Preservation of Forests" (1873); idem, *Report upon Forestry* (1878), 76n, 268, 309; idem, *The Elements of Forestry* (Cincinnati, 1882), 26; Hough to GPM, 28 Oct. 1873, 9 Apr. 1878. Hough visited Marsh in 1881 (Hough, "Experimental Plantation of the Eucalyptus near Rome," *American Journal of Forestry* 1 [1882–83]: 402–13; GPM, *Earth as Modified* [1884 ed.], 383n). Marsh did not reciprocate Hough's admiration; "I hope Congress may do something for Forestry, though I do not expect much from Dr. Hough" (GPM to C. S. Sargent, 25 Mar. 1880). See Herbert A. Smith et al., *A National Plan for American Forestry*, 73 Cong. 1 sess., Sen. Doc. 12 (Washington, 1933), 1:743–44; H. A. Smith, "The Early Forestry Movement in the United States," *Agricultural History* 12 (1938): 326–46; Ralph R. Widner, ed., *Forests and Forestry in the American States* (Missoula, Mont.: Natl. Assoc. of State Foresters, 1968), 3, 27; Michael Williams, *Americans and Their Forests* (1989), 370–77, 490–91.

29. SFB to GPM, 21 Dec. 1872; Frederick Starr, "American Forests; Their Destruction and Preservation," in *Rept. Commissioner of Agriculture . . . 1865*, 39 Cong. 1 sess., H. Ex. Doc. 136, 210–34 at 221–29; I. A. Lapham et al., *Report on the Disastrous Effects of the Destruction of Forest Trees . . . in the State of Wisconsin* (Madison, 1867); N. S. Shaler, *Notes on the Investigations of the Kentucky Geological Survey during the Years 1873, 1874, and 1875* (Frankfort, 1877); Livingstone, *Nathaniel Southgate Shaler*, 195; idem, "Nature and Man in America" (1980): 371; N. H. Egleston, *Arbor Day: Its History and Observance* (Washington, D.C., 1896), 9–10; G. B. Emerson to GPM, 11 Dec. 1875; A. D. Rodgers, *Bernhard Eduard Fernow* (1951), 153; B. E. Fernow, "Do Forests Influence Rainfall?" *Garden and Forest* 1 (1888): 489–90; J. A. Warder to GPM, 13 Dec. 1877; C. S. Sargent, "George Perkins Marsh," *Nation* 35 (1882): 136; idem, *A Few Suggestions on Tree-Planting* (Boston, 1876), 3–18; idem, "The Protection of Forests," *North American Review* 135 (1882): 386–401 at 401; Ida Hay, *Science and the Pleasure Ground: A History of the Arnold Arboretum* (1995), 54, 322–23; F. V. Hayden to Susan Edmunds, 5 Jan. 1872 (UVM); Gifford Pinchot, *Breaking New Ground* (1947), xvi–xvii.

30. Thomas R. Cox et al., *This Well-Wooded Land: Americans and Their Forests from Colonial Times to the Present* (Lincoln: University of Nebraska Press, 1985), 144–47; Nicolai Cikovsky, "'The Ravages of the Axe': The Meaning of the Tree Stump in Nineteenth-Century American Art," *Art Bulletin* 61 (1971): 611–26 at 613; A. F. Hawes, "What Some Vermonters Are Doing in Forestry," *Vermont Agricultural Bull.* no. 9 (1911): 16–17; R. W. Judd, *Common Lands, Common People* (1997), 97–98; Robin Winks, *Frederick Billings* (1991), 293–99; U.S. Na-

tional Park Service, *Marsh-Billings National Historical Park: Land Use History* (1994), 59–63, and *Draft General Management Plan* (1998), 35–36.

31. GPM, *Earth as Modified* (1874 ed.), viii. See Reclus to GPM, various dates, 1868–70, and to CCM, 14 Sept. 1884; GPM to Reclus, 20 Nov. 1869; GPM, "Preliminary Notice" [for Reclus's *La Terre*] (ca. 1870) MS; Reclus, *A New Physical Geography* [La Terre] (London, 1874), 519; Joël Cornuault, "Élisée Reclus & George Perkins Marsh. Quelques notes," *Les Cahiers Élisée Reclus*, no. 20, Jan. 1999; K. R. Olwig, "Historical Geography and the Society/Nature 'Problematic': The Perspective of J. F. Schouw, G. P. Marsh and E. Reclus" (1980): 39–41. Reclus chose to counterpoise Marsh by eulogizing human creative powers (J. R. Whitaker, "World View of Destruction and Conservation of Natural Resources" [1940], 147–49).

32. Gerolamo Boccardo to GPM, 10 Apr. 1868; Marcus Hall, "Restoring the Countryside: George Perkins Marsh and the Italian Land Ethic" (1998), 6–10. *Man and Nature* was twice translated into Italian (*L'Uomo et la natura*, Florence, 1869, 1872). The 1872 Italian translation was reprinted, with a long introduction by Fabienne O. Villani, in Milan in 1988. A French translation begun in the 1860s, under Reclus's guidance, was lost when the translator was killed in the Franco-Prussian War (GPM to S. Dana Horton, 15 Oct. 1869 [CCM copy, with note by her to S. G. Brown]).

33. Hugh Cleghorn to GPM, 6 Mar. 1868; Dietrich Brandis to GPM, dated Rome, "Friday"; Yule to GPM, 15 Jan. 1878; A. J. Stuart, *Extracts from 'Man and Nature' . . . with some notes on forests and rainfall in Madras* (1882); J. C. Brown, *Forests and Moisture* (1877); Graeme Wynn, "Pioneers, Politicians and the Conservation of Forests in Early New Zealand," *Journal of Historical Geography* 5 (1979): 171–88 at 181–85; Keiichi Takeuchi, "Nationalism and Geography in Modern Japan, 1880s to 1920s," in David Hooson, ed., *Geography and National Identity* (Oxford: Blackwell, 1994), 104–11 at 107–9; J. M. Powell, "Enterprise and Dependency: Water Management in Australia," in Tom Griffiths and Libby Robin, eds., *Ecology and Empire* (1997), 102–21 at 107. See Whitaker, "World View," 149–53; C. J. Glacken, "Changing Ideas of the Habitable World" (1956), 83.

34. GPM to SFB, 12 Oct. 1881, BSI (also see GPM to John Bigelow, 3 Sept. 1863, in Bigelow, *Retrospections of an Active Life* [1909–13], 2:51); C. E. Norton to J. B. Harrison, in *Garden & Forest* 2 (1889): 333; C. S. Sargent to Robert Underwood Johnson, 25 Nov. 1908, quoted in S. B. Sutton, *Charles Sprague Sargent and the Arnold Arboretum* (1970), 77–78. See Hay, *Science and the Pleasure Ground*, 103;

D. J. Pisani, "Forests and Conservation, 1865–1890" (1986), 352. My 1965 reprint of the first (1864) edition of *Man and Nature* incorporates as footnotes many Marsh additions through his last (1884) edition.

35. Max Oelschlaeger, *The Idea of Wilderness* (1991), 337, 106; John Ball to Charles Darwin, 25 June 1874, Darwin Papers Cal. 9512, MS.DAR.160:34. The American surgeon William Williams Keen brought Marsh's observation on gestural expression in *Lectures on the English Language* (35n; see Chapter 10 above) to Darwin's notice, as relevant to *The Descent of Man, Selection with Relation to Sex* (New York, 1872), 52 (Keen to Darwin, 26 Sept. 1873, Darwin Papers Cal. 9072).

36. Charles Darwin, *On the Various Contrivances by Which British and Foreign Orchids are Fertilized by Insects, and on the Good Effects of Intercrossing* (London, 1862); idem, "On the Formation of Mould," *Trans. Geolog. Soc.* 5 (1837); idem, *The Formation of Vegetable Mould through the Action of Worms* [1881], *Works of Charles Darwin* (London: Pickering, 1989), 26:138–39; M&N, 90; 1884 ed., 129n, 403n.

37. *LEL* 1885 ed., 236–37; F. W. Farrar (1870) quoted and Müller cited in S. G. Alter, *Darwinism and the Linguistic Image* (1999), 92–93, 83 (also 11, 56–59, 73–80, 100–4, 122, 135); O&H, 51–54, 84–85. See also Hans Aarsleff, *The Study of Language in England* (1967), 103–10; Gillian Beer, "Darwin and the Growth of Language Theory" (1989) in her *Open Fields: Science in Cultural Encounter* (Oxford: Clarendon Press, 1996), 95–114.

38. GPM, *Earth as Modified* (1874 ed.), 10n; Darwin, *The Descent of Man, Selection with Relation to Sex*, 2d ed. [1877], *Works*, vols. 21–22 (1989), 21:52–53; idem, *On The Origin of Species by Means of Natural Selection*, 33–41, 80 (quote); M&N, 42n39.

39. Darwin, *On the Origin of Species* (1860), 69; M&N, 247. Marsh later noted that the yellow-horned poppy (*Glaucium luteum*), "unknown to modern science, but . . . described by Pliny and Dioscorides," had sprung to life after more than 1,500 years, with the removal for smelting of mining refuse near Mount Laurium, Greece (M&N, 1884 ed., 288n). Darwin showed keen interest in and experimented with the vitality of seeds (*On the Origin of Species*, 6th ed., 1876, *Works* [1988], 16:339–41), but only once wrote of longevity beyond the span of a few decades, in referring to seeds from the graves of ancient Gauls which had lately germinated (Darwin to *Gardiners' Chronicle*, before 19 Dec. 1855, in his *Correspondence*, 5:532–33).

40. Darwin, *Descent of Man*, 70, 102. Like Marsh, the philologist Max Müller stressed language as a uniquely human trait (*Lectures on the Science of Lan-*

guage, cited in Gillian Beer, "'The Face of Nature': Anthropomorphic Elements in the Language of *The Origin of the Species*" [1986], 218–19).

41. Patrick Geddes to Lewis Mumford, 13 Nov. 1920, in Frank G. Novak, Jr., ed., *Lewis Mumford and Patrick Geddes: The Correspondence* (London: Routledge, 1995), 80; Mumford first mentions Marsh in *Sticks and Stones: A Study of American Architecture and Civilization* (1924; rev. ed., N.Y.: Dover, 1955), 201; Lewis Mumford, *The Brown Decades* (1931), 78; W. L. Thomas, Jr., *Man's Role in Changing the Face of the Earth* (1956); B. L. Turner II et al., *Earth as Transformed by Human Action* (1990); W. B. Meyer, *Human Impact on the Earth* (1996), 21, 216. See also Mumford, *Sketches from Life, the Autobiography: The Early Years* (New York: Dial, 1982), 408; R. Guha and J. Martinez-Alier, *Varieties of Environmentalism* (1997), 187, 197, 222n10. In 1923 a British geographer termed *Man and Nature* "the only book dealing with" human impacts on nature (R. L. Sherlock, "The Influence of Man as an Agent in Geographical Change," *Geographical Journal* 61 [1923]: 258–73 at 258). If Marsh was "little if at all known" to American geographers (J. O. M. Broek, "Agrarische Opnemingen [Surveys] in de Vereenigde Staten," *Tijdschrift voor Economische Geographie* 24 [1933]: 232–44 at 238n), soil scientists were familiar with *Man and Nature* (W. C. Lowdermilk, "Civilization and Soil Erosion," *Journal of Forestry* 33 [1935]: 554–60; H. S. Person, *Little Waters* [Washington, D.C.: Soil Conservation Service, 1935], 18, 29–32, 41–74). In 1940 an American geographer acclaimed Marsh's seminal influence (Whitaker, "World View," 144–47, 153, 161–62).

42. P. S. Jennison, *The History of Woodstock, Vermont, 1890–1983* (1983), 140–43; U.S. National Park Service, Marsh-Billings National Historical Park, *Conservation Stewardship Workshop: Findings and Recommendations* (1995), *Draft General Management Plan* (1998), 7–13, 33–36.

43. *M&N*, 13–14; Burton V. Barnes et al., *Forest Ecology*, 4th ed. (New York: Wiley, 1988), 357–60; Timothy O'Riordan, "Special Report: The Earth as Transformed by Human Action," *Environment* 30:1 (1988): 25–28; "Human-Dominated Ecosystems," *Science* 277 (1997): 485–525. Early twentieth century holistic ecologists like Frederic Clements were more narrowly biotic than was Marsh; see Chapter 18 below.

44. S. L. Udall, *The Quiet Crisis* (1963), 69–82; Glacken, "Changing Ideas of the Habitable World," 83. However, American historians "purblind" to environmental contexts long ignored Marsh's insights (A. W. Crosby, "The Past and Present of Environmental History" [1995]: 1180–81, 1185). See my "George Perkins Marsh and the American Geographical Tradition" (1953) and

Daniel W. Gade, "The Growing Recognition of George Perkins Marsh," *Geographical Review* 73 (1983): 341–44.

45. *M&N*, 45–46.

46. GPM, *Human Knowledge* (1847), 4.

47. Pisani, "Forests and Conservation," 358–59.

48. GPM to Harriet Preston, 12 Apr. 1882. Preston had sent Marsh her translation of Virgil's *Georgics*; Marsh urged her to translate Voss's "Der siebzigste Geburtstag" (1771) (Johann Heinrich Voss, *Sämmtliche poetische Werke* [Leipzig, 1835], 99–103).

49. Ralph Waldo Emerson, "The American Scholar" (1837), in *The Portable Emerson* (New York: Viking, 1946), 23–46 at 43.

50. *LEL*, 637, 678.

15. Florence and Unfinished Italy

1. George Washington Wurts quoted in CCM to GPM, 26 July 1869.

2. Edouard Drouyn de Lhuys to Costantino Nigra, quoted in Arrigo Solmi, *The Making of Modern Italy* (1925), 332. See Marziano Brignoli, "Ricordo della convenzione di settembre nel centenario della stipulazione," *Risorgimento* 16 (1964): 144–63; J. W. Bush, *Venetia Redeemed: Franco-Italian Relations 1864–1866* (1967), 13–20.

3. GPM, 27 Sept., 25 Oct. 1864, DD 105, 108; Cj, 17–24 Sept., 15 Oct. (Giuseppe Baruffi quoted on Turin handbill), 1 Dec. 1864, 15:61–62, 16:1–3, 17:51–52; GPM to C. E. Norton, 16 Jan. 1865 (CCM copy). H. F. d'Ideville, *Journal d'un diplomate* (1873), 24 Oct. 1864, 2:228–30, reports the royal palace placarded with "Castel da vendere, e Re da pendere. Vendite di sudditi" (Castle for sale, King for hanging. Auction of subjects). The Turin debacle is detailed in P. D. Pasolini, *Giuseppe Pasolini* (1915), 2:42–48; Henry G. Elliot, *Some Revolutions and Other Diplomatic Experiences*, ed. Gertrude Elliot (London: John Murray, 1922), 175–78.

4. GPM to Seward, 19 Sept., 25 Oct. 1864, DD 102 & 108; see also 10 Apr. 1867, 21 Dec. 1881, DD 174 & 1010; Trauth, *IADR*, 167–68.

5. GPM to Norton, 16 Jan. 1865; to R. H. Dana, Jr., 22 Apr. 1867, Dana MSS.

6. GPM to Seward, 16 Jan. 1865, DD 111; GPM to HP, 6 Nov. 1864, PAA. Marsh later accused Tuscans of supporting the papacy and Napoleon III (GPM to W. M. Evarts, 1 Dec. 1880, DD 928).

7. GPM to FM, 5 Aug. 1855, FC; to SFB, 2 Aug. 1865, BSI.

8. The Giardino had passed from the Soderini to the Bourbon del Monte family in the seventeenth century, and in 1861 to Emilio Forini, who leased it to Marsh, and after Marsh's death to Marsh's Icelandic scholar colleague, Willard Fiske. See Gargano Gargani, *Il giardino già dei Soderini di Firenze, attualmente di Emilio Forini, presso San Salvo* (Florence, 1878); Theodore Stanton, "The Villa Forini," N.Y. *Semi-Weekly Tribune*, 23 Jan. 1885 (reprinted as "The Italian Home of George P. Marsh," *Vermont Alumni Weekly* 14:3 [1934]: 43–44); D. W. Fiske to CCM, 5 Nov. 1883. The remnant "Giardino," now the Villa Arrivabene (the family of Forini's son-in-law), restored in the 1970s, is owned by local sector 12 of the city of Florence (Laura Lucchesi and Stefano Bertocci, *Villa Il Giardino* [1985], 35–41).

9. GPM to Seward, 30 Jan. 1865, DD 112; Seward to GPM, 27 Feb. 1865, DI 126, pp. 213–14; GPM to Seward, private, 21 Mar. 1865, Seward Papers; GPM to Solomon Foot, 21 Mar. 1865, MV. A later request to take medical leave in Paris for Caroline's sake was "reluctantly granted" by President Johnson (3 May 1866, DI 143); but Marsh's Paris stay was too brief to need to avail himself of it.

10. GO to GPM, 2 Apr., 16 Oct. 1859; CCM to LW, 7 & 19 Dec. 1859, WP; CCM to Silas A. Crane, 19 Aug. 1860.

11. U.S. Army, George Ozias Marsh Certif. of Discharge, 9 Aug. 1862, UVM; Mrs. M. Brakeley to GPM, 14 May 1865; J. A. C. Gray to GPM, 1 Mar. 1865.

12. Maria Buell Hickok to CCM, 18 May 1865; GFE to GPM, 24 Apr. 1865.

13. GPM to GO, 1 Dec. 1855, CFP; CCM to LW, 7 Dec. 1859, WP.

14. J. A. C. Gray to GPM, 18 May 1865; Mrs. M. Brakeley to GPM, 14 May 1865.

15. GPM to Seward, 21 Mar. 1865, Seward Papers; to Norton, 29 Mar. 1865 (CCM copy).

16. GPM to LW, 29 Oct. 1869, WP; to SFB, 9 Feb. 1867, BSI; to Solomon Foot, 21 Mar. 1865, VHS MSS no. 64; to CBT, 24 Apr. 1868 (CCM copy).

17. Solomon Foot to GFE, 11 Mar. 1865; GFE to GPM, 8 June 1865; GPM to D. H. Wheeler, 4 Jan. 1866; GFE to GPM, 10 May, 2 July 1866. Marsh himself had been puffed for appointment to Foot's Senate seat ("Vermont and Her Senators," *Nation* 2 [1866], 700–10). Marsh was proposed as envoy to France on W. L. Dayton's death in 1864, and Grant would have sent him there or to London in 1869, but Marsh declined both as too expensive (J. H. Alexander to

W. H. Seward, 21 Dec. 1864, U.S. Dept. of State Appointment Papers; CCM to GPM, 5 Aug. 1869).

18. Justin Morrill and GFE to Hamilton Fish, 24 Mar. 1869, U.S. Dept. of State Appointment Papers; Ulysses S. Grant quoted in CCM to GPM, 9 Sept. 1869; GFE to GPM, 29 Mar. 1869. See GFE to Jacob Collamer, 3 May 1864, Collamer Papers; Solomon Foot to GFE, 11 Mar. 1865; J. R. Young, *Around the World with General Grant* (New York, 1879), 1:264-66; L. D. White, *The Republican Era, 1869-1901* (1958), 6.

19. GPM to Seward, 18 Nov. 1865, 18 Jan. 1868, DD 127 & 199 (also nos. 111, 187); to T. A. Trollope, 30 Nov. 1867, in Trollope, *What I Remember* (1888), 449. The sale of church lands, at least in Piedmont, succeeded better than Marsh thought, many peasants adeptly circumventing papal sanctions (Alfonso Bogge and Modesto Sibona, *La vendita dell'asse ecclesiastico in Piemonte dal 1867 al 1916* [Milan: Banca Commerciale Italiana, 1987]). On landholding and agriculture in Piedmont, see Emilio Sereni, *History of the Italian Agricultural Landscape* (1961), 267-72, 304-13; Valerio Castronovo, *Economia e società in Piemonte dall'unita al 1914* (Milan: Banca Commerciale Italiana, 1969), 5-82.

20. E. L. Godkin, "The Pope and the Catholic Nations," *Nation* 5 (7 Nov. 1867): 132-34.

21. GPM to SFB, 7 Apr. 1870, BSI; to Seward, 6 Jan. 1869, DD 239. See Christopher Seton Watson, *Italy from Liberalism to Fascism, 1870-1925* (London: Methuen, 1967), 18-29; Denis Mack Smith, *Victor Emanuel, Cavour, and the Risorgimento* (1971), 80-81; idem, "Francesco De Sanctis" (1991), 253-56; Trauth, *IADR*, 158; J. A. Davis, *Conflict and Control: Law and Order in Nineteenth-Century Italy* (1988), 188-91; Nunzio Pernicone, *Italian Anarchism, 1864-1892* (1993), 33.

22. GPM, MS, Milan, 1864; to Hamilton Fish, 6 Oct. 1874, DD 520; to W. M. Evarts, 10 Dec. 1878, DD 802 (also DD 111 & 865). See Jessie White Mario, "Right and Wrong in Italy," *Nation* 9 (1869): 48-49; GPM to John Bigelow, 22 Apr. 1865, in Bigelow, *Retrospections of an Active Life* (1909), 1:510; Stephen G. Hughes, *Crime, Disorder and the Risorgimento: The Politics of Policing in Bologna* (Cambridge: Cambridge University Press, 1994), 230-53; Davis, *Conflict and Control*, 176-91.

23. GPM to Seward, 20 June, 11 July 1866, DD 143 & 146; Mack Smith, *Victor Emanuel*, 309-16; Geoffrey Wawro, *The Austro-Prussian War: Austria's War with Prussia and Italy in 1866* (Cambridge: Cambridge University Press, 1996), 70-72, 83-123. Marsh had a copy of, but left no reaction to, La Marmora's face-saving *Un po più di luce sugli eventi politici e militari dell'anno 1866* (Florence, 1873).

24. Mack Smith, *Victor Emanuel*, 303, 317-29; Bush, *Venetia Redeemed*.

25. GPM to Seward, 31 July 1866, DD 150; to Lewis Richmond, 21 Dec. 1881, DD 1010; to R. H. Dana, Jr., 22 Apr. 1867, Dana MSS; Mack Smith, "Francesco De Sanctis," 257; GPM to Seward, 2 Feb. 1864, DD 83. See Trauth, *IADR*, 161–62.

26. GPM to Seward, 7 Oct., 2 Nov., 20 Dec. 1867, 24 June 1868, DD 191, 192, 214; Garibaldi to GPM, 20 Jan. 1868; H. N. Gay, "Garibaldi's American Contacts and His Claims to American Citizenship" (1932), 18–19. On the 1867 imbroglio, see Terrence Murphy, "Lord Acton and the Free Church Policy of Baron Ricasoli," *Journal of Ecclesiastical History* 32 (1981): 321–35; F. J. Coppa, *The Origins of the Italian Wars of Independence* (1992), 130–34; Mack Smith, *Victor Emanuel*, 344–52.

27. E. Gryzanowski, "The Political Crisis in Italy," *Nation* 11 (1870): 418–19.

28. GPM to Seward, 12 Oct. 1870, DD 309; to Carrie Marsh Crane, 7 Sept. 1870; to F. P. Nash, 28 Mar. 1877; to Seward, 23 Nov. 1870, DD 323. See also DD 299–301, 321, 329; E. E. Y. Hales, *Pio Nono* (1954), 314–20.

29. GPM to SFB, Sept. 21, 1871, BSI. See Coppa, *Origins*, 138–45; idem, *Cardinal Giacomo Antonelli and Papal Politics in European Affairs* (1990), 152–69; Denis Mack Smith, *Modern Italy* (1997), 90–92; Anthony Rhodes, *The Power of Rome in the Twentieth Century* (1983), 16–32; Eamon Duffy, *Saints & Sinners: A History of the Popes* (1997), 223–34. Not until 1929 did the Lateran Accord that set up Vatican City at last settle territorial issues, though Church-State relations remained troubled even after fascism.

30. GPM to Seward, 28 Nov. 1866, DD 168; to Rufus King, 25 Nov. 1866, U.S. PR, Misc. Corr. 2:279–81; Guy W. Moore, *The Case of Mrs. Surratt* (Norman: University of Oklahoma Press, 1954), 75–77; Trauth, *IADR*, 36–37.

31. GPM to Seward, 4 Aug. 1862, 20 Mar. 1866, DD 48 & 105; to Hamilton Fish, 16 Nov. 1871, DD 384 (Diplomatic Post Rec., RG 84); Trauth, *IADR*, 42–48; Adrian Cook, *The Alabama Claims: American Politics and Anglo-American Relations, 1865–1872* (Ithaca: Cornell University Press, 1975), 239; Pia G. Celozzi Baldelli, *Arbitrati e politica di potenza: Gli Stati Uniti dopo la guerra di secessione* (Rome: La Sapienza editrice, 1990), 356–58.

32. GPM to Seward, 6 May 1864, DD 92; Seward to GPM, 3 Sept. 1864, DI 110, p. 202; GPM to Fish, 28 June 1871, DD 361. On the negotiations, see DD 122, 142, 166, 171, 178, 200, 212, 221, 348, 357; DI 103, 158.

33. U.S., *Commercial Relations . . . with Foreign Countries . . . 1882 and 1883,* 1:550–54; GPM to Evarts, 17 Dec. 1877, DD 720; to George Opdyke, 15 May 1868.

34. GPM to F. W. Behn, 21 Jan. 1873, U.S. PR, Misc. Corr. 4:123–24; GPM to Hamilton Fish, 12 Dec. 1871, DD 391 (Diplomatic Post Rec., RG 84, vol. 5); Trauth, *IADR*, 94–97. See DD 214, 236, 237; DI 207.

35. Trauth, *IADR*, 159–60, 171; GPM to J. G. Blaine, 25 & 28 May 1881, DD 963, 967.

36. GPM press release, 7 Jan. 1873, in GPM to Fish, 9 Jan. 1873, DD 434; Trauth, *IADR*, 87–94; Davis, *Conflict and Control*, 189; Josiah Strong, *Our Country: Its Possible Future and Its Present Crisis*, rev. ed. [1891], ed. Jurgen Herbst (Cambridge: Harvard University Press, 1963), 59–88; L. W. Levine, *The Opening of the American Mind* (1996), 80. See Fish to GPM, 16 Dec. 1872, DI 363, pp. 409–10.

37. GPM to Seward, 6 June, 24 & 25 Oct. 1864, DD 95, 107, 108; H. R. Marraro, "Spezia: An American Naval Base," *Military Affairs* 7 (1943): 203–8; GPM to SFB, 8 Jan. 1866, BSI; L. P. di Cesnola to GPM, 17 Oct., 8 & 30 Dec. 1870; GPM to Evarts, 28 Aug. 1880, DD 908; GPM to B. O. Duncan, Consul at Naples, 18 Apr. 1880, U.S. Dept. State Papers, Italy: Letters to Consuls, vol. 5; SFB cited in Evarts to GPM, 11 Aug. 1880, DI 731. See SI, *Ann. Rept. 1867*, 39, *Ann. Rept. 1870*, 25; GPM, DD 405 & 688. On Cesnola's Cypriote artifacts, see Ellen Herscher, "Cyprus in Context: Harvard's Semitic Museum Unveils Its Cesnola Collection," *Archaeology* 50:4 (1997): 68–71.

38. GPM to D. H. Wheeler, 14 Feb. 1866 (also 9 & 16 Nov. 1864); W. W. Story to J. R. Lowell, 11 Feb. 1853, in Henry James, *William Wetmore Story and His Friends* (1903), 1:254–57; GPM to H. A. Homes, 17 Mar. 1879; Preston Powers to GPM, 25 Feb. 1865; GPM to Preston Powers, 29 Feb. 1865; Wurts to Fish, 11 Aug. 1874, DD 505; Fish to GPM, 3 Sept. 1874, DI 441, pp. 466–67 (also DD 478, 522, 625).

39. GPM to Blaine, 17 June 1881, DD 976; to Reverend Moorehead, 22 Mar. 1864. See DD 816 & 858.

40. GPM to Fish, 12 Sept. 1870, DD 303, in U.S., *Foreign Relations . . . 1870*, 450–52; G. W. Wurts to CCM, 1 Oct. 1884. The incident caused a flurry in Washington; Hamilton Fish thought he might have to transfer Marsh to Madrid (T. E. Vermilyen to Fish, 4 Mar. 1872, Fish Corr.; CM to GPM, 22 & 30 Jan., 12 & 17 Feb. 1872; GPM to W. T. Sherman, 5 Aug. 1872; SP; GPM to John Bigelow, 2 Mar. 1872, in Bigelow, *Retrospections*, 2:19–20).

41. Trollope, *What I Remember*, 488–90; Garibaldi to GPM, 20 Jan. 1868; W. J. Stillman, *Autobiography of a Journalist* (1901), 2:390–91. C. F. Adams was U.S. Minister to Britain (1861–68), James Russell Lowell to Spain (1877–80) and to Britain (1880–85).

42. Trauth, *IADR*, 155–57; GPM quoted in Cornelia Underwood to Levi

Underwood, 5 Dec. 1873, in Tom Daniels, "In Italy with Mr. and Mrs. George Perkins Marsh" (1979), 193.

43. GPM to Charles Sumner, 15 Apr. 1865, Sumner Papers; to C. D. Drake, 30 Dec. 1867; GPM, "Were the States Ever Sovereign?" (1865) and "State Sovereignty" (in five parts; 1865).

44. GPM to Sumner, 1 Aug. 1864, Sumner Papers; to Bigelow, 22 Apr. 1865, in Bigelow, *Retrospections*, 2:510.

45. GPM to Drake, 30 Sept. 1867; to Dana, 21 Dec. 1865, 8 Oct. 1866, Dana MSS.

46. GPM to Godkin, 22 Nov. 1865, Godkin Papers; to Sumner, 17 Feb. 1868, Sumner Papers; to C. D. Drake, 20 May 1868. See White, *Republican Era*, 7–8, 366–67.

47. GPM to SFB, 9 Feb. 1867, BSI; to Godkin, 20 Sept. 1870 (also 24 Oct. 1865), Godkin Papers.

48. GPM, "Notes on the New Edition of Webster's Dictionary" (1866–67); Godkin to GPM, 8 Aug. 1867; Lowell to Godkin, 16 Oct. 1866, in J. R. Lowell, *Letters* (1894), 1:372; George Washington Moon, "Criticisms Originally Published in America, on the Hon. George P. Marsh's Strictures," in Moon, *Bad English Exposed: A Series of Criticisms on the Errors and Inconsistencies of Lindley Murray and Other Grammarians* (London, 1879), 48–116 at 106–7. See also "Moon's English," letters in the *Nation* 3 (1866): 314–15.

49. Godkin to GPM, 6 Oct. 1870; GPM, "A Cheap and Easy Way to Fame" (1865). Authorship of Marsh's *Nation* essays is given in Daniel Haskell, ed., *The Nation: Index of Titles and Contributors, Vols. 1–105* (New York: N.Y. Public Library, 1951–53). One essay attributed to Marsh, "'Protection to Naturalized Citizens,'" *Nation* 3 (1866): 115–16, is not by him; it is a letter, not in his style, dated 22 June 1866 from Berlin, which Marsh never visited.

50. GPM, "Thoughts and Aphorisms" (1875), 7; GPM, "Old English Literature" (1865); James Murray to GPM, 26 May 1870, 22 Oct. 1879; *A New English Dictionary on Historical Principles*, ed. James A. H. Murray, 10 vols. (Oxford, 1888–1928), 1:iii; Phonetic Alphabet bill in U.S. Senate, 29 Mar. 1867, U.S. CG, 40 Cong. 1 sess., 429; GPM to E. M. Gallaudet, 14 Oct. 1870.

51. GPM to H. C. Lea, 4 Jan., 22 June, 1 Aug., 20 Dec. 1870 (all UPa); Lea to GPM, 20 Dec. 1869, 4 Aug. 1870, 23 Jan. 1871 (UPa); GPM to Nash, 22 & 27 Dec. 1870; Lea to Nash, Aug. 1887; Henry Charles Lea, *A History of the Inquisition of the Middle Ages* (1888; rev. ed., New York: Macmillan, 1922), 1:iv. Through Villari, Italy's Minister of Education after 1875, Lea gained access to Vatican archives (Edward Scully Bradley, *Henry Charles Lea: A Biography* [Phila-

delphia, 1931]). See J. H. McDaniels, "Francis Philip Nash," *Hobart College Bull.* 9:4 suppl. (1911): 1–23; Anne Jacobson-Schutte, "Palazzo del Sant'Uffizio: The Opening of the Roman Inquisition's Central Archive," American Historical Association *Perspectives* 37:5 (May 1999): 25–28.

52. GPM to Yule, dated only 1869. See Henry Yule, tr. and ed., *The Book of Ser Marco Polo the Venetian, Concerning the Kingdoms and Marvels of the East* (London, 1875), and GPM review, "Book of Marco Polo" (1875). For help on handedness, Marsh got effusive thanks from Paolo Lioy in the 2d edition of his *Sulla legge della produzione dei sessi, coll'aggiunta di una lettera al Signor Marsh sul lato destro e sinistro* (Milan, 1873), v–xxi.

53. GPM to R. H. Dana, Jr., 22 Apr. 1867, Dana MSS. See Stanton, "Villa Forini"; Lucchesi and Bertocci, *Villa Il Giardino*.

54. GPM to CCM, 29 Oct. 1868 (CCM copy). See Elizabeth Adams Daniels, *Jessie White Mario: Risorgimento Revolutionary* (Athens: Ohio University Press, 1972).

55. W. H. Seward, 27 July 1871, in his *Travels around the World*, ed. Olive R. Seward (New York, 1873), 725; Trollope, *What I Remember*, 465.

56. GPM to CBT (CCM copy), 24 Apr. 1868; to SFB, 14 Mar. 1868, BSI.

57. GPM to SFB, 26 Dec. 1867, BSI; J. M. Sims to GPM, 6 Sept. 1866, 17 Aug. 1870. See Seale Harris, *Woman's Surgeon: The Life Story of J. Marion Sims* (New York: Macmillan, 1950).

58. GPM to S. Dana Horton, 19 Aug. 1869 (CCM copy). The *drei kalten Heiligen* were saints Servaz, Pankraz, and Bonifaz, before whose birthdays in mid-May it was held climatically unsafe to plant certain flowers. On the Three Kings (relics of the Magi), see Patrick J. Geary, *Living with the Dead in the Middle Ages* (Ithaca: Cornell University Press, 1994), 251–55.

59. GPM to SFB, 8 Jan. 1866, BSI; to Henry Yule, 21 July 1869; to C. E. Norton, 4 Nov. 1871, Norton Papers.

60. GPM to George Bancroft, 12 Sept. 1871, Bancroft MSS; to Yule, 21 Mar., 15 May 1871; to SFB, 2 Feb. 1874, BSI.

61. GPM to General W. Cotton, 9 Mar. 1874; to Sarah Butler Wister, 20 Jan. 1874. "Read this morning with Mr M the first Ode of Horace from a beautiful edition which he gave me last night. . . . The wonderful poems are the same, but how changed the two faces that bend over them!" (CCM, Diary, MS, 22 Nov. 1873).

62. GPM to SFB, 2 Feb. 1872, BSI; to CBT, 1 Jan. 1880 (CCM copy).

63. GPM to Fish, 24 June 1872, DD 407 (also DD 345 & 363); to Yule,

17 Dec. 1871; Rhodes, *Power of Rome*, 52–62; W. J. Stillman, *The Union of Italy, 1815–1895* (1899), 359–60.

64. GPM to Carrie M. Crane, 2 Oct. 1870; to CCM, 27 Dec. 1871; Fish to GPM, 3 July 1872, DI 346, p. 395.

16. Last Watersheds: Rome, Cravairola, Vallombrosa

1. GPM to Fish, 6 Oct. 1874, DD 520; to Evarts, 23 Apr. 1877, 23 Oct. 1880, DD 662, 920.

2. GPM to Evarts, 20 Feb., 6 Mar. 1878, DD 741, 744; to GFE, 29 Jan. 1878. The State Department had "a terrific struggle with itself over . . . whether to call the new king, 'King Humbert' simply, or 'King Humbert *first*.' . . . After a due amount of incubation, however, it was decided to call him 'King' simply, leaving it to posterity to decide whether he is King Humbert 'first,' or 'last'" (GPM to GFE, 17 Feb. 1877; GFE to GPM, 29 Jan. 1878).

3. GPM to Evarts, 26 Jan. 1880, DD 865. On King Umberto, see Richard Drake, *Byzantium for Rome: The Politics of Nostalgia in Umbertian Italy, 1878–1900* (Chapel Hill: University of North Carolina Press, 1980), xiv–xxii; Denis Mack Smith, *Italy and Its Monarchy* (New Haven: Yale University Press, 1983), 71–74 (with an acid appraisal of the deified Vittorio Emanuele's deficiencies, 63–67).

4. GPM to Evarts, 26 Jan. 1880, 30 Nov. 1878, DD 865 & 800. See Denis Mack Smith, "Francesco De Sanctis" (1991), 261–63; J. A. Davis, *Conflict and Control: Law and Order in Nineteenth-Century Italy* (1988), 291.

5. GPM to Evarts, 23 Apr. 1877, DD 662 (also DD 798, 800, 802, 1023). See also W. J. Stillman, *The Union of Italy, 1815–1895* (1899), 364–68; Trauth, *IADR*, 176n44; Nunzio Pernicone, *Italian Anarchism* (1993), 133, 155; Floriana Colao, *Il delitto politico tra Ottocento e Novecento: Da 'delitto fittizio' a 'nemico dello stato'* (Milan: Giuffrè, 1986); Denis Mack Smith, *Modern Italy* (1997), 97–106.

6. GPM to F. P. Nash, 14 July 1876; to E. L. Noyes, 27 Feb. 1881, in U.S., Dept. of State Papers, Italy: Letters to Consuls, vol. 5; to C. E. Norton, Apr. 1881. See Emilio Sereni, *History of the Italian Agricultural Landscape* (1961), 331–33.

7. GPM to Evarts, 1 Dec. 1880, to Frelinghuysen, 1 July 1881, DD 928 & 981 (also DD 897 & 920); to Sarah Wister, 23 Mar. 1879 (CCM copy).

8. GPM to Evarts, 23 July 1877, 1 July 1881, DD 777 & 981; Prince Alexander Gortchakov, 1878, cited in Mack Smith, *Modern Italy*, 113. See Trauth, *IADR*, 172–74; R. J. B. Bosworth, *Italy and the Wider World* (1996), 19–25.

9. GPM to Evarts, 1 July 1881, DD 981; to Nash, 14 July 1876. See Ester De Fort, *La scuola elementare dall'Unità alla caduta del fascismo* (Bologna: Il Mulino, 1996), 68–111.

10. GPM to GFE, 30 Dec. 1877; to Nash, 3 Aug. 1877; to W. T. Sherman, 19 Aug. 1876, SP; confid. to Fish through GFE, 27 Oct. 1876.

11. GPM to Yule, 20 Oct. 1878; to Seward, 21 Mar. 1864, DD 87; to Frelinghuysen, 10 Jan. 1882, DD 1015; Trauth, *IADR*, 176–77.

12. GPM to Frelinghuysen, 30 Jan. 1882, DD 1016; *M&N*, 286n4.

13. GPM to GFE, 29 Jan. 1878.

14. GPM to C. D. Drake, 20 Feb. 1869; to CCM, 2 Oct. 1868 (CCM copy). On Wurts, see H. S. Sanford to GPM, 17 June 1865; G. W. Wurts to GPM, 23 Oct. 1865. Caroline E. Crane, pers. comm., summer 1951, recalled Wurts.

15. GPM to R. H. Dana, Jr., 11 June 1868, Dana MSS; GFE to GPM, 11 June 1868.

16. GPM to Evarts, 23 Jan. 1879, DD 811; Eugene Schuyler to GPM, 2, 17, & 19 Dec. 1879; GPM to Schuyler, 4 Dec. 1879, in U.S. Dept. of State Papers, Italy: Letters to Consuls, vol. 5; Schuyler, *American Diplomacy* (1886), 172–73.

17. GPM to Fish, 15 & 25 Sept. 1874, DD 512, 513.

18. Swiss counsellor Delarageaz, 1868, quoted in Paul Guggenheim, "Fixation de la frontière de l'Alpe de Cravairola" (1954), 471 [my translation].

19. GPM to Yule, 24 Sept 1874; to CCM, 14 & 17 Sept. 1874 (CCM copies), 10 Sept. 1874, CFP.

20. Antonio Malintoppi, "Diritto ed equità nell'arbitrato per l'Alpe Cravairola" (1975), 517–18. Campo was willing to buy the alp, but the Italian communes did not want to sell it.

21. GPM, "Decision of Arbitration . . . of the Italian-Swiss frontier at . . . Alpe de Cravairola" (1874).

22. GPM, *Earth as Modified by Human Action* (1874 ed.), 1877 printing, app. 656n30; Karl Culmann, *Rapport au Conseil Fédéral sur les Torrents des Alpes Suisses* (1865), 7–13. This description of Campo and the Rovana is implicit in Marsh's "Decision of Arbitration," 2030–31.

23. GPM to Fish, 25 Sept. 1874, DD 513; "Decision of Arbitration," 2033.

24. "Decision of Arbitration," 2044–47. Marsh recognized that assigning Cravairola to the Swiss would have worked only if the Italians sold their land rights; for "the extension of the sovereignty, institutions, laws, and administration of the Swiss over the Alp, while its proprietors and occupants remained citizens of a foreign country, would occasion continued jealousies and dis-

sensions and therefore prove prejudicial, rather than beneficial, to the peace, harmony and good will of the two countries" (GPM to Fish, 25 Sept. 1874, DD 513; "Decision of Arbitration," 2034).

25. GPM to Yule, 24 Sept. 1874; to Fish, 15 & 25 Sept. 1874, DD 512, 513; Fish to GFE (re presents to Marsh), U.S., 44 Cong. 1 sess., Sen. Misc. Doc. 16 (13 Dec. 1875); GPM cited by Swiss envoy to Italy Jean Baptiste Pioda, 19 June 1875, in Guggenheim, "Fixation de la frontière," 508n. Noting that Marsh's decision had given "much joy," the Comune of Crodo sent condolences on Marsh's death eight years later (Mayor Guglielmi to CCM, 26 July 1882).

26. Guggenheim, "Fixation de la frontière," 483, 510–14.

27. Malintoppi, "Diritto ed equità," 521–23. On watersheds see Chapter 1 above.

28. GPM, "Watershed" (1878); M&N, 222. On Barton Pond, see M&N, 302; on Val di chiana, Chapter 13 above, on Lake Thun and the Aar, M&N, 341–42.

29. Stephen B. Jones, *Boundary-Making: A Handbook for Statesmen, Treaty Editors and Boundary Commissioners* (Washington, D.C.: Carnegie Endowment for International Peace, 1945), 101–4; A. O. Cukwurah, *The Settlement of Boundary Disputes in International Law* (1967), 41–45, 81–83.

30. Max Huber, "The Island of Palmas (or Miangas) . . . Arbitral Award," *American Journal of International Law* 22 (1928): 867–912 at 894.

31. "Introduction au Tome III," *Recueil des arbitrages internationaux* (1954), 3:li (see Guggenheim, "Fixation de la frontière"); Cukwurah, *Settlement of Boundary Disputes*, 201–3, 231–32; A. L. W. Munkman, "Adjudication and Adjustment— International Judicial Decision and the Settlement of Territorial and Boundary Disputes," *British Yearbook of International Law* 46 (1972–73): 1–116 at 1, 5–7.

32. The nonconformist preacher Charles Marsh was martyred in 1555. GPM's remark about heresy appears in *BFP* 17 Jan. 1845, but not in the published version of his *NES* address.

33. GPM, "The Catholic Church and Modern Civilization" (1867), 231; GPM to Bancroft, 3 Aug. 1869, Bancroft MSS. See E. E. Y. Hales, *Pio Nono* (1954), 255–62; Trauth, *IADR*, 167–69.

34. Anthony Rhodes, *The Power of Rome in the Twentieth Century* (1983), 9–17; GPM to Evarts, 8 Feb. 1878, DD 736.

35. GPM, "Catholic Church and Modern Civilization"; Godkin to GPM, 10 Feb. 1874.

36. [GPM,] *Mediaeval and Modern Saints and Miracles NOT Ab Uno e Societate Jesu* (1876) [hereafter *MMSM*], 122; Harper's to GPM, 11 Jan. 1881; review,

Harper's New Monthly Mag. 53 (1876): 787. See GPM to H. C. Lea, 16 Aug. 1874; A. B. Crane to GPM, various dates, 1875–76, CFP; GPM-Harper's contract, 20 Jan. 1876, cancellation, 25 Jan. 1882, statement, 8 Jan. 1884, Harper & Bros. Archives. A modern reviewer rates Marsh's book a "strongly anti-Jesuit but scholarly attack upon the Church's cynical exploitation of ignorance and superstition" (William L. Vance, *America's Rome*, 2 vols., 2: *Catholic and Contemporary Rome* [New Haven: Yale University Press, 1989], 15).

37. GPM to H. C. Lea, 19 Nov. 1876, UPa; *MMSM*, 129n.

38. *MMSM*, 18–21, incorporating GPM's handwritten corrections; *LEL*, 1885 ed., 194n.

39. *MMSM*, 20, 214, 215 (GPM handwritten insert).

40. *MMSM*, 137–38.

41. *MMSM*, 149–51, 141–42, 294–95.

42. *MMSM*, 98, 104, 106.

43. *MMSM*, 104, 200–201, 119.

44. GPM to Nash, 26 Apr. 1877; to W. J. Knapp, 29 Jan. 1880. The Marsh essays mentioned are all in *Johnson's New Universal Cyclopaedia*, q.v.

45. *M&N*, 312; GPM to GFE, 4 Aug. 1862. See Sereni, *History of the Italian Agricultural Landscape*, 264–72, 298–312.

46. *M&N*, 311–25. Salinization and other effects of irrigation continue to degrade huge areas (W. B. Meyer, *Human Impact on the Earth* [1996], 76–77). In common with most, Marsh at first ascribed malaria and other diseases to "miasma." But by the 1870s he was aware that germs were "the true seeds of infection and death" in the spread of contagious disease (*Earth as Modified by Human Action*, 1874 ed., 147n, 1884 ed., 145n; see also Meyer, *Human Impact*, 110, 125).

47. *M&N*, 321–22n47.

48. GPM to Gen. W. Cotton, 9 Mar. 1874; GPM, *Irrigation: Its Evils, the Remedies, and the Compensations* (1874), 3–4; handwritten slip pasted in Marsh's annotated copy, p. 12; GPM, "Irrigation," *Johnson's New Universal Cyclopaedia* (1876), 2:1311–13.

49. GPM, *Irrigation*, 15–16. See James Watts to GPM, 21 May 1873; J. A. Kasson to GPM, 30 Mar. 1874; William Gilpin, *Mission of the North American People* (Philadelphia, 1873), 71–76, 89–90, for classic faith in the American "garden"; H. N. Smith, *Virgin Land: The American West as Symbol and Myth* (1950), 35–43; Wallace Stegner, *Beyond the Hundredth Meridian: John Wesley Powell and the Second Opening of the West* (Boston: Houghton Mifflin, 1953), 1–8.

50. GPM, *Irrigation*, 4–6.

51. Ibid., 15–17, 19; GPM, "Irrigation," 1313.

52. SFB to GPM, 24 Feb. 1874; GPM to SFB, 2 Feb. 1874. See J. W. Powell, *Report on the Lands of the Arid Region of the United States* . . . 2d ed. (Washington, D.C., 1879), 83–84; Stegner, *Beyond the Hundredth Meridian*, 301–55.

53. Arthur A. Ekirch, *Man and Nature in America* (New York: Columbia University Press, 1963), 78–79; Lawrence B. Lee, "Environmental Implications of Government Reclamation in California," *Agricultural History* 44 (1975): 223–37 at 225.

54. C. S. Sargent to GPM, 23 Jan. 1879; GPM to Sargent, 14 & 16 May 1879, 25 June 1882; Sargent, "George Perkins Marsh," *Nation* 35 (1882): 136. Marsh's animadversion on Ruskin bore on Ruskin's claim "that the cross-sectional area of the stem of a tree, and the sum of the cross-sectional areas of all the branches are equal." Marsh thought he "had seen cases where the latter quantity was the greatest," and believed "the assertion of the proposition by Ruskin is prime evidence that it is not true, for in regard to the facts of nature, Ruskin is, as he said of Raphael, always in the wrong" (to Sargent, 16 May 1879). On eucalyptus and malaria, GPM to H. C. Lea, 19 Mar. 1875; *Earth as Modified by Human Action*, 1884 ed., 383n. Eucalyptus was extolled for absorbing humidity, exhaling moisture, producing an antiseptic volatile oil, and dropping foliage on decaying organic matter (Kenneth Thompson, "The Australian Fever Tree in California: Eucalyptus and Malaria Prophylaxis," *Annals of the Association of American Geographers* 60 [1970]: 230–44).

55. GPM to Sargent, 30 Aug. 1879; *M&N*, 1867 printing, app. 555n21.

56. GPM, *Earth as Modified by Human Action*, 1884 ed., 320n, 35n; *M&N*, 37n.

57. GPM, *Earth as Modified* (1884), 605n, 587n, 473n, 46. For Marsh on urban climate, see Lewis Mumford, "The Natural History of Urbanization," in W. L. Thomas, Jr., ed., *Man's Role in Changing the Face of the Earth* (1956), 382–98 at 398.

58. GPM to Nash, 6 Feb. 1875; to Yule, 15 Feb. 1875 (*litera longa* is the "long capital I" from which the hanged thief dangles in Plautus's *Pot of Gold*). See A. J. Johnson and GPM, Publisher's Agreement, 2 Feb. 1874; GPM to Johnson, Dec. 1875, 28 Feb., 5 Mar. 1878; Johnson to GPM, 4 Apr. 1878. Marsh's encyclopedia articles are grouped under GPM, *Johnson's New Universal Cyclopaedia* in the bibliography.

59. GPM to Yule, 25 Jan. 1875, 10 July 1876.

60. *Nation* 20 (1875): 134, report of lecture by Prof. Edmund Singleton Holden, astronomer-librarian at the U.S. Naval Observatory (*On the Number of Words Used in Speaking and Writing* [Philadelphia, 1876]); *LEL* (1860), 128 ff.; GPM to Nash, 25 Mar. 1875; F. A. March, "Weisse's Origin of the English Language,"

Nation 28 (1879): 153–54; see L. W. Levine, *The Opening of the American Mind* (1996), 80. Marsh's published rejoinder to Holden, "Milton's and Shakespere's Vocabularies" (1875), was more temperate than his letter to Nash.

61. GPM to Nash, 6 Feb., 27 Dec. 1875; to Eugenio Canvazzi (author of *Vocabolario di agricoltura*, Bologna, 1871), 10 Feb. 1875; GPM, "Biography of a Word" (1881).

62. GPM, *Human Knowledge* (1847), 42, amplified in *O&H* (1862), 415.

63. GPM, "Report of a Committee . . . for the Erection of a Monument over the Grave of Ethan Allen" (1858); GPM to J. N. Pomeroy, 21 May 1872; Pomeroy to GPM, 21 July 1873, citing *BFP & Times*, 7 July 1873. See *BFP & Times*, 24 Apr. 1873; Daniel Robbins, *The Vermont State House* (1980), 126.

64. GPM to GFE, 9 Feb., 25 Apr. 1879, in Robert C. Winthrop, *Addresses and Speeches on Various Occasions* (Boston, 1886), 4:140–43; GPM letters & R. C. Winthrop et al., Memorial to Congress, 26 Apr. 1880, in F. L. Harvey, *History of the Washington National Monument* (1903), 122–23, 97–98; Louis Torres, *"To the Immortal Name and Memory of George Washington": The United States Corps of Engineers and the Construction of the Washington Monument* (Washington, D.C.: GPO, 1985), 55–58, 78, 80, 85; Pamela Scott and Antoinette J. Lee, *Buildings of the District of Columbia* (New York: Oxford University Press, 1993), 100–102.

65. Amy Yule to CCM, 2 July 1876; GPM to George Bancroft, 15 Apr. 1877, Bancroft MSS; Domenico Carutti di Cantogno, "Cenno necrologio del socio straniero Perkins Marsh," Atti della Reale Accademia dei Lincei, 3d ser. *Transunti* 7 (1882–83): 172–73; Guido Quazza, *L'Utopia di Quintino Sella: La politica della scienza* (Turin: Comitato di Torino dell'Istituto per la storia del Risorgimento Italiano, 1992), 531–32. The Marsh-Sella correspondence is at UVM and at the Fondazione Sella, Biella, Piedmont; letters thus far published are in *Epistolario di Quintino Sella*, ed. Guido and Marisa Quazza, vol. 3, 1870–71 (Rome: Istituto per la storia del Risorgimento Italiano, ser. 2: *Fonti*, vol. 30, 1991), 85, 175–76, 223, 228.

66. GPM, "Tiber," *Johnson's New Universal Cyclopaedia* (1878), 4:853; GPM to W. T. Sherman, 22 Mar. 1877, SP. See John Agnew, *Rome* (Chichester: Wiley, 1995), 14–44. Marsh noted that modern Rome, like many European capitals, was but recently, if yet, supplied with pure water (GPM, "The Aqueducts of Ancient Rome" [1881]).

67. GPM to SFB, 30 June 1881, BSI; to David H. Wheeler, 4 Jan. 1866 (CCM copy).

68. Henry James, "The After-Season at Rome," *Nation* 16 (1873): 399–400; GPM to GFE, 15 Feb. 1879; to SFB, 21 Feb. 1872; to Carrie M. Crane,

16 Dec. 1872. Vance, *America's Rome*, 2:261–63, documents Marsh's indictment of American kowtowing to royalty.

69. Eleanor Clark, *Rome and a Villa* (Garden City, N.Y.: Doubleday, 1952), 16. See Otto Wittmann, Jr., "The Italian Experience (American Artists in Italy, 1830–1875)," *American Quarterly* 4 (1952): 3–15 at 4, 13.

70. CCM to Susan Edmunds, 15 Dec. 1872; Matthew Arnold to Fanny Arnold, 17 Apr. 1873, in Arnold, *Letters* (1895), 1:112; *The Roman Journals of Ferdinand Gregorovius, 1852–1874*, ed. Friedrich Althouse (London: G. Bell, 1907), 30 Mar. 1873, 439; Michelet story in GPM to George Bancroft, 14 Mar. 1871, Bancroft MSS.

71. Henry James, *William Wetmore Story and His Friends* (1903), 1:257–58; GPM to CBT, 1 Jan. 1880.

72. GPM to LW, 2 Jan. 1877 (CCM copy); to Yule, 25 Dec. 1872; to Sherman, 27 Feb. 1872, SP; to D. G. Mitchell, n.d., in Mitchell, *American Lands and Letters: Leather-stocking to Poe's "Raven"* (1907), 43.

73. T. A. Trollope, *What I Remember* (1888), 448; GPM to Evarts, 5 Mar. 1878, DD 743. In 1893 America's heads of diplomatic missions were elevated from "Envoy Extraordinary & Minister Plenipotentiary" to "Ambassador Extraordinary & Plenipotentiary."

74. Rhodes, *Power of Rome*, 48; F. J. Coppa, *Cardinal Giacomo Antonelli* (1990), 129; GPM to T. Adolphus Trollope, 30 Nov. 1867, in Trollope, *What I Remember*, 449.

75. CCM to LW, 30 Dec. 1877; GPM to LW, 2 Jan. 1877 (CCM copy); GPM to GFE, 30 Dec. 1877; GPM to H. C. Lea, 29 Apr. 1880; GPM to LW, 25 Feb. 1880.

76. GPM to LW, 10 May 1879; to CBT, 1 Jan. 1880.

77. CCM to Edward Hungerford, 3 Sept. 1882, Karen Perez Coll., UVM.

78. GPM to C. B. Hartman (Ohio), 13 Mar. 1882; to N. J. Bowditch (Massachusetts Board of Health), 20 May 1870, in U.S. Dept. of State Papers, Italy: Letters to Consuls, vols. 6 & 2.

79. GPM to J. S. Pike, 12 May 1864, Pike Papers; GPM, "Dry Wines," *American Register* (Paris), 1 Sept. 1877, 4; GPM to GFE, 29 Jan. 1878; GFE to GPM, 12 Oct. 1877.

80. Carrie M. Crane to GPM, 2 Feb. 1872; GPM to Carrie Crane, 19 Nov. 1869 (CCM copy); to Nash, 24 Aug. 1875. On the drowning, see GPM to Giuseppe Pasolini, 10, 14, 16, & 29 May, 10 June 1875.

81. GPM to LW, 21 Oct. 1876, WP; to Yule, 17 June 1876; Elizabeth Green Crane to A. B. Crane, 20 Nov. 1881, CFP.

82. GPM to Nash, Sept. 1877; to H. A. Homes, Sept. 1852, in *LL*, 293; to Fish, 20 Apr. 1876, DD 602. See C. C. Andrews to GPM, 26 Apr. 1876, Andrews Papers; GPM to Andrews, 22 Jan. 1877.

83. GPM to Sarah Wister, 23 Mar. 1879 (CCM copy); to CBT, 1 Jan. 1880. After Marsh's death Carlo went to America with Caroline, was expelled from several schools, and disappeared into the Midwest, resurfacing occasionally to cadge funds.

84. Billings twice remodeled the Marsh house, in 1869–70 "in the fashionable Stick Style with a French roof and a broad veranda," and more elaborately in 1885, by the architect Hudson Holly (R. W. Winks, *Frederick Billings* [1991], 302–3). Peter S. Jennison terms Holly's mansion "an understated Queen Anne style" (1980 introduction to reprint of H. S. Dana, *The History of Woodstock* [1889], n.p.).

85. GPM to CBT, 1 Jan. 1880.

86. GPM to LW, 25 Jan. 1880; to Nash, 26 Apr. 1877; A. B. Crane to GPM, 1 Nov. 1880; CCM to LW, 15 Mar. 1881, WP; GPM to Yule, 18 Sept. 1876.

87. Washington correspondent, Boston *Daily Transcript*, 11 Jan. 1873; Mrs. E. E. Evans to GPM, 16 Mar. 1873, citing her reply in Boston *Evening Transcript*, 27 Feb. 1873; GPM to GFE, 15 Feb. 1879. See J. R. Lowell to W. J. Hoppin, 22 Nov. 1881, in *New Letters of James Russell Lowell*, ed. M. A. De Wolfe Howe (New York: Harper, 1932), 259.

88. GPM to M. F. Force, 5 May 1877; to GFE, 21 Apr. 1877; GFE to GPM, 10 May 1877.

89. GFE to GPM, 29 Mar., 1 July 1880, 13 Apr. 1881; M. F. Force to GPM, 2 Nov. 1881; GFE to GPM, 9 Mar. 1881, 29 Jan. 1878. See Selig Adler, "The Senatorial Career of George Franklin Edmunds, 1866–1891," Ph.D. thesis, University of Illinois, 1934; Warren F. Austin, "Address," in *George F. Edmunds Centenary Exercises, 1828–1928* (Burlington: Historical Society, 1928), 1–28; Emily E. Briggs, *The Olivia Letters: Being Some History of Washington City for Forty Years* (New York, 1906); Adam Badeau, *Grant in Peace . . . a Personal Memoir* (Hartford, Conn., 1887), 536–37.

90. CCM to LW, 15 Mar. 1881, WP; Edward Crane cited in CCM to S. G. Brown, 25 Oct. 1884; GPM to LW, 15 Mar. 1881, WP.

91. GPM to GFE, Apr. 1881; Charles Marsh estate, 14 Apr. 1874, Vt. Court of Probate, Dist. of Hartford, 32:140. After all his fears, Marsh left his wife more than $54,000 (GPM, will and estate inventory, Probate Office Burlington no. 4676, Chittenden County, 60:301–4, 384, 532; 61:11).

92. GPM to M. H. Buckham, 4 Apr., 23 Sept. 1881. See GPM to J. R.

Lowell, 19 Jan. 1882, Lowell Papers; University of Vermont, *Catalogue of the Library of George Perkins Marsh* (1892), v–viii. GPM, "Fireproof Construction in Italy," *Johnson's New Universal Cyclopaedia* (1876), 2:113–14, faults American tinderbox building.

93. CCM Diary, 22 Nov. 1873.

94. GPM to C. E. Norton, 16 Oct. 1865 (CCM copy); GPM to R. C. Winthrop, 8 Nov. 1872, Winthrop Papers, and to W. T. Sherman, 25 Dec. 1872, SP (on Greeley); to M. F. Force, 5 May 1877; to GFE, 22 Mar. 1877.

95. GPM, *Speech on the Tariff Question* (1846), 1010; to Sumner, 17 Feb. 1868, Sumner Papers; to H. C. Lea, 6 Nov. 1876, UPa; to Sherman, 18 Mar. 1876, SP; to Sarah Wister, 5 Apr. 1880 (CCM copy). On the moral decline, see L. D. White, *The Republican Era, 1869–1901* (1958), 365–80.

96. GPM to Giuseppe Pasolini, 29 May 1875; GFE in the Senate, 14 Jan. 1887, U.S. *Cong. Record*, 42 Cong. 2 sess., 645–46; William Letwin, *Law and Economic Policy in America: The Evolution of the Sherman Antitrust Act* (1965; reprint, Westport: Greenwood, 1980), 94–95.

97. GPM to Sherman, 25 Dec. 1872, SP; to GFE, 19 May 1877.

98. GPM to F. P. Nash, 9 Aug. 1879, 28 Mar. 1877; to Pasolini, 11 Sept. 1876.

99. GPM to C. S. Sargent, 14 Nov. 1880.

100. W. W. Story, *Vallombrosa* (1881), 6, 8, 28, 69, 101; James, *William Wetmore Story*, 2:330–37; Story to GPM, 12 July 1882. On Marsh's donation of *Pinus monophylla* seeds, see *Nuova Rivista Forestale* 2 (1879), cited in Marcus Hall, "Restoring the Countryside: George Perkins Marsh and the Italian Land Ethic" (1998), 14n34. Marsh held Di Bérenger's *Dell' antica storia e giurisprudenza forestale in Italia* (Treviso, 1859–63) "the most learned work ever published on the social history of the forest" (marginal note, UV, *Catalogue of the Library of George Perkins Marsh*, 237). The monastic buildings reverted to the church in the 1940s and have been much restored; the forestry school removed to Florence in 1913, and its former buildings now (1998) house a biogenetic institute.

101. Frelinghuysen to GPM, 7 July telegram, 18 July 1882, DI 828, 2:206–9; GPM to GFE, 24 Sept. 1876; GPM quoted in CCM, "Last Days of George P. Marsh," MS; G. W. Wurts to CCM, 10 July 1882; Schuyler, *American Diplomacy*, 173–74. See W. J. Stillman, "The Late George P. Marsh," and "A Diplomatic Intrigue," *Nation* 35 (1882): 304–5, 529–30. At his death Marsh was preparing a dispatch "opposing this measure from which he considered he would be the greatest and first sufferer" (Wurts to Frelinghuysen, 15 Sept. 1882, DD 1044).

102. GPM to C. S. Sargent, 20 July 1882. Marsh was especially struck by the inversion of the natural order of succession, in this case, from beech to fir (*M&N*, 276n245, 1884 ed., 354n). See Mitchell, *American Lands and Letters*, 36; Donald P. Duncan, "Forest Practice at Vallombrosa in Central Italy," *Journal of Forestry* 44 (1946): 347–53; Henry Kernan, "The Trees of Vallambrosa [*sic*]," *American Forests* 60:7 (1954): 14–16, 44–45.

103. GPM quoted in CCM, "Last Days of George P. Marsh." The vista Marsh enjoyed was by 1998 obscured by forest growth; for a view of the valley below one had to climb to the former Hermitage, then "Paradisino," high above the monastery, dedicated to Milton in 1925. Milton himself never visited Vallombrosa, a name he took from Ariosto's *Orlando Furioso* to lend *Paradise Lost* an aptly picturesque aura. To forestall the monastery's dissolution in the 1860s, the monks elaborated this fiction, claiming (but not showing) letters from Milton, and an organ used by the poet kept in reverential disrepair (Edward Chaney, "Milton's Visit to Vallombrosa: A Literary Tradition" [1991], in his *The Evolution of the Grand Tour: Anglo-Italian Cultural Relations since the Renaissance* [London: Frank Cass, 1998], 278–313). Chaney terms Milton's "thick autumnal leaves that strow the brooks in Vallombrosa" poetic license, as Vallombrosa's trees were mostly conifers, and even its beeches kept their leaves after October. Also dubious are William Beckford's autumnal leaves, even less likely Elizabeth Barrett Browning's dead leaves thick in the spring (pp. 286–87, 298). But fallen beech leaves would in fact remain through the winter and beyond. Marsh noted none in July.

104. Mary E. Crane to LW, 27 July 1882, WP; CCM, "Last Days of George P. Marsh."

105. Trollope, *What I Remember*, 530; Wurts to Frelinghuysen, 30 July 1882, DD 1033; H. Nelson Gay, *The Protestant Burial Ground in Rome: A Historical Sketch* (London: Macmillan, 1913); CCM to S. G. Brown [ca. 1883]. On leaving Vallombrosa, Caroline confided to Marsh's niece that "my dear husband's age and increasing weakness . . . made life almost a burthen, & death so desirable for him," adding that "for years the feeling that such a change might come at any time has been so constantly present . . . as to throw a shadow over every present enjoyment" (CCM to Maria Buell Hungerford, 2 Sept. 1882, Karen Perez Coll., UVM).

17. Retrospect: Forming a Life

1. GPM to CCM, 1839, in *LL*, 30; to F. P. Nash, 9 Aug. 1879.

2. GPM to SFB, 2 July 1855, BSI.

3. GPM to Lieber, 4 Jan. 1860; GPM quoted in *LL*, 424.

4. GPM to C. C. Jewett, 8 Feb. 1851, in *LL*, 203; to Sarah Wister, 23 Mar. 1879 (CCM copy).

5. Trauth, *IADR* (1958), xvii.

6. GPM to SFB, 9 Feb. 1867, BSI; to H. S. Dana, 15 Feb. 1861; GPM quoted in Harry L. Koopman, "George P. Marsh," An Address in the Marsh Library of the University of Vermont, before the New England College Librarians, May 1926, MS, UVM, 5. The book may have been *Histoire de Foulques Fitz-Warin* (Paris, 1840), to which Marsh refers in his additions to Hensleigh Wedgwood's *Dictionary of English Etymology* (1862), 153.

7. GPM, "A Cheap and Easy Way to Fame" (1865); "Thoughts and Aphorisms" (1875), 4–6; to CCM, 1839, in *LL*, 31.

8. GPM to CCM, 30 Mar. 1855; to FM, 29 Mar. 1855, GMM. Marsh slightly varies James Thomson's "expressive Silence, muse His praise" (*A Hymn on the Seasons*, 1730).

9. GPM, "A Cheap and Easy Way to Fame."

10. *LEL*, 291.

11. Wurts to CCM, 1 Oct. 1884; CCM to Samuel Gilman Brown, 5 Mar. 1883. Marsh's additions to Wedgwood's *Dictionary* are laden with Arabic and Turkish matter (e.g., Arsenal, 50; Borough, 113; Carboy, 160–61; Dub, 244–45).

12. *LEL* 1885 ed., 527n; GPM to Nash, 9 Apr. 1879.

13. GPM to S. Dana Horton, 15 Oct. 1869 (CCM copy); GPM, "Notes on the New Edition of Webster's Dictionary; IX. Form and Arrangement of Dictionaries," *Nation* 3 (27 Dec. 1866): 516; *LEL*, 448n; GPM, "Lexicon, Dictionary, Thesaurus, Vocabulary, Glossary," *Johnson's New Universal Cyclopaedia* (1876), 2:1752.

14. GPM to Nash, 3 Aug., 8 June 1877.

15. GPM to Nash, Sept. 1877; to LW and FW, 13 Oct. 1862, WP; to CCM, 14 Feb. 1855.

16. GPM to Nash, 14 July 1876; GPM, "England Old and New" (1859/60), MS, 5; GPM and CCM to LW and FW, 25 July 1862, WP; M. K. Cayton, *Emerson's Emergence* (1989), 148–49.

17. GPM to HP, 7 Aug. 1863, PAA.

18. GPM, "Thoughts and Aphorisms," 2–3; GPM to Nash, 3 Aug. 1877; CCM to S. G. Brown, n.d.

19. George Allen to GPM, 19 June 1844; T. D. Seymour Bassett review of my 1958 biography of Marsh, *New England Quarterly* 32 (1960): 249–51 at 251.

20. GPM to Giuseppe Pasolini, 16 May 1875; to LW, 25 July 1862, WP; F. A. March, "George Perkins Marsh" [review of *LL*], *Nation* 47 (1888): 213–15 at 214; Cayton, *Emerson's Emergence*, 137, 218–21, 232–34.

21. Cornelia Underwood to Levi Underwood, 5 Dec. 1973, in Tom Daniels, "In Italy with Mr. and Mrs. George Perkins Marsh" (1979): 192; Susan Edmunds to CCM, 6 July 1883. For Caroline's reciprocal dependence on Marsh, see her poem "To —— ——," in CCM, *Wolfe of the Knoll* (1860), 324–27.

22. CCM to GPM, 15 May 1856; GPM to HP, PAA, 31 Mar. 1863; Dedication to CCM in *LEL*, 1885 ed., iii.

23. On Marsh's devotion to his Buell relations, see CCM to Maria Buell Hungerford, 2 Sept. 1882 (Karen Perez Coll., UVM).

24. GPM to Nash, Sept. 1877; to Homes's son, Sept. 1852, in *LL*, 293; Carrie M. Crane to GPM, 2 Feb. 1872.

25. GPM to Miss Estcourt, Aug. 1855, in *LL*, 376. I am unable to identify Seymour.

26. Lucy W. Mitchell to family, 16 Dec. 1876, John K. Wright Papers; O. W. Holmes, Jr., to GPM, 23 Apr. 1860; Norton to CCM, 5 Aug. 1882; GPM to LW, 9 Nov. 1875; Cornelia Underwood to Levi Underwood, 3 Nov. 1873, in Daniels, "In Italy with . . . Marsh," 192. See [T. D. Seymour], "Lucy Myers Mitchell," *Critic* 15 (1888): 176; Richard H. Dana, *Hospitable England in the Seventies: The Diary of a Young American, 1875–1876* (Boston: Houghton Mifflin, 1921), 243–46.

27. SFB to GPM, 4 Feb. 1854, in W. H. Dall, *Spencer Fullerton Baird* (1915), 315–16; CBT to GPM, 13 Jan. 1862 (misdated 1861); Humphrey Sandwith to GPM, 13 March 1880.

28. SFB to GPM, 5 Feb. 1854 (differs slightly from letter dated 4 Feb. in Dall, note 27 above); D. H. Wheeler, "Recollections of George P. Marsh," MS, UVM.

29. GPM, *NES* (1845), 20; *MMSM* (1876), 204; "Study of Nature" (1860): 59; GPM to C. D. Drake, 15 Feb. 1845.

30. GPM to S. Dana Horton, 15 Oct. 1869 (CCM copy). Marsh here refers to Mlle. Nélie Jacquemart (1841–1912), Paris portrait painter married to Edouard André; their elegant home is now the Musée Jacquemart-André. Marsh was later elected, along with Lucretia Mott and Lydia Maria Child, an honorary member of the Massachusetts Society for the Universal Education of Women (Hannah E. Stevenson to GPM, 8 July 1877). See Ishbel Ross, *Child of Destiny: The Life Story of the First Woman Doctor* (New York: Harper, 1949), 136; Nancy Ann Sahli, *Elizabeth Blackwell, M.D.: A Biography* (New York: Arno Press, 1982); Eliza-

beth Blackwell, *Pioneer Work for Women* (London, 1895), 101; Margaret Armstrong, *Fanny Kemble: A Passionate Victorian* (New York: Macmillan, 1938), 278.

31. GPM to Carrie M. Crane, 4 Oct. 1869; to Sarah Wister, 23 Mar. 1879 (CCM copy); Carl N. Degler, *At Odds: Women and the Family in America from the Revolution to the Present* (New York: Oxford University Press, 1980), 308–11.

32. GPM, "The Education of Women" (1866): 165; Horace Bushnell, *Women's Suffrage, the Reform against Nature* (New York, 1869), 110–13, 143. Unlike most states, New Jersey's 1776 constitution did not disfranchise women, whose right to vote was made explicit in 1790. The growing electoral influence of women alarmed the state legislature in 1807 into confining suffrage to free white males; colored women too had voted, and white male anxieties over color and gender coalesced ("Women Vote in New Jersey," in Gerda Lerner, *The Female Experience* [Indianapolis: Bobbs-Merrill, 1977], 323–29). Slavery in New Jersey was partly abolished in 1804, reinstated a few years later, abolished again in theory but not practice in 1846, and finally ended by the Thirteenth Amendment in 1865 (Arthur Zilversmit, *The First Emancipation: The Abolition of Slavery in the North* [Chicago: University of Chicago Press, 1967], 184–99, 215–22).

33. GPM, "Female Education in Italy" (1866).

34. Ibid.; R. W. Emerson to Paulina W. Davis, Sept. 1850, quoted in Cayton, *Emerson's Emergence*, 281–82n6.

35 GPM, "Oration" [New Hampshire] (1856), 83–84. Marsh was unaware of the extent to which American women, white as well as black, did heavy farm work (Degler, *At Odds*, 363–64).

36. GPM, handwritten addition to *MMSM*, 216. See A. L. Cardoza, *Aristocrats in Bourgeois Italy* (1997), 147–48.

37. GPM, *NES*, 19–20; "Female education in Italy." In Italy women were objects of both passion *and* reverence. They "may be angels until they marry, but then almost immediately become very far from angelic. After they have produced sons, they evolve, through matriarchy, to the angelic status once more. At their death, all the bells of Heaven ring, and ever afterwards their names are only mentioned in hushed tones of profound veneration" (E. R. Vincent, preface to Massimo D'Azeglio, *Things I Remember* [1966], ix). See also Mauro Moretti, "Pasquale Villari e l'istruzione femminile: dibattiti dell'opinione e iniziative di reforma," in Simonetta Soldani, ed., *L'educazione delle donne: Scuole e modelli di vita femminile nell'Italia dell'Ottocento* (Milan: Franco Angeli, 1989), 497–530; Silvia Franchini, *Élites ed educazione femminile nell'Italia dell'Ottocento: L'Istituto della SS. Annunziata di Firenze* (Florence: Leo S. Olschki, 1993), 343–86; John Perlman and Dennis Shirley, "When Did New England Women Acquire Literacy?" *Wil-*

liam & Mary Quarterly 3d ser. 48 (1991): 50–67; W. J. Gilmore, *Reading Becomes a Necessity of Life: Cultural Life in Rural New England, 1780–1835* (1989), 42–50.

38. GPM to Horace Greeley, 7 May 1869, FC; GPM, "Education of Women," 166. Even anticlerical liberals kept Italian women in their place (Alice Kelikian, "Science, Gender, and Moral Ascendancy in Liberal Italy," *Journal of Modern Italian Studies* 1 [1995]: 377–89).

39. GPM to C. E. Norton, Apr. 1881; to Nash, 14 July 1876. See Collegio Ferretti (Protestant Orphanage for Girls), *Report of the Executive Committee for the Year 1882* (Florence, 1883); Emily Bliss Gould to GPM, 23 Jan., 8 & 30 Aug. 1869.

40. Cj, 23 Feb. 1864, 13:56–57.

41. GPM to CCM, 24 Jan. 1855; to Professor Haynes, 21 Mar. 1881; to Henry Yule, 4 Apr. 1879; to Caroline Estcourt, 28 Mar. 1851, in *LL*, 212.

42. GPM, "Study of Nature," 59–60; GPM to C. C. Andrews, 5 July 1875, Andrews Papers; GPM, "Aphorisms" (ca. 1875), MS; GPM, Jan. 1845, quoted in *LL*, 94n.

43. Davide Lévi, (pen-name "Julius") author of *Democrazia e papismo* (Milan, 1863), was highly simpatico with Marsh. Marsh letters to Lévi (4 July 1864; 19 Feb. 1869) are in the library of the Museo del Risorgimento in Turin. On Lévi's visits to the Marshes, Cj, 4:40, 8:55, 12:43 13:64–65, 14:60, 16:22. See also Elena Loewenthal, "Vita ebraica a Torino fra '800 e '900," *Studi Piemontese* 14:1 (1985): 117–23; Clara M. Lovett, *The Democratic Movement in Italy, 1830–1876* (Cambridge: Harvard University Press, 1982).

44. GPM, "Address [delivered before the American Colonization Society]" (1856), 13, 17; "England Old and New," 11; CCM to Maria Buell Hickok, 29 Nov. 1845; *LL*, 73n; GPM to Lieber, 27 Apr. 1860; E. Q. Putnam to Eugene Schuyler, 30 May 1880 (copy to GPM).

45. *M&N*, 46, 28–29; GPM, "Study of Nature," 46–47; William James, "On a Certain Blindness in Human Beings"; in his *Talks to Teachers on Psychology; and to Students on Some of Life's Ideals* (New York, 1899), 229–64 at 257. Marsh instanced Ruskin's contempt for the "humble-life" pictures of Gerrit Dou and Bartolomé Murillo.

46. *MMSM*, 191–93.

47. *M&N*, 58n9. This admonition is omitted from 1874 and later editions. The Vermont of Marsh's time had still been one of "servants," not of Dorothy Canfield Fisher's "hired hands" (*Vermont Tradition* [1953], 323–31).

48. GPM to William Tecumseh Sherman, 27 Feb. 1872; to SFB, 10 Apr. 1854, in *LL*, 357; to J. G. Saxe, 7 Feb. 1851.

49. GPM to C. E. Norton, Apr. 1881.

50. GPM, "Aphorisms", MS; *MMSM*, 184–86, 114 (handwritten addition), 204; "Legend, " *Johnson's New Universal Cyclopaedia* (1876), 2:1715. Marsh termed "the reading of flashy novels the principal agency in the intellectual degradation" of England and America (to Norton, Apr. 1881).

51. *MMSM*, 152.

52. GPM to George Bancroft, 10 Jan. 1868 (alluding to Alessandro Manzoni's famed 1821 ode "Il cinque maggio," on Napoleon as history's arbiter), Bancroft MSS; GPM to Robert C. Winthrop, 2 Jan. 1878 (CCM copy).

53. GPM to Nash, 9 Aug. 1879.

54. GPM to Nash, 5 June 1879; to Seward, 21 Mar. 1864, DD 87; *LEL* (1860), 8. Marsh shared Caroline's view of many Germans as "learned, arrogant and unprincipled" (Cj, 21 Dec. 1864, 16:65).

55. GPM to Nash, 5 June 1879; to R. H. Dana, Jr., 21 Dec. 1865, Dana MSS; to Nash, 26 Sept. 1874; GPM, *The Camel* (1856), 169.

56. GPM to Norton, 17 Aug. 1881, Norton Papers; GPM, *Human Knowledge* (1847), 33–34.

57. *LEL*, 313–14. The Yale classicist James Hadley's "On Formation of Indo-European Futures" ([1859], in his *Essays Philosophical and Critical* [New York, 1873], 184–98), the sole study of the topic at the time, ventured no such surmises. Noting that many languages lack future tenses, and that "Marsh's reasoning would be considered totally unfounded today, even by those of us who acknowledge the extent to which . . . language change is influenced by culture," a modern linguist rates Marsh's "an absurd suggestion, but a good try" (Suzanne Fleischman to the author, 24 Mar. 1999; see also her *The Future in Thought and Language: Diachronic Evidence from Romance* [Cambridge: Cambridge University Press, 1982], 22–24, 29–30, 45).

58. GPM to C. S. Sargent, 14 Feb. 1880.

59. *VHG*, 1 (1867), Preface, iii, also quoted in *Abby Hemenway's Vermont*, ed. Brenda C. Morrissey (Brattleboro: Stephen Greene, 1972), 5; GPM, *Human Knowledge*, 30–31.

60. Cj, 12 Jan. 1863, 7:83.

61. *LEL*, 449–50.

62. GPM to Harriet Preston, 12 Apr. 1882 (CCM copy); *M&N*, 379n123. Marsh here paraphrases Jacques Babinet's 1856 complaint against "Bernard de Palissy, who, 150 years ago, came and took away from me, a humble academician of the 19th century, this discovery."

63. GPM to Yule, 11 May 1868.

64. GPM to Sargent, 2 June 1879; GPM, "Thoughts and Aphorisms," 6; *Earth as Modified by Human Action* (1874), 546n. The literary thief was Tiberius Cornelius Winkler (*Zand en Duinen* [Dockum, 1865]).

65. *LEL,* 278n; GPM to Sargent, 16 May 1879. His friend's note, not in *Nature* but in the *Journal of Forestry,* is cited by Marsh in *Earth as Modified by Human Action* (1884 ed.), 83n. Marsh's inventive squirrels resemble Thoreau's squirrels as trimmers of pines and harvesters of cones and acorns (H. D. Thoreau, *Faith in a Seed* [1850-61/1993], 28–33, 128–37).

66. GPM to SFB, 2 July 1855, BSI. Marsh was amused by one seventeenth-century printer who, when "Master" and "Mister" were commonly conflated, had saved space by abbreviating "Abbot's master-piece" to his "Mr.-piece" (*LEL,* 431–32).

67. *LEL* 1885 ed., 525–26n. Marsh himself was guilty of retranslating from German a translation from English, Edward Daniel Clarke's 1801 depiction of Mediterranean mud flats deposited by the Nile (*M&N* 1864 ed. 433n; Clarke's original text in my 1965 ed., 367n105).

68. *LEL* 1885 ed., 490–91n, 211–12n; Jules Michelet, *Histoire de France* (Paris, 1835–67), 5 (1841): 158; GPM, "Italian Language and Literature" (1876), 1332, 1330; *LEL* 1885 ed., 22; 1861 ed., 40.

69. *LEL* 1861 ed., 80–83; 1885 ed., 84n, 67n.

70. GPM to Carrie M. Crane, 4 Oct. 1869 (CCM copy); *LEL* 1885 ed., 22.

71. *LEL* 1885 ed., 527n.

72. *O&H* (1862), 412–13.

73. GPM to Nash, 28 Mar. 1877, 9 Apr. 1879; *LEL* 1885 ed., 322n.

74. GPM to Henry Yule, 28 Apr. 1868.

75. *LEL* 1861 ed., 466n, 12–14.

76. *LEL* 1885 ed., 17n; *O&H,* 27–28; *M&N,* 89n74.

77. GPM to Yule, 16 July 1874. See also GPM, "Index, Concordance, Digest, Table of Contents," *Johnson's New Universal Cyclopaedia* (1876), 2:1136.

78. *LEL* 1861 ed., 219–20, 679.

79. GPM, "Study of Nature," 36–37; "England Old and New," 7; Charles Darwin, *On the Origin of Species* (1860), 424.

80. GPM to Sarah Wister, 7 Feb. 1879.

81. GPM to George Ticknor, 20 Feb. 1860.

82. *LEL,* 15–16; GPM, "Speech on the Tariff Question" (1846), 1013. On presumed American ingenuity, see R. M. Elson, *Guardians of Tradition: American Schoolbooks of the Nineteenth Century* (1964), 167–69; Arthur M. Schlesinger, "What Then Is the American, This New Man?" *AHR* 48 (1943): 225–44 at 230.

83. GPM to H. C. Lea, 19 Nov. 1876, UPa; GPM, *Human Knowledge*, 14. Typically, Marsh adds a "By the way" on a fifteenth-century precursor of uncanny talents, whom Crichton perhaps emulated; but because this nonpareil linguist, swordsman, musician, master of all arts, and theologian's "knowledge, of course, came from the Evil One," he was foretold to be carried off by devils at the age of 33 (GPM, "Crichton," *Johnson's New Universal Cyclopaedia* [1878], 4:1584). Marsh was unaware that this prodigy was Ferdinand of Cordova (ca. 1420–1480), an advocate of papal temporal rights and emissary from the Spanish court to Pope Alexander VI.

84. "Study of Nature," 44. Marsh's essay is a set of reflections on Humboldt's *Views of Nature* and on his life; see R. B. Stein, *John Ruskin and Aesthetic Thought in America* (1967), 159–63; L. D. Walls, *Seeing New Worlds* (1995), 78–107.

85. GPM, "Preliminary Notice" [to Reclus's *La Terre*] (ca. 1870), MS. See my "George Perkins Marsh on the Nature and Purpose of Geography," *Geographical Journal* 126 (1960): 413–17.

86. *M&N*, 15, 4–5; *O&H*, 27. Though the Marsh cousins were mutually inspired by German scholarship, James like George railed against "unintelligible" Teutonic metaphysicians (Walls, *Seeing New Worlds*, 31).

87. *M&N*, 225n164, 17n14 (1884 ed., 276n, 13–14n).

88. *M&N*, 225n164. See also GPM, *Camel*, 167–69.

89. GPM, "Physical Science in Italy" (1868), 420; A. E. Housman, "The Application of Thought to Textual Criticism" (1922), in his *Collected Poems and Selected Prose*, ed. Christopher Ricks (London: Penguin, 1988), 325–29.

90. GPM, "Study of Nature," 40–44, 57. See Walls, *Seeing New Worlds*, 83, 92 (on Schiller), 129.

91. GPM, *Human Knowledge*, 38; *Address Delivered before the Graduating Class of the U.S. Military Academy at West Point* (1860), 7.

18. Prospect: Reforming Nature

1. William F. Murison, "Prescient Prophet of Ill Consequences" [review of GPM, *M&N*], *Landscape Architecture* 56 (1965): 74–75. A more recent verdict is that *Man and Nature* "reads today like any classic from another time. It disconcertingly blends errors that have been exploded with insights that have endured" (W. B. Meyer, *Human Impact on the Earth* [1996], 5).

2. C. J. Glacken, *Traces on the Rhodian Shore* (1967); GPM, *M&N*, 465; "The Study of Nature" (1860), 34. On the nineteenth-century personification of

nature as female, see Gillian Beer, "'The Face of Nature': Anthropomorphic Elements in the Language of *The Origin of Species*" (1986), 230–33; Carolyn Merchant, *Ecological Revolutions: Nature, Gender, and Science in New England* (1989), 202–5.

3. Shirley A. Briggs, "Rachel Carson: Her Vision and Her Legacy," in G. J. Marco et al., eds., *Silent Spring Revisited* (Washington, D.C.: American Chemical Society, 1987), 3–11. The Carson/Marsh comparison is underscored by S. B. Sutton, *Charles Sprague Sargent and the Arnold Arboretum* (1970), 77–78.

4. *M&N*, 51n53.

5. Michael Williams, *Americans and Their Forests* (1989), 381. F. V. Hayden, "Report on Nebraska Territory," in *Report of the Commissioner of the General Land Office, 1867*, 152–205 at 155–60, misreads Marsh as a rainmaker. See A. D. Rodgers, *Bernhard Eduard Fernow* (1951), 148–52; H. N. Smith, "Rain Follows the Plough" (1946–47): 176–82; idem, *Virgin Land* (1950), 26–29, 128 ff.; Richard G. Lillard, *The Great Forest* (New York: Knopf, 1947), 258–60; R. H. Grove, *Green Imperialism* (1995), 470–71; D. J. Pisani, "Forests and Conservation, 1865–1890" (1985–86): 348–49; Nancy Langston, *Forest Dreams, Forest Nightmares* (1995), 144–48.

6. G. G. Whitney, *From Coastal Wilderness to Fruited Plain* (1994), 158, 264; Meyer, *Human Impact*, 5.

7. *M&N*, 20–28; GPM, MS reply to Daniel Draper, "Has Our Climate Changed?" *Popular Science Monthly* 1 (1872): 665–74; Yule to GPM, 15 Jan. 1879; GPM, *Irrigation: Its Evils, the Remedies, and the Compensations* (1873); James Watts to GPM, 21 May 1873; J. A. Kasson to GPM, 30 Mar. 1874; John A. Warder (American Nurserymen's Assoc.) to GPM, 13 Dec. 1877.

8. R. F. Nash, *Wilderness and the American Mind* (1967), 122–81; Stephen Fox, *The American Conservation Movement: John Muir and His Legacy* (1985), 139–47; Clayton R. Koppes, "Efficiency, Equity, Esthetics: Shifting Themes in American Conservation," in Donald Worster, ed., *The Ends of the Earth: Perspectives in Modern Environmental History* (New York: Cambridge University Press, 1988), 230–51; John Passmore, *Man's Responsibility for Nature* (London: Duckworth, 1974), 24. Richard N. L. Andrews, *Managing the Environment, Managing Ourselves: A History of American Environmental Policy* (New Haven: Yale University Press, 1999), 149–53, blames the Pinchot era's stress on utility, neglecting Marsh's "other environmental goals and values," for much present angst.

9. Richard Hofstadter, *Social Darwinism in American Thought*, rev. ed. (Boston: Beacon Press, 1955), 99–100; Holmes Rolston III, "Can and Ought We to Follow Nature?" *Environmental Ethics* 1 (1979): 7–30; William James, *The Will to*

Believe (New York, 1896), 43. For these and John Stuart Mill's analogous repro-
bation of nature, see my "Awareness of Human Impacts" (1990), 123.

10. S. P. Hays, *Conservation and the Gospel of Efficiency* (1959), 189–98, 271–
73; K. J. Gregory and D. E. Walling, eds., *Human Activity and Environmental
Processes* (Chichester: Wiley, 1987), 3; C. J. Glacken, "Changing Ideas of the
Habitable World" (1956), 81–88.

11. Sigmund Freud, *Civilization and Its Discontents* [1929], 3d ed., tr. Joan
Riviere (London: Hogarth Press, 1946), 53–54.

12. Paul Fejos and C. Warren Thornthwaite quoted, J. C. Bugher, W. A.
Albrecht, Michael Graham, and James Malin cited, in my "Awareness of Human Impacts," 124–25. The gloomy minority at "Man's Role" were Lewis
Mumford, Kenneth Boulding, F. F. Darling, Paul Sears, and above all Carl
Sauer, who lamented that "the worry of the earlier part of the century . . . that
we might not use our natural resources thriftily [had] given way to easy confidence in the capacities of technologic advance without limit" ("The Agency of
Man on Earth," in W. L. Thomas, Jr., *Man's Role in Changing the Face of the Earth*
[1956], 49–69 at 66; see Michael Williams, "Sauer and 'Man's Role in Changing the Face of the Earth,'" *Geographical Review* 77 [1987]: 218–31 at 230).

13. Frederic E. Clements, *Plant Communities: An Analysis of the Succession of
Vegetation* (Washington, D.C.: Carnegie Institution, 1916); Paul Sears, *Deserts
on the March* (Norman: University of Oklahoma Press, 1935); Donald Worster,
Nature's Economy: A History of Ecological Ideas (Cambridge: Cambridge University
Press, 1977), 209–42.

14. Kirkpatrick Sale, *The Conquest of Paradise: Christopher Columbus and the Co-
lumbian Legacy* (New York: Knopf, 1990), 68–69; Max Oelschlaeger, *The Idea of
Wilderness: From Prehistory to the Age of Ecology* (1991), 17.

15. Aldo Leopold, "The Land Ethic," in his *A Sand County Almanac and
Sketches Here and There* (New York: Oxford University Press, 1949), 201–26 at
224–26; Susan L. Flader, *Thinking Like a Mountain: Aldo Leopold and the Evolution of
an Ecological Attitude toward Deer, Wolves, and Forests* (Lincoln: University of Nebraska Press, 1978), 34–35, 270–71; Charles Birch and John E. Cobb, Jr., *The
Liberation of Life* (Cambridge: Cambridge University Press, 1981), 273–74, 282–
83; Neil Evernden, *The Natural Alien* (Toronto: University of Toronto Press,
1985), 15–16; Paul W. Taylor, *Respect for Nature: A Theory of Environmental Ethics*
(Princeton: Princeton University Press, 1986), 258.

16. Eugene P. Odum, *Fundamentals of Ecology* [1953], 3d ed. (Philadelphia:
Saunders, 1971; Odum's *Ecology: A Bridge between Science and Society* [Sunderland,
Mass.: Sinauer Associates, 1997], 299, 319, reiterates *Man and Nature's* seminal

role); Donald Worster, *The Wealth of Nature: Environmental History and the Ecological Imagination* (New York: Oxford University Press, 1993), 156–62; Bill McKibben, *The End of Nature* (Harmondsworth: Penguin, 1990); Michael G. Barbour, "Ecological Fragmentation in the Fifties," in William Cronon, ed., *Uncommon Ground: Toward Reinventing Nature* (New York: W. W. Norton, 1995), 233–55; Mark Sagoff, *The Economy of the Earth: Philosophy, Law, and the Environment* (Cambridge: Cambridge University Press, 1990); Frank Oldfield, "Man's Impact on the Environment: Some Recent Perspectives," *Geography* 68 (1983): 245–56 at 248. "The whole of popular ecology," concludes a British biogeographer, "is warped, dated, . . . suffused with a desire for 'stability' and 'safety'" (Philip Stott, "Biogeography and Ecology in Crisis: The Urgent Need for a New Metalanguage," *Journal of Biogeography* 25 [1998], 1–2).

17. Hays, *Conservation*, 347.

18. John S. Collis, *The Triumph of the Tree* (London: Cape, 1950), 246; P. G. Anson in *Landscape* 3 (1953–54): 2; E. G. D. Murray, "The Place of Nature in Man's World," *American Scientist* 42 (1954): 142.

19. W. L. Thomas, Jr., "Introductory," in his *Man's Role*, xxi–xxxviii at xxii.

20. Robert C. Mitchell, "Public Opinion and Environmental Politics in the 1970s and 1980s," in Norman Vig and M. E. Kraft, eds., *Environmental Policy in the 1980s* (Washington, D.C.: Congressional Quarterly Press, 1984), 51–74 at 56–58; idem, *Public Opinion on Environmental Issues* (Washington, D.C.: GPO, 1980), 3, 41; S. P. Hays, *Beauty, Health, and Permanence: Environmental Politics in the United States, 1955–1985* (1987), 32–33; Stephen Cotgrove, *Catastrophe or Cornucopia: The Environment, Politics and the Future* (New York: John Wiley, 1982), 31; Noel Malcolm, "Green Thoughts in a Blue Shade," *Spectator* (London), 7 Jan. 1989, 6.

21. *M&N*, 42–43.

22. Ghillean Prance, "Flora," in B. L. Turner et al., *The Earth as Transformed by Human Action* (1990), 387–91; P. M. Vitousek et al., "Human Domination of Earth's Ecosystems" (1997).

23. Only one of ten key issues listed by an environmental conference at MIT in 1970 was so cited in Sweden in 1982 (Václav Smil, *Energy, Food, Environment: Realities, Myths, Options* [Oxford: Clarendon Press, 1987], 208–9). And nuclear waste excepted, none of the key issues raised at Rio in 1992 had been emphasized at Stockholm just a decade earlier (Gilbert F. White, "Emerging Issues in Global Environmental Policy," *Ambio: A Journal of Human Environment* 25 [1996], 58–60).

24. The 1995 Madrid working group barely agreed to call human influence on global climate "discernable" rather than "appreciable," "notable," "mea-

surable," or "detectable" (Intergovernmental Panel on Climate Change, Working Group I, "Summary for Policy Makers," in J. T. Houghton et al., eds., *Climate Change 1995: The Science of Climate Change* [Cambridge: Cambridge University Press, 1996], 3–7 at 5; see also 413–17, 438–39).

25. Kai Erikson, *A New Species of Trouble: Explorations in Disaster, Trauma, and Community* (New York: Norton, 1994), 150–51.

26. Ulrich Beck, "Risk Society and the Provident State," in Scott Lash et al., *Risk, Environment and Modernity* (1996), 27–43 at 31; Erikson, *New Species of Trouble*, 203–25; Robert W. Fri, "Using Science Soundly: The Yucca Mountain Standard," *RFF [Resources for the Future] Review* (Summer 1995): 15–18, summarizing U.S. Natl. Research Council, *Technical Bases for Yucca Mountain Standards* (1995).

27. Martin J. Pasqualetti and K. David Pijawka, "*Un*siting Nuclear Power Plants," *Professional Geographer* 48 (1996): 57–69. On declining faith in science, see Nicholas Rescher, *Unpopular Essays on Scientific Progress* (Pittsburgh: University of Pittsburgh Press, 1980); Maurie J. Cohen, "Risk Society and Ecological Modernisation," *Futures* 29 (1997): 105–19 at 107–8, 111–13.

28. World Commission on Environment and Development, *Our Common Future* (Oxford: Oxford University Press, 1987), 342; Lowenthal, "Awareness of Human Impacts," 127.

29. *M&N*, 465.

30. Hays, *Beauty*, 172–74; Edith Efron, *The Apocalyptics: Politics, Science, and the Big Cancer Lie* (New York: Simon & Schuster, 1984).

31. David Ehrenfeld, *Beginning Again: People and Nature in the New Millennium* (New York: Oxford University Press, 1993), 177–79; on pigeonholing Marsh as a utilitarian, Larry Anderson, "Nothing Small in Nature," *Wilderness* (Summer 1990): 64–68 at 68.

32. From 1955 to 1987 the Science Citation Index had 413 references to Thoreau, 248 to Muir, 68 to Marsh (I have updated Robin MacDowell, "Thoreau in the Current Scientific Literature," *Thoreau Society Bull.* 143 [Spring 1982]: 2, & 172 [Summer 1985]: 3–4). Most of the citations are in hard-core science journals. Lawrence Buell, *The Environmental Imagination: Thoreau, Nature Writing . . .* (1995), 180, 363, 542n58, typifies this stance, whose advance in academe and in religion is charted in R. F. Nash, *The Rights of Nature* (1989), 87–160.

33. Cecelia Tichi, *New World, New Earth: Environmental Reform in American Literature from the Puritans through Whitman* (1979), 220; Buell, *Environmental Imagination*, 306–7. Another critic admits that "Marsh foresaw the twentieth century better than Thoreau," and then laments that "Marsh's vision, urbanity and ob-

servations . . . were not matched by a succinct and aphoristic style" (Edward Hoagland, "The Unknown Thoreau," *Nation* [7 June 1993]: 768–70).

34. H. D. Thoreau, "Walking" [1862], in his *Walden and Other Writings* (New York: Modern Library, 1937), 593–632 at 613; idem, *The Maine Woods* [1846] (London: Eyre & Spottiswoode, 1950), 277; John G. Blair and Augustus Trowbridge, "Thoreau on Katahdin," *American Quarterly* 12 (1960): 508–17; L. D. Walls, *Seeing New Worlds: Henry David Thoreau and Nineteenth-Century Natural Science* (1995), 112.

35. Thoreau, "Walking," 617–18, 602, 631–32; Leo Marx, "The Full Thoreau," *New York Review of Books*, 15 July 1999, 44–48.

36. Thoreau, *Journal*, ed. Bradford Torrey and F. H. Allen (Boston: Houghton Mifflin, 1906), 14:306–7, 3 Jan. 1861. Thoreau, like Marsh, is the victim of today's anachronistic dualism setting poetry, imagination, and intuition against science, hard facts, and objective reality. Thoreau lauded *un*scientific knowledge of nature, but his own scientific interests kept growing (Walls, *Seeing New Worlds*, 113–14, 132, 221; Robert D. Richardson, "Thoreau and Science," in Robert J. Scholnik, ed., *American Literature and Science* [Lexington: University of Kentucky Press, 1992], 110–25 at 119–20). Like Marsh, he "saw no polarity between poetry and nature's economy" (Gary Paul Nabhan, "Foreword," in H. D. Thoreau, *Faith in a Seed* [1993], xii). See also Philip and Kathryn Whitford, "Thoreau: Pioneer Ecologist and Conservationist," in Walter Harding, ed., *Thoreau, a Century of Criticism* (Dallas: Southern Methodist University Press, 1954), 192–205; E. A. Schofield and R. C. Baron, eds., *Thoreau's World and Ours* (Golden, Colo.: North American Press, 1993), 39–76, 172–80.

37. R. F. Nash, *Wilderness and the American Mind*, 131, 105; Frederick Turner, *John Muir: From Scotland to the Sierra* (New York: Viking, 1985), 97, 308, 310, 383–84. On Muir's early friendship and later break with Gifford Pinchot, see Char Miller, *Gifford Pinchot: The Evolution of an American Conservationist* (Milford, Pa.: Grey Towers Press, 1992).

38. Buell, *Environmental Imagination*, 477n53; Cj, 25 June 1864, 15:25.

39. References to Thoreau are in *M&N*, 267n235; 1867 printing, 552n5, 554n12, 558–59n35, 565n60, 566n65; all incorporated in *Earth as Modified by Human Action* (1874). Thoreau's *Excursions* (1863, including "Autumnal Tints") and *Maine Woods* are in Marsh's list of works consulted (*Earth as Modified*, 1877 printing, xxiii); *Man and Nature* also refers to *A Week on the Concord and Merrimack Rivers*. Marsh also owned Thoreau's *Walden* and *Cape Cod* (University of Vermont, *Catalogue of the Library of GPM* [1892], 670).

40. *M&N*, 61; GPM to Sargent, 16 May 1879.

41. Meyer, *Human Impact*, 6. Like several others, Meyer overstates the contrast of Thoreauvian wildness and Marshian domestication in my 1958 biography (p. 272).

42. *M&N*, 203–4, 249; *Earth as Modified*, 1884 ed., 320n. For Marsh's Adirondack role, see J. R. Ross, "Man over Nature: Origins of the Conservation Movement" (1975): 55; Karl Jacoby, "Class and Environmental History: Lessons from 'the War in the Adirondacks,'" *Environmental History* 2 (1997): 324–42.

43. Mumford, "Prospect" (1956), in Thomas, *Man's Role*, 1141–52 at 1150. The merging of romanticism and science, rather than their oft-assumed opposition, is stressed in Walls, *Seeing New Worlds*. That Frederick Law Olmsted "must have been familiar with" *Man and Nature* seems certain to Anne Whiston Spirn ("The Authority of Nature: Conflict and Confusion in Landscape Architecture," in *Nature and Ideology: Natural Garden Design in the Twentieth Century*, ed. Joachim Wolschke-Bulmahn [Washington, D.C.: Dumbarton Oaks, 1997], 249–61 at 259n45). Spirn sees Olmsted's The Fens and the Riverway in Boston (1881) as landscapes embodying Marsh's idea of becoming "a co-worker with nature in the reconstruction of the damaged fabric" of lands laid waste by human improvidence or malice (*M&N*, 35; see p. 298 above).

44. The belletrist Herbert Quick in 1925 referred to Marsh's "forgotten poems, some equally forgotten scientific books" (quoted in J. R. Whitaker, "World View of Destruction and Conservation of Natural Resources" [1940], 161n30); Carl Sauer called Marsh a "forgotten scientist" ("Theme of Plant and Animal Destruction in Economic History" [1938], in his *Land and Life* [1963], 145–54 at 147–48); Lewis Mumford termed *Man and Nature* "quite forgotten even by geographers until I resurrected it in *The Brown Decades* [1931]" (to Babette Deutsch, 30 Oct. 1960, in his *My Works and Days* [New York: Harcourt Brace Jovanovich, 1979], 177). See also Milburn McCarty, "Forgotten Dartmouth Men: Greatest Scholar—George Perkins Marsh," *Dartmouth Alumni Mag.* 28:4 (1936), 18, 68; Robert J. Wheel, "The Forgotten Vermonter: George Perkins Marsh" (M.A. thesis, St. Michael's College, Winooski, Vt., 1955).

45. Fox, *American Conservation Movement*, 109. Typical is Stephen Jay Gould's put-down: "The conservation movement was born, in large part, as an elitist attempt by wealthy social leaders to preserve wildness as a domain for patrician leisure and contemplation" (S. J. Gould, *Eight Little Piggies: Reflections on Natural History* [London: Cape, 1993], 50). Today's crusaders for environmental justice cast yesterday's crusades as elitist in origin, leadership, and interests, as plots against the poor, and as Western ploys to keep the Third World underdeveloped (Ramachandra Guha and Juan Martinez-Alier, *Varieties of Environmentalism*

[1997], xvii–xviii, 3–21, 31–45). However, the mainstream continues to honor Marsh. His "legacy runs deep" in the American Society for Environmental History, whose founders termed him the "primal environmental historian" (James E. Sherow to the author, 31 Mar. 1997).

46. R. W. Judd, *Common Lands, Common People* (1997), 2–3, 8, 49–56, 73–77, 84–89, 90–97, 147, 265–66. Among supportive data kindly sent me by Richard Judd (13 Nov. 1997), the sole "ecological" item predating Marsh is an 1851 essay lauding hillside forests "for binding [soil] to the earth, preserving it from wearing and washing away under heavy rains and snows" (Edmund H. Bennett, "Ornamental and Forest Trees," *Transactions . . . Agric. Soc. Commonwealth Massachusetts . . . 1851*, 380).

47. Wallace Stegner, "It All Began with Conservation," *Smithsonian* (Apr. 1990): 35–43 at 39.

48. Grove, *Green Imperialism*, 486, 475, 470–71.

49. Ibid., 430–31, 445, citing Donald Butter, *Topography of Oudh* (1839), and Edward Green Balfour, "Notes on the Influence Exercised by Trees in Inducing Rain and Preserving Moisture" (1849). Pierre Poivre in Mauritius went no further than to say cleared lands would be burnt and dessicated by the sun (Grove, *Green Imperialism*, 185–88; see also 2, 10–15, 63, 204, 252–53, 259–62, 428, 470–72). "A handful of British, Dutch and French scientists" whose environmental concerns were at most marginal "is a slender basis" for Grove's claim (Peter Coates, *Nature: Western Attitudes since Ancient Times* [Cambridge: Polity Press, 1998], 104).

50. Marsh's influence vis-à-vis John Croumbie Brown is later acknowledged by Grove ("Scotland in South Africa: John Croumbie Brown and the Roots of Settler Environmentalism," in Tom Griffiths and Libby Robin, *Ecology and Empire* [1997], 139–53 at 147). Marsh encouraged Brown to publish his *Hydrology in South Africa* (1875). Brown's compilation, *Forests and Moisture* (1877), is laden with extracts (pp. 21–23, 45–46, 57, 64, 74–75, 88–92, 106–17, 125, 144–48, 160–61, 167–72, 216–17, 223–35, 283–85) from *Man and Nature*.

51. John MacKenzie, "Empire and the Ecological Apocalypse" (1997), in Griffiths and Robin, *Ecology and Empire*, 215–28 at 221. Marsh's "mistaken" primacy is likewise ascribed to "Yankee imperialism" in William Beinart and Peter Coates, *Environment and History: The Taming of Nature in the USA and South Africa* [London: Routledge, 1995], 2, 44–45.

52. Langston, *Forest Dreams, Forest Nightmares*, 104–9; S. J. Pyne, *Vestal Fire: An Environmental History* (1997), 195; James C. Scott, *Seeing Like a State: How Cer-*

tain Schemes to Improve the Human Condition Have Failed (New Haven: Yale University Press, 1998), 11–21.

53. T. L. Whited, *The Struggle for the Forest in the French Alps and Pyrenees, 1860–1940* (1994), 88, 102–4, 112–17. No mention of Marsh, however, has been found in French forestry sources (Tamara Whited to the author, 10 June 1996; Jean-Pierre Feuvrier to the author, 16 Sept. 1996).

54. GPM, *Address . . . Agric. Soc. Rutland* (1847), 18.

55. *M&N*, 381. To be sure, corporate interests exaggerate reform costs; while "most environmental initiatives of the past seemed expensive and questionable at the time, . . . every one of them appears a bargain in retrospect" (Gregg Easterbrook, *A Moment on Earth: The Coming Age of Environmental Optimism* [New York: Penguin, 1996], 175, 210).

56. A. W. Crosby, "The Past and Present of Environmental History" (1995): 1189; Murison, "Prescient Prophet of Ill Consequences," 75.

57. James C. Malin, *The Contriving Brain and the Skillful Hand in the United States* (Lawrence, Kans.: privately printed, 1955). "That eventually all ecosystems will have to be managed" is becoming received wisdom (Richard Gallagher and Betsy Carpenter, "Human-Dominated Ecosystems," *Science* 277 [1997]: 485).

58. *M&N*, 36.

59. Meyer, *Human Impact*, 5–6; Worster, *Wealth of Nature*, 170; *M&N*, 91–92.

60. Yi-Fu Tuan, *Escapism* (Baltimore: Johns Hopkins University Press, 1998), 6. Espousing this "pathetic fallacy," Harold Nicolson "could scarcely believe that [his] swans were being sincere in their indifference" to the outbreak of the Second World War (quoted in *Escapism*, 87). It remains unclear who should speak on behalf of voiceless nature.

61. Emile Durkheim, *The Elementary Forms of Religious Life* [1912], tr. Karen E. Fields (New York: Free Press, 1995), 213–14, 351–52, 372, 379.

62. *M&N*, 201–2, 36.

63. W. M. Davis, "Biographical Memoir of George Perkins Marsh" (1909): 80.

64. GPM, "Physical Science in Italy" (1868), 420.

65. Brian Wynne, "May the Sheep Safely Graze? A Reflexive View of the Expert–Lay Knowledge Divide" (1996), in Lash et al., *Risk, Environment and Modernity*, 44–83; Rolf Lidskog, "Scientific Evidence or Lay People's Experience? On Risk and Trust with Regard to Modern Environmental Threats," in Maurie J. Cohen, ed., *Risk in the Modern Age* (London: Macmillan, 2000), 196–224.

66. Marsh as construed by Langston, *Forest Dreams, Forest Nightmares*, 102.

67. *M&N*, 44–45; *NES* (1845), 12–13.

68. Thomas Carlyle, "The Hero as Divinity" (1840), in his *On Heroes, Hero-Worship, & the Heroic in History* (Berkeley: University of California Press, 1997), 3–36 at 26. Other Carlyle texts come closer to Marsh's position that all human lives are historically consequential.

69. GPM to Harriet Preston, 12 Apr. 1882.

70. Peter Wild, "Book Find" [review of GPM, *M&N*], *Orion Nature Quarterly* 1:1 (1982): 54–55.

BIBLIOGRAPHY

THE MOST IMPORTANT SOURCE FOR ANY STUDY OF Marsh's life is the Marsh manuscript collection in the Bailey/Howe Library of the University of Vermont. The bulk of this collection, tens of thousands of letters to and from Marsh, along with diaries, notebooks, and drafts of speeches, documents, and books, was given to the University of Vermont between 1946 and 1950 by Caroline Crane Marsh's grandniece, the late Caroline E. Crane, of Scarsdale, New York. Of material still in Miss Crane's and her sister Aurelia Crane's possession in 1950, Caroline Marsh's Italian diaries are of special value; these with other letters and papers were given to me to be added to the University of Vermont collection. Marsh's diplomatic dispatches in the National Archives are vital both for his diplomatic career in Turkey and in Italy, and for his views on politics, religion, science, and culture.

Other major manuscript holdings are Marsh letters at the Vermont Historical Society, the Dartmouth College Archives, the Spencer F. Baird correspondence at the Smithsonian Institution, and the Wislizenus Papers (now joined to the UVM Marsh Papers) loaned to me by Marsh's grandnephew, the late Dr. Marsh Pitzman of St. Louis, the Hiram Powers Papers at the Archives of American Art, and the Crane Family Papers at the New York Public Library. Data in Italy, manuscript and published, throw much light on Marsh's last two decades there, including material bearing on his *Man and Nature*.

Of published materials, Marsh's own extensive works and his great library, bought and given to the University of Vermont by Frederick Billings, and catalogued by H. L. Koopman, are indispensable. Caroline Crane Marsh's *Life and Letters* (1888) of her husband, which ill-health prevented her from carrying beyond 1861, incorporates valuable insights as well as extracts from many letters no longer to be found. Simultaneously with my own earlier biography appeared Sister Mary Philip Trauth's *Italo-American Diplomatic Relations, 1861–1882* (1958), a study of Marsh's mission to Italy, and in 1982 Jane and Will Curtis and Frank Lieberman's *World of George Perkins Marsh*, a popular pictorial sketch.

Manuscript Collections and Archives

C. C. Andrews Papers. Minnesota Historical Society.

Spencer F. Baird Correspondence. Smithsonian Institution.

George Bancroft MSS. Massachusetts Historical Society.

Park Benjamin Collection. Columbia University, Rare Books and MSS Library.

Burlington, Vermont. Town Records.

———. Grand List, 1825–52 [destroyed, order of City Clerk].

———. Highway Rate Bill for District Number Two, 1831. UVM.

———. Land List and Real Estate Assessment, 1837.

———. Land Records (Real Estate Records), 1831–63.

Chittenden County (Vt.). Court Records, 1825–58. Formerly Burlington; transferred to Vt. Public Records Div., Dept. of Admin., Montpelier.

———. Probate Office. Wills of Ozias Buell, Wyllys Lyman, and George Perkins Marsh.

John M. Clayton Papers. Library of Congress.

Jacob Collamer Papers. Formerly Woodstock, Vt.; now UVM (*VH*, 57 [1989], 243–44).

Columbia College, Minutes of the Trustees, vol. 5. Columbia University Archives.

Crane Family Papers. New York Public Library.

R. H. Dana, Jr., MSS. Massachusetts Historical Society.

Dartmouth College Archives.

Charles Darwin Papers. Cambridge University Library.

Charles Stewart Daveis Correspondence. Columbia University Rare Books and MSS Library.

C. S. Daveis Papers. Library of Congress.

Edward Everett Papers. Massachusetts Historical Society.

BIBLIOGRAPHY

C. C. Felton MSS. Harvard College Archives.

Hamilton Fish Correspondence. Library of Congress.

Willard Fiske Papers. Columbia University Library.

Charles Folsom MSS. Boston Public Library.

Ford Collection. New York Public Library.

Albert Gallatin Papers. New York Historical Society.

Galloway–Maxcy–Markoe Papers. Library of Congress.

E. L. Godkin Papers. Harvard University Library.

Asa Gray Collection. Harvard University Herbarium.

Harper & Brothers Archives, New York.

Hartford (Vt.), District of. Court of Probate. Will of Charles Marsh.

Rush C. Hawkins Papers. Annmary Brown Memorial Library, Brown University.

Isaac Hayes Papers. American Philosophical Society.

Joseph Henry Papers. SI.

Franklin B. Hough Papers. New York State Library, Albany.

Franklin B. Hough Papers. U.S. National Agricultural Library.

C. C. Jewett Papers. Boston Public Library.

Miner Kilbourne Kellogg Papers. Archives of American Art, SI.

Knollenberg Collection. Yale University Library.

Library of Congress, Letter Book of the Librarian of Congress, 1843–49.

Francis Lieber Correspondence. Library of Congress.

Francis Lieber MSS. Henry E. Huntington Library, San Marino, California.

James Russell Lowell Papers. Harvard University Library.

Marsh MSS. Vermont Historical Society.

Marsh Papers. Dartmouth College Archives.

Caroline Crane Marsh Journals, Turin, Italy [cited as Cj]. 17 notebooks, 1861–1865, UVM. [An emended typescript made for CCM in the 1880s is in the Crane Family Papers at the New York Public Library.]

————. "Last Days of George P. Marsh," UVM.

George P. Marsh Collection. UVM.

S. F. B. Morse MSS. Library of Congress.

Norcross MSS. Massachusetts Historical Society.

Charles Eliot Norton Papers. Harvard University Library.

Karen Perez Collection. UVM [letters to and from Marsh's niece Maria Buell Hungerford, donated August 1998].

Elisha Perkins letter and account book, 1788–1816. Yale University, Harvey Cushing/John Hay Whitney Medical Library.

J. S. Pike Papers. Calais Free Library, Calais, Maine.

Hiram Powers MSS. New York Historical Society.

Hiram Powers Papers. Archives of American Art, SI.

Ricasoli Archives. Brolio, Tuscany.

W. H. Seward Papers. University of Rochester Library, Manuscripts Collection.

W. T. Sherman Papers. Library of Congress.

Smithsonian Institution Archives, Correspondence Received.

E. G. Squier Collection. Library of Congress.

Charles Sumner Papers. Harvard University Library.

Benjamin Tappan, Senate Journal. Library of Congress.

United States, Department of State. National Archives. See "Other Sources," below, for diplomatic and other MSS.

University of Vermont Records. Trustees Minutes, vol. 3 (1829–65).

Vermont State Archives. Office of the Secretary of State, Montpelier.

Vermont State Papers. In Vermont State Archives.

Windsor County (Vt.). Court Records, 1825. Woodstock.

Robert C. Winthrop Papers. Massachusetts Historical Society.

Wislizenus Papers. In Marsh Collection, UVM.

Woodstock (Vt.). Land Records, vols. 1–2, 1793–96.

John K. Wright Papers. Dartmouth College Archives.

Newspapers (only long runs cited)

Burlington, Vt. *Free Press* (after 1848, *Daily Free Press*; after 1868, *Daily Free Press & Times*), 1827–85.

    Sentinel (after 1829–30, *Northern Sentinel*; after 1844, *Sentinel and Democrat*), 1825–50.

Montpelier, Vt. *Watchman & State Gazette* (after 1837, *Watchman & State Journal*), 1835–44.

New York *Times*, 1861–82.

    Tribune (weekly, semiweekly, daily), 1850–85.

Woodstock, Vt. *Observer*, 1820–25.

Published Material by George Perkins Marsh

Address Delivered before the Agricultural Society of Rutland County, Sept. 30, 1847. Rutland, Vt., 1848.

"Address [delivered before the American Colonization Society," Jan. 15, 1856], *Thirty-ninth Annual Report* (1856), 10–17. [Also in *African Repository* 32 (1856): 40–47.]

BIBLIOGRAPHY

Address Delivered before the Burlington Mechanics Institute, April 5, 1843. BFP extra, June 2, 1843.

Address Delivered before the Graduating Class of the U.S. Military Academy at West Point, June, 1860. New York, 1860.

Address, Delivered before the New England Society of the City of New-York, Dec. 24, 1844 [cited as *NES*]. New York, 1845. Greek epigraph before title, "Iomen eis athenas" (Let Us Go to Athens).

"Agriculture in Italy," *Nation* 2 (1866): 183–84.

"American Diplomacy," N.Y. *World*, June 30, 1860.

"American Heraldry," N.Y. *World*, July 2, 1860.

The American Historical School: A Discourse Delivered before the Literary Societies of Union College. Troy, N.Y., 1847.

"The American Image Abroad," *BDFP*, Sept. 21, 1854.

"American Representatives Abroad," N.Y. *World*, June 21, 1860.

An Apology for the Study of English; Delivered on Monday, Nov. 1, 1858; Introductory to a Series of Lectures in the Post-Graduate Course of Columbia College, New York. New York, 1859. Also in *Inaugural Addresses of Theodore W. Dwight, and of George P. Marsh, in Columbia College, New York*. New York, 1859, 57–93. [Rev. as ch. 1 of *Lectures on the English Language* (1860), q.v.]

"The Aqueducts of Ancient Rome," *Nation* 32 (1881): 147–48. [Review of Rodolfo Lanciani, *Topografia di Roma antica; I commentarii di Frontino intorno le acque e gli aquedotti* (1880).]

"Biographical Sketch of the Author [J. G. Biernatzki]," in Caroline C. Marsh, *The Hallig: or, The Sheepfold in the Waters* (1856), q.v., 17–24.

"The Biography of a Word," *Nation* 32 (1881): 88–89.

"Boccardo's Dictionary of Political Economy," *Nation* 22 (1876): 65–66.

"The Book of Marco Polo" [review of Henry Yule's *Ser Marco Polo*], *Nation* 21 (1875): 135–37, 152–53.

"The Camel," in *Ninth Annual Report of the Smithsonian Institution for 1854*, 98–122. 33 Cong. 2 sess., Misc. Doc. 24. Washington, 1855.

The Camel; His Organization, Habits and Uses, Considered with Reference to His Introduction into the United States. Boston, 1856.

"The Catholic Church and Modern Civilization," *Nation* 5 (1867): 229–31.

"The 'Catholic Party' of Cesare Cantù and American Slavery," *Nation* 2 (1866): 564–65.

"A Cheap and Easy Way to Fame," *Nation* 1 (1865): 778.

A Compendious Grammar of the Old-Northern or Icelandic Language: Compiled and Translated from the Grammars of [Rasmus Christian] Rask. Burlington, Vt., 1838.

"Cutting Metals with a 'Burr,'" letter to N.Y. *Tribune*, April 27, 1881.

"Decision of Arbitration . . . concerning the definite fixing of the Italian-Swiss frontier at the place called: Alpe de Cravairola, pronounced by the Umpire, Sept. 23, 1874," in *History and Digest of the International Arbitrations to which the United States Has Been a Party*, ed. John Bassett Moore, 6 vols. (Washington, D.C., 1898), 2:2027–49. [In French in Guggenheim, "Fixation de la frontière," q.v., 497–507.]

"The Desert: I. The Ship of the Desert; or, A Discourse of Camels, and Herein of Their Furniture, Diet, and Drivers," *American Whig Review* 16 (1852): 39–51. [Part II unpublished.]

A Dictionary of English Etymology, by Hensleigh Wedgwood. Vol. 1 (A–D) [London, 1859]; *with notes and additions, by GPM.* New York, 1862.

[Diplomatic Correspondence as Minister to Italy, 1861–82], in U.S. Dept. of State. Papers Relating to Foreign Affairs, 1861–83. [1861–66 dispatches reprinted in Howard R. Marraro, ed., *L'Unificazione Italiano vista dai diplomatici statunitensi*, vol. 4. Istituto per la Storia del Risorgimento Italiano, Bibl. scientia, *Fonti*, vol. 62, Rome, 1971.]

[Diplomatic Correspondence as Minister to Turkey, 1849–53], in part in the following publications, q.v.: United States, *Francis Dainese; [Jonas King]; Kossuth and Captain Long; Martin Koszta.*

"Dry Wines," *American Register* (Paris), 1 Sept. 1877, p. 4.

The Earth as Modified by Human Action. 1874, 1884. See *Man and Nature.*

"The Education of Women," *Nation* 3 (1866): 165–66.

"The Excommunication of Noxious Animals by the Catholic Church," *Nation* 2 (1866): 763–64.

"Female Education in Italy" [review of Anna Maria Mozzoni, *Un Passo avanti nella Cultura Femminile* (Milan, 1866)], *Nation* 3 (1866): 5–7.

"The Future of Italy," N.Y. *World*, July 9, 1860.

"The Future of Turkey," *Christian Examiner* 65 (1858): 401–19.

"A Glossary of Later and Byzantine Greek, by E. A. Sophocles" [review], N.Y. *World*, June 28, 1860.

The Goths in New-England; A Discourse Delivered at the Anniversary of the Philomathesian Society of Middlebury College, Aug. 15, 1843. Middlebury, Vt., 1843.

"The Grammar of English Grammars, by Goold Brown.—A Treatise on the English Language, by Simon Kerl" [review], N.Y. *World*, June 14, 1860.

Human Knowledge: A Discourse Delivered before the Massachusetts Alpha of the Phi Beta Kappa Society, at Cambridge, Aug. 26, 1847. Boston, 1847. [In *LL*, 432–52.]

Irrigation: Its Evils, the Remedies, and the Compensations. (Rome, July 24, 1873.) 43 Cong. 1 sess., Sen. Misc. Doc. 55. Washington, Feb. 10, 1874.

"The Italian Cause and Its Sympathizers." N.Y. *World*, July 7, 1860.

"Italian Language and Literature," in *Johnson's New Universal Cyclopaedia*, q.v., 2 (1876), 1330–36.

"Italian Nationality," N.Y. *World*, June 14, 1860.

"The Italian Question" (abstract of lecture, 23 Mar. 1860), *BDFP*, March 26, 1860.

Johnson's New Universal Cyclopaedia . . . 4 vols. New York, 1874–78.

> The following articles: Amat (Felix de Torres), 4:1560; Castanheda, de (Fernão Lopez), 4:1573; Catalan Language and Literature, 4:1573–74; Crichton (James), 4:1584; D'Esclot . . . (Bernat), 4:1591–92; Fireproof Construction in Italy, 2:113–14; Fréjus, Col de, Tunnel of, or Tunnel of Mont Cenis, 2:331–32; Fresco, or Fresco-Painting, 2:339; Fucino, or Celano, Lake and Tunnel of, 2:355–56; Genoa, 2:471; Girgenti, 2:556; Improvisation, 2:1129–30; Index, Concordance, Digest, Table of Contents, 2:1135–36; Index Librorum Prohibitorum, 2:1136–37; Inundations and Floods of Rivers, 2:1273–75; Irrigation, 2:1311–13; Italian Language and Literature, 2:1330–36; Jacme (Jayme or Jaume) En I., 2:1356; Legend, 2:1714–15; Lexicon, Dictionary, Thesaurus, Vocabulary, Glossary, 2:1751–52; Lombardini (Elia), 3:98; Lopes, or Lopez (Fernão), 3:112; Lull (Ramon), 3:147; March (Ausias), 3:294; Mulberry, 3:657–58; Muntaner Ramon, 3:668; Olive, 3:946–47; Po, 3:1298–99; Pontine Marshes, 3:1332; Romansch . . . , 3:1704–5; St. Gothard, Tunnel of, 4:18–19; Sicilian Vespers, 4:265; Sicilies, The Two, 4:265–67; Sicily, Island of, 4:267–69; Straw, Manufacture of, 4:591–92; Tiber, 4:853–54; Velvet, 4:1712–13; Watershed, 4:1299–1300; Well, 4:1345–46.

"The Late General Estcourt," *National Intelligencer* (Washington, D.C.), Feb. 7, 1856. [In *LL*, 474–79.]

Lectures on the English Language. First Series. New York, 1860. "Revised and enlarged" eds., New York and London, 1861 [cited as *LEL*], 1864, 1872 (all identically paginated). 1885 rev. ed., New York. [For ch. 1, see "An Apology . . ." above.]

Lectures on the English Language. Edited, truncated, and expanded by William George Smith as *The Students' Manual of the English Language.* London: John

Murray, 1862; later editions to 1880. [This work replaced Marsh's first two chapters with two of Smith's own on the "origin, affinities and constituent elements" of English. Reprinted as vol. 6 of *Origins of American Linguistics, 1643–1914.* Edited by Ray Harris. 13 vols. London: Routledge/Thoemmes Press, 1997.]

Man and Nature; or, Physical Geography as Modified by Human Action. New York and London, 1864 (reprint with new Appendix, 1867). [Reprint with introduction, annotations, and many later additions, ed. David Lowenthal. Cambridge, Mass.: Harvard University Press, 1965. Cited as *M&N.*]

    The Earth as Modified by Human Action; a New Edition of Man and Nature. New York and London, 1874 (reprint with new Appendix, 1877); reprint, St. Clair Shores, Mich.: Scholarly Press, 1970; New York: Arno, 1970; and North Stratford, N.H.: Ayer, 1976.

    The Earth as Modified by Human Action; a Last Revision of Man and Nature. New York, 1884.

    Cheloviek i priroda, ili, o vliianii chelovieka na fiziko-geograficheskikh uslovi i prirody. Translated by N. A. Neviedomskii. St. Petersburg: Izd. N. Poliakova i Ko, 1866.

    L'Uomo e la natura; ossia, La superficie terrestre modificata per opera dell'uomo. Florence, 1872. [Reprint with introduction, ed. Fabienne O. Vallino. Milan: Franco Angeli, 1988.]

"Marsh, James," in George Ripley and C. A. Dana, eds., *New American Cyclopaedia,* New York, 1859–61, 9:216–17.

Mediaeval and Modern Saints and Miracles [cited as *MMSM*]. [Pen name used: Not Ab Uno e Societate Jesu.] New York, 1876. [Reset reprint (New York: Harper & Row, 1969) with some but not all of Marsh's corrections and additions, and with his name as author on the title page. The reprint carries no introduction and omits the date of the original.]

Memorial of George P. Marsh, of Vermont, Asking an Appropriation for the Compensation of His Services as Minister Resident to the Ottoman Porte, under the Act of August 11, 1848, imposing judicial duties on the Minister, and of his services under a special mission to the government of Greece. Dec. 1, 1854. 33 Cong. 2 sess., Sen. Misc. Doc. 8 (with GPM to Solomon Foot, Jan. 3, 1855). Washington, 1855.

"Milton's and Shakespere's Vocabularies," *Nation* 20 (1875): 274–75 [letter from Rome, March 23, 1875].

"Monumental Honors," *Nation* 1 (1865): 491–92.

New Dictionary by the Philological Society of London. [Notice] Burlington, Aug. 8, 1859.

"Notes on Mr. Hensleigh Wedgwood's Dictionary of English Etymology, and on Some Words Not Discussed by Him," *Transactions of the Philological Society* [London] (1865), 187–206; "Postscript to Notes . . . ," 307–12. Also issued as a separate pamphlet by the Philological Society.

"Notes on the New Edition of Webster's Dictionary," *Nation* 3 (1866): 125–27, 147–48, 186–87, 225–26, 268–69, 288–89, 369, 408–9, 515–17; 4 (1867): 7–9, 108–10, 127–28, 312–13, 373, 392–93, 516–17; 5 (1867): 7–8, 88–89, 208–9.

"Notes on Vesuvius, and Miscellaneous Observations on Egypt (from Letters to the Smithsonian Institution from a Traveller in the East)," *American Journal of Science and Arts* 2d ser. 13 (1852): 131–34.

"Old English Literature," *Nation* 1 (1865): 778.

"Old Northern Literature," *American Whig Review* 1 (1845): 250–57.

"Oration" [before the New Hampshire State Agricultural Society, Oct. 10, 1856], in N.H. State Agric. Soc. *Transactions . . . 1856* (Concord, N.H., 1857), 35–89.

"Oriental Christianity and Islamism," *Christian Examiner* 65 (1858): 95–125.

"The Oriental Question," *Christian Examiner* 64 (1858): 393–420.

"The Origin and History of the Danish Sound and Belt-Tolls," *Hunt's Merchants' Mag.* 10 (1844): 218–32, 303–8. Translated from Johan Friderich Wilhelm Schlegel, *Danmark's og Hertugdommenes Statsret* (1827), ch. 6.

The Origin and History of the English Language, and of the Early Literature It Embodies [cited as *O&H*]. New York, 1862; rev. eds., 1871, 1885, 1892 (all eds. have same pagination); reprint of 1871 ed., Ann Arbor: Gryphon Books, 1971.

"The Origin of the Italian Language," *North American Review* 105 (1867): 1–41.

"The Origin, Progress, and Decline of Icelandic Historical Literature, by Peter Erasmus Mueller . . . Translated, with Notes . . . ," *American Eclectic* 1 (1841): 446–68; 2 (1841): 131–46.

"Our English Dictionaries," *Christian Review* 101 (1860): 384–415.

"[Petition, Dec. 12, 1837] To His Excellency the Governor of Vermont" [with 22 others], *Vermont Chronicle*, Dec. 20, 1837. [Against Quebecois insurgency.]

"Physical Science in Italy" [review of Gerolamo Boccardo's *Fisica del globo*], *Nation* 7 (1868): 420–21.

"The Principles and Tendencies of Modern Commerce; with Special Reference to the Character and Influence of the Traffic between the Christian States and the Oriental World," *Hunt's Merchants' Mag.* 33 (1855): 147–68.

"The Proposed Revision of the English Bible," *Nation* 11 (1870): 238–39, 261–63, 281–82.

"Pruning Forest Trees," *Nation* 1 (1865): 690–91.

Railroad Commissioner of the State of Vermont. *Third Annual Report to the General Assembly, 1858.* Burlington, 1858.

———. *Special Report, Oct. Sess., 1858.* Burlington, 1858.

———. *Fourth Annual Report, 1859.* Burlington, 1859.

Remarks on Slavery in the Territories of New Mexico, California and Oregon; Delivered in the House of Representatives, Aug. 3d, 1848 (CG, 30 Cong. 1 sess., App. 1072–76). Burlington, Vt., 1848.

Reply to Mr. Brodhead's Remarks in the Senate on the Bill for the Relief of George P. Marsh, as Reported in the [*Congressional*] *Globe* of April 26, 1856. Washington, May 1, 1856.

"Report of a Committee under the Act Providing for the Erection of a Monument over the Grave of Ethan Allen" [with John N. Pomeroy], *Vermont Senate Journal* 1858, app., 319–22.

"Report of the Commissioners on the Plan of the State House" [with Norman Williams and John Porter], April 1, 1857, *Journal of the House of Representatives of the State of Vt.,* Oct. Sess. 1857. Montpelier, 1857, App., 469–71.

Report on Durazzo Library, June 7, 1844. 28 Cong. 1 sess., H. Rep. 553. [Purchase inexpedient for Library of Congress.]

Report [on Memorial of] James A. Stevens. May 24, 1844. 28 Cong. 1 sess., H. Rep. 510. [On water resistance experiments.]

Report [on Memorial of] Joshua Dodge. April 26, 1848. 30 Cong. 1 sess., H. Rep. 589. [On services as U.S. tobacco agent in Germany and Italy.]

Report [on Memorial of] P. J. Farnham and Jed Frye. March 3, 1849. 30 Cong. 2 sess., H. Rep. 142. [Demands reparation for British seizure of bark *Jones* at St. Helena.]

Report [on Memorial of the National Institute for the Promotion of Science]. June 7, 1844. 28 Cong. 1 sess., H. Rep. 539. [To nationalize its holdings].

Report [on Petition of E. H.] Holmes and [W.] Pedrick. March 29, 1844. 28 Cong. 1 sess., H. Rep. 389. [On compensation for rawhide cutting machine.]

Report [on Petition of] Elisha H. Holmes. [1], Jan. 30, 1846. 29 Cong. 1 sess., H. Rep. 160. [2], Feb. 16, 1847. 29 Cong. 2 sess., H. Rep. 62. [On dredging machine patent.]

Report on Spirit Ration in Navy. Jan. 28, 1845. 28 Cong. 2 sess., H. Rep. 73.

Report, on the Artificial Propagation of Fish. Made under the Authority of the Legislature of Vt. Burlington, 1857.

[Report on the Education of the Deaf and Dumb], in Vermont, *Journal of the General Assembly, 1824,* 23–30.

"'The River', from the Swedish of Tegnér" [GPM, tr.], *American Whig Review* 2 (1845): 357. Revised version in Hemenway, *Poets and Poetry of Vermont* (1860), q.v., 338–39.

Speech on the Bill for Establishing the Smithsonian Institution; Delivered in the House of Representatives of the U. States, April 22, 1846 (CG, 29 Cong. 1 Sess., App., pp. 850–55). Washington, 1846. [Reprinted in Rhees, *Smithsonian Institution . . . Origin and History* (1879), q.v., 410–28.]

Speech on the Mexican War, Delivered in the House of Representatives of the U.S., Feb. 10, 1848 (CG, 30 Cong. 1 sess., 331–33). Washington, 1848. [In *LL*, 453–71.]

"Speech on the Question of the Annexation of Texas," Jan. 20, 1845. CG, 28 Cong. 2 sess., App., 14: 314–19.

Speech on the Tariff Bill; Delivered in the House of Representatives of the United States, on the 30th of April, 1844 (CG, 28 Cong. 1 sess., 594–97). St. Albans, Vt., 1844.

Speech on the Tariff Question; Delivered in the House of Representatives of the U.S., June 30th, 1846 (CG, 29 Cong. 1 sess., App., 1009–14). Washington, 1846.

[Speech on Turkish Missions], *BDFP*, October 1, 2, & 3, 1854.

"State Sovereignty" [entitled "The Sovereignty of the States" after the first installment], *Nation* 1 (1865): 554–56, 648–50, 715–16, 776–77, 810–12.

"Statistics of the Mont Cenis Tunnel," *Nation* 5 (1867): 259–60.

"The Study of Nature," *Christian Examiner,* 68 (1860): 33–62.

"Summary of the Statistics of Sweden," *Hunt's Merchants' Mag.* 24 (1851): 194–99.

"Swedish Literature: 1. Olof Rudbeck the Elder and His Atlantica; 2. The Life and Works of the Painter Hörberg," *American Eclectic* 1 (1841): 63–81, 313–32.

"Thoughts and Aphorisms," in T. Adolphus Trollope, ed., *In Memoriam: A Wreath of Stray Leaves to the Memory of Emily Bliss Gould ob: 31st Aug. 1875.* Rome, 1875, 1–7.

"Translations from the German [Claudius's 'Rhine-Wine Song'; Matthisson's 'The Gnomes and The Fairies']," *American Whig Review* 2 (1845): 256–58.

"Trieste, and the Participation of Austria in the Commerce of the World, 1832–41; Translated from the Austrian Lloyd's Journal," *Hunt's Merchants' Mag.* 10 (1844): 495–521.

"The Two Dictionaries," N.Y. *World,* June 15, 1860.

"Vermont," in John Bouvier, *A Law Dictionary, adapted to the Constitution and Laws of the United States of America*, 12th ed. (Philadelphia 1868), 2:636–37.

"The War and the Peace," *Christian Examiner* 67 (1859): 260–82.

"Waterglass," *BDFP*, July 13, 1857.

"Watershed," *Johnson's New Universal Cyclopaedia*, q.v., 4 (1878), 1299–1300.

"Were the States Ever Sovereign?" *Nation* 1 (1865): 5–8.

"What American Diplomacy Might Do," N.Y. *World*, July 2, 1860.

"'Whole and Halfe' from ye High-Dutche of [Johann Gabriel] Seidl," in Hemenway, *Poets and Poetry of Vermont* (1860), q.v., 337–38.

Selected Unpublished Papers by George Perkins Marsh

"Aphorisms" [in notebook; additional to "Thoughts and Aphorisms," q.v. above]. ca. 1875.

"Books and other things sent to Constantinople" [in 1850 notebook].

"Broussa." ca. 1855.

[Campagna of Rome]. 1881.

Casa d'Angennes [Turin, Italy], First Lease Inventory, July 6, 1861.

"Constantinople and the Bosphorus." ca. 1855.

"Domestic Life and Arts of the Greeks and Romans." Burlington Lyceum, March 12, 1832.

"Economy of the Forest." With letter to Asa Gray, May 9, 1849.

"England Old and New." Anniversary discourse given somewhere in New England, ca. 1859/60.

[Independence Day Oration. Burlington, July 4, 1829 (?)]. (In praise of American institutions.)

[Independence Day Oration. Burlington, July 4, 1843]. (On British institutions and aggressions.)

[Independence Day Oration. Woodstock, July 4, 1857 (?)]. (On changes in the American scene since the Revolution.)

"Indian Corn and the Pellagra." ca. 1875.

"Italian Independence." Lecture, Burlington, Mar. 23, 1860. Abstract as "The Italian Question," *BDFP*, Mar. 26, 1860.

[Italo-Swiss Boundary Arbitration Report]. 1874.

"List of Engravings prepared for Mr. Healy by his friend George P. Marsh." Washington, May 1846.

"Martin Koszta and the American Legation at Constantinople." Draft of letter to Mr. Benedict, signed "Mediterranean." 1853.

"Messina to Catania." ca. 1855.

Morse's New Atlas.

"Office-Holders Turned Out." 1854.

"The Old World and the New." Written for the *North American Review*, 1865 or 1866.

"The Past, the Present and the Future of New England and Her Offspring and the Influence They Are Probably to Have on the World's Future." Speech at Forefathers' Day, Middlebury, Vt., Dec. 2, 1859.

"Persia Treaty Negotiations." 1852.

Petition from Citizens of Burlington for Erection of a New State House, Oct. 17, 1832. Vt. State Archives, 66:39.

Phraseology of Paul Louis Courier. Collection of organized notes. N.d.

Preliminary Notice [for an English edition of Élisée Reclus's *La Terre*]. ca. 1870.

"Relations between Commerce and Civilization." Post 1870.

Report to the University of Vermont on the condition of the library, Nov. 27, 1855.

"Supplementary Statement of Complaint" (about Dubois's Hotel). April 4, 1850.

Other Selected Sources

Adams, John Quincy. *Memoirs of John Quincy Adams, Comprising Portions of His Diary from 1795 to 1848*. Edited by Charles Francis Adams. 12 vols. Philadelphia, 1874–77.

Allard, Dean Conrad, Jr. *Spencer Fullerton Baird and the U.S. Fish Commission: A Study in the History of American Science*. New York: Arno Press, 1978.

Alter, Stephen G. *Darwinism and the Linguistic Image: Language, Race, and Natural Theology in the Nineteenth Century*. Baltimore: The Johns Hopkins University Press, 1999.

Ames, Winslow. "The Vermont State House," *Journal of the Society of Architectural Historians* 23 (1964): 193–99.

Arnold, Matthew. *Letters of Matthew Arnold 1848–1888*. Edited by G. W. E. Russell. 2 vols. New York, 1895.

Bailey, Harold L. "Vermont's State Houses; Being a Narration of the Battles over the Location of the Capitol and Its Construction," *Vermont Quarterly* 12 (1944): 135–56.

Bailey, Richard W. *Nineteenth-Century English*. Ann Arbor: University of Michigan Press, 1996.

[Baird, S. F.] *Report of the [United States] Commissioner [of Fish and Fisheries] for 1872 and 1873*. 42 Cong. 3 sess., Sen. Misc. Doc. 74. Washington, D.C., 1874.

Bassett, T. D. Seymour. "500 Miles of Trouble and Excitement: Vermont Railroads, 1848–1861," *VH* 49 (1981): 133–54.

———. "The George Perkins Marsh Papers," *Dartmouth College Library Bull.* n.s. 10:1 (1969): 9–14.

———. *The Growing Edge: Vermont Villages, 1840–1880.* Montpelier: VHS, 1992.

Baugh, Albert C., and Thomas Cable. *A History of the English Language.* 4th ed. London: Routledge, 1993.

Beck, Richard. "George P. Marsh and Old Icelandic Studies," *SS* 17 (1943): 195–203.

Beer, Gillian. "'The Face of Nature': Anthropomorphic Elements in the Language of *The Origin of Species*" (1986), in Ludmilla Jordanova, ed., *Languages of Nature: Critical Essays on Science and Literature.* London: Free Association Books, 1986, 212–43.

Bigelow, John. *Retrospections of an Active Life.* 5 vols. New York: Baker & Taylor, 1909–13.

Billington, Ray Allen. *The Protestant Crusade, 1800–1860: A Study of the Origins of American Nativism.* New York: Macmillan, 1938.

Boromé, Joseph A. *Charles Coffin Jewett.* Chicago: American Library Association, 1951.

Bosworth, R. J. B. *Italy and the Wider World, 1860–1960.* London: Routledge, 1996.

Brown, Ernest Francis. *Raymond of the Times.* New York: W. W. Norton, 1951.

Brown, John Croumbie, compiler. *Forests and Moisture; or, Effects of Forests on Humidity of Climate.* Edinburgh, 1877.

Brown, Samuel Gilman. *A Discourse Commemorative of the Hon. George Perkins Marsh, LL.D.* Burlington, 1883.

———. *The Works of Rufus Choate, with a Memoir of His Life.* 2 vols. Boston, 1862.

Bruce, Robert V. *The Launching of Modern American Science, 1846–1876.* New York: Knopf, 1987.

Buell, Lawrence. *The Environmental Imagination: Thoreau, Nature Writing, and the Formation of American Culture.* Cambridge: Harvard University Press, 1995.

Burlington *Free Press. Index, 1848–1861.* Montpelier: Historical Records Survey, 1940.

Bush, John W. *Venetia Redeemed: Franco-Italian Relations, 1864–1866.* Syracuse, N.Y.: Syracuse University Press, 1967.

Cardoza, Anthony L. *Aristocrats in Bourgeois Italy: The Piedmontese Nobility, 1861–1930.* New York: Cambridge University Press, 1997.

Carlisle, George W. F. H. *Diary in Turkish and Greek Waters.* C. C. Felton, ed. Boston, 1855.

Carpi, Leone. *Il risorgimento italiano: Biografi storico-politiche d'illustre italiani contempo-rane.* 4 vols. Milan, 1884–88.

Carroll, Daniel B. *The Unicameral Legislature of Vermont.* Montpelier: VHS, 1932.

Cayton, Mary Kupiec. *Emerson's Emergence: Self and Society in the Transformation of New England, 1800–1845.* Chapel Hill: University of North Carolina Press, 1989.

Coleman, Peter J. *Debtors and Creditors in America: Insolvency, Imprisonment for Debt, and Bankruptcy, 1607–1900.* Madison: State Historical Society of Wisconsin, 1974.

Conant, H. J. "Imprisonment for Debt in Vermont: A History," *Vermont Quarterly* 19 (1951): 67–80.

Coppa, Frank J. *Cardinal Giacomo Antonelli and Papal Politics in European Affairs.* Albany, N.Y.: SUNY Press, 1990.

————. *The Origins of the Italian Wars of Independence.* London: Longman, 1992.

Coulson, Thomas. *Joseph Henry: His Life and Work.* Princeton: Princeton University Press, 1950.

[Crane, Elizabeth Green]. *Caroline Crane Marsh: A Life Sketch.* N.d., n.p.

Crawford, Martin. *The Anglo-American Crisis of the Mid-Nineteenth Century: The Times and America, 1850–1862.* Athens: University of Georgia Press, 1987.

Cronon, William. *Changes in the Land: Indians, Colonists, and the Ecology of New England.* New York: Hill and Wang, 1983.

Crosby, Alfred W. "The Past and Present of Environmental History," *AHR* 100 (1995): 1177–89.

Cukwurah, A. O. *The Settlement of Boundary Disputes in International Law.* Manchester: Manchester University Press, 1967.

[Culmann, Karl]. *Rapport au Conseil Fédéral sur les Torrents des Alpes Suisses inspectés en 1858, 1859, 1860 et 1863.* Translated by H. F. Bessard. Lausanne, 1865.

Curti, Merle. "Austria and the United States, 1848–1852: A Study in Diplomatic Relations," *Smith College Studies in History* 11 (1926): 141–206.

————. *The Growth of American Thought.* New York: Harper & Bros, 1943.

————. *Probing Our Past.* New York: Harper & Bros., 1955.

————. *The Roots of American Loyalty.* New York: Columbia University Press, 1946.

Curtis, Jane, Will Curtis, and Frank Lieberman. *The World of George Perkins Marsh, America's First Conservationist and Environmentalist.* Woodstock, Vt.: Countryman Press, 1982.

Dall, William H. *Spencer Fullerton Baird: A Biography.* Philadelphia: Lippincott, 1915.

Dana, Henry Swan. *History of Woodstock, Vermont.* Boston, 1889. Reprint, without family genealogies, with unpaginated Introduction and Epilogue by Peter S. Jennison; Taftsville, Vt.: Countryman Press, 1980.

Daniels, Robert V., ed. *The University of Vermont: The First Two Hundred Years.* Hanover, N.H.: University of Vermont Press, 1991.

Daniels, Tom. "In Italy with Mr. and Mrs. George Perkins Marsh," *VH* 47 (1979): 191–95.

Darwin, Charles. *On the Origin of Species by Means of Natural Selection.* New York: D. Appleton, 1860.

Davis, John A. *Conflict and Control: Law and Order in Nineteenth-Century Italy.* London: Macmillan, 1988.

Davis, John A., and Paul Ginsborg, eds. *Society and Politics in the Age of the Risorgimento.* Cambridge: Cambridge University Press, 1991.

Davis, William Morris. "Biographical Memoir of George Perkins Marsh 1801–1882," Natl. Acad. of Sciences *Biographical Memoirs* 6 (1909): 71–80 (read April 18, 1906).

D'Azeglio, Massimo. *Things I Remember (I miei ricordi,* 1873). Translated by E. R. Vincent. London, 1966.

De Mare, Marie. *G. P. A. Healy: American Artist.* New York: David McKay, 1954.

De Mauro, Tullio. *Storia linguistica dell'Italia unita.* Bari: Editori Laterza, 1963.

Dowty, Alan. *The Limits of American Isolation: The United States and the Crimean War.* New York: New York University Press, 1971.

Duffy, Eamon. *Saints & Sinners: A History of the Popes.* New Haven: Yale University Press, 1997.

Duffy, John J., and H. Nicholas Muller III. *An Anxious Democracy: Aspects of the 1830s.* Westport, Conn.: Greenwood Press, 1982.

Dupree, A. Hunter. *Science in the Federal Government: A History of Policies and Activities to 1940.* Cambridge: Harvard University Press, 1957.

Durden, Robert F. "James S. Pike: President Lincoln's Minister to the Netherlands," *NEQ* 29 (1956): 341–64.

———. *James Shepherd Pike: Republicanism and the American Negro, 1850–1882.* Durham: Duke University Press, 1957.

Dwight, Timothy. *Travels in New-England and New-York* (1821–22). 4 vols. London, 1823.

Ekirch, Arthur A. *The Idea of Progress in America, 1815–1860.* New York, 1944; reprint, New York: AMS Press, 1969.

Eldridge, Charles William. "Journal of a Tour through Vermont to Montreal and Quebec in 1833," *VHS Proceedings* n.s. 2 (1931): 53–82.

Ellis, Joseph J. *American Sphinx: The Character of Thomas Jefferson*. New York: Knopf, 1997.

Elson, Ruth Miller. *Guardians of Tradition: American Schoolbooks of the Nineteenth Century*. Lincoln: University of Nebraska Press, 1964.

Falnes, Oscar J. "New England Interest in Scandinavian Culture and the Norsemen," *NEQ* 10 (1937): 211–42.

Ferris, Norman B. *Desperate Diplomacy: William H. Seward's Foreign Policy, 1861*. Knoxville: University of Tennessee Press, 1976.

Feuer, Lewis S. "James Marsh and the Conservative Transcendentalist Philosophy: A Political Interpretation," *NEQ* 31 (1958): 3–31.

Finberg, Luisa Spencer. "The Press and the Pulpit: Nativist Voices in Burlington and Middlebury, 1853–1860," *VH* 61 (1993): 156–75.

Fisher, Dorothy Canfield. *Vermont Tradition: The Biography of an Outlook on Life*. Boston: Little, Brown, 1953.

Fleming, Robin. "Picturesque History and the Medieval in Nineteenth-Century America," *AHR* 100 (1995): 1061–94.

Fowler, H. D. *Camels to California: A Chapter in Western Transportation*. Stanford, Calif.: Stanford University Press, 1950.

Fox, Stephen. *The American Conservation Movement: John Muir and His Legacy*. Madison: University of Wisconsin Press, 1985.

Franchot, Jenny. *Roads to Rome: The Antebellum Protestant Encounter with Catholicism*. Berkeley: University of California Press, 1994.

Freidel, Frank. *Francis Lieber, Nineteenth Century Liberal*. Baton Rouge: Louisiana State University Press, 1947.

Fryd, Vivien Green. *Art and Empire: The Politics of Ethnicity in the United States Capitol, 1815–1860*. New Haven: Yale University Press, 1992.

————. "Hiram Powers's *Greek Slave*: Emblem of Freedom," *American Art Journal* 14:4 (1982): 31–39.

Gay, H. Nelson. "Garibaldi's American Contacts and His Claims to American Citizenship," *AHR* 38 (1932): 1–19.

Gilman, M. D. *The Bibliography of Vermont; or, A List of Books and Pamphlets Relating in Any Way to the State, with Biographical and Other Notes*. Burlington, 1897.

Gilmore, William J. *Reading Becomes a Necessity of Life: Cultural Life in Rural New England, 1780–1835*. Knoxville: University of Tennessee Press, 1989.

Glacken, Clarence J. "Changing Ideas of the Habitable World," 70–92, in W. L. Thomas, Jr., *Man's Role in Changing the Face of the Earth*, q.v.

————. *Traces on the Rhodian Shore: Nature and Culture in Western Thought from Ancient*

Times to the End of the Eighteenth Century. Berkeley: University of California Press, 1967.

[Godkin, E. L.] "American Ministers Abroad," *Nation* 4 (1867): 132–34.

Goldfrank, David M. *The Origins of the Crimean War.* London: Longman, 1994.

Goode, G. Brown. "The Genesis of the U.S. National Museum." A Memorial of G. Brown Goode, Together with a Selection of His Papers . . . , *SI Ann. Rept. 1897*, 83–192. Washington, D.C., 1901.

———, ed. *The Smithsonian Institution, 1846–1896.* Washington, D.C., 1897.

Goode, James M. *The Outdoor Sculpture of Washington, D.C.: A Comprehensive Historical Guide.* Washington, D.C.: Smithsonian Institution Press, 1974.

Goodman, Paul. *Towards a Christian Republic: Antimasonry and the Great Tradition in New England, 1826–1836.* New York: Oxford University Press, 1988.

Griffiths, Tom, and Libby Robin, eds. *Ecology and Empire: Environmental History of Settler Societies.* Edinburgh: Keele University Press, 1997.

Grøndal, Benedikt, ed. *Breve fra og til Carl Christian Rafn, med en Biografi.* Copenhagen, 1869.

Grove, Richard H. *Green Imperialism: Colonial Expansion, Tropical Island Edens and the Origins of Environmentalism, 1600–1860.* Cambridge: Cambridge University Press, 1995.

Guggenheim, Paul. "Fixation de la frontière de l'Alpe de Cravairola, 23 septembre 1874," 464–514, in A. de La Pradelle, Jacques Politis, and André Salomon, eds., *Receuil des arbitrages internationaux*, III. Paris: Éditions Internationales, 1954.

Guha, Ramachandra, and Juan Martinez-Alier. *Varieties of Environmentalism: Essays North and South.* London: Earthscan, 1997.

Guyot, Arnold. *The Earth and Man: Lectures on Comparative Physical Geography in Its Relation to the History of Mankind.* Translated by C. C. Felton. Boston, 1849.

Hale, Edward Everett. [Review of Marsh's *Lectures on the English Language*], *Christian Examiner* 69 (1860): 1–18.

Hales, E. E. Y. *Pio Nono, a Study of European Politics and Religion in the Nineteenth Century.* London: Eyre & Spottiswoode, 1954.

Hall, Marcus. "Restoring the Countryside: George Perkins Marsh and the Italian Land Ethic (1861–1882)," *Environment and History* 4 (1998): 1–14.

Hamilton, Holman. *Zachary Taylor: Soldier in the White House.* Indianapolis: Bobbs-Merrill, 1951.

Harris, Michael H., ed. *The Age of Jewett: Charles Coffin Jewett and American Librarianship, 1841–1868.* Littleton, Colo.: Libraries Unlimited, 1975.

Harvey, F. L. *History of the Washington National Monument and the Washington National Monument Society*. 57 Cong. 2 sess., Sen. Doc. 224. Washington, 1903.

Hay, Ida. *Science in the Pleasure Ground: A History of the Arnold Arboretum*. Boston: Northeastern University Press, 1995.

Hays, Samuel P. *Beauty, Health, and Permanence: Environmental Politics in the United States, 1955–1985*. New York: Cambridge University Press, 1987.

————. *Conservation and the Gospel of Efficiency: The Progressive Conservation Movement, 1890–1920*. Cambridge: Harvard University Press, 1959.

Hearder, Henry. *Italy in the Age of the Risorgimento, 1790–1870*. London: Longman, 1988.

Hemenway, Abby Maria, ed. *Poets and Poetry of Vermont*. Boston & Brattleboro, 1860.

————, ed. *Vermont Historical Gazetteer* [cited as *HVG*]. 5 vols. Various places, 1862–82.

Henry, Joseph. *The Papers of Joseph Henry*. Edited by Marc Rothenberg. Vol. 7, *1847–1849: The Washington Years*. Washington: Smithsonian Institution, 1996.

Hill, Ralph Nading. *The Winooski: Heartway of Vermont*. New York: Rinehart, 1949.

Holt, Michael F. *The Political Crisis of the 1850s*. New York: Norton, 1978.

————. *The Rise and Fall of the American Whig Party: Jacksonian Politics and the Onset of the American Civil War*. New York: Oxford University Press, 1999.

Hough, Franklin B. "On the Duty of Governments in the Preservation of Forests," Amer. Assn. for the Advancement of Science *Proc.* 22, Part 2 (1873): 1–10.

————. *Report upon Forestry*. Washington, D.C., 1878.

Hughes, J. Donald. *Pan's Travail: Environmental Problems of the Ancient Greeks and Romans*. Baltimore: Johns Hopkins University Press, 1994.

Humphreys, Sexson E. "Le relazioni diplomatiche fra gli Stati Uniti e l'Italia del Risorgimento (1847–1871)." Dottore in lettere, Università degli Studi, Rome, 1945. MS, Biblioteca Alessandrina.

Huxley, T. H. *Man's Place in Nature* [1863]. Ann Arbor: University of Michigan Press, 1959.

Ideville, Henry d'. *Journal d'un diplomate en Italie*, I: 1859–62. 2d ed., Paris, 1872. II: 1862–66. Paris, 1873.

James, Henry. *William Wetmore Story and His Friends . . .* 2 vols. Boston: Houghton Mifflin, 1903.

Jennison, Peter S. *The History of Woodstock, Vermont, 1890–1983* Woodstock: Countryman Press, 1983.

Judd, Richard W. *Common Lands, Common People: The Origins of Conservation in Northern New England*. Cambridge: Harvard University Press, 1997.

Kark, Ruth. *American Consuls in the Holy Land, 1834–1914*. Jerusalem: Hebrew University, 1994.

Kirkland, Edward C. *Men, Cities and Transportation: A Study in New England History, 1820–1900*. 2 vols. Cambridge: Harvard University Press, 1948.

Klay, Andor. *Daring Diplomacy: The Case of the First American Ultimatum*. Minneapolis: University of Minnesota Press, 1957.

Kliger, Samuel. "George Perkins Marsh and the Gothic Tradition in America," *NEQ* 19 (1946): 524–31.

Koopman, Harry L., compiler. *Bibliography of George Perkins Marsh*. Burlington, 1892. [Reprint from UV, *Catalogue of the Library of George Perkins Marsh*, q.v.]

Krapp, George Philip. *The English Language in America*. 2 vols. New York: Ungar, 1925. [Reprint 1960.]

Lane, Harlan. *When the Mind Hears: A History of the Deaf*. New York: Random House, 1984.

Lane-Poole, Stanley. *The Life of the Right Honourable Stratford Canning, Viscount de Redcliffe*. 2 vols. London, 1888.

Langston, Nancy. *Forest Dreams, Forest Nightmares: The Paradox of Old Growth in the Inland West*. Seattle: University of Washington Press, 1995.

Lanman, Charles. *Haphazard Personalities; Chiefly of Noted Americans*. Boston, 1886. [On Marsh, 91–108.]

———. *Letters from a Landscape Painter*. Boston, 1845.

Lash, Scott, Bronislaw Szerszynski, and Brian Wynne, eds. *Risk, Environment and Modernity: Towards a New Ecology*. London: Sage, 1996.

Leopold, Richard W. *Robert Dale Owen: A Biography*. Cambridge: Harvard University Press, 1940.

Levine, Lawrence W. *The Opening of the American Mind: Canons, Culture, and History*. Boston: Beacon Press, 1996.

Lindsay, Julian I. *Tradition Looks Forward: The University of Vermont: A History, 1791–1904*. Burlington: University of Vermont, 1954.

Livingstone, David. *Nathaniel Southgate Shaler and the Culture of Science*. Tuscaloosa: University of Alabama Press, 1987.

———. "Nature and Man in America: Nathaniel Southgate Shaler and the Conservation of Natural Resources," *Transactions of the Institute of British Geographers* n.s. 5 (1980): 369–82.

Lowell, James Russell. *Letters*. Edited by Charles Eliot Norton. 2 vols. New York, 1894.

Lowenthal, David. "Awareness of Human Impacts: Changing Attitudes and Emphases" (1990), 121–35, in B. L. Turner et al., *The Earth as Transformed by Human Action*, q.v.

———. "G. P. Marsh and Scandinavian Studies," *SS* 29 (1957): 41–52.

———. "George Perkins Marsh and the American Geographical Tradition," *Geographical Review* 43 (1953): 207–13.

———. "George Perkins Marsh on the Nature and Purpose of Geography," *Geographical Journal* 126 (1960): 413–17.

———. *George Perkins Marsh: Versatile Vermonter*. New York: Columbia University Press, 1958.

———. "Introduction" to Marsh, *Man and Nature* (1864; 1965), q.v., x–xxix.

———. *The Past Is a Foreign Country*. Cambridge: Cambridge University Press, 1985.

———. *The Vermont Heritage of George Perkins Marsh*. Woodstock, Vt.: The Woodstock Historical Society, 1960.

Lucchesi, Laura, and Stefano Bertocci. *Villa Il Giardino: Una dimora signorile nella campagna di San Salvi*. Florence: Salimbeni, 1985.

Ludlum, David. *Social Ferment in Vermont, 1791–1850*. New York: Columbia University Press, 1939; reprint, Montpelier: VHS, 1948.

Lund, John M. "Vermont Nativism: William Paul Dillingham and U.S. Immigration Legislation," *VH* 63 (1995): 15–29.

McDonnell, Mark J., and Stewart T. A. Pickett, eds. *Humans as Components of Ecosystems: The Ecology of Subtle Human Effects and Population Areas*. New York: Springer Verlag, 1993.

Mack Smith, Denis. "Francesco De Sanctis: The Politics of a Literary Critic," 251–70, in Davis and Ginsborg, *Society and Politics in the Age of the Risorgimento*, q.v.

———. *Modern Italy: A Political History*. New Haven: Yale University Press, 1997.

———. *Victor Emanuel, Cavour, and the Risorgimento*. London: Oxford University Press, 1971.

———, ed. *The Making of Italy, 1796–1870*. New York: Macmillan, 1968.

Maiden, Martin, and Mair Parry, eds. *The Dialects of Italy*. London: Routledge, 1997.

Malintoppi, Antonio. "Diritto ed equità nell'arbitrato per l'Alpe Cravairola," in *Studi in onore di Gaetano Morelli*. Istituto di Diritto Internazionale e Straniero della Università di Milano, *Comunicazione e studi*, vol. 14. Milan: Giuffrè, 1975, 501–24.

Marraro, Howard R. *American Opinion on the Unification of Italy, 1846–1861*. New York: Columbia University Press, 1932.

——. *L'Unificazione Italiano vista dai diplomatici statunitensi*. 4 vols. Istituto per la Storia del Risorgimento Italiano, Bibl. scientia, *Fonti*, vols. 49, 51, 56, 62. Rome, 1963–71.

Marsh, Caroline Crane [Mrs. George P. Marsh]. *The Hallig: or, The Sheepfold in the Waters. A Tale of Humble Life on the Coast of Schleswig*. Translated from the German of [J. G.] Biernatzki. Boston, 1856.

——, compiler. *Life and Letters of George Perkins Marsh*. Vol. I. New York, 1888 [cited as *LL*]. [Vol. II never completed.]

——. *Wolfe of the Knoll, and Other Poems*. New York, 1860.

Marsh, James. *Coleridge's American Disciples: The Selected Correspondence of James Marsh*. Edited by John J. Duffy. Amherst: University of Massachusetts Press, 1973.

——, ed. *Aids to Reflection, by S. T. Coleridge . . . Together with a Preliminary Essay* [xiii–xlvi], and Additional Notes . . . Burlington, 1829. Excerpted in Perry Miller, ed., *The Transcendentalists: An Anthology* (Cambridge: Harvard University Press, 1950), 34–39.

Matthews, Jean V. *Rufus Choate: The Law and Civic Virtue*. Philadelphia: Temple University Press, 1980.

Mencken, H. L. *The American Language: An Inquiry into the Development of English in the United States*. 3d ed. New York: Knopf, 1923. And *Supplement II*. New York: Knopf, 1948.

Merchant, Carolyn. *Ecological Revolutions: Nature, Gender, and Science in New England*. Chapel Hill: University of North Carolina Press, 1989.

Merlone, Rinaldo. *Dalla Rivoluzione francese alle soglie del Duemila: due secoli di musica, cooperazione, associazionismo in un comune della pianura torinese*. Turin: Comune di Piòbesi Torinese, 1996.

Meyer, William B. *Human Impact on the Earth*. New York: Cambridge University Press, 1996.

Mitchell, Donald Grant. *American Lands and Letters; Leather-stocking to Poe's "Raven"*. Vol. 15 of his *Works*. New York: Scribner, 1907. ["George P. Marsh," 35–45.]

Montesquieu, Charles de Secondat, Baron de. *The Spirit of Laws* [1748]. Translated by Thomas Nugent. Chicago: Encyclopedia Britannica, 1952.

Moriondi, Carlo. *Questi Piemontese*. Turin: Risveglio, 1990.

Müller, Friedrich Max. *Lectures on the Science of Language: First Series*. Rev. ed. New York, 1874.

Muller, H. Nicholas, III, and Samuel B. Hand, eds. *In a State of Nature: Readings in Vermont History*. Montpelier: VHS, 1982.

Mumford, Lewis. *The Brown Decades: A Study of the Arts in America, 1865–1895* (1931). Rev. ed., New York: Dover, 1955.

Nash, Bennett Hubbard, and Francis Philip Nash. "Notice of George Perkins Marsh," Amer. Assn. for the Advancement of Science *Proc.* 18 (1882–83): 447–57.

Nash, Roderick Frazier. *The Rights of Nature: A History of Environmental Ethics*. Madison: University of Wisconsin Press, 1989.

————. *Wilderness and the American Mind*. New Haven: Yale University Press, 1967.

Nicolson, Marjorie. "James Marsh and the Vermont Transcendentalists," *Philosophical Review* 34 (1925): 28–50.

Oehser, Paul H. *Sons of Science: The Story of the Smithsonian Institution and Its Leaders*. New York: Henry Schuman, 1949.

Oelschlaeger, Max. *The Idea of Wilderness: From Prehistory to the Age of Ecology*. New Haven: Yale University Press, 1991.

Olwig, Kenneth Robert. "Historical Geography and the Society/Nature 'Problematic': The Perspective of J. F. Schouw, G. P. Marsh and E. Reclus," *Journal of Historical Geography* 6 (1980): 29–45.

Paine, Caroline. *Tent and Harem: Notes of an Oriental Trip*. New York, 1859.

Pasolini dall'Onda, P. D. *Giuseppe Pasolini, 1815–1876: Memorie raccolte da suo figlio*. 2 vols. 4th ed. Turin, 1915.

Pernicone, Nunzio. *Italian Anarchism, 1864–1892*. Princeton: Princeton University Press, 1993.

Pescanti Botti, Renata. *Donne del risorgimento italiano*. Milan: Casa Editrice Ceschina, 1966.

Pinchot, Gifford. *Breaking New Ground*. New York: Harcourt Brace, 1947.

Pisani, Donald J. "Forests and Conservation, 1865–1890," *Journal of American History* 72 (1985–86): 340–59.

Pollak, Martha D. *Turin, 1564–1680: Urban Design, Military Culture, and the Creation of the Absolutist Capital*. Chicago: University of Chicago Press, 1991.

Poore, Benjamin Perley. *Perley's Reminiscences of Sixty Years in the National Metropolis*. 2 vols. Philadelphia, 1886.

Pyne, Stephen J. *Vestal Fire: A Environmental History, Told through Fire, of Europe and*

Europe's Encounter with the World. Seattle: University of Washington Press, 1997.

Rafn, C. C., ed. *Antiquitates Americanae; sive, Scriptores septentrionales rerum ante-Columbianarum in America ... fra det 10de til det 14de Aarhundrede.* Copenhagen, 1837.

Rhees, W. J., ed. *The Smithsonian Institution: Documents Relative to Its Origin and History.* SI. Misc. Coll., vol. 17. Washington, D.C., 1879.

———, ed. *The Smithsonian Institution: Journals of the Board of Regents, Reports of Committees, Statistics, Etc.* [cited as *SIJ*]. SI Misc. Coll., vol. 18. Washington, D.C., 1879.

Rhodes, Anthony. *The Power of Rome in the Twentieth Century: The Vatican in the Age of Liberal Democracies, 1870–1922.* London: Sidgwick and Jackson, 1983.

Rice, John L. "Dartmouth College and the State of New Connecticut, 1776–1782," *Connecticut Valley Historical Soc. Papers and Proc.* 1 (1876–81): 176–206.

Richardson, Leon Burr. *History of Dartmouth College.* 2 vols. Hanover, N.H.: Dartmouth College Publ., 1932.

Rivinus, E. F., and E. M. Youssef. *Spencer Baird of the Smithsonian.* Washington, D.C.: SI Press, 1992.

Robbins, Daniel. *The Vermont State House: A History & Guide.* [Montpelier]: Vermont Council on the Arts, 1980.

Rodgers, Andrew Denny, III. *Bernhard Eduard Fernow: A Story of North American Forestry.* Princeton, N.J., 1951; reissue New York: Hafner, 1968.

Ross, John R. "Man over Nature: Origins of the Conservation Movement," *American Studies* 16 (1975): 49–62.

Roth, Randolph A. *The Democratic Dilemma: Religion, Reform and the Social Order in the Connecticut River Valley of Vermont, 1791–1850.* New York: Cambridge University Press, 1987.

———. "The Other Masonic Outrage: The Death and Transfiguration of Joseph Burnham," *Journal of the Early Republic* 14 (1994): 35–69.

Sallares, Robert. *The Ecology of the Ancient Greek World.* London: Duckworth, 1991.

Salomone, A. William. "The Nineteenth-Century Discovery of Italy: An Essay in American Cultural History. Prolegomena to a Historiographical Problem," *AHR* 73 (1968): 1359–91.

Sauer, Carl O. *Land and Life: A Selection from the Writings of Carl Ortwin Sauer.* Edited by John Leighly. Berkeley: University of California Press, 1963.

Saveth, Edward N. *American Historians and European Immigrants, 1875–1925.* New York: Columbia University Press, 1948.

Schuyler, Eugene. *American Diplomacy and the Furtherance of Commerce.* New York, 1886.

Sereni, Emilio. *History of the Italian Agricultural Landscape* (1961). Translated by R. Burr Litchfield. Princeton: Princeton University Press, 1997.

Sherman, Michael, ed. *A More Perfect Union: Vermont Becomes a State, 1777–1816.* Montpelier: VHS, 1991.

Shirley, John M. *The Dartmouth College Causes and the Supreme Court of the United States.* St. Louis, 1877.

Sicard, Roch-Ambroise. *Cours d'instruction d'un sourd-muet de naissance, pour servir à l'education des sourds-muets.* Paris, An VIII (1800); 2d ed., 1803.

Slotten, Hugh Richard. *Patronage, Practice, and the Culture of American Science: Alexander Dallas Bache and the United States Coast Survey.* New York: Cambridge University Press, 1994.

Smith, Henry Nash. "Rain Follows the Plough; The Notion of Increased Rainfall for the Great Plains, 1844–1880," *Huntington Library Quarterly* 10 (1946–47): 169–93.

———. *Virgin Land: The American West as Symbol and Myth.* Cambridge: Harvard University Press, 1950.

Smithsonian Institution. *Annual Reports of the Board of Regents.* Washington, D.C., 1847–.

Solmi, Arrigo. *The Making of Modern Italy.* London: Ernest Benn, 1925.

Squier, E. G., and E. H. Davis. *Ancient Monuments of the Mississippi Valley; Comprising the Results of Extensive Original Surveys and Explorations.* SI Contributions to Knowledge, vol. I. Washington, D.C., 1848.

Stein, Roger B. *John Ruskin and Aesthetic Thought in America, 1840–1900.* Cambridge: Harvard University Press, 1967.

Stillman, William James. *The Autobiography of a Journalist.* 2 vols. Boston: Houghton Mifflin, 1901.

———. *The Union of Italy, 1815–1895.* Cambridge, Eng., 1899.

Stilwell, Lewis D. *Migration from Vermont.* Montpelier: VHS, 1948.

Stock, Leo F., ed. *Consular Relations between the United States and the Papal States: Instructions and Despatches.* Washington, D.C.: American Catholic Historical Association, 1945.

———. *United States Ministers to the Papal States: Instructions and Despatches 1848–1868.* Washington, D.C.: Catholic University Press, 1933.

Story, William Wetmore. *Vallombrosa*. Edinburgh, 1881.

Stuart, A. J. *Extracts from 'Man and Nature' . . . with some notes on forests and rainfall in Madras*. Madras, 1882.

Sutton, S. B. *Charles Sprague Sargent and the Arnold Arboretum*. Cambridge: Harvard University Press, 1970.

Tamagnone, Michele. *Piobesi nei dodici secoli della sua storia*. Turin: Piobesi Torinese, 1985.

Taylor, Isaac. *Words and Places; or, Etymological Illustrations of History, Ethnology, and Geography*. 4th rev. ed. London, 1873.

Thirgood, J. V. *Man and the Mediterranean Forest: A History of Resource Depletion*. London: Academic Press, 1981.

Thomas, William L., Jr., ed. *Man's Role in Changing the Face of the Earth*. Chicago: University of Chicago Press, 1956.

Thompson, Zadock. *Natural History of Vermont*. Burlington, 1853.

Thoreau, H. D. *Faith in a Seed: The Dispersal of Seeds and Other Late Natural History Writings* (1850–61). Edited by Bradley P. Dunn. Washington, D.C.: Island Press, 1993.

Tichi, Cecelia. *New World, New Earth: Environmental Reform in American Literature from the Puritans through Whitman*. New Haven: Yale University Press, 1979.

Tocqueville, Alexis de. *Democracy in America* (1835–40). 2 vols. New York: Vintage, 1954.

Trauth, Sister Mary Philip. *Italo-American Diplomatic Relations, 1861–1882: The Mission of George Perkins Marsh, First American Minister to the Kingdom of Italy* [cited as *IADR*]. Washington, D.C.: Catholic University of America Press, 1958. [Reprint, Westport, Conn.: Greenwood Press, 1980.]

Trollope, T. Adolphus. *What I Remember*. New York, 1888.

Turner, B. L., II, et al., eds. *The Earth as Transformed by Human Action*. New York: Cambridge University Press, 1990.

Udall, Stewart L. *The Quiet Crisis*. New York: Holt, Rinehart & Winston, 1963.

United States. *Biographical Directory of the American Congress 1774–1949*. 81 Cong. 2 sess., H. Doc. 607. Washington: GPO, 1950.

———. *Congressional Globe*, 1843–.

———. *Foreign Relations of the United States, 1870*. 41 Cong. 3 sess., H. Ex. Doc. 1, pp. 448–52. Washington, 1871.

———. *Francis Dainese*. Feb. 26, 1857. 34 Cong. 3 sess., H. Ex. Doc. 82.

———. *House Journal*, 1843–61.

———. *[Jonas King] Communications . . . Relative to the Case of the Reverend Mr. King*.

May 24, 1854. 33 Cong. 1 sess., Sen. Ex. Doc. 67. *Further Correspondence . . .* Dec. 19, 1854. 33 Cong. 2 sess., Sen. Ex. Doc. 9.

————. *Kossuth and Captain Long*. Feb. 20, 1852. 32 Cong. 1 sess., H. Ex. Doc. 78.

————. *[Marsh Claim for Extra Compensation]*. Feb. 20, 1855, 33 Cong. 2 sess., Sen. Ex. Doc. 40. May 23, 1856, 34 Cong. 1 sess., H. Rep. 166. Dec. 22, 1857, 35 Cong. 1 sess., Sen. Rep. 2. March 12, 1858, 35 Cong. 1 sess., H. Rep. 168. April 6, 1860, 36 Cong. 1 sess., H. Rep. 350. "Act for the Relief of George Perkins Marsh," 12 Stat. at Large 857, June 13, 1860.

————. *[Martin Koszta.]* March 2, 1854. 33 Cong. 1 sess., Sen. Ex. Doc. 40.

————, National Park Service. *Marsh-Billings National Historical Park. Conservation Stewardship Workshop* (1993): *Findings and Recommendations.* Boston, [1995].

————, ————. *Draft General Management Plan.* Woodstock, Vt., April 1998.

————, ————. *Land Use History.* Olmsted Center for Landscape Preservation, Cultural Landscape Publ., no. 4. Boston, 1994.

————, State Department. Papal States, see Stock, Leo F., ed.

————. State Department Papers, National Archives

————, ————. Appointment and Recommendation Papers.

————, ————. Belgium. Brussels Legation. Post Records, vol. 7, 1861–63.

————, ————. Italy: Diplomatic Instructions, vols. 1–2, 1838–94.

————, ————. Italy: Diplomatic Despatches, vols. 10–19, 1861–83.

————, ————. Italy: Post Records [hereafter PR], Miscellaneous Correspondence: Letters to Consuls and Others. Vols. 1–6, 1861–82. Serial nos. 20–25.

————, ————. Italy: PR, Consular and Miscellaneous Letters Received. Vols. 1–9, 1861–82. Serial nos. 39–47.

————, ————. Turkey: Diplomatic Instructions, vol. 1, 1825–59.

————, ————. Turkey: Diplomatic Despatches, vol. 12, 1849–53.

————, ————Turkey: PR, Miscellaneous Correspondence of the Legation of the U.S. Letters Sent, 1849–59; Received, vol. 2, 1850–54.

University of Vermont. *Catalogue of the Library of George Perkins Marsh.* Burlington: The University, 1892.

Varg, Paul A. *New England and Foreign Relations, 1789–1850.* Hanover, N.H.: University Press of New England, 1983.

Vermont. Directory and Rules of the House of Representatives, for the Present Session. Montpelier, 1835.

————. General Assembly Journal, 1823–35.

————. House Journal, 1836–83.

————. Joint Assembly Journal, 1857.

————. Laws of Vermont, 1835–60.

————. Railroad Commissioner. Annual Reports, 1856–62.

————. *Records of the Council of Censors.* Paul S. Gillies and D. Gregory Sanford, eds. Montpelier: Secretary of State, 1997.

————. Senate Journal, 1857–60.

————. Supreme Court, Vt. Reports (Records of the Supreme Court of the State of Vermont), vols. 1–28, 1828–55.

————. Supreme Executive Council. *Records of the Council of Safety and Governor and Council of the State of Vermont.* Edited by E. P. Walton. 8 vols. Montpelier, 1873–80.

Vermont: A Guide to the Green Mountain State. American Guide Series. Boston: Houghton Mifflin, 1937.

Vitousek, Peter M., et al. "Human Domination of Earth's Ecosystems," *Science* 277 (1997): 494–99.

Walker, Joseph T., Jr. "'Old Pewter': A Biographical Sketch of Captain Alden Partridge of Norwich, Vermont," *VH* 33 (1965): 313–25.

Walls, Laura Dassow. *Seeing New Worlds: Henry David Thoreau and Nineteenth-Century Natural Science.* Madison: University of Wisconsin Press, 1995.

Webster, Daniel. *The Papers of Daniel Webster. Diplomatic,* vol. 2: 1850–52. Edited by Kenneth E. Shewmaker and Kenneth R. Stevens. Hanover, N.H.: University Press of New England, 1987.

Wessels, Tom. *Reading the Forested Landscape: A Natural History of New England.* Woodstock, Vt.: Countryman Press, 1997.

Wheeler, David Hilton. "Recollections of George P. Marsh," MS, UVM [reprinted from *Christian at Work*].

Whitaker, J. R. "World View of Destruction and Conservation of Natural Resources," *Annals of the Association of American Geographers* 30 (1940): 143–62.

White, Leonard D. *The Republican Era, 1869–1901: A Study in Administrative History.* New York: Macmillan, 1958.

Whited, Tamara Louise. *The Struggle for the Forest in the French Alps and Pyrenees, 1860–1940.* Ann Arbor: University Microfilms, 1994.

Whitney, Gordon G. *From Coastal Wilderness to Fruited Plain: A History of Environmental Change in Temperate North America, 1500 to the Present.* Cambridge: Cambridge University Press, 1994.

Williams, Michael. *Americans and Their Forests: A Historical Geography.* Cambridge: Cambridge University Press, 1989.

BIBLIOGRAPHY

Winks, Robin W. *Frederick Billings: A Life.* New York: Oxford University Press, 1991.

Wunder, Richard P. *Hiram Powers: Vermont Sculptor, 1805–1873.* 2 vols. Newark: University of Delaware Press, 1991.

Addendum

Works listed below reached me after this book was in page proof.

Albers, Jan. *Hands on the Land: A History of the Vermont Landscape.* Cambridge: MIT Press, 1999. Reviews Abenaki Indian landscape impacts (pp. 56–66) and Marsh's ecological insights (pp.182–88).

Dorman, Robert L. *A Word for Nature: Four Pioneering Environmental Advocates, 1845–1913.* Chapel Hill: University of North Carolina Press, 1998. Summarizes my 1958 biography (pp. 3–45, 227–29). Stresses Marsh's aloof elitism (pp. 19, 36); likens *Man and Nature* to Darwin's *On the Origin of Species* (p. 43).

Krech, Shepard, III. *The Ecological Indian: Myth and History* (New York: Norton, 1999). Reinforces the view that environmental damage is not peculiar to modern technology but has occurred throughout human tenancy of the earth; see pp. 296 and 506n13 above.

Tyrrell, Ian. *True Gardens of the Gods: Californian-Australian Environmental Reform, 1860–1930.* Berkeley: University of California Press, 1999. On eucalyptus's hygienic value, notes Marsh's debt to Australia's Ferdinand von Mueller, both cited to foster irrigation in California. Though *Man and Nature* was long "practically ineffectual" (p. 18), it is now "almost obligatory in a work on American environmental history to begin with [Marsh's] achievement" (p. 17); see p. 415 above.

Winter, James. *Secure from Rash Assault: Sustaining the Victorian Environment.* Berkeley: University of California Press, 1999. The call to action in *Man and Nature* was uniquely ignored in England, where reviewers (*Atheneum; Edinburgh Review*) termed Marsh blind to God's benign design and oblivious to progress in science and technology enabling advanced peoples to repair exploitative environmental damage (pp. 29–34).

Illustration Credits

1, 5, 6, 7: H. S. Dana, *History of Woodstock, Vt.;* 4: Stephen Greene Press, Brattle-boro, Vt.; 2, 3, 15: University of Vermont Library; 8, 9: Billings Family Archives, Woodstock, Vt.; 12: Library of Congress LC-US262-61104; 13, 27, 28, 34: British Library; 16: Frederick H. Meserve Coll., New York; 17: Martha D. Pollak; 18–20: Archivio Storico della Città di Torino, Collezione Simeom (no. 18, D 371, courtesy Martha D. Pollak; nos. 19, 20, D 332, D 608, courtesy Guido Gentile); 29: Rinaldo Merlone; 10, 11, 14, 21–26, 30–33: author's collection.

INDEX